Fifth Edition

MATHEMATICS FOR ELEMENTARY TEACHERS

AN ACTIVITY APPROACH

Fifth Edition

MATHEMATICS FOR ELEMENTARY TEACHERS

An Activity Approach

Albert B. Bennett, Jr.
University of New Hampshire

L. Ted Nelson
Portland State University

Boston Burr Ridge, IL Dubuque, IA Madison, WI New York San Francisco St. Louis
Bangkok Bogotá Caracas Lisbon London Madrid
Mexico City Milan New Delhi Seoul Singapore Sydney Taipei Toronto

McGraw-Hill Higher Education

A Division of The ***McGraw-Hill*** *Companies*

MATHEMATICS FOR ELEMENTARY TEACHERS:
AN ACTIVITY APPROACH, FIFTH EDITION

Published by McGraw-Hill, an imprint of The McGraw-Hill Companies, Inc., 1221 Avenue of the Americas, New York, NY 10020. Copyright © 2001, 1998, 1992, 1985, 1979 by The McGraw-Hill Companies, Inc. All rights reserved. No part of this publication may be reproduced or distributed in any form or by any means, or stored in a database or retrieval system, without the prior written consent of The McGraw-Hill Companies, Inc., including, but not limited to, in any network or other electronic storage or transmission, or broadcast for distance learning.

Some ancillaries, including electronic and print components, may not be available to customers outside the United States.

This book is printed on recycled, acid-free paper containing 10% postconsumer waste.

1 2 3 4 5 6 7 8 9 0 QPD/QPD 0 9 8 7 6 5 4 3 2 1 0

ISBN 0–07–232653–0

Vice president and editor-in-chief: *Kevin T. Kane*
Publisher: *JP Lenney*
Senior sponsoring editor: *William K. Barter*
Developmental editor: *Erin Brown*
Marketing manager: *Mary K. Kittell*
Media technology project manager: *Steve Metz*
Project manager: *Vicki Krug*
Production supervisor: *Enboge Chong*
Designer: *K. Wayne Harms*
Cover design and artwork: *Rokusek Design*
Illustration concept: *Rebecca Bennett*
Senior photo research coordinator: *Carrie K. Burger*
Supplement coordinator: *Tammy Juran*
Compositor: *GAC–Indianapolis*
Typeface: *10/12 Caslon 224 Book*
Printer: *Quebecor Printing Book Group/Dubuque, IA*

The credits section for this book begins on page 301 and is considered an extension of the copyright page.

www.mhhe.com

About the Authors

It was at the University of Michigan that Albert Bennett and L. Ted Nelson and their families first met. Bennett and Nelson had been invited to participate in a National Science Foundation sponsored program of graduate studies in mathematics. Ten years later while on sabbaticals at the University of Oregon they collaborated in writing lessons to actively involve prospective teachers in learning the mathematical concepts they would be teaching. These lessons eventually led to the publication of the first editions of *Mathematics for Elementary Teachers: A Conceptual Approach* and *Mathematics for Elementary Teachers: An Activity Approach.*

Albert Bennett completed his undergraduate and masters degrees at the University of Maine in Orono. He taught mathematics at Gorham State College and became active in the summer mathematics institutes that were sponsored by the Association of Teachers of Mathematics in New England. An early bias that was reflected in his teaching of these institutes was the need to encourage intuition in the teaching and learning of mathematics. He received his doctorate in mathematics from the University of Michigan in 1966 and joined the mathematics faculty at the University of New Hampshire to teach mathematics to prospective teachers. There he organized a mathematics lab and started writing laboratory activities for teachers. In the next few years his efforts led to the publication of Fraction Bars, Decimal Squares, and articles and textbooks for elementary and middle school teachers. These publications support methods of using models and concrete materials in the teaching of mathematics.

Ted Nelson is professor of mathematics and education at Portland State University. He taught junior and senior high school mathematics after graduating from St. Cloud State University, and then continued mathematical studies at Bowdoin College and the University of Michigan, where he received his doctorate in 1968. After serving four years as the first mathematics department chair at Southwest Minnesota State University he moved to Oregon to follow his interest in teaching mathematics to teachers. Currently, his main goal is to continue development of three lab-based courses for prospective elementary teachers and eight additional lab-based courses for middle school teachers. His teaching and curriculum efforts led to a faculty achievement award for outstanding university teaching in 1988. Over the past fifteen years he has written curriculum materials and given workshops designed to bring more concrete materials, visual models, and problem-solving investigations into the elementary and middle school mathematics curriculums.

JUST FOR FUN ACTIVITIES AND ELEMENTARY CLASSROOM IDEAS

1.1	The Peg-Jumping Puzzle	7
1.2	Fibonacci Numbers in Nature	13
1.3	Algebraic Expressions Game	18
	Elementary Classroom Idea: Act Out the Problem	19
2.1	Attribute Identity Game	27
2.2	Coordinate Games	34
2.3	Pica-Centro	39
	Elementary Classroom Idea: Math Communication	40
3.1	Mind-Reading Cards and the Game of Nim	47
3.2	Force Out	54
3.3	Cross-Numbers for Calculators	59
3.4	Calculator Games and Number Tricks	67
	Elementary Classroom Idea: Visualizing Basic Operations	68
4.1	Number Chart Primes and Multiples	78
4.2	Star Polygons	84
	Elementary Classroom Idea: Odd-Even Class Models	87
5.1	Games for Negative Numbers	95
5.2	Fraction Games	103
5.3	Fraction Games for Operations	110
	Elementary Classroom Idea: Integer Balloon	111
6.1	Decimal Games	123
6.2	Decimal Games for Operations	133
6.3	Game of Interest	141
6.4	Golden Rectangles	148
	Elementary Classroom Idea: Base-Ten Decimal Model	150
7.1	Simulated Racing Game	158
7.2	Page Guessing	165
7.3	Cryptanalysis	175
	Elementary Classroom Idea: Student-Centered Data Collection	177
8.1	Probability Games	185
8.2	Trick Dice	197
	Elementary Classroom Idea: Racetrack Probability	198
9.1	Tangram Puzzles	207
9.2	The Game of Hex	214
9.3	Instant Insanity	220
9.4	Snowflakes	228
	Elementary Classroom Idea: Paper-Folding Symmetries	229
10.1	Centimeter Racing Game	238
10.2	Pentominoes	244
10.3	Soma Cubes	251
	Elementary Classroom Idea: Folding Boxes	252
11.1	Line Designs	261
11.2	Paper Puzzles	269
11.3	Enlarging Drawings	278
	Elementary Classroom Idea: Analyzing Shapes	280

CONTENTS

ACTIVITY SETS

Preface *xi*

PROBLEM SOLVING

1.1 Seeing and Extending Patterns with Pattern Blocks 1
1.2 Geometric Number Patterns with Color Tile 8
1.3 Solving Story Problems with Algebra Pieces 14

SETS, FUNCTIONS, AND REASONING

2.1 Sorting and Classifying with Attribute Pieces 21
2.2 Graphing Spirolaterals 28
2.3 Logic Problems for Cooperative Learning Groups 36

WHOLE NUMBERS

3.1 Models for Numeration with Multibase Pieces 42
3.2 Adding and Subtracting with Multibase Pieces 49
3.3 Multiplying with Base-Ten Pieces 55
3.4 Dividing with Base-Ten Pieces 60

NUMBER THEORY

4.1 Models for Even Numbers, Odd Numbers, Factors, and Primes 70
4.2 Models for Greatest Common Factor and Least Common Multiple 79

INTEGERS AND FRACTIONS

5.1 Black and Red Tile Model for Integers 89
5.2 Fraction Bar Model for Equality and Inequality 97
5.3 Computing with Fraction Bars 104

DECIMALS: RATIONAL AND IRRATIONAL

6.1 Decimal Squares Model 115
6.2 Operations with Decimal Squares 124
6.3 A Model for Introducing Percent 134
6.4 Irrational Numbers on the Geoboard 142

STATISTICS

7.1 Randomness, Sampling, and Simulation in Statistics 151
7.2 Scatter Plots: Looking for Relationships 159
7.3 Statistical Distributions: Observations and Applications 166

PROBABILITY

8.1 Probability Experiments 179
8.2 Multistage Probability Experiments 187

GEOMETRIC FIGURES

9.1 Figures on Rectangular and Circular Geoboards 200
9.2 Regular and Semiregular Tessellations 209
9.3 Models for Regular and Semiregular Polyhedra 215
9.4 Creating Symmetric Figures: Pattern Blocks and Paper Folding 221

MEASUREMENT

10.1 Measuring with Metric Units 232
10.2 Areas on Geoboards 239
10.3 Models for Volume and Surface Area 245

MOTIONS IN GEOMETRY

11.1 Locating Sets of Points in the Plane 255
11.2 Drawing Escher-Type Tessellations 263
11.3 Devices for Indirect Measurement 270

Answers to Puzzlers 283
Answers to Selected Activities 285
Credits 301
Index 303
Material Cards 309

Material Cards

1. Rectangular Grid (2.2)
2. Isometric Grid (2.2)
3. Attribute-Game Grid (2.1)
4. Two-Circle Venn Diagram (2.1)
5. Three-Circle Venn Diagram (2.1, 9.1)
6. Pica-Centro Recording Sheet (2.3)
7. Coordinate Guessing and Hide-a-Region Grids (2.2)
8. Table of Random Digits (7.1, 7.3, 8.1)
9. Two-Penny Grid (8.1)
10. Three-Penny Grid (8.1)
11. Geoboard Recording Paper (9.1)
12. Grids for Game of Hex (9.2)
13. Perpendicular Lines for Symmetry (9.4)
14. Metric Measuring Tape (10.1)
15. Centimeter Racing Mat (10.1)
16. Pentomino Game Grid (10.2)
17. Attribute Label Cards (2.1)
18. Logic Problem Clue Cards and People Pieces (Problem 1) (2.3)
19. Logic Problem Clue Cards (Problems 2 and 3) (2.3)
20. Logic Problem Clue Cards (Problems 4 and 5) (2.3)
21. Object Pieces for Logic Problem 5 (2.3)
22. Mind-Reading Cards (3.1)
23. Decimal Squares* (6.1, 6.2)
24. Decimal Squares* (6.1, 6.2)
25. Decimal Squares* (6.1, 6.2)
26. Decimal Squares* (6.1, 6.2)
27. Rectangular Geoboard Template (6.4, 9.1, 10.2)
28. Algebra Pieces (1.3)
29. Algebraic Expression Cards (1.3)
30. Simulation Spinners (8.2)
31. Trick Dice (8.2)
32. Metric Ruler, Protractor, and Compass (9.1, 10.1, 10.3, 11.1, 11.2, 11.3)
33. Circular Geoboard Template (9.1)
34. Regular Polyhedra (9.3)
35. Regular Polyhedra (9.3)
36. Cube Patterns for Instant Insanity (9.3)
37. Pentominoes (10.2)
38. Prism, Pyramid, and Cylinder (10.3)
39. Hypsometer-Clinometer (11.3)
40. Interest Gameboard (6.3)

*Decimal Squares® is a registered trademark of Scott Resources.

Preface

The primary purpose of *Mathematics for Elementary Teachers: An Activity Approach* is to engage prospective elementary and middle school teachers in mathematical activities that will enhance their conceptual knowledge, introduce them to important manipulatives, and model the kind of mathematical learning experiences that they will be expected to provide for their students.

The National Council of Teachers of Mathematics' *Principles and Standards for School Mathematics* ("Standards 2000") and its predecessor, *Curriculum and Evaluation Standards for School Mathematics,* strongly assert that students learn mathematics well only when they construct their own mathematical thinking. Information can be transmitted from one person to another but mathematical understanding and knowledge come from within the learner as that individual explores, discovers, and makes connections.

The National Council of Teachers of Mathematics' *Professional Standards for Teaching Mathematics* presents a vision of mathematics teaching that "redirects mathematics instruction from a focus on presenting content through lecture and demonstration to a focus on active participation and involvement. Mathematics instructors do not simply deliver content; rather, they facilitate learners' construction of their own knowledge of mathematics."

This book provides instructors with activities and materials to actively engage students in mathematical explorations. It provides prospective elementary and middle school teachers the opportunity to examine and learn mathematics in a meaningful way.

Reasons for Mathematical Activities

Current views about the learning of mathematics are summarized in *Everybody Counts: A Report to the Nation on the Future of Mathematics Education* (National Research Council, pp. 58–59):

> Effective teachers are those who can stimulate students to *learn* mathematics. Educational research offers compelling evidence that students learn mathematics well only when they *construct* their own mathematical understanding. To understand what they learn, they must enact for themselves verbs that permeate the mathematics curriculum: "examine," "represent," "transform," "solve," "apply," "prove," "communicate." This happens most readily when students work in groups, engage in discussion, make presentations, and in other ways take charge of their own learning.

Mathematics for Elementary Teachers: An Activity Approach is a book of inductive activities for prospective teachers. It provides the instructor with the resources to make students' mathematical activity the focus of attention. The instructor is then free to interact with small groups, pose questions, and encourage reflective thinking and class discussions. The book enables students to experience mathematics directly by using models that embody concepts and promote mathematical thinking.

This book reflects the beliefs that

- Students who learn mathematics constructively through the appropriate use of manipulatives, models, and diagrams are more likely to develop a solid conceptual basis.

- A concrete approach diminishes the mathematical anxiety often accompanying a more abstract approach.
- Tactile and visual approaches provide mental images that can be more easily retained than symbols and called forth as needed in mathematical thinking.
- Prospective teachers who learn mathematics by being actively involved in doing mathematics will be more likely to teach in the same manner.

Features

Mathematics for Elementary Teachers: An Activity Approach contains 11 chapters with 34 activity sets and accompanying materials to provide a self-contained mathematics laboratory. Here are the special features of this book.

Active Learning Each activity set uses physical materials or visual models to provide a context for understanding. The questions and activities in each activity set are sequentially developed to encourage discovery and to proceed with depth into the topic.

Cooperative Learning The activity sets can be done individually or in small groups. In particular, Activity Set 2.3, Logic Problems for Cooperative Groups, is designed for small-group interaction.

Individual Reflections Throughout the activity sets, students are encouraged to describe patterns, discuss their thinking, and write explanations of their reasoning.

Manipulative Kit and Material Cards The Manipulative Kit contains 9 common manipulatives on color card stock together with storage envelopes. The Activity Book also has 40 additional material cards with manipulatives, models, grids, templates, game mats, and other materials to be used in the activities.

Pedagogy The activity sets demonstrate ways that manipulatives and visual models can be used in classrooms to promote conceptual understanding and mathematical thinking. There are concrete or visual models for teaching the following:

reasoning
numeration
whole numbers
fractions
integers
decimals
percents
patterns and sequences
algebraic expressions
indirect measurement
even and odd numbers
factors and primes
greatest common divisor
least common multiple
geometric relationships
length, area, and volume
statistics and probability
algebra word problems
metric measurements

Just for Fun Each activity set is followed by a Just for Fun activity. These are related to the topics of the activity sets and often are recreational or artistic.

Classroom Ideas At the end of each chapter a Suggested Classroom Activity is given for students to try with children. All of the models and manipulatives are appropriate for use with children.

Readings for More Classroom Ideas Many creative teachers have written about activities they have successfully used in their classrooms. Also, the National

Council of Teachers of Mathematics has sponsored many booklets that discuss reform issues for teachers. There is a selected list of readings at the end of each chapter. Additional sources can be found on the web site (described below).

Puzzlers Brain teasers are interspersed throughout the text to add variety and provide practice in problem solving.

Answer Section Answers to all puzzlers and selected (★) activities are in the back of the Activity Book.

Supplements

Activity Book Instructor's Manual The *Instructor's Manual for Mathematics for Elementary Teachers: An Activity Approach* contains answers for all activity sets, puzzlers, and Just for Fun activities. There is a set of sample test questions, with answers, for each chapter of the activity book. There are blackline masters of various grids used throughout the book.

Companion Text The text, *Mathematics for Elementary Teachers: A Conceptual Approach,* Fifth Edition, is a companion volume to this activity book. Like the activity book, the text contains 11 chapters and 34 sections. Each of these sections corresponds to an activity in the activity book. The text also contains a one-page math activity at the beginning of each section that uses the same Manipulative Kit as the Activity Book.

Web Site (http://www.mhhe.com/math/ltbmath/bennett_nelson/)

This new web site created for the Fifth Edition of the Activity Book and its companion text offers a variety of resources to both instructors and students. From this site you can:

- Download color masters for transparencies of the manipulatives and black and white masters for a variety of grids and dot paper.
- Access an extended bibliography and list of website sources.
- Explore 14 open-ended computer investigations by downloading the Mathematics Investigator software to generate data.
- Download Logo instructions and exercises.
- Use the Bulletin Board to communicate with others about mathematical ideas or questions that arise as you study the material.

Class Formats

The familiar chapter headings in *Mathematics for Elementary Teachers: An Activity Approach* will be convenient for those who wish to use the activities with the companion text or another mathematics textbook. Many of the activity sets are independent and may be selected out of sequence.

Mathematics for Elementary Teachers: An Activity Approach can be effectively used in several types of class formats:

- **A lab course** based on the activity sets with outside readings from a reference text or journals

- **A combination lab and recitation course** in which the activities are used as starting points, followed by discussions or lectures based on extensions of the ideas raised in the investigations
- **A traditional lecture/recitation course** with the activity sets as supplemental and used for outside assignments

To Prospective Teachers

In this course you are a student. However, this is an important time for you to begin to think like the teacher you are planning to become. This is the time to begin your thinking about the nature of mathematics and about teaching and learning mathematics. You can do this by monitoring your own thoughts and feelings as you explore and investigate, by keeping a personal journal during your course, by observing the way your classmates learn and "grapple" with math, by noting how you might change or adapt an activity to use in an elementary or middle school classroom, and by deciding why it is important that children study mathematics.

This book was written to help you experience the mathematical ideas you will be teaching and the way you will be expected to teach them to your students in the twenty-first century. You will construct models, use manipulatives, work cooperatively, explore, investigate, discover, make conjectures, and form conclusions. The models and manipulatives that accompany this book can be used with children. We encourage you to try appropriate explorations with the children you know and, by observing and asking questions, try to understand how they are thinking about mathematics. Eleanor Duckworth used her observations of children learning in her book, *The Having of Wonderful Ideas and Other Essays on Teaching and Learning,* where she makes these comments on teaching and learning:[1]

> So what is the role of teaching, if knowledge must be constructed by each individual? In my view, there are two aspects to teaching. The first is to put students into contact with phenomena related to the area to be studied—the real thing, not books or lectures about it—and to help them notice what is interesting; to engage them so they will continue to think and wonder about it. The second is to have the students try to explain the sense they are making, and, instead of explaining things to students, to try to understand their sense. These two aspects are, of course, interdependent: When people are engaged in the matter, they try to explain it and in order to explain it they seek out more phenomena that will shed light on it.

Acknowledgments:

The authors would like to thank the students and instructors who have used the previous editions of this text, along with the instructors who reviewed the fifth edition of this text and its companion text, *Mathematics for Elementary Teachers, A Conceptual Approach*, Fifth Edition. The authors are sincerely grateful for the helpful suggestions and continued support.

[1]New York: Teachers College Press, Columbia University, 1987, p. 123.

CHAPTER 1

PROBLEM SOLVING

From the earliest grades, the curriculum should give the students opportunities to focus on regularities in events, shapes, designs, and sets of numbers. Children should begin to see that regularity is the essence of mathematics. The idea of a functional relationship can be intuitively developed through observations of regularity and work with generalizable patterns.[1]

ACTIVITY SET 1.1

SEEING AND EXTENDING PATTERNS WITH PATTERN BLOCKS

Purpose To recognize, describe, construct, and extend geometric patterns

Materials Pattern blocks from the Manipulative Kit

Activity Human beings are pattern-seeking creatures. Babies begin life's journey listening for verbal patterns and looking for visual patterns. Scientists in search of extraterrestrial intelligence send patterned signals into the universe and listen for incoming patterns on radio telescopes. Mathematics is also concerned with patterns. Many mathematicians and educators involved in reforming mathematics teaching and learning at the elementary and middle school levels are suggesting that the notion of mathematics as the study of number and shape needs to be expanded. Some suggest that "mathematics is an exploratory science that seeks to understand every kind of pattern."[2]

In this first activity set colored geometric shapes called pattern blocks will be used to recognize, study, and extend geometric patterns. The set of pattern blocks consists of six different figures: a green triangle, an orange square, a red trapezoid, a blue rhombus, a white rhombus, and a yellow hexagon.

[1]*Curriculum and Evaluation Standards for School Mathematics* (Reston, VA: National Council of Teachers of Mathematics, 1989), 60.
[2]Lynn A. Steen, *On the Shoulders of Giants: New Approaches to Numeracy* (Washington, DC: National Academy Press, 1990), 1–8.

1. The pattern block figures shown here form the first five figures of a sequence. Use your green triangles to construct the sixth and seventh figures that you think extend the given pattern.

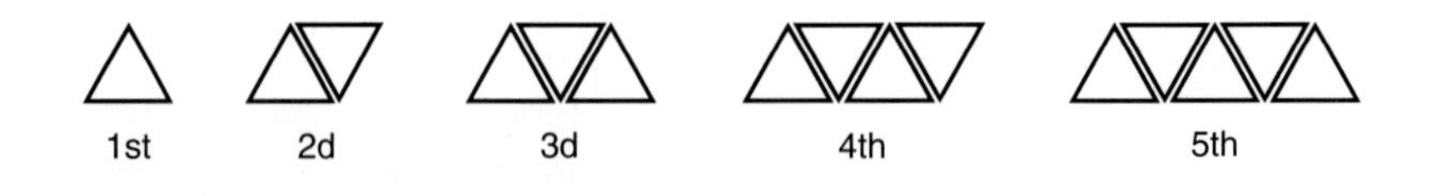

★ **a.** Describe in writing at least three ways that the seventh figure in your sequence differs from the sixth figure.

★ **b.** Describe in writing what the 15th figure in this sequence would look like so that someone reading your description could build the same figure. Give your written description to another person and ask him or her to build a figure according to your instructions.

2. Use your pattern blocks to construct the sixth and seventh figures of the sequence below.

a. Describe in writing how new figures are created as this sequence is extended.

b. Will the 10th figure in an extended sequence have a green triangle or a blue rhombus on the right end? Explain your reasoning.

c. How many triangles and how many rhombuses are in the 25th figure of the extended sequence? Explain how you arrived at your answer.

d. Write a statement that will enable readers to determine the number of triangles and rhombuses in any figure they choose.

3. The pattern block sequence started below uses three different types of pattern blocks. Use your pattern blocks to build the next two figures in the extended sequence.

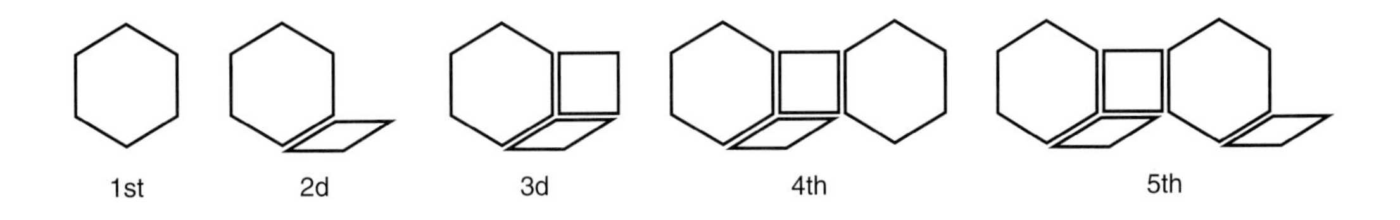

★ **a.** Describe in writing how new figures are created as this sequence is extended.

★ **b.** What pattern block will be on the right end of the 17th figure in this sequence? Explain how you arrived at your answer.

★ **c.** Determine the number of hexagons, squares, and rhombuses in the 20th figure of the sequence without building the figure. Explain how you thought about it.

★ **d.** Repeat part c for the 57th figure in the sequence.

4. Use your pattern blocks to build the sixth and seventh figures of the sequence here.

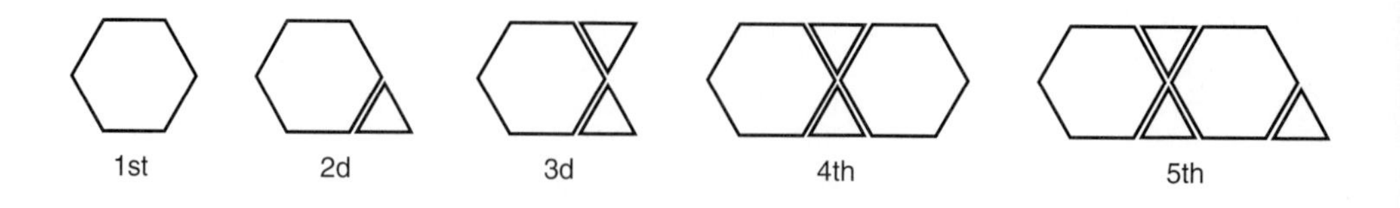

a. Determine the number of triangles and hexagons in the 10th figure of the extended sequence. Do the same for the 15th figure.

b. Devise a way to determine the number of triangles and hexagons in any given figure in this sequence. Write an explanation of your procedure so that readers can use it to determine the number of triangles and hexagons in any figure they choose.

5. The third and fourth figures of a sequence are given below. Use your pattern blocks to construct the first, second, and fifth figures in this sequence. Sketch diagrams of your figures in the space provided.

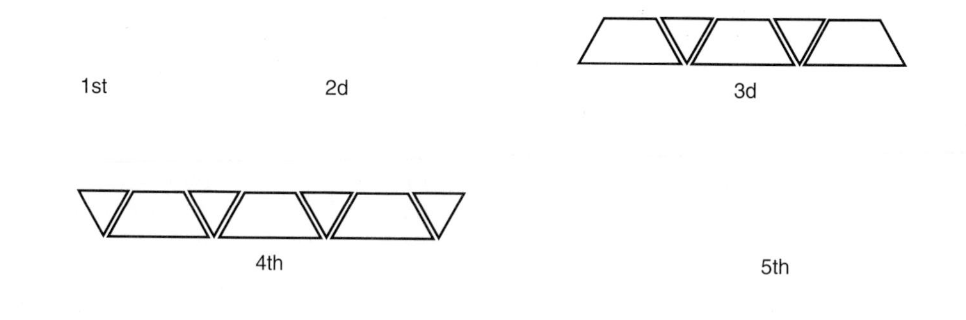

a. Describe how the odd-numbered figures differ from the even-numbered figures.

b. Sketch the missing figures for the sequence on the next page. Explain how you can determine the number of hexagons in any even-numbered figure of the sequence, then explain it for any odd-numbered figure.

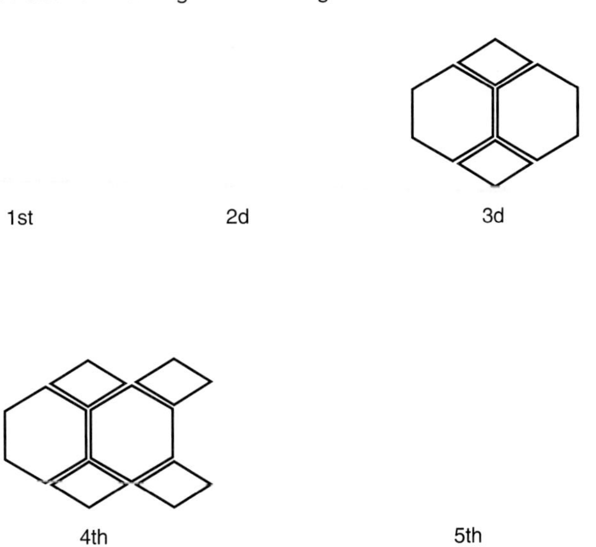

★ **6.** The third term of a sequence is shown below. Create more than one sequence for which the given figure is the third term and sketch diagrams of the first, second, and fourth figures in the space provided. Write a rule for extending each pattern you create so that the reader is able to build the next few figures in the sequence. (You may wish to use the colored tiles from your Manipulative Kit for this activity.)

Sequence I

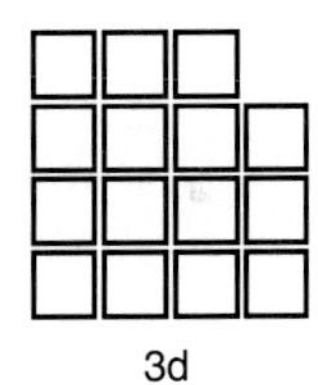

1st 2d 3d 4th

Sequence II

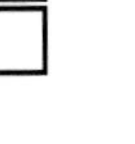

1st 2d 3d 4th

Sequence III

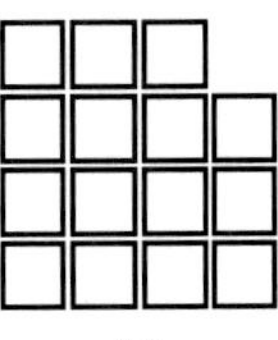

1st 2d 3d 4th

7. Sequences I and II begin repeating in the fifth figure and sequence III begins repeating in the sixth figure. Build the next two figures in each sequence with your pattern blocks. For the 38th figure in each sequence determine which pattern block is at its right end and how many of each type of pattern block it contains. Describe how you reached your conclusion in each case.

Sequence I

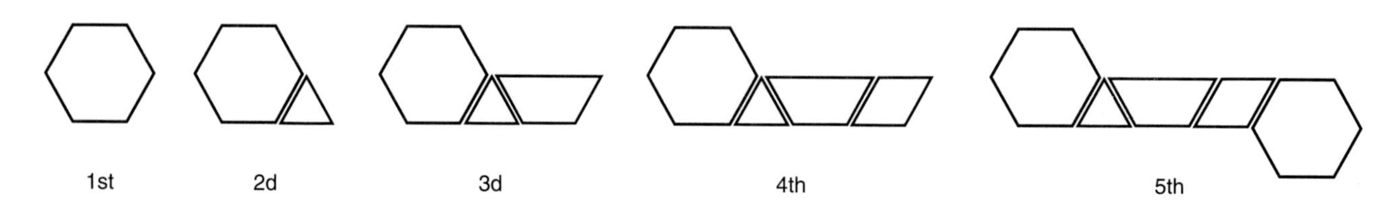

★ *Sequence II*

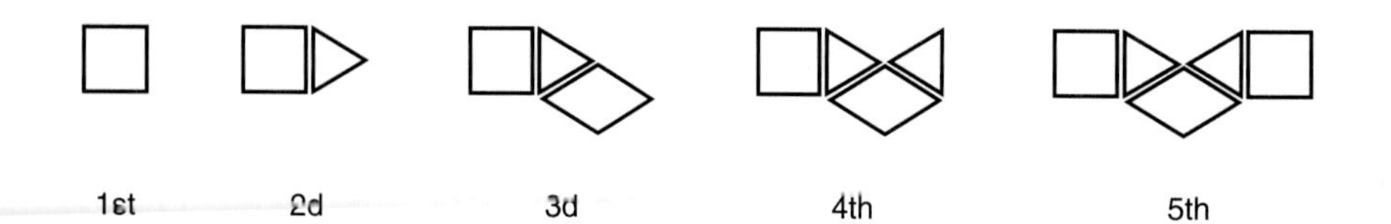

Sequence III

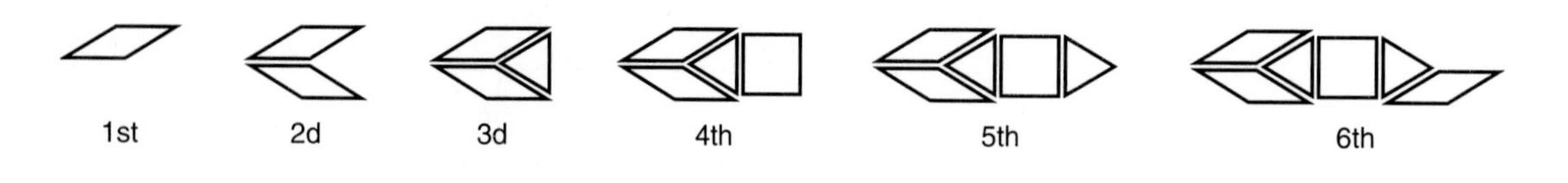

8. Devise your own sequence of figures with pattern blocks. Pose at least three questions about your sequence. Ask another person to build your sequence and answer your questions.

PUZZLER

By moving exactly two toothpicks, get the cherry out of the cup without changing the size and shape of the cup.

JUST FOR FUN

THE PEG-JUMPING PUZZLE

The official peg-jumping puzzle is shown here. A block of wood has seven holes in a row, with three movable black pegs in the holes at one end and three movable red pegs in the holes at the other end. The hole in the center is empty.

The challenge of the peg-jumping puzzle is to interchange the pegs between the right and left sides so that the black pegs move to the positions occupied by the red pegs and vice versa. There are two legal moves: any peg can move to an adjacent empty hole, and a peg of one color can jump a single peg of another color if there is a hole to jump into. (You cannot jump over two or more pegs.)

A simple model for this puzzle uses black and red tiles from the Manipulative Kit on a puzzle grid drawn like the one here.

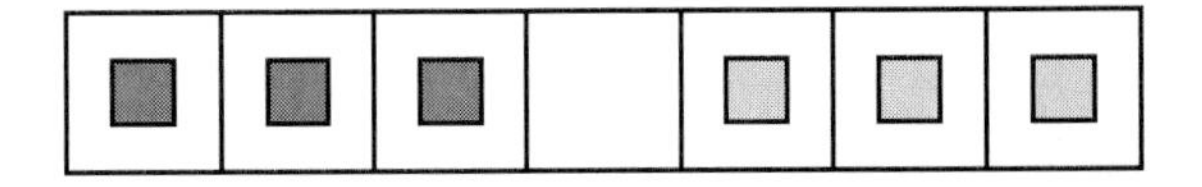

★ 1. Use three problem-solving strategies as you investigate this puzzle: using a model, simplifying the problem (start with one red and one black peg and three holes), and making a table (to record the numbers of moves).

2. What is the least number of moves required to solve the puzzle?

ACTIVITY SET 1.2

GEOMETRIC NUMBER PATTERNS WITH COLOR TILE

Purpose To use geometric patterns to represent number patterns and provide visual support for extending number sequences

Materials Colored tiles from the Manipulative Kit

Activity How long would it take you to find the sum of the counting numbers from 1 to 100?

$$1 + 2 + 3 + 4 + \cdots + 49 + 50 + 51 + \cdots + 97 + 98 + 99 + 100$$

Karl Friedrich Gauss

Karl Friedrich Gauss (1777–1855), one of the greatest mathematicians of all time, was asked to compute such a sum when he was 10 years old. As was the custom, the first student to get the answer was to put his or her slate on the teacher's desk. The schoolmaster had barely stated the problem when Gauss placed his slate on the table and said, "There it lies."

No one knows for sure how the young Gauss obtained the sum so quickly. It is possible, however, that he, like many other creative thinkers, made a mental calculation by thinking of this number problem in a pictorial or visual way. Can you think of a picture or diagram that represents Gauss' sum?

Often in mathematics visual information can give valuable insights into numerical questions. Visual images can also help us remember mathematical ideas and concepts. In the following activities, geometric patterns will be used to generate number sequences. The visual information in the patterns will aid you in making numerical generalizations.

1. Find a pattern in the following sequence and use your tiles to construct the sixth and seventh figures of the sequence.

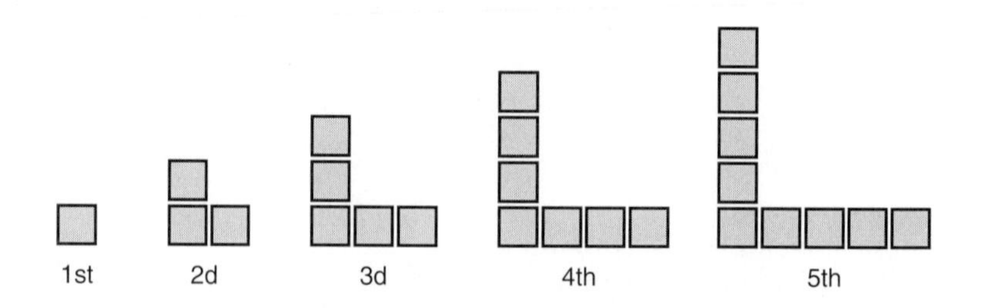

★ a. By counting the number of tiles in each figure, we can see that the first seven figures represent the sequence of odd numbers 1, 3, 5, 7, 9, 11, and 13. Use your tiles to build the tenth figure in this sequence. Determine the tenth odd number by counting the tiles in the tenth figure.

★ **b.** Write a sentence or two describing precisely how you would build the 20th figure in this sequence. How many tiles would be needed? What is the 20th odd number?

★ **c.** Write a sentence or two describing what the 50th figure would look like and how many tiles it would contain.

★ **d.** Write a statement that will enable readers to determine the number of tiles in any figure they choose from this sequence.

2. The first three terms of the number sequence represented by the sequence here are 7, 12, and 17. Build the fourth figure. Sketch this figure, and beneath it record the number of tiles needed to build it.

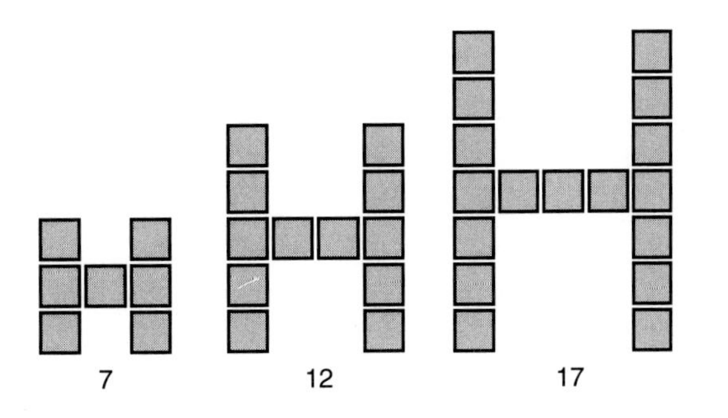

a. Write directions for constructing the eighth figure in the sequence. Ask someone to build the figure by following your directions.

b. How many tiles are in the eighth figure?

c. Determine the number of tiles in the 15th figure. (Imagine how you would construct that figure.)

d. Describe in words what the 50th figure would look like and how many tiles it would contain.

e. Write a procedure using words or an algebraic expression to determine the number of tiles for any figure, given the figure number.

3. Here are three sequences and the number sequences they represent. Build the fourth figure in each sequence, and record the fourth number in each number sequence. Determine the 10th number in each number sequence by imagining how you would construct the 10th figure in each geometric pattern. Write a procedure using words or an algebraic expression that would enable the reader to determine the number of tiles in any figure, given the figure number.

a.

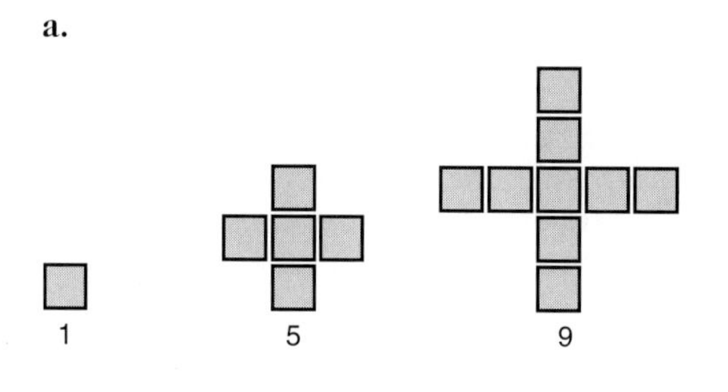

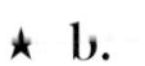
★ b.

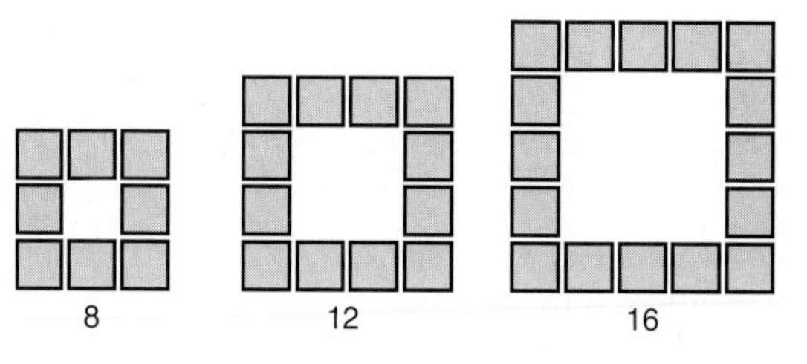

c.

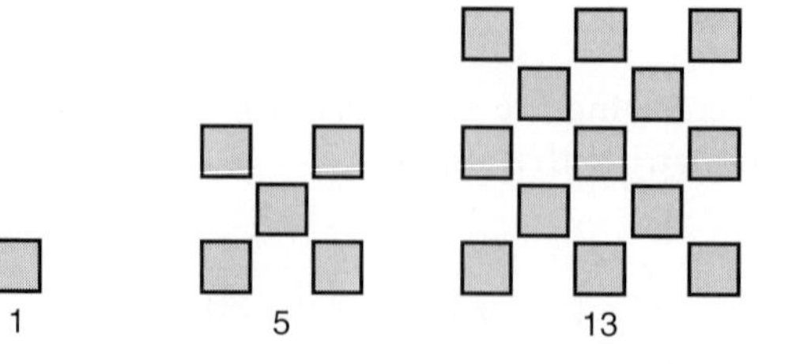

4. Using red and blue tiles, construct the fourth and fifth patterns in this sequence of rectangles and sketch your results. Determine the fourth and fifth terms of the corresponding number sequence.

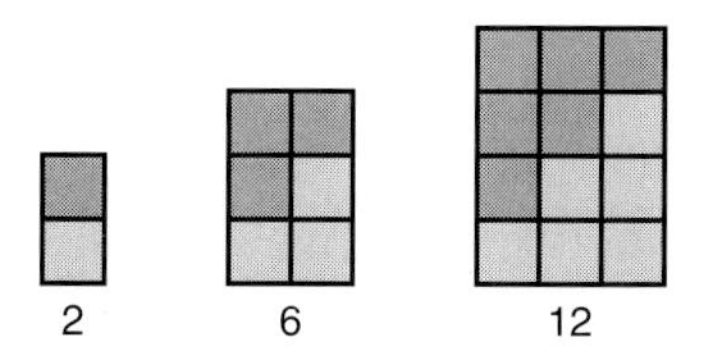

a. Describe the 10th rectangle in the sequence, including height, width, total number of tiles, and number of red tiles and blue tiles.

b. Explain how you can determine the number of red tiles and the number of blue tiles in the 50th rectangle in the sequence.

5. The four figures in the following sequence have a stairstep pattern. Use your tiles to build the fifth figure in this sequence.

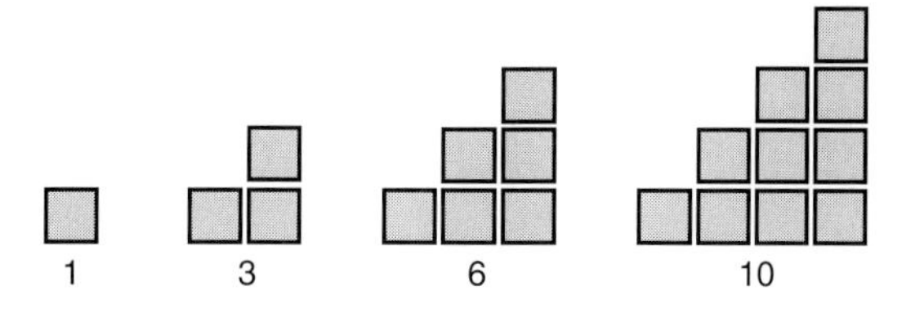

★ **a.** What number does the fifth stairstep represent? The 10th stairstep?

★ **b.** Use the results from activity 4 above to help you determine the number of tiles in the 50th stairstep.

★ **c.** Which stairstep corresponds to Gauss' sum $1 + 2 + 3 + 4 + \ldots + 99 + 100$?

★ **d.** What was Gauss' answer? Explain how you reached your conclusion.

★ e. Write a paragraph explaining how the sum of consecutive whole numbers from 1 to any given number can be obtained.

6. Suppose young Gauss had been asked to compute the sum

$$2 + 4 + 6 + 8 + 10 + 12 + 14 + \ldots + 78 + 80$$

How might he have computed this sum quickly? Devise a method of your own to compute the sum. (*Hint:* One way is to build a stairstep sequence similar to that in activity 5, except for the height of the steps.)

PUZZLER

Use 12 toothpicks to make this figure of squares. By repositioning exactly 3 toothpicks, form a figure with exactly 3 squares, all the same size.

These toothpicks form five squares of the same size. Reposition exactly two toothpicks to get exactly four squares of the same size.

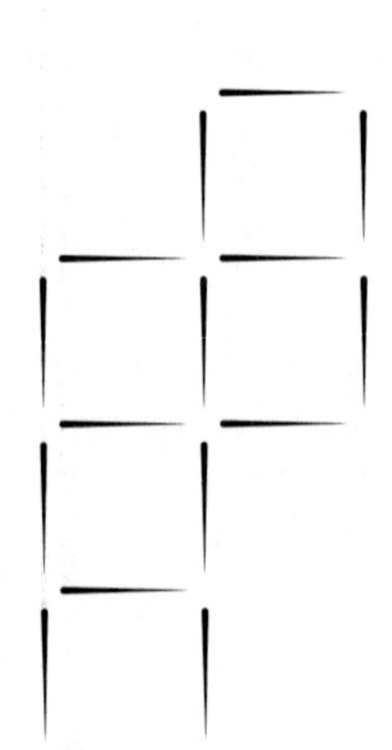

JUST FOR FUN

FIBONACCI NUMBERS IN NATURE

The Fibonacci numbers 1, 1, 2, 3, 5, 8, 13, . . . occur in nature in a variety of unexpected ways. Following the first two numbers of this sequence, each number is obtained by adding the previous two numbers. What are the next five numbers in this sequence?

★ **1. Daisies:** Field daisies often have 21, 34, 55, or 89 petals. If you are playing the game "loves me, loves me not" with a daisy, which numbers of petals will result in a yes answer? The centers of daisies have clockwise and counterclockwise spirals. The numbers of these spirals are also Fibonacci numbers.

2. Sunflowers: The seeds of the sunflower form two spiral patterns, one proceeding in a clockwise direction and one in a counterclockwise direction. The numbers of spirals in the two directions are consecutive Fibonacci numbers. In the drawing, there are 34 counterclockwise and 55 clockwise spirals. In larger sunflowers, there are spirals of 89 and 144. Find a sunflower and count its spirals.

3. Cones and Pineapples: Pine, hemlock, and spruce cones have clockwise and counterclockwise spirals of scalelike structures called bracts. Notice the two spirals that pass through the bract marked with an X in the pinecone diagram. The numbers of these spirals are almost always Fibonacci numbers. For example, the white pinecone has five and eight spirals.

The sections of a pineapple are also arranged in spirals that represent Fibonacci number patterns. Count the spirals on a pineapple from upper left to lower right and from lower left to upper right.

ACTIVITY SET 1.3

SOLVING STORY PROBLEMS WITH ALGEBRA PIECES

Purpose To use algebra pieces as a visual model for representing and solving algebra story problems

Materials Algebra pieces are on Material Card 28.

Activity The Greek algebra of the Pythagoreans (ca. 540 B.C.) and Euclid (ca. 300 B.C.) was not symbolic with letter for variables, as we think of algebra, but geometric. An arbitrary number was expressed as a line segment. To express the sum of two arbitrary numbers, the Greeks joined the two segments end to end. The product of two numbers was represented as the area of a rectangle with the two segments as sides.

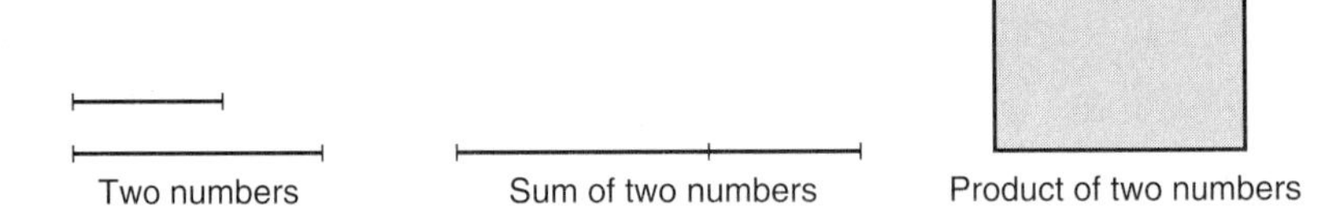

In this activity set, we will represent and solve algebra story problems using a geometric model similar to that of the Greeks. Our model will consist of two types of pieces: a variable piece and a unit piece. The variable piece will be used to represent an arbitrary line segment or an arbitrary number. Each unit piece will represent a length of 1 unit.

Unit Variable piece

The algebra pieces can be placed together to form different expressions.

An arbitrary number

Twice a number

3 more than a number

2 more than 3 times a number

The product of a number times itself

1. Form the rectangle shown here with your algebra pieces.

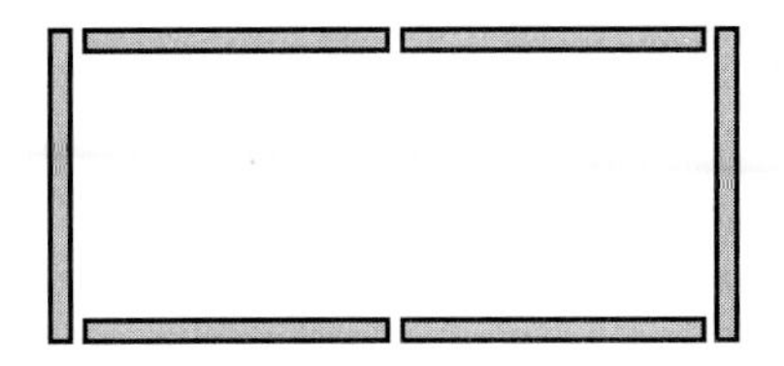

★ **a.** Describe, in a sentence, the relationship between the length and the width of this rectangle. (Remember that the algebra pieces represent line segments so they have length but no width.)

★ **b.** If the perimeter of this rectangle is 54 units, explain how you can determine the length and the width of the rectangle using only your algebra-piece model.

★ **c.** Use your algebra pieces to alter the rectangle so that its length is 3 more units than twice its width. Draw a sketch of your new rectangle.

★ **d.** If this new rectangle has a perimeter of 90 units, explain how you can determine its length and its width using only your algebra-piece model.

2. If a variable piece represents a whole number, then two consecutive whole numbers are represented by the algebra pieces in figure a. Figure b illustrates the sum of the two consecutive whole numbers.

(a) (b)

a. Represent four consecutive whole numbers with your algebra pieces. Sketch them here.

b. Use your model to determine four consecutive whole numbers, which have a sum of 58. Explain your solution by referring to the algebra-piece model for the sum of four consecutive whole numbers.

3. Use your algebra pieces to represent the sides of the triangle described as follows:

The second side of the triangle is twice the length of the first side. The third side of the triangle is 6 units longer than the second side.

★ a. Draw a sketch of your algebra-piece model.

★ b. Use your model to determine the length of each side of the triangle if the perimeter of the triangle is 66 units. Using your model, explain how you arrived at your conclusion.

4. A box contains $2.25 in nickels and dimes. There are three times as many nickels as dimes. If one variable piece represents the *number of dimes* in the box, explain why the variable pieces shown here represent the total number of nickels and dimes.

a. Suppose each dime is replaced by its value in nickels so that there are only nickels in the box. Explain why the total number of coins is now represented by the following pieces.

b. The total amount of money in the box is $2.25. Explain how to use this information to determine the number of nickels represented by each variable piece in the last diagram.

c. How can you use the above information to determine the original number of dimes and nickels in the box?

Activities 5–11: Use your algebra pieces to represent the information in each story problem. Sketch an algebra-piece model for each problem. Explain how each solution can be arrived at using the algebra pieces.

★ **5.** Two pieces of rope differ in length by 7 meters. End to end, their total length is 75 meters. How long is each piece? (*Hint:* Let a variable piece represent the length of the shorter rope, and let the unit piece be 1 meter.)

6. There are 3 boys on the school playground for every girl on the playground. Altogether there are 76 children. How many are boys? (*Hint:* Let one variable piece represent the number of girls on the playground.)

★ **7.** Andrea has a collection of nickels, and Greg has a collection of dimes. The number of nickels Andrea has is four times the number of dimes that Greg has. Andrea has 80 cents more than Greg. How much money does Greg have? (*Hint:* Let one variable piece represent the number of dimes Greg has.)

8. The length of a rectangle is 5 feet more than three times its width. Determine the length and the width if the perimeter is 130 feet.

★ **9.** Three-fifths of the students in a class are women. If the number of men in the class were doubled and the number of women were increased by 9, there would be an equal number of men and women. How many students are there? (*Hint:* Represent the number of women by three variable pieces and the number of men by two variable pieces.)

10. The sum of three numbers is 43. The first number is 5 more than the second number and the third number is 8 less than twice the first number. What are the three numbers? (*Hint:* Let one variable piece represent the second number, construct the first and third numbers, and add.)

PUZZLER

You and your friend each have some pennies. If you give your friend 1 penny, then you and your friend will have the same number of pennies. If your friend gives you 1 penny, then you will have twice as many pennies as your friend. How many pennies do each of you have?

JUST FOR FUN

ALGEBRAIC EXPRESSIONS GAME[3]

(2 teams)

In this game, players match algebraic expressions to word descriptions. Copies of algebraic expression cards appear on Material Card 29. There are two decks of 12 cards, one for each team. Each student has an expression card similar to the samples below.

Play: The two teams play simultaneously. On team A, the game begins with player 1 saying, "I have $n + 1$. Who has two less than a number?" Then player 2 (on the same team), who has the card with $n - 2$, says, "I have $n - 2$. Who has one more than two times a number?" The game continues in this manner.

On team B, player 1 is the player with the $n + 2$ card, and player 2 is the player with an algebraic expression matching the word description on player 1's card.

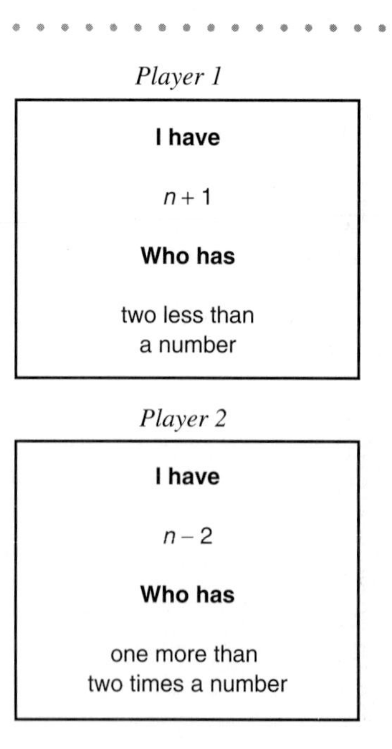
Player 1
I have
n + 1
Who has
two less than
a number
Player 2
I have
n – 2
Who has
one more than
two times a number

[3]T. Giambrone, "I HAVE . . . WHO HAS . . . ?" *The Mathematics Teacher* 73 (October 1980), 504–506.

Each deck of 12 cards is circular. That is, the last card in the deck calls the first card. Therefore, all the cards in the deck should be used. If there are fewer than 12 students on a team, some students can use more than one card. If there are more than 12 students on a team, two students can share a card, or more cards can be made.

Objective: The two teams compete against each other to complete the cycle of cards in the shortest amount of time.

Variation: The game can be changed so that word descriptions are matched to algebraic expressions. Two sample cards are shown here. In this case player 1 says, "I have two more than three times a number. Who has $4x - 1$?" Player 2 says, "I have one less than four times a number. Who has $3x + 5$?" Make a deck of these cards and play this version of the game.

I have

two more than three times a number

Who has

$4x - 1$

I have

one less than four times a number

Who has

$3x + 5$

IDEAS FOR THE ELEMENTARY CLASSROOM

SUGGESTED CLASSROOM ACTIVITY: ACT OUT THE PROBLEM

Many problems can be solved by having students act out the problem. The handshake problem is an excellent example. Ask two students to show the class what one handshake looks like. Then choose a group of five students and ask them to determine how many handshakes there will be if each of the five students shakes hands with every other person in the group. Let the entire class be consultants to find ways to systematically count handshakes. Extend the handshake problem to the entire class.

Another example is from the peg-jumping puzzle (Activity Set 1.1, Just for Fun). Put seven chairs in a row, as shown. Select three boys to sit at one end and three girls to sit at the other, with an empty chair in the middle. See what strategies the students use to solve the problem.

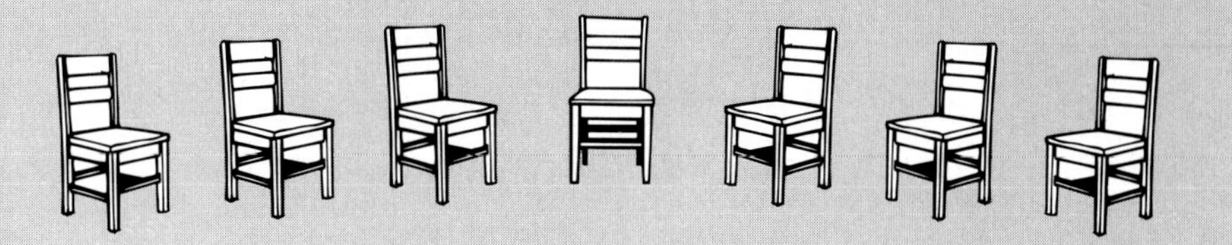

Readings for More Classroom Ideas

Arcidiacono, Michael J., and E. Maier. "Picturing Algebra," in *Math and the Mind's Eye* (Unit ME9). Salem, OR: The Math Learning Center, Box 3226, 1993.

Bennett, A., E. Maier, and L. Ted Nelson. "Seeing Mathematical Relationships." (Unit ME1). *Math and the Mind's Eye.* Salem, OR: The Math Learning Center, Box 3226, 1988.

Berman, Barbara, and F. Friederwitzer. "Algebra Can Be Elementary . . . When It's Concrete." *Arithmetic Teacher* 36 (April 1989): 21–24.

Burns, Marilyn. *Problem-Solving Lessons: Grades 1–6.* Sausalito, CA: Math Solutions Publications, 1996.

Burns, Marilyn. *Writing in Math Class: A Resource of Grades 2–8.* Sausalito, CA: Math Solutions Publications, 1996.

Carey, Deborah A. "The Patchwork Quilt: A Context for Problem Solving." *Arithmetic Teacher* 40 (December 1992): 199–203.

Coburn, Terrence G., et al. *Patterns,* in the *Curriculum and Evaluation Standards for School Mathematics Addenda Series: Grades K–6.* Reston, VA: National Council of Teachers of Mathematics, 1993.

Cox, Pam, and Bridges, Linda. "Algebra Activities for All: Calculating Human Horsepower." *The Mathematics Teacher* 92 (March 1999): 225–228.

Cramer, Kathleen, and Thomas Post. "Making Connections: A Case for Proportionality." *Arithmetic Teacher* 40 (February 1993): 342–346.

Farivar, S., and N. M. Webb. "Helping and Getting Help-Essential Skills for Effective Group Problem Solving." *Arithmetic Teacher* 41 (May 1994): 521–525.

Gill, Alice J. "Multiple Strategies: Product of Reasoning and Communication." *Arithmetic Teacher* 40 (March 1993): 380–387.

Jensen, Rosalie, and O'Neil, David R. "Let's Do It: Classical Problems for All Ages." *Arithmetic Teacher* 29 (January 1982): 8–12.

Kenney, Patricia A., Zawojewsky, Judith S., and Silver, Edward A. "The Thinking of Students: Marcy's Dot Pattern." *Mathematics Teaching in the Middle School* 3 (May 1998): 474–477.

Loewen, A. C. "Lima Beans, Paper Cups, and Algebra." *Arithmetic Teacher* 38 (April 1991): 34–37.

Norman, F. Alexander. "Figurate Numbers in the Classroom." *Arithmetic Teacher* 38 (March 1991): 42–45.

Schifter, Deborah. *Reasoning About Operations: Early Algebraic Thinking in Grades K–6,* in *Developing Mathematical Reasoning in Grades K–12,* 1999 Yearbook, edited by L. V. Stiff and R. R. Curcio, VA: National Council of Teachers of Mathematics, 1999, 62–81.

Silverman, F., K. Winograd, and D. Strohauer. "Student Generated Story Problems." *Arithmetic Teacher* 39 (April 1992): 6–12.

Van de Walle, John A., and Holbrook, Helen. "Patterns, Thinking, and Problem Solving." *Arithmetic Teacher* 34 (April 1987): 6–12.

Van de Walle, John A., and Thompson, Charles S. "Let's Do It: Promoting Mathematical Thinking." *Arithmetic Teacher* 32 (February 1985): 7–13.

CHAPTER 2

SETS, FUNCTIONS, AND REASONING

A climate should be established in the classroom that places critical thinking at the heart of instruction. . . . Children need to know that being able to explain and justify their thinking is important and that how a problem is solved is as important as its answer. . . . Manipulatives and other physical models help children relate processes to their conceptual underpinnings and give them concrete objects to talk about in explaining and justifying their thinking.[1]

ACTIVITY SET 2.1

SORTING AND CLASSIFYING WITH ATTRIBUTE PIECES

Purpose To use attribute pieces in games and activities for sorting and classifying, reasoning logically, formulating and verifying hypotheses, and introducing set terminology and operations

Materials Attribute pieces from the Manipulative Kit, attribute label cards from Material Card 17, attribute game grid on Material Card 3, and the two-circle and three-circle Venn diagrams on Material Cards 4 and 5, respectively

Activity "Mathematics is reasoning. One cannot do mathematics without reasoning. The standard does not suggest, however, that formal reasoning strategies be taught in grades K–4. At this level, the mathematical reasoning should involve the kind of informal thinking, conjecturing, and validating that helps children to see that mathematics makes sense.

[1]*Curriculum and Evaluation Standards for School Mathematics* (Reston, VA: National Council of Teachers of Mathematics, 1989), 29.

"Manipulatives and other physical models help children relate processes to their conceptual underpinnings and give them concrete objects to talk about in explaining and justifying their thinking. Observing children interact with objects in this way allows teachers to reinforce thinking processes and evaluate any possible misunderstandings."[2]

In the late 1960s, sets of geometric figures called attribute pieces became a popular physical model for activities that promote logical thinking. These attribute pieces are also well suited for introducing ideas and terminology related to sets. The 24 attribute pieces used in this activity set vary in shape, color, and size. There are four shapes (triangle, square, hexagon, and circle), three colors (red, blue, and yellow), and two sizes (large and small). Each attribute piece differs from every other piece in at least one of the attributes of shape, color, or size. Many of the games and activities in this activity set can be adapted for use with children.

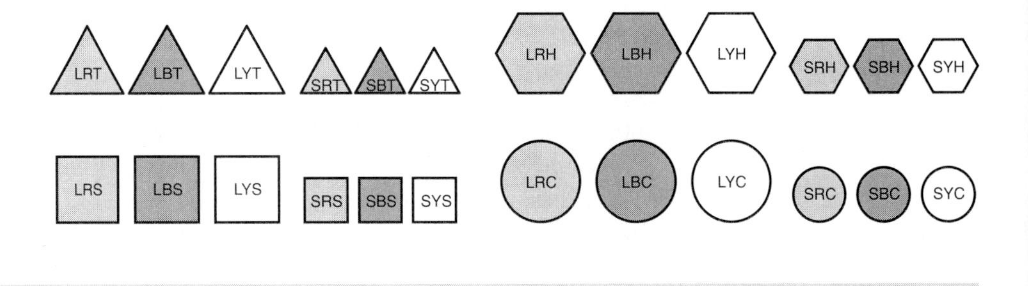

1. Place your *small blue square, small red square,* and *small yellow circle* on the indicated squares of the grid below. In the first row of the grid, the small blue square differs from the small red square in exactly one attribute—color. In the third column, the small red square differs from the small yellow circle in exactly two attributes—color and shape. Use your remaining small attribute pieces to fill the grid so that adjacent pieces in rows differ in exactly one attribute and adjacent pieces in columns differ in exactly two attributes. Record your results on the grid.

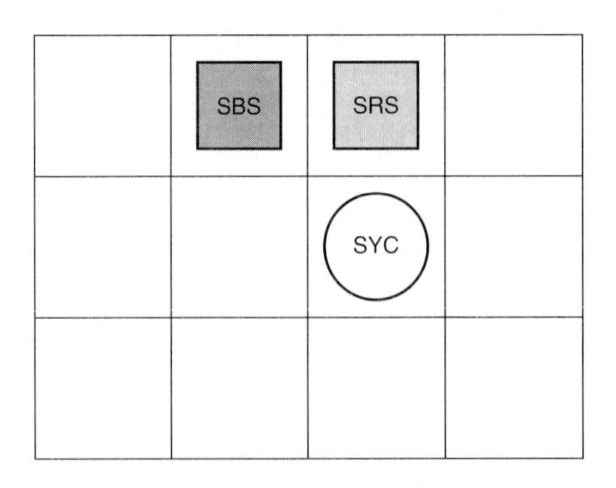

[2]NCTM's *Curriculum and Evaluation Standards for School Mathematics,* 29–30.

2. Attribute Guessing Game (2 players): To play this game, display all 24 attribute pieces on a flat surface. One player thinks of a specific attribute piece. The other player tries to determine that piece by asking questions that can be answered yes or no. Pieces that are ruled out by questions may be physically separated from the rest. The score is the number of questions needed to identify the piece. Low score wins. Players alternate roles.

★ **a.** If the player trying to guess the piece restricts all questions to the attributes of color, size, and shape, what is the minimum number of guesses necessary to ensure the identification of a randomly chosen piece? List the questions you would ask in order to identify a piece in the minimum number of guesses.

b. If the player guessing is allowed to use additional attributes of the pieces (like number of sides, size of angles, opposite sides parallel), can the minimum number of guesses be reduced? Explain.

3. Attribute Grid Game (2 or more players or teams, or solitaire): To begin this game, place any attribute piece on the center square of the grid (Material Card 3). The *large red hexagon* was used to start the game on the grid shown on page 24. The players take turns placing an attribute piece on the grid according to the following conditions: the piece that is played must be placed on a square that is adjacent, by row, column, or diagonal, to a piece that has already been played; the adjacent attribute pieces in the rows must differ in one attribute (color, size, or shape); the adjacent pieces in the columns must differ in two attributes; and the adjacent pieces in the diagonals must differ in three attributes. A player's score on each turn is the total number of attributes in which the piece played differs from *all* adjacent attribute pieces.

Examples: The first piece played on the grid below was a *small blue square,* which differs in three attributes from the large red hexagon. The first player scored 3 points. The second player played the *small red hexagon.* This piece differs from the small blue square in two attributes and from the large red hexagon in one attribute. This player also scored 3 points. The game ends when no more pieces can be played on the grid.

Solitaire Version: Place an attribute piece on the center square. Try to place as many pieces as possible on the grid according to the rules. Record your results. Is it possible to place all 24 pieces on the grid?

Attribute-Game Grid

Rows, 1 difference (1 point)
Columns, 2 differences (2 points)
Diagonals, 3 differences (3 points)

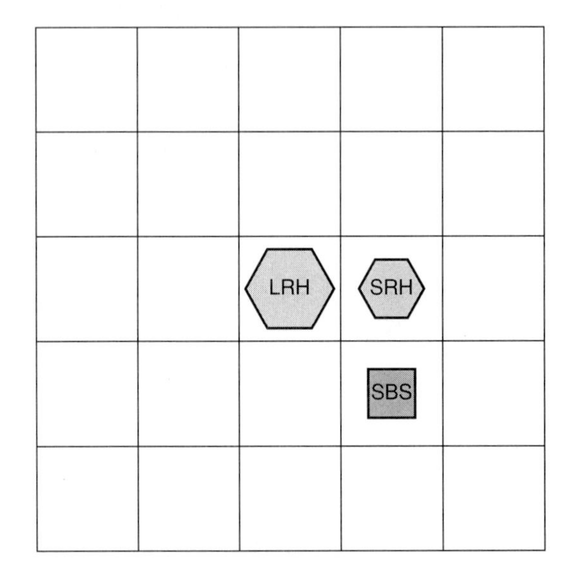

4. Label your two-circle Venn diagram (Material Card 4) with the attribute labels (from Material Card 17) RED and TRIANGULAR, as shown in the diagram below. In this diagram the *large red circle* is placed in the circle labeled RED and the *small blue triangle* is placed in the circle labeled TRIANGULAR. The *small red triangle* possesses both the red attribute and the triangular attribute, so it is placed in the intersection of the two circles. The *large yellow hexagon,* having neither attribute, is placed outside both circles. Place the same four pieces on your diagram. Distribute your remaining attribute pieces on the correct regions of the diagram.

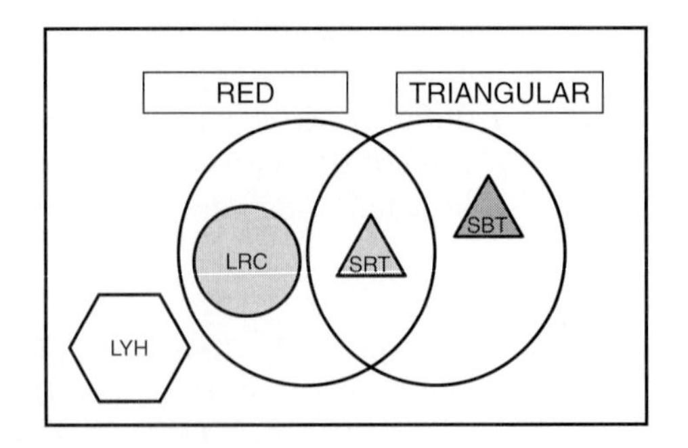

a. The attribute pieces that lie in the intersection of the two circles are described as being RED *and* TRIANGULAR. This set is called the *intersection* of the set of red pieces and the set of triangular pieces. Referring to your diagram, complete the following list of pieces in the intersection of the two sets.

Red *and* Triangular: SRT, ____________________

★ **b.** The attribute pieces that lie in the RED circle or the TRIANGULAR circle or both circles are described as being RED *or* TRIANGULAR. This set is called the *union* of the set of red pieces with the set of triangular pieces. Referring to your diagram, complete the following list of pieces in the union of the two sets.

Red *or* Triangular: LRC, SRT, SBT, ____________________

5. Label your two-circle Venn diagram as shown here. Distribute your attribute pieces in the appropriate regions.

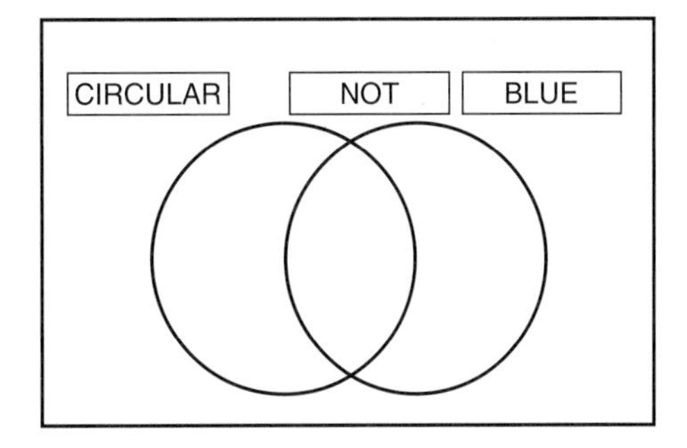

★ **a.** Record the set of attribute pieces in the *intersection* of the following two sets.

Circular *and* Not Blue: ____________________

b. Record the set of attribute pieces in the *union* of the following two sets.

Circular *or* Not Blue: ____________________

c. Using only words found on the attribute label cards, describe the pieces that are outside both circles in this diagram.

6. Mathematicians invent symbols for the purpose of abbreviating and representing ideas. In this activity the symbols NR, NL, T, NY, and NB will represent the NOT RED, NOT LARGE, TRIANGULAR, NOT YELLOW, and NOT BLUE sets of attribute pieces, respectively. The symbol ∪ will represent the *union of sets,* and the symbol ∩ will represent the *intersection of sets.* Use your labels and two-circle Venn diagram to determine and record the following sets of attribute pieces. (Each question calls for a different diagram.)

a. NR ∩ NL:

★ **b.** Using only words found on the attribute label cards, describe the pieces that are outside both circles in part a.

★ **c.** T ∪ NY:

d. NB ∪ NY:

7. Label your three-circle Venn diagram (Material Card 5) with the attribute labels shown in the diagram. There are eight different regions within the rectangle. Seven of the regions are enclosed by one or more circles, and one region is outside all three circles. Distribute the attribute pieces on the correct regions of your diagram.

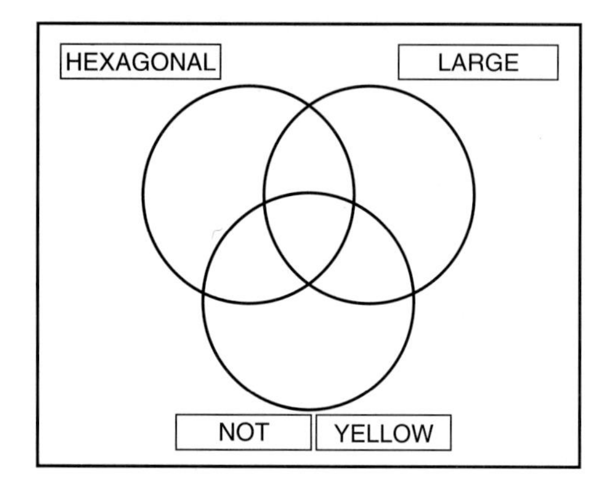

a. Record your solution by writing the abbreviation for each piece on the correct region of the diagram above.

b. The pieces LBH, LRH, and LYH are in the region H ∩ L. The pieces LBH and LRH are in the region (H ∩ L) ∩ NY. List the pieces in each of the following regions.

(1) H ∩ NY:

★ (2) (H ∩ NY) ∩ L:

(3) (NY ∩ L) ∪ H:

(4) (H ∩ L) ∪ H:

★ (5) H ∩ (NY ∪ L):

(6) (H ∩ NY) ∪ L:

PUZZLER

These three containers are all labeled incorrectly. But it is true that one contains two nickels, another two pennies, and another one nickel and one penny. How is it possible to determine the correct labels for all three containers by selecting one coin from one container?

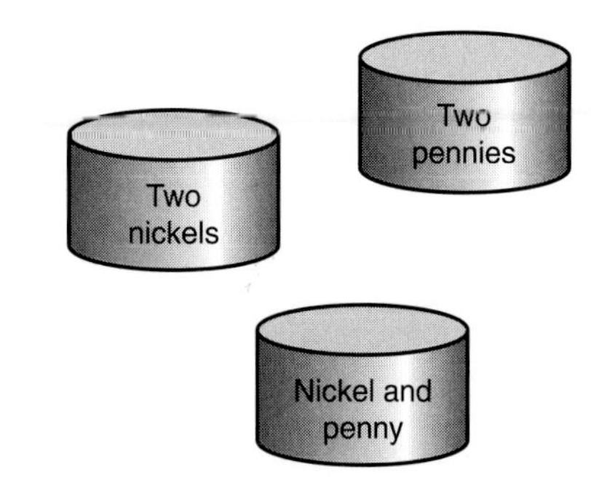

JUST FOR FUN

ATTRIBUTE IDENTITY GAME (2 teams or 2 players)

Draw three large circles and label them *X*, *Y*, and *Z*. Each circle represents a set of attribute pieces. The labeling team (or player) decides what the sets are to be, but does not tell the guessing team (or player). The descriptions of the sets should be written down. The guessing team selects an attribute piece, and the labeling team places that piece on the correct region of the diagram. The guessing team continues selecting pieces for placement until its members believe they have correctly identified sets *X*, *Y*, and *Z*. The labeling team answers yes or no to each guess and records the number of guesses. If part or all of a proposed identification is incorrect, the guessing team continues selecting pieces for placement until its members are able to guess all three sets.

After the guessing team correctly identifies the three sets, the teams reverse roles and play again. The winner is the team requiring the *fewest number of pieces (placed on the diagram) and guesses* to make an identification of all three sets.

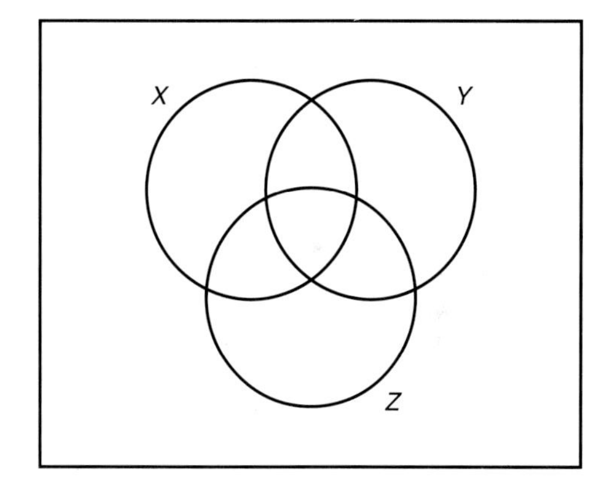

ACTIVITY SET 2.2

GRAPHING SPIROLATERALS

Purpose To draw, analyze, predict, and write conjectures about graphical patterns

Materials Rectangular and isometric grid paper on Material Cards 1 and 2

Activity "Mathematics is an exploratory science that seeks to understand every kind of pattern—patterns that occur in nature, patterns invented by the human mind, and even patterns created by other patterns. To grow mathematically, children must be exposed to a rich variety of patterns appropriate to their own lives through which they can see variety, regularity, and interconnections."[3]

In this activity visual patterns that are created by graphing number sequences will be analyzed. You will make and test predictions and write conjectures based on the data you have collected.

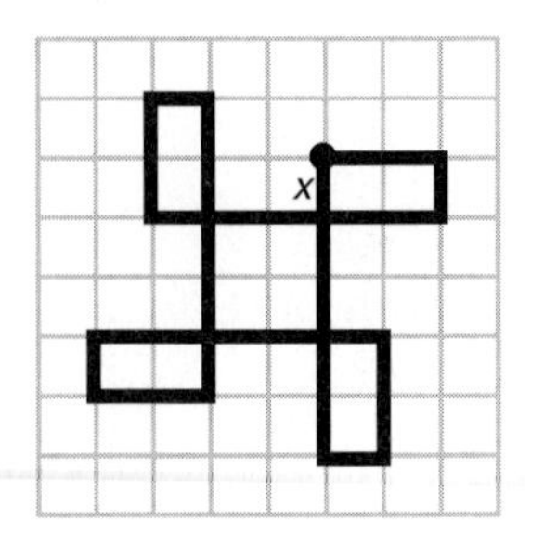

Spirolaterals are graphs of number sequences. The spirolateral shown here is the graph of the sequence 2, 1, 5. Segments of length 2 units, 1 unit, and 5 units were drawn, starting at point *x:* 2 units to the east, 1 unit to the south, 5 units to the west, 2 units to the north, 1 unit to the east, and so on, repeating the number sequence with a 90-degree clockwise turn each time. After the sequence 2, 1, 5 has been repeated four times this spirolateral returns to the starting point and further repetitions will simply retrace the figure. Any spirolateral that returns to its starting point and begins repeating after the sequence has been used one or more times is *completed.*

The number of numbers in a sequence is called the *order of the spirolateral.* The spirolateral generated by the sequence 2, 1, 5 has order three.

★ **1.** Graph the spirolaterals generated by the following consecutive number sequences. Begin at the given starting points on the grid below and continue repeating the sequence until the spirolateral completes or you can see that it will not complete.

a.	1	e.	1, 2, 3, 4, 5
b.	1, 2	f.	1, 2, 3, 4, 5, 6
c.	1, 2, 3	g.	1, 2, 3, 4, 5, 6, 7
d.	1, 2, 3, 4	h.	1, 2, 3, 4, 5, 6, 7, 8

[3]Lynn A. Steen. *On the Shoulders of Giants: New Approaches to Numeracy* (Washington, DC: National Academy Press, 1990): 8.

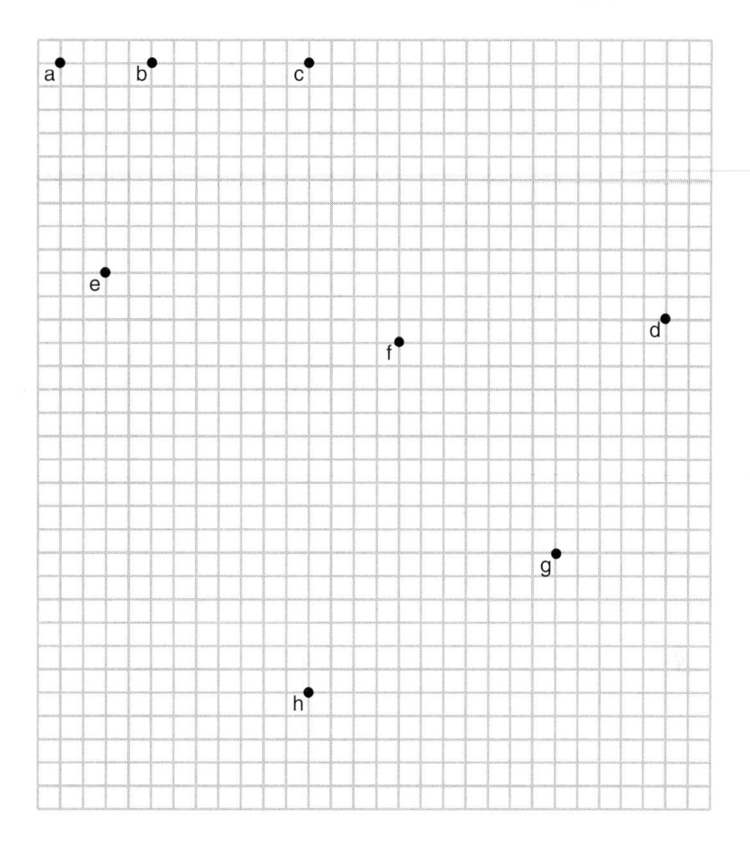

2. Use the information from the graphs you drew in activity 1 to complete the following table. Then use the data from the table to form conjectures about spirolaterals.

	ORDER	COMPLETE (yes, no)	NUMBER OF PASSES (number of times through sequence to complete)
a.	1	________	________
b.	2	________	________
c.	3	________	________
d.	4	________	________
e.	5	________	________
f.	6	________	________
g.	7	________	________
h.	8	________	________

Your preliminary conjectures:

★ **3.** Using the information gathered in the preceding table, and your preliminary conjectures, make a prediction about whether or not the spirolaterals of orders 9, 10, 11, and 12 listed below can be completed, and the number of passes needed when they can. Record your predictions in the table. Then test your prediction by sketching those sequences on the grid provided (start each spirolateral at the given letter on the grid) and record the actual results.

SEQUENCE	ORDER	COMPLETE Prediction	COMPLETE Actual	NUMBER OF PASSES Prediction	NUMBER OF PASSES Actual
i. 1, 2, 3, 4, 5, 6, 7, 8, 9	9	____	____	____	____
j. 1, 2, 3, 4, 5, 6, 7, 8, 9, 10	10	____	____	____	____
k. 1, 2, 3, 4, 5, 6, 7, 8, 9, 10, 11	11	____	____	____	____
l. 1, 2, 3, 4, 5, 6, 7, 8, 9, 10, 11, 12	12	____	____	____	____

j

i

k

l

4. The numbers used to generate spirolaterals do not have to be consecutive. Predict which of the sequences listed below can be completed, and the number of passes needed when they can. Record your predictions in the table. Then test your prediction by graphing the sequences on the grid. Record the actual data you obtain.

	SEQUENCE	ORDER	COMPLETE		NUMBER OF PASSES	
			Prediction	Actual	Prediction	Actual
a.	3	1	____	____	____	____
b.	4, 7	2	____	____	____	____
c.	2, 5, 4	3	____	____	____	____
d.	4, 2, 1, 5	4	____	____	____	____
e.	3, 5, 1, 3, 7	5	____	____	____	____

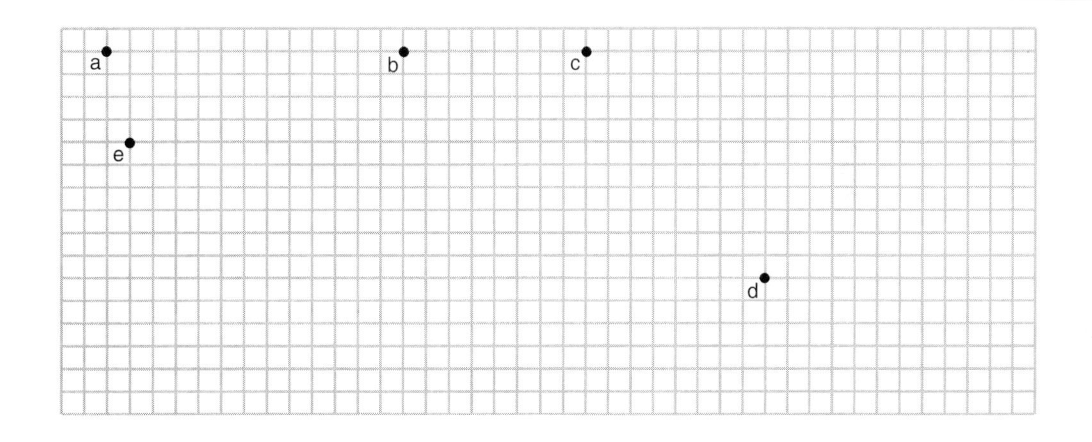

★ 5. In order for a spirolateral to be completed, the total movement in the east direction must equal the total movement in the west direction. Similarly, the total movement in the north direction must equal the total movement in the south direction. For the spirolateral 2, 4, 3, 2, 6, 1, the amount of movement in each direction is shown in the table below. The sum of the numbers in both the east and west directions is 11 and the sum in both the north and south directions is 7. The table also shows that this spirolateral is completed after two passes, because the spirolateral has returned to its starting position.

	E	S	W	N
	2	4	3	2
	6	1	2	4
	3	2	6	1
TOTALS	11	7	11	7

Complete the following tables for the indicated sequences of numbers to check completeness and number of passes.

a. (1, 2, 6)

E	S	W	N

b. (1, 2, 1, 4)

E	S	W	N

c. (5, 2, 1, 7, 3, 4)

E	S	W	N

d. (1, 1, 2, 2, 3, 3)

E	S	W	N

e. (8, 7, 6, 5, 4, 3, 2, 1)

E	S	W	N

f. (1, 2, 3, 4, 5, 4, 3, 2, 1)

E	S	W	N

g. Based on the data you have collected, form and state a conjecture about which sequences of numbers have spirolaterals that can be *completed.* (You may test your conjecture by drawing more spirolaterals using the grid on Material Card 1.)

h. For sequences that can be completed, form and state a conjecture about the number of times the sequences must be repeated (number of passes) before the spirolateral is completed.

6. Spirolaterals can also be graphed on isometric grids. The graph of 1, 2, 3, 4 is shown on the next page. Starting at point x, each turn is 120 degrees clockwise. Graph a few number sequences on isometric grid (Material Card 2) and form some conjectures about which ones can be completed. The three directions on an isometric grid can be labeled east (E), southwest (SW), and northwest (NW). Try analyzing these spirolaterals using a table with three columns labeled E, SW, and NW.

7. You can graph your telephone number, house number, social security number, etc., using spirolaterals. Using a table like the one below, you can give the letters in words number values and graph the resulting number sequence. For example, the name Tam, or the initials T.A.M. would translate into the sequence 2, 1, 4 and its graph on a rectangular or isometric grid would look as follows. Use your own initials to form a spirolateral.

1	2	3	4	5	6	7	8	9
A	B	C	D	E	F	G	H	I
J	K	L	M	N	O	P	Q	R
S	T	U	V	W	X	Y	Z	

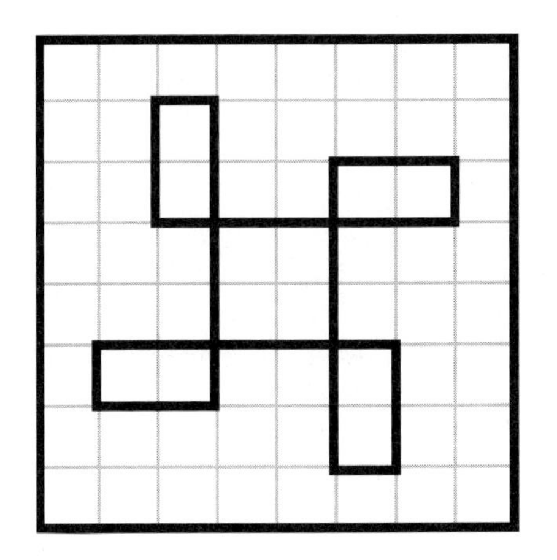

PUZZLER

In the first diagram, a block is standing on end on the table and an identical block is on its side on the floor. The distance between the highest parts of both blocks is 32 inches.

In the second diagram, the blocks are in reversed position and the distance between them is 28 inches.

How tall is the table?

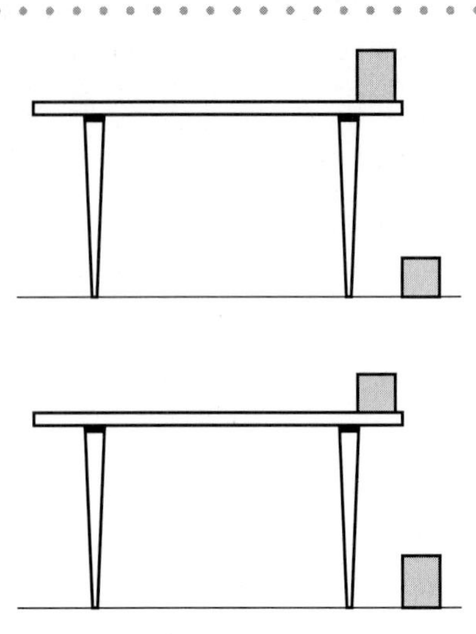

JUST FOR FUN

Both of these games use coordinate grids from Material Card 7.

COORDINATE GUESSING

(2 players or teams)

The object of this game is to determine the coordinates of the opponent's target. Each player uses a pair of rectangular grids such as the ones shown here (Material Card 7). Each player draws three targets on one grid and records guesses as to the opponent's targets on the other grid. A target consists of three adjacent intersection points of the grid (in rows, columns, or diagonals) whose coordinates are integers.

The players should not be able to see each other's grids. They take turns guessing three points at a time by giving the coordinates of each point. When a guessed point is on a target, the opponent must acknowledge this immediately. The game ends when one player has identified all three of the opponents' targets.

Variation: To speed up the game, you may wish to agree to inform your opponent whenever a guess is within 1 unit (horizontally or vertically) of a target.

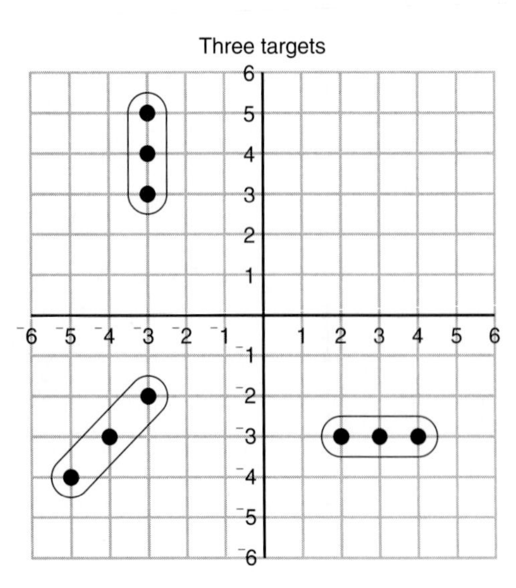

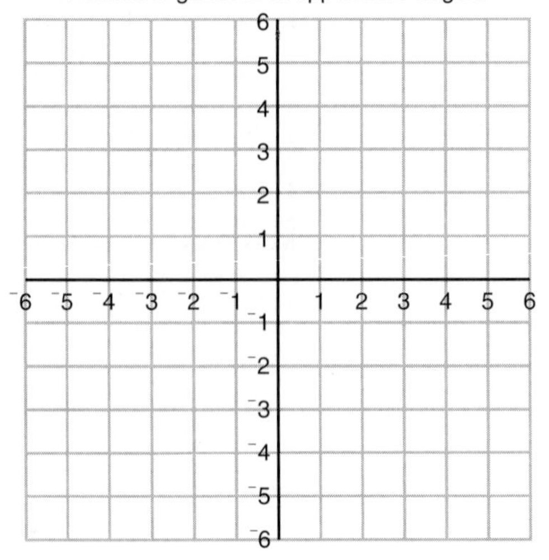

HIDE-A-REGION (2 players or teams)

This game is similar to Coordinate Guessing, but players search for the opponent's rectangle. Each player uses a pair of rectangular grids (Material Card 7). The game begins with the players agreeing on the area for the rectangles. Each player then sketches a rectangle on one grid, with the coordinates of the vertices being integers. (The players should not see each other's grids.) The players take turns guessing points by giving coordinates. Only one guess is allowed on each turn. Each player must tell whether the opponent's guess is an exterior, interior, vertex, or boundary point. Both players keep a record of their opponent's guesses and their own guesses. The first player to locate all four vertices of the opponent's rectangle is the winner.

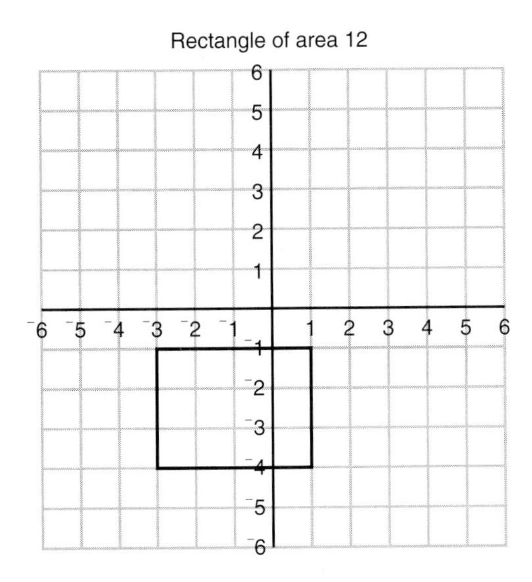

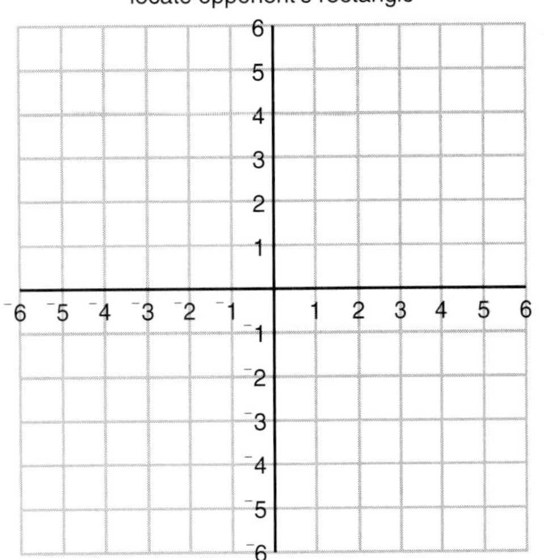

Variation: The players may agree to switch the shape of the region but keep all other rules the same. For example, instead of a rectangle, the region could be a right triangle, triangle, or parallelogram.

ACTIVITY SET 2.3

LOGIC PROBLEMS FOR COOPERATIVE LEARNING GROUPS

Purpose To have cooperative learning groups use deductive reasoning to solve logic problems

Materials The logic problem clue cards and other materials for the five cooperative logic problems are on Material Cards 18, 19, 20, and 21.

Activity "In a cooperative learning situation, students' achievements of their shared goals are positively correlated; students perceive that they can reach their learning goals if and only if the other students in the learning group also reach their goals. Thus, students seek outcomes that are beneficial to all group members. When students work on mathematics assignments, they discuss the material with their group, explain how to complete the work, listen to each other's explanations, encourage each other to understand the solutions, and provide academic help and assistance."[4]

The logic problems in this activity set are designed to be solved cooperatively by a group of four people. Each member of the group will have a clue card that contains statements not found on any other person's card. No member of the group is allowed to read another member's card. The group is to work as a team to solve the problem cooperatively. When the activity begins, members of the group may read their clues to the group. The tasks of the group members are to (1) identify the problem that is to be solved; (2) agree on a process (Will there be one group recorder? Will the solution be arrived at mentally, or will diagrams, charts, or models be used?); (3) listen carefully when clues are read; and (4) make sure that all members of the group understand each step toward the solution.

Inductive reasoning is the process of forming conclusions on the basis of observations and experiments; it involves making an educated guess. *Deductive reasoning* is the process of deriving conclusions from given statements. The cooperative logic problems in this activity set will focus on deductive reasoning.

Cooperative Logic Problems

Form a team of four people. Follow these guidelines as you solve the problems:

- Each team member takes one clue card. (If there are not enough people to form a group of four, some members may hold more than one card.)
- Each person may read the clues aloud but may not show the clue card to another person.
- The team is to identify the problem and agree on a method for solving it.
- The team members are to solve the problem cooperatively by assisting each other.
- The problem is solved when all team members understand the solution and agree that the problem has been solved.

[4]David Johnson and Roger Johnson, "Cooperative Learning in Mathematics Education," in *New Directions for Elementary School Mathematics,* 1989 Yearbook (Reston, VA: National Council of Teachers of Mathematics, 1989), 235.

After a problem has been solved, take a few minutes to record the solution in the blanks provided below. In the space for comments, record thoughts and feelings you had about the problem while they are fresh in your mind. What aids (models, diagrams, charts) did the team use in the solution? How did you feel about working with a team? Looking back and reflecting on the problem, can you see another approach to solving it?

1. **Nine in a Row:** The clue cards and people pieces are on Material Card 18. Use one set of clue cards and one set of people pieces in your cooperative group. Help each other arrange the people pieces as the clues are read.

    _____ _____ _____ _____ _____ _____ _____ _____ _____
    Shortest Tallest

    Comments:

2. **The Tennis Team:** The clue cards (Logic Problem 2) are on Material Card 19. Once you understand the problem, cooperatively devise a way to represent the problem so that you can solve it together.

    Team members for the road trip _____ _____ _____ _____

    Comments:

★ **3. The Prizes:** The clue cards (Logic Problem 3) are on Material Card 19. Devise a way to represent this problem that is acceptable to the entire group.

________	________	________	________
Math prize	English prize	French prize	Logic prize

Comments:

★ **4. Who's Who:** The clue cards (Logic Problem 4) are on Material Card 20.

________	________	________
Pilot's name	Copilot's name	Engineer's name

Comments:

5. Five-House Puzzle: The clue cards (Logic Problem 5) are on Material Card 20, and sets of objects referred to by the clues are on Material Card 21. Use one set of objects for your group.

House color	________	________	________	________	________
Nationality	________	________	________	________	________
Car make	________	________	________	________	________
Beverage	________	________	________	________	________
Pet	________	________	________	________	________

Comments:

JUST FOR FUN

PICA-CENTRO (2 players)

In this game, one player chooses a three-digit number and the other player tries to discover it through logical deduction.

Preparation: Suppose that the chooser is player A and the guesser is player B. Player A begins by selecting a three-digit number, all of whose digits are different. (Zero may be used.) Player A records that number on a slip of paper and keeps it from the view of player B.

Player B uses the pica-centro recording sheet (Material Card 6) to record player B's guesses and player A's responses.

Play: Player B begins the process of deducing the identity of the number that player A has chosen by picking a three-digit number and recording that number on the recording sheet. A pica is a digit that is correct but not in the correct position; a centro is a digit that is both correct and in the correct position. Thus, player A might respond

- "One pica and no centros," which means that one digit has been correctly guessed, but it is in the wrong position
- "No picas and one centro," which means that the guess contains 1 correct digit and it is in the correct position
- "One pica and one centro," which means 1 correct digit in the wrong position and 1 correct digit in the correct position
- "No picas and no centros," which means that all digits are incorrect

Player B uses the clues to figure out other three-digit numbers to try. After each of player B's choices, player A responds with the appropriate number of picas and centros.

Example: Suppose player A chooses 471 and player B's first guess is 123. Player A responds, "You have one pica and no centros," and player B records that on the first line of the recording sheet, as shown below. One logical second choice for player B is 321. Player A responds, "You have no picas and one centro." Player B now deduces that either the 1 or the 3 is correct and in the correct position, but that 2 is not a digit in the number. (Why can player B deduce this?) Player B next chooses 541. Player A responds, "You have one pica and one centro," which player B records on the sheet. Play continues in this manner until player B has deduced player A's number.

Ending: After player B has deduced player A's number, there will be a zero in the pica column and a 3 in the centro column. The players' roles are then reversed, with player B choosing the number and player A deducing the number. The winner is the player who requires the fewest number of turns to identify the number.

Variations: Four-digit, five-digit, or larger numbers can be used, as long as the players agree on the number of digits before playing.

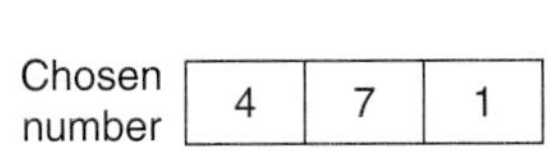

Chosen number	4	7	1

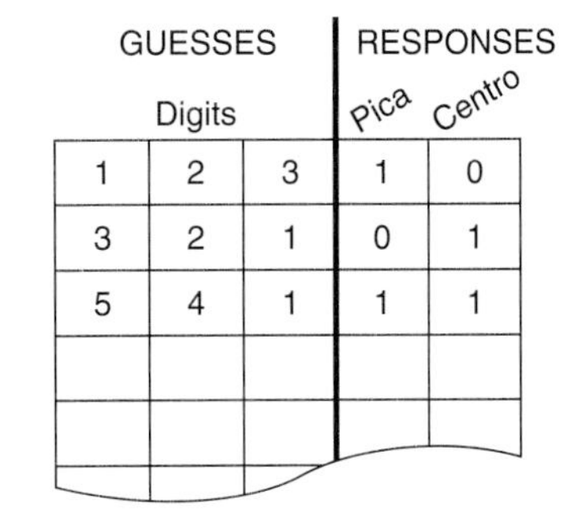

GUESSES			RESPONSES	
Digits			Pica	Centro
1	2	3	1	0
3	2	1	0	1
5	4	1	1	1

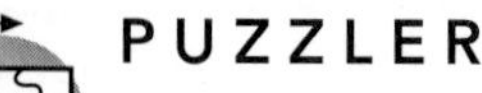

PUZZLER

Divide this parcel of land into two parts that each have the same size and shape (i.e., two congruent pieces).

IDEAS FOR THE ELEMENTARY CLASSROOM

SUGGESTED CLASSROOM ACTIVITY: MATH COMMUNICATION

Two students sit on opposite sides of a barrier so that neither one can see the objects in front of the other.

Communication and Reasoning: Give each student a complete set of attribute pieces. One student selects a single piece and places it against the barrier. The other student attempts to identify that piece by asking questions that can be answered yes or no and by sorting his or her complete set of pieces to keep a record of which pieces have been ruled out (as in the Attribute Guessing Game). The students then reverse roles.

Communication and Visualization: Give each student the same number of cubes. One student uses the cubes to build a structure and then gives verbal instructions to the other student as to how it is constructed. The other student builds according to the instructions. When the student is finished building, the barrier is removed and the structures are compared. If the structures are not identical, the students discuss statements that led to misunderstanding or confusion. The students then reverse roles.[5]

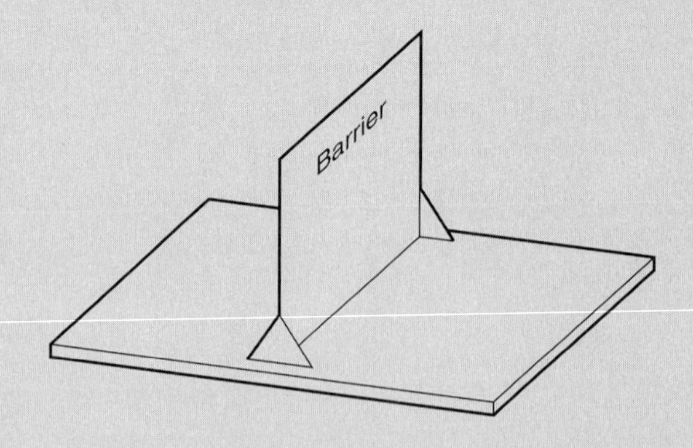

[5]See Glenda Lappan and Pamela Schram's "Communication and Reasoning: Critical Dimensions of Sense Making in Mathematics," in *New Directions for Elementary School Mathematics*, 1989 Yearbook (Reston, VA: National Council of Teachers of Mathematics, 1989), 14–30.

Readings for More Classroom Ideas

Cook, Marcy, "Ideas, Create a House Number." *Arithmetic Teacher* 36 (January 1989): 19–24.

Cruishank, Douglas E., "Sorting, Classifying, and Logic." *Arithmetic Teacher* 21 (November 1974): 588–598.

Edwards, Thomas G. "Students as Researchers: An Inclined-Plane Activity." *Mathematics Teaching in the Middle School* 1 (November-December, 1995): 532–535.

Goodman, Jan M. *Group Solutions: Cooperative Logic Activities, Grades K–4. GEMS Series.* Berkeley, CA: Lawrence Hall of Science, 1992.

Lappan, Glenda, and Pamela Schram. *Communication and Reasoning: Critical Dimensions of Sense Making in Mathematics,* in *New Directions for Elementary School Mathematics,* 1989 Yearbook, edited by P. Trafton and A. Schulte. Reston, VA: National Council of Teachers of Mathematics, 1989, 14–30.

Malloy, Carol E. *Developing Mathematical Reasoning in the Middle Grades: Recognizing Diversity,* in *Developing Mathematical Reasoning in Grades K–12,* 1999 Yearbook, edited by L. V. Stiff and R. R. Curcio. Reston, VA: National Council of Teachers of Mathematics, 1999: 13–21.

Odds, Frank C. "Spirolaterals." *The Mathematics Teacher* 66 (February 1973): 121–124.

Phillips, Elizabeth, et al. *Patterns and Functions,* in the *Curriculum and Evaluation Standards for School Mathematics Addenda Series: Grades 5–8.* Reston, VA: National Council of Teachers of Mathematics, 1991.

Russell, Susan Jo. *Mathematical Reasoning in the Elementary Grades,* in *Developing Mathematical Reasoning in Grades K–12,* 1999 Yearbook, edited by L. V. Stiff and R. R. Curcio. Reston, VA: National Council of Teachers of Mathematics, 1999: 1–12.

Schwartz, James E. "'Silent Teacher' and Mathematics as Reasoning." *Arithmetic Teacher* 40 (October 1992): 122–124.

Scott, Thomas L. "A Different Attribute Game." *Arithmetic Teacher* 28 (March 1981): 47–48.

Shaw, J., et al. "Cooperative Problem Solving: Using K-W-D-L as an Organizational Technique." *Teaching Children Mathematics* 3 (May 1997): 482–486.

Warman, Michele, "Fun with Logical Reasoning." *Arithmetic Teacher* 29 (May 1982): 26–30.

Welchman-Tischler, Rosamond. "Making Mathematical Connections." *Arithmetic Teacher* 39 (May 1992): 12–18.

Whitin, David J. "Bring On the Buttons." *Arithmetic Teacher* 36 (January 1989): 4–6.

CHAPTER 3

WHOLE NUMBERS

Children come to understand number meanings gradually. To encourage these understandings, teachers can offer classroom experiences in which students first manipulate physical objects and then use their own language to explain their thinking. This active involvement in, and expression of, physical manipulations encourages children to reflect on their actions and to construct their own number meanings. In all situations, work with number symbols should be meaningfully linked to concrete materials.[1]

ACTIVITY SET 3.1

MODELS FOR NUMERATION WITH MULTIBASE PIECES

Purpose To use base-five number pieces and visual diagrams for other bases to introduce the concepts of grouping and place value in a positional numeration system

Materials Base-five number pieces from the Manipulative Kit

Activity The concepts underlying positional numeration systems—grouping, regrouping, and place value—become much clearer when illustrated with concrete models, and many models have been created for this purpose. Models for numeration fall roughly into two categories or levels of abstractness. *Grouping models*, such as bundles of sticks and Dienes blocks, show clearly how groupings are formed; *abacus-type models*, such as chip-trading or place-value charts, use color and/or position to designate different groupings.

[1]*Curriculum and Evaluation Standards for School Mathematics* (Reston, VA: National Council of Teachers of Mathematics, 1989), 38.

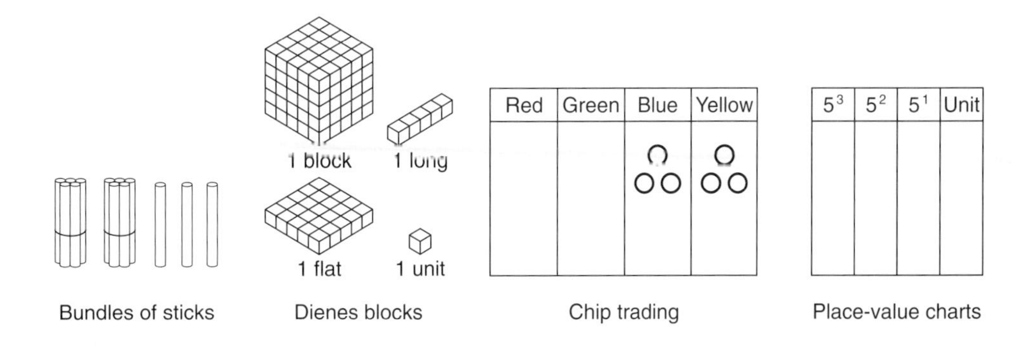

The base number pieces shown here are in the Manipulative Kit. They are called base-five number pieces because each piece, other than the unit, has 5 times as many unit squares as the preceding piece.

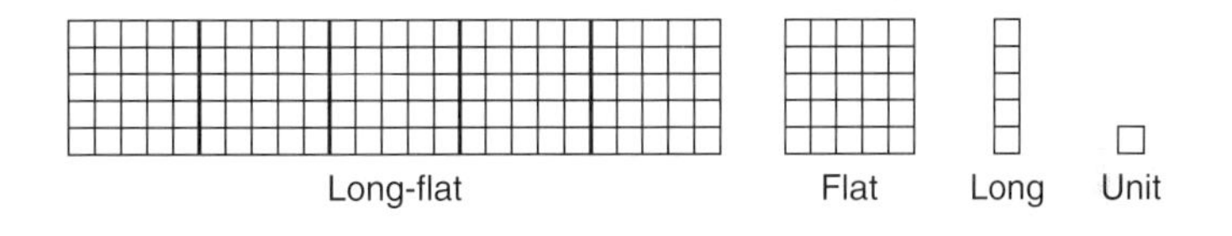

1. Use your base-five pieces to form the collection consisting of 8 units, 6 longs, and 5 flats, as shown in figure a. Exchange 5 units for 1 long, 5 longs for 1 flat, and 5 flats for 1 long-flat. This results in the collection shown in figure b. The two collections are *equivalent* because they both contain the same number of unit squares, but the second collection has fewer base-five pieces. (The first collection has 19 pieces, and the second has 7 pieces.)

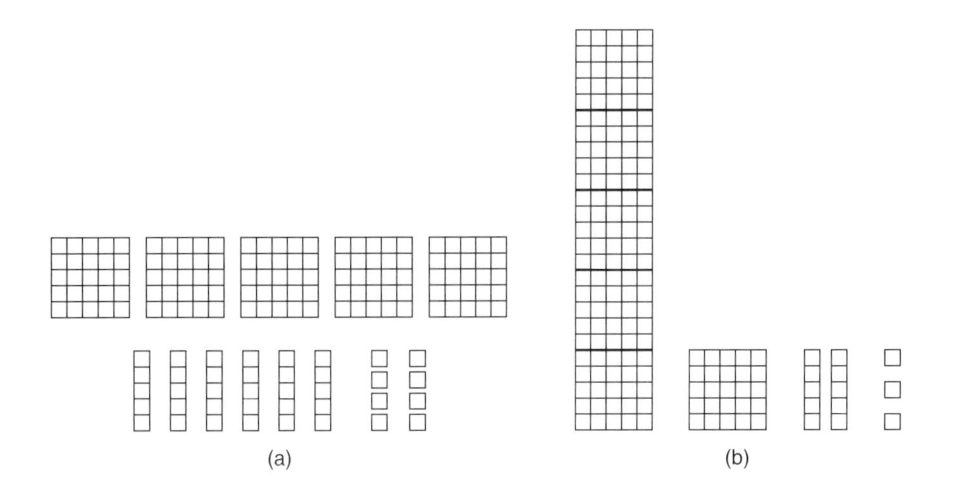

Use your base-five pieces to form each of the following collections. Make exchanges until you have an equivalent collection with the least number of base-five pieces. (This is called the *minimal collection.*) Record the number of long-flats, flats, longs, and units in your minimal collection in the table below.

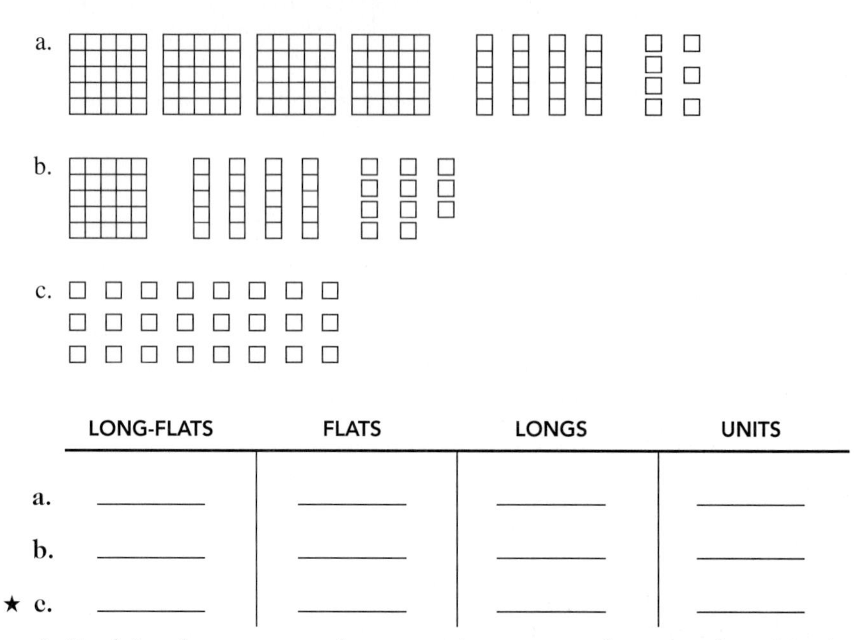

	LONG-FLATS	FLATS	LONGS	UNITS
a.	______	______	______	______
b.	______	______	______	______
★ c.	______	______	______	______

d. Explain why you never have to write a numeral greater than 4 in the table.

2. Using the base-five pieces, you can represent a collection of 28 units by a minimal collection of 1 flat, 0 longs, and 3 units. This is recorded in the table below.

Using base-five pieces to aid visualization, supply the missing entries in the following table:

	NO. OF UNIT SQUARES	LONG-FLATS	FLATS	LONGS	UNITS
a.	28	0	1	0	3
b.	31	______	______	______	______
★ c.	126	______	______	______	______
d.	200	______	______	______	______
e.	______	0	1	2	3
★ f.	______	3	3	0	3
g.	______	4	4	4	4

3. Minimal collections of base-five pieces can be recorded without using tables. For example, the first entry in the table above can be written as 103_{five}. In doing this, we must agree that the positions of the digits from right to left represent the numbers of units, longs, flats, and long-flats. This method of writing numbers is called *positional numeration,* and 103_{five} is called a *base-five* *numeral.* Write the base-five numerals for each of the other entries in the table in problem 2.

a. 103_{five} b. ★ c. d.

e. ★ f. g.

4. The base-five pieces that represent the numeral 2034_{five} are shown here. There are a total of 269 unit squares in these pieces.

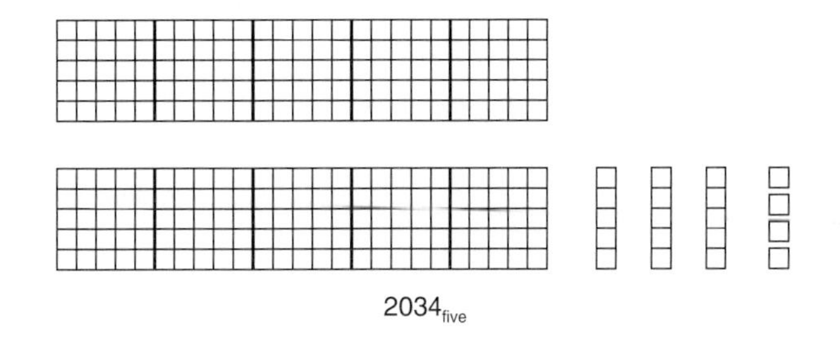

2034_{five}

Represent the following numbers with your base-five pieces, and determine the total number of unit squares in each.

	BASE-FIVE NUMERAL	TOTAL NUMBER OF UNIT SQUARES
a.	2304_{five}	________
★ b.	1032_{five}	________
c.	2004_{five}	________

5. Here are the first three number pieces for base seven. Draw the first three number pieces for base three and base ten in the space provided below.

Base seven

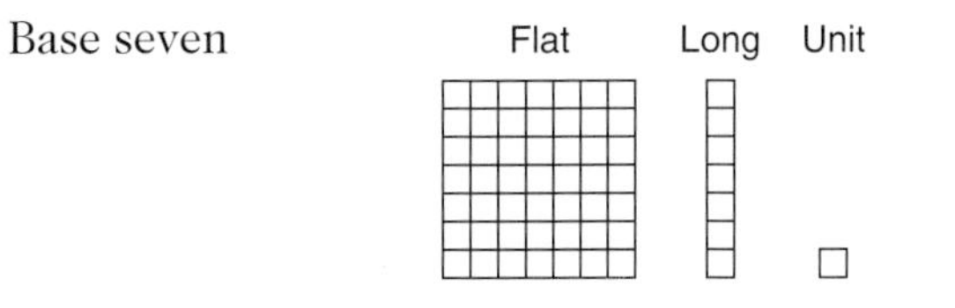

★ **a.** Base three

b. Base ten

6. Draw a diagram of the collection of base pieces representing each of the following numbers. Then determine the total number of unit squares in each collection.

a. 122_{three} Total number of unit squares ________

★ **b.** 425_{seven} Total number of unit squares ________

c. 157_{ten} Total number of unit squares ________

7. For each of the bases in parts a, b, and c, make a sketch of the minimal collection of base pieces that represents the collection of unit squares shown here. Then write the base numeral below the sketch.

a. Base five

★ **b.** Base nine

c. Base ten

JUST FOR FUN

The numbers 1, 2, 4, 8, 16, 32, and so on are called *binary numbers*. Following 1, each number is obtained by doubling the previous number. There are a variety of applications of binary numbers. Many of these applications provide solutions to games and puzzles, such as the intriguing mind-reading cards and the ancient game of Nim.

MIND-READING CARDS

With the five cards in Fig. 1, you can determine the age of any person who is not over 31. Which of these cards is your age on? If you selected card 4 and card 16, then you are 20 years old. If you selected cards 1, 2, and 16, then you are 19 years old.

1. The number in the upper left-hand corner of each of these cards is a binary number. Write your age as the sum of the fewest possible binary numbers. On which cards do the binary numbers that sum to your age appear? On which cards does your age appear?

★ **2.** Write the number 27 as a sum of binary numbers. On which cards do these binary numbers occur? On which cards does 27 occur?

3. If someone chooses a number that is only on cards 1, 2, and 8, what is this number?

★ **4.** Explain how the mind-reading cards work.

★ **5.** The mind-reading cards can be extended to include greater numbers. The next card is card 32 (see Fig. 2). To extend this system to six cards, more numbers must be placed on the first five cards. For example, $33 = 32 + 1$, so 33 must be put on card 1 as well as on card 32. On which cards should 44 be placed?

6. Write the additional numbers that must be put on cards 1, 2, 4, 8, 16, and 32 in Fig. 2 to extend this system to six cards.

7. Cut out the mind-reading cards from Material Card 22 and use them to intrigue your friends or students. Ask someone to select the cards containing his or her age, or to pick a number less than 64 and tell you which cards it is on. Then, amaze them by revealing that number.

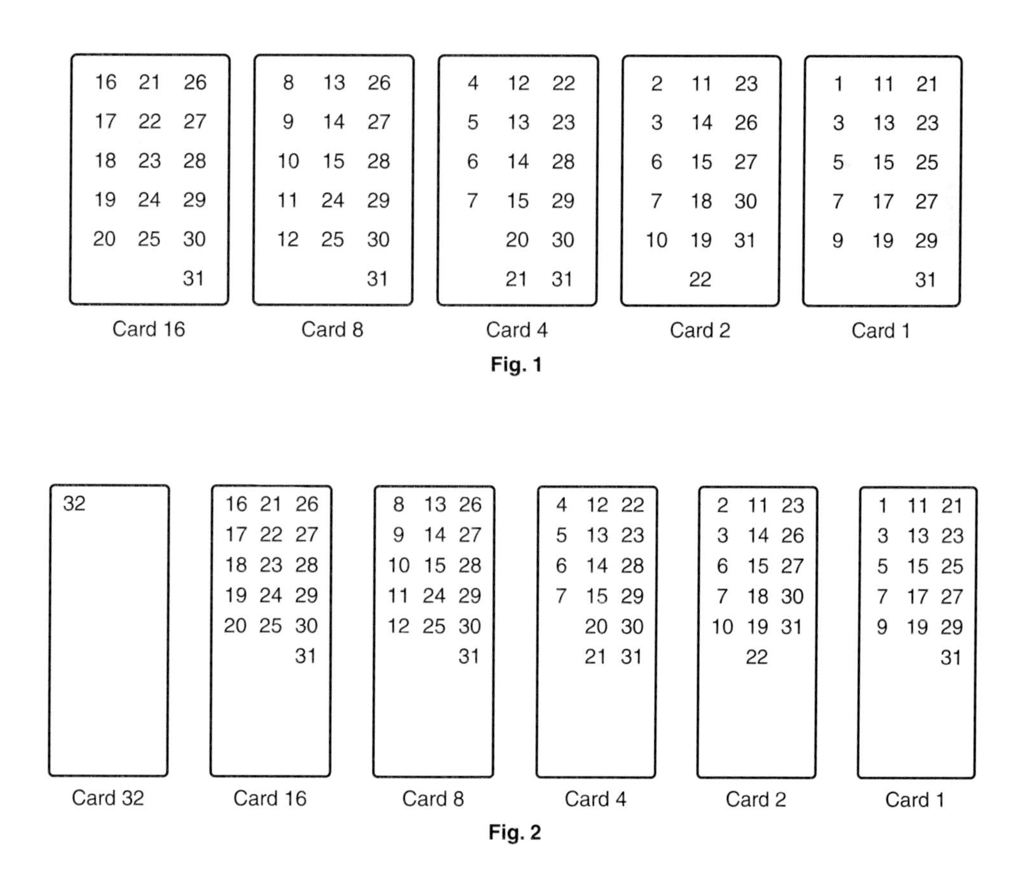

Fig. 1

Fig. 2

GAME OF NIM (2 players)

The game of Nim is said to be of ancient Chinese origin. In its simplest form, it is played with three rows of toothpicks or markers. On a turn, a player may take any number of toothpicks from any one row. The winner of the game is the player who takes the last toothpick or group of toothpicks. Try a few rounds of this game with a classmate.

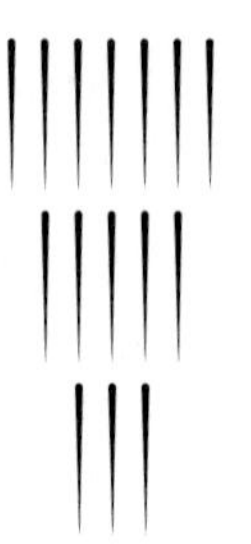

With a little practice, it is easy to discover some situations in which you can win. For example, if it is your opponent's turn to play and there is a 1-2-3 arrangement, as shown here, you will be able to win. Sketch three more situations in which you will win if it is your opponent's turn to play.

Winning Strategy: A winning strategy for the game of Nim involves binary numbers.

1. On your turn, group the remaining toothpicks in each row by binary numbers, using the largest binary numbers possible in each row, as shown in this example.

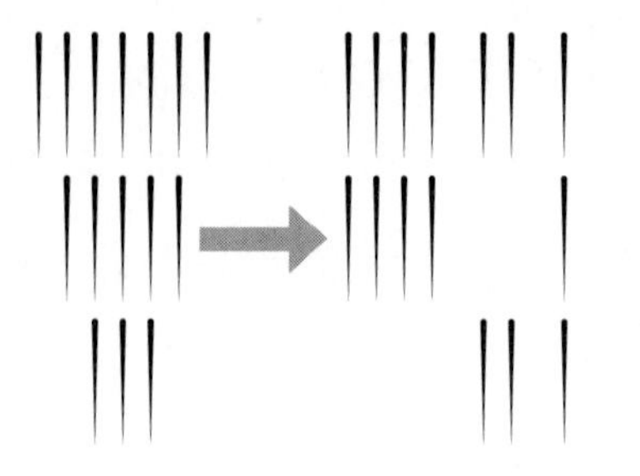

2. Then remove toothpicks so that there is an *even number* of each type of pile left. In the example above, you will see an even number of 4s, an even number of 2s, and an odd number of 1s. So 1 toothpick must be taken from one of the rows. Show that the 1-2-3 arrangement above has an even number of binary groups.

★3. Assume it is your turn and you have the arrangement shown here. Cross out toothpick(s) so that your opponent will be left with an even number of binary groups.

★4. After the markers in each row have been grouped by binary numbers, there will be either an *even situation* (an even number of each type of binary group) or an *odd situation* (an odd number of at least one type of binary group). Experiment to see whether one turn will always change an even situation to an odd situation. Is it possible in one turn to change an even situation into another even situation?

Generalized Nim: Play the game of Nim with any number of rows and any number of toothpicks in each row. Try the strategy of leaving your opponent with an even situation each time it is his or her turn. Is this still a winning strategy?

ACTIVITY SET 3.2

ADDING AND SUBTRACTING WITH MULTIBASE PIECES

Purpose To use multibase number pieces to visually illustrate addition and subtraction

Materials A pair of dice and base-five number pieces from the Manipulative Kit

Activity The operations of addition and subtraction are inverses of each other. *Addition* is explained by putting together sets of objects, and *subtraction* is explained by taking away a subset of objects from a given set. This dual relationship between subtraction and addition can be seen in the trading-up game, which introduces addition, and the trading-down game, which introduces subtraction.

Algorithms for addition and subtraction are step-by-step procedures for adding and subtracting numbers. One of the best ways to gain insight into the addition and subtraction algorithms is by thinking the process through using manipulatives that represent other number bases. The base-five number pieces will be used in the following activities.

Addition

1. **Trading-Up Game** (2 to 5 players): Use your base-five pieces to play. On a player's turn, two dice are rolled. The total number of dots facing up on the dice is the number of base-five units the player wins. At the end of a player's turn, the player's pieces should be traded (regrouped), if necessary, so that the total winnings are represented by the fewest number of base-five pieces (minimal collection). The base-five numeral for the minimal collection should be recorded. The first player to get 1 long-flat wins the game.

   Example: On the player's first turn, shown below, the 8 units won were traded for a minimal collection and the corresponding base-five numeral was recorded.

   On the player's second turn, the 8 additional pieces won were added to the player's collection. After trading to get a minimal collection, the player recorded the base-five numeral for the total collection.

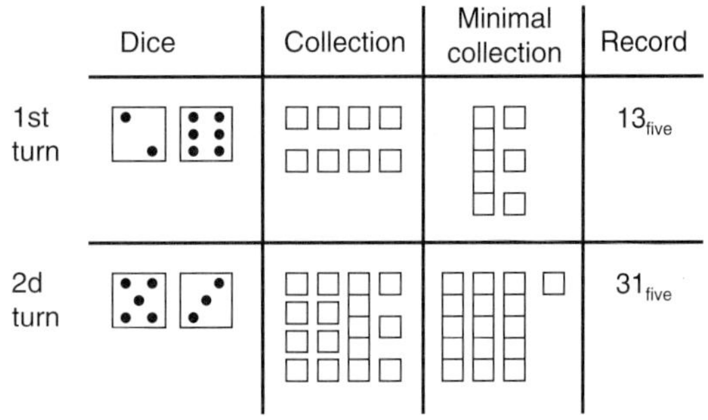

	Dice	Collection	Minimal collection	Record
1st turn				13_{five}
2d turn				31_{five}

2. Answer the following questions for the trading-up game described in activity 1.

a. What is the maximum number of longs and units you can win in one roll of the two dice?

★ **b.** By comparing the maximum collection for one turn (part a) to the long-flat, determine the fewest number of rolls in which a game can be won. Explain your reasoning.

c. Determine the maximum number of turns it can take to obtain a long-flat. Explain how you arrived at your conclusion. (*Hint:* Compare the least possible winnings from one roll of the dice to the long-flat.)

★ **d.** If your total is 421_{five} after several turns, will it be possible for you to win on your next roll? Explain why or why not.

e. Suppose your total score is 333_{five}. What is the fewest number of turns in which you can reach 1000_{five}? Explain how you reached your conclusion. (*Hint:* Use your base-five pieces to represent 333_{five}.)

3. Numbers written as base-five numerals can be added with base-five pieces. For example, $324_{\text{five}} + 243_{\text{five}}$ can be added by representing each number with base-five pieces, combining the collections, and trading (regrouping) to find the minimal collection, as shown below.

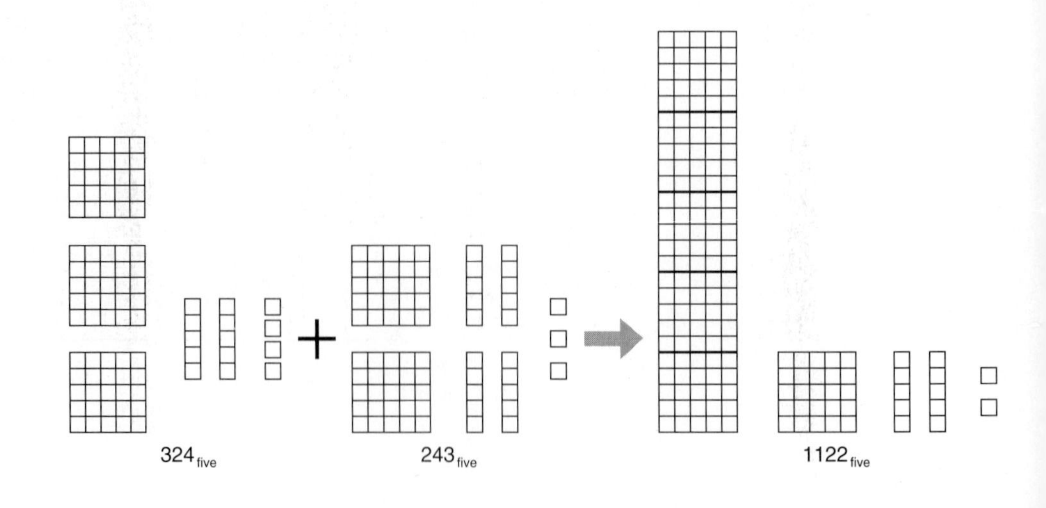

Use base-five pieces to represent each of the following sums. Then regroup to obtain the minimal collection, and record the base-five numeral for this collection.

★ **a.** $43_{five} + 24_{five} =$ **b.** $313_{five} + 233_{five} =$

★ **c.** $304_{five} + 20_{five} + 120_{five} + 22_{five} =$ **d.** $1000_{five} + 100_{five} + 10_{five} =$

4. Numbers in other bases can be added mentally by visualizing the appropriate base number pieces or by making sketches of the appropriate base pieces. For example, in order to add $1221_{three} + 122_{three}$, the following diagrams show what regrouping must be done to obtain 2 long-flats, 1 flat, and 2 longs.

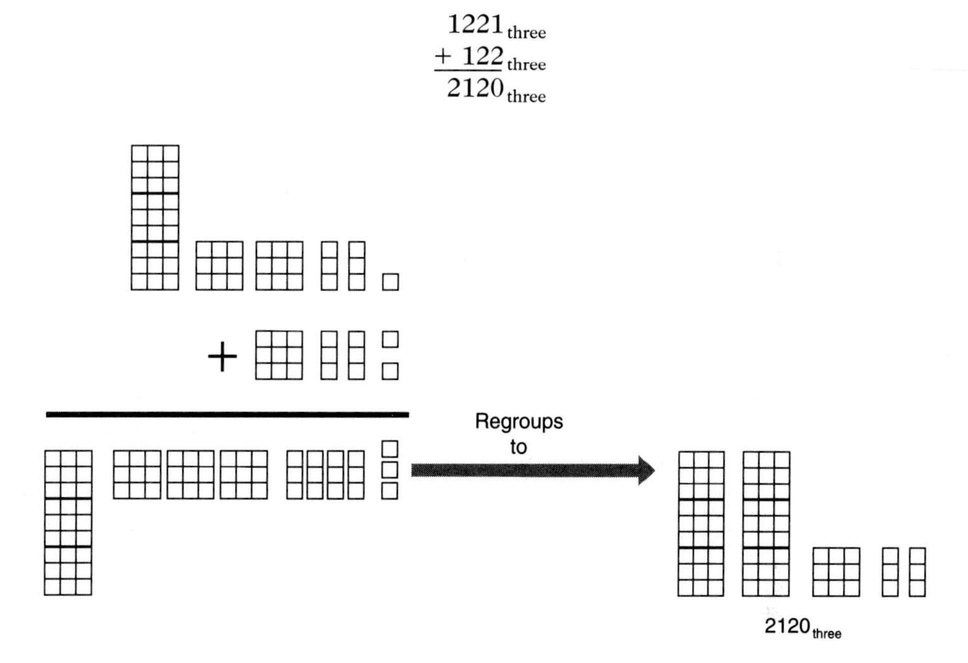

Make sketches of base number pieces representing the following sums and their regroupings (as above). Record the numeral for each sum.

★ **a.**
$$\begin{array}{r} 201_{three} \\ +\ 102_{three} \\ \hline \end{array}$$

b.
$$\begin{array}{r} 2312_{four} \\ +\ 203_{four} \\ \hline \end{array}$$

★ **c.**
$$\begin{array}{r} 255_{six} \\ +\ 134_{six} \\ \hline \end{array}$$

d.
$$\begin{array}{r} 11011_{two} \\ +\ 10101_{two} \\ \hline \end{array}$$

Subtraction

5. Trading-Down Game (2 to 5 players): Use your base-five pieces to play. Each player begins the game with 1 long-flat. On a player's turn, two dice are rolled. The total number of dots facing up on the dice is the number of units the player discards. The object of the game is to get rid of all base-five pieces. The first player to do so wins the game. On some turns the player will have to trade (regroup) in order to discard the exact number of pieces. At the end of a player's turn, the remaining pieces should be a minimal collection, and the base-five numeral should be recorded.

Example: On the player's first turn, shown below, the long-flat was traded for other base-five pieces so that 1 long and 2 units could be discarded. The numeral representing the remaining minimal collection was recorded.

On the player's second turn, 1 long was traded for units so that the correct number of pieces could be discarded. The base-five numeral for the remaining collection was recorded.

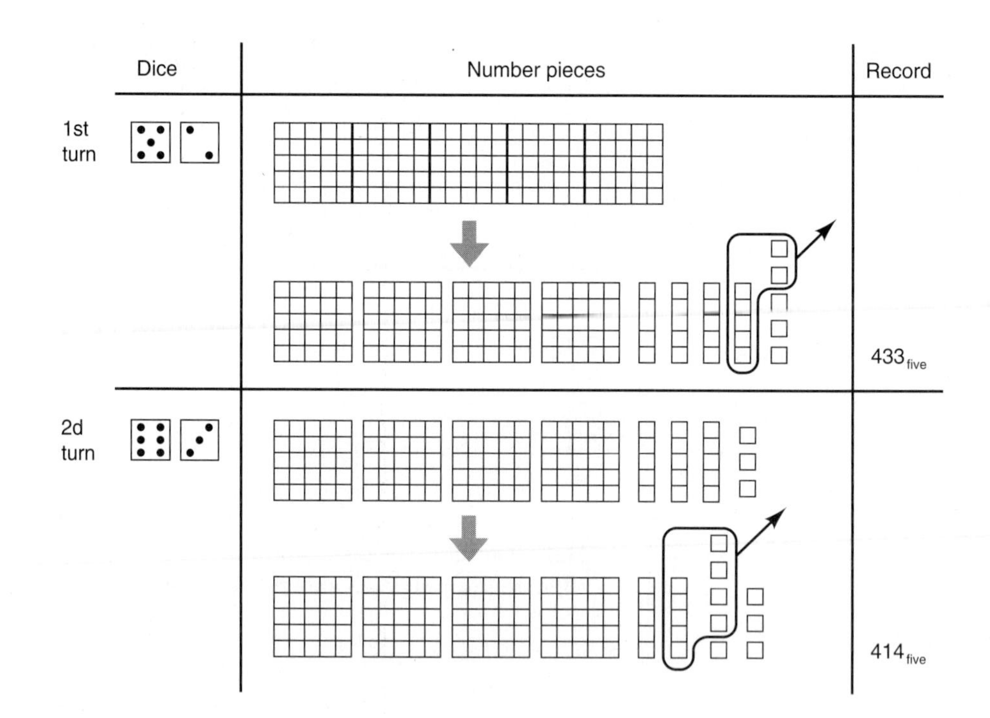

6. Answer the following questions for the trading-down game described in activity 5.

   a. Determine the fewest number of rolls in which a player can discard all pieces. Explain your conclusion.

★ b. If a player's total was reduced from 313_{five} to 242_{five} in one turn, what did the player roll on that turn? (*Hint:* Use your base number pieces to represent both scores.)

7. Here are three methods that students have been observed using to determine $232_{\text{five}} - 143_{\text{five}}$ with their base-five pieces. Work through each method using your base number pieces. Then briefly explain the strategy or technique you believe was used in each method.

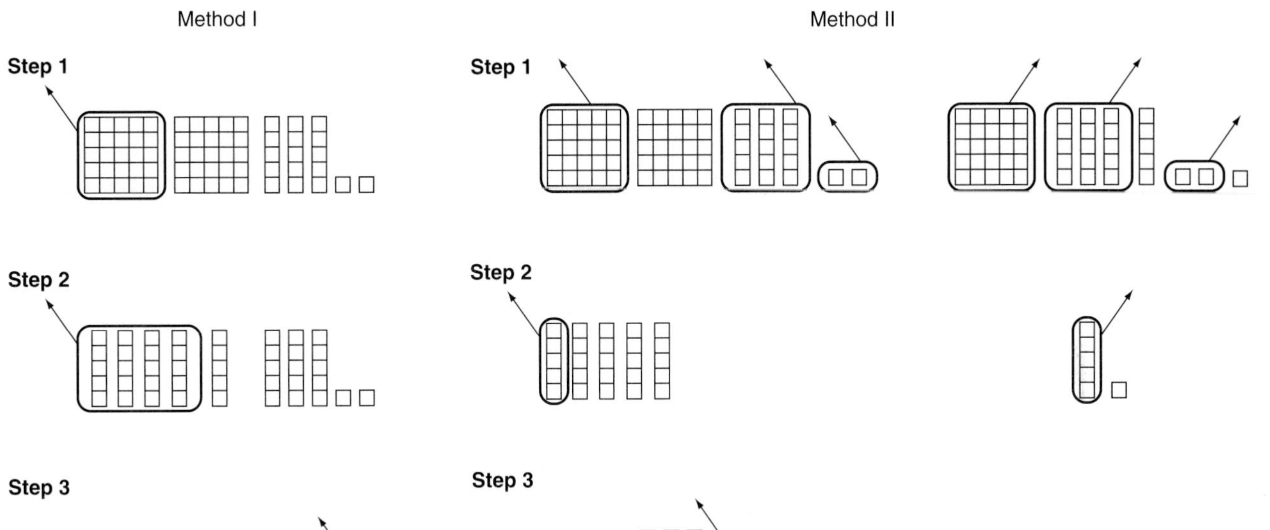

Step 3

Step 3

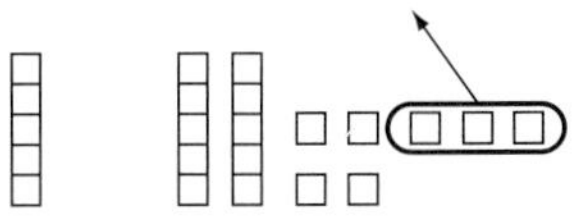

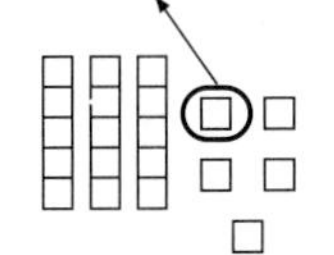

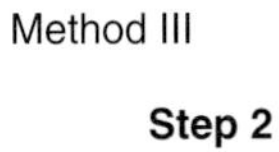

Method III

Step 1

Step 2

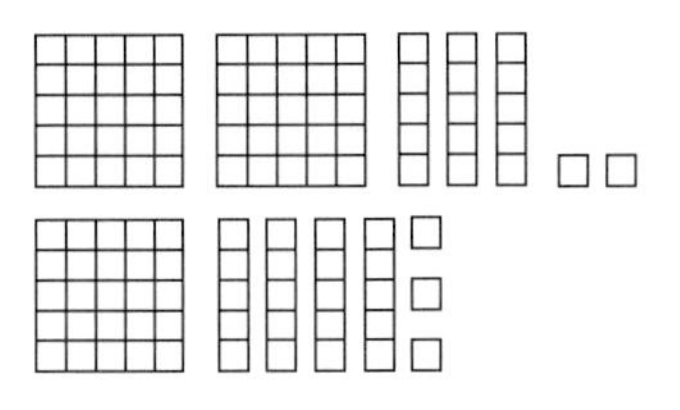

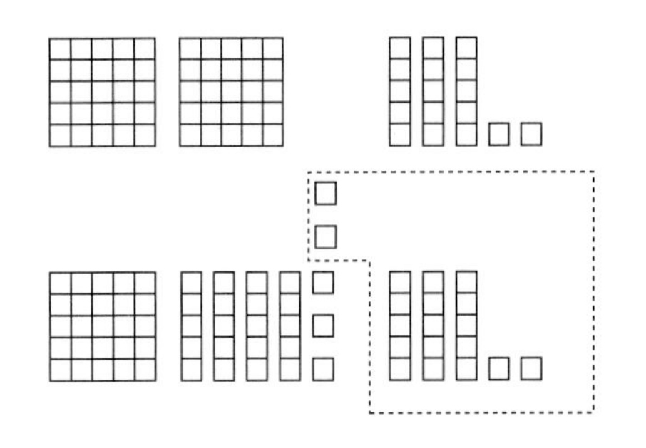

8. Methods I and II in activity 7 are forms of the *take away* model for subtraction, and method III is called the *comparison* model. Use your base number pieces and one or more of these subtraction models to determine the following differences.

★ a. $1142_{five} - 213_{five}$ b. $2331_{five} - 124_{five}$ ★ c. $4112_{five} - 143_{five}$

9. Determine each of the following differences by drawing diagrams of the base number pieces that represent the numerals.

★ a. $1221_{three} - 122_{three}$ b. $2312_{four} - 203_{four}$

★ c. $1010_{two} - 101_{two}$ d. $101_{ten} - 11_{ten}$

JUST FOR FUN

FORCE OUT (2 players)

The whole numbers from 1 to 21 are listed as shown below. The players take turns crossing off one, two, or three consecutive numbers, starting with 1. For example, player A might cross off 1 and 2, player B might cross off 3, 4, and 5, and so forth. The loser is the player who is forced to cross off the last number, 21. Play the game a few times. Can you develop a strategy for winning?

1 2 3 4 5 6 7 8 9 10 11 12 13 14 15 16 17 18 19 20 21

★ ***Winning Strategy:*** With a little practice, you will see some "winning numbers." Suppose it is your turn to move and the numbers from 1 to 15 have been crossed off. If you cross off the number 16, you can win the game no matter what your opponent does. Why? What number preceding 16 must you cross off in order to guarantee that you will get to cross off 16? Work backward in this manner to develop a winning strategy. Can you always win if you start? Can you always win if your opponent starts?

Toothpick Variation: Twenty-one toothpicks are spread in a row, as shown below. One of the toothpicks is bent. The players take turns selecting 1, 2, 3, or 4 toothpicks from the row. The loser is the player who is forced to select the bent toothpick. Play the game a few times. Can you develop a strategy for winning?

| | | | | | | | | | | | | | | | | | | | >

Other Variations: An arbitrary whole number is selected. The players take turns mentally subtracting 1, 2, 3, 4, or 5 from that number. The loser is the player who is forced to reach zero. Describe a strategy for winning this game. (*Hint:* Start with a number like 30 and play the game a few times.)

PUZZLER

Three cardboard disks are numbered on both sides. When the disks are dropped onto a table, the sum of the three visible numbers will be 6, 7, 8, 9, 10, 11, 12, or 13. What number is written on the back of each disk?

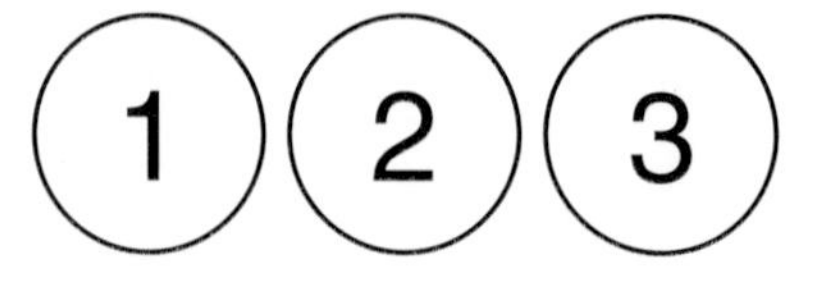

ACTIVITY SET 3.3

MULTIPLYING WITH BASE-TEN PIECES

Purpose To use base-ten pieces to build a visual model for multiplication

Materials Base-ten pieces from the Manipulative Kit

Activity Multiplication is often thought of as repeated addition. For example,

$$4 \times 8$$

can be thought of as 4 copies of 8, or

$$8 + 8 + 8 + 8$$

If repeated addition is illustrated with 4 sets of 8 tiles, the tiles can be pushed together to form a rectangular array of squares with dimensions 4 by 8.

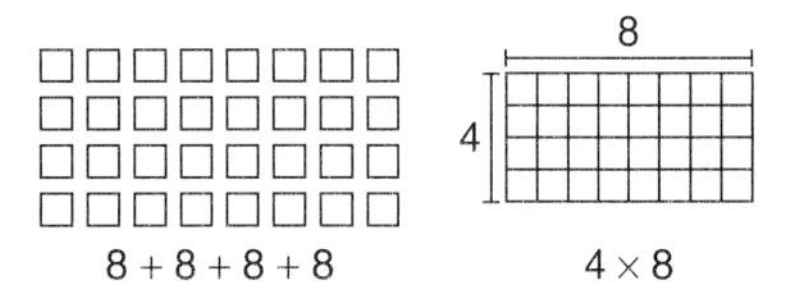

So, the conceptual model for multiplication will be a rectangle. The two numbers being multiplied are the dimensions of the rectangle, and the number of squares in the rectangle is the product. In this activity set, base-ten pieces will be used to illustrate the paper-and-pencil multiplication algorithm.

1. The product 7×12 can be represented with base-ten pieces as 7 copies of 12. When the pieces are pushed together, they form a rectangle with dimensions 7 $\times$ 12. By regrouping the base pieces to form a minimal collection of 8 longs and 4 units, we can determine that the product is 84.

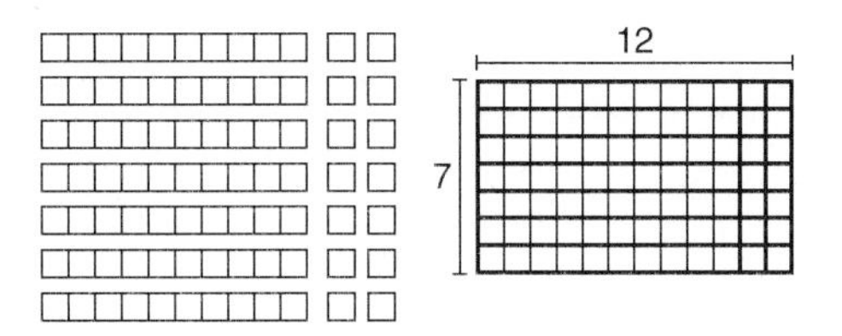

Determine the following products by forming rectangles with your base-ten pieces and then regrouping to obtain the minimal collection. Sketch each rectangle, and write the numeral representing the minimal collection beneath it.

a. 6×14

b. 7×23

★ 2. Explain how it is possible, using base-ten number pieces, to multiply a one-digit number by a two-digit number without knowing any multiplication facts. For example, think of representing 7×64 with your base-ten number pieces. Draw a diagram to illustrate your explanation.

3. The product 12×13 is represented below with base-ten pieces. Several exchanges can be made without affecting the dimensions of the rectangle. Ten longs can be exchanged for 1 flat, and groups of 10 adjacent units can be exchanged for longs. The resulting rectangle is composed of 1 flat, 5 longs, and 6 units, so the product of 12 and 13 is 156.

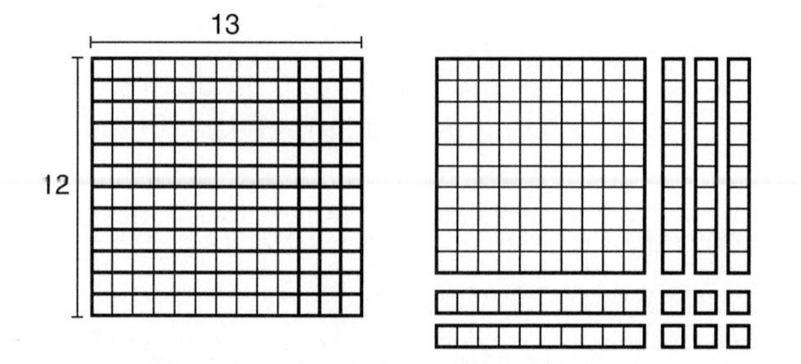

Use your base number pieces to determine these products. Make exchanges when possible, and use as few base pieces as you can when building your rectangle. Sketch each rectangle, and record the minimal set of base-ten pieces in the rectangle.

a. 13×13

★ b. 21×23

c. 17×12

★ **4.** Explain how it is possible, using base-ten number pieces, to multiply a two-digit number by a two-digit number without knowing any multiplication facts. Draw a diagram to illustrate your explanation for a specific product.

5. The rectangle model for multiplication corresponds very closely to the usual paper-and-pencil algorithm for multiplication. Use your base-ten number pieces to form the rectangle representing 23×24.

★ **a.** Here is a paper-and-pencil procedure that uses four partial products for multiplication. Make a diagram of the base-ten number piece rectangle representing 23×24. Clearly match each partial product with the corresponding region of your diagram.

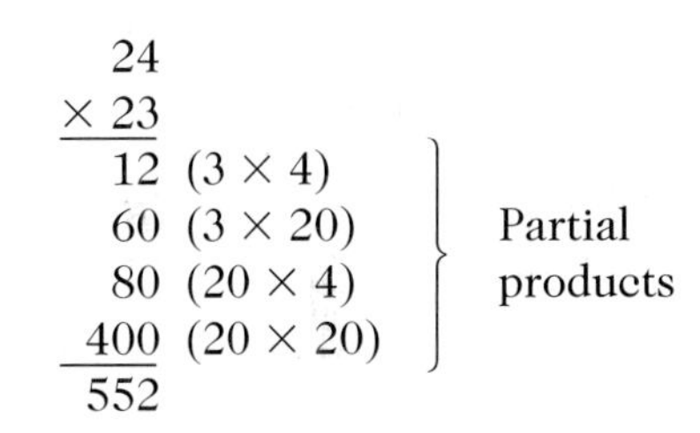

$$\begin{array}{rll} & 24 & \\ \times & 23 & \\ \hline & 12 & (3 \times 4) \\ & 60 & (3 \times 20) \\ & 80 & (20 \times 4) \\ & 400 & (20 \times 20) \\ \hline & 552 & \end{array} \left.\vphantom{\begin{array}{l}1\\1\\1\\1\end{array}}\right\} \text{Partial products}$$

b. The following procedure uses two partial products for multiplication. Make another sketch of your number piece rectangle, and match the parts of the rectangle with the corresponding partial products.

$$\begin{array}{rll} & 24 & \\ \times & 23 & \\ \hline & 72 & (3 \times 24) \\ & 480 & (20 \times 24) \\ \hline & 552 & \end{array} \left.\vphantom{\begin{array}{l}1\\1\end{array}}\right\} \text{Partial products}$$

6. Once you become familiar with the base-ten number piece model, it is easy to sketch diagrams of rectangles in order to compute products. For example, to compute 33×41, you would outline the rectangle as in the first sketch below and fill it with flats, longs, and units as in the second sketch. The product is obtained by counting the number pieces (12 hundreds, 15 tens, and 3 units) and adding (1353).

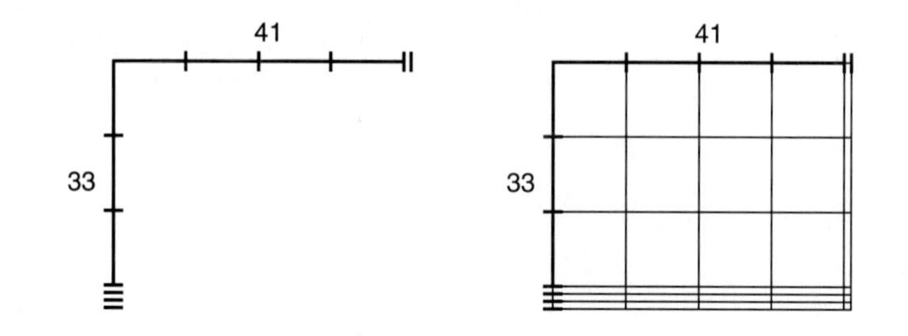

For each of the following products,

(1) Draw a number piece sketch like the one above.

(2) Determine the product from the sketch.

(3) Adjacent to each sketch, compute the product using the usual paper-and-pencil algorithm, and match the partial products to the corresponding parts of your sketch.

a. 10×23

★ **b.** 22×43

c. 45×45

JUST FOR FUN

CROSS-NUMBERS FOR CALCULATORS

Sit back, relax, and exercise your calculator's multiplication key with this cross-number puzzle. Not all answers are obtainable on a calculator, however. Clues for some of the numbers have been omitted, but no square is left blank.

1.	2.			3.	4.		5.	
6.			7.					
8.			9.				10.	11.
	12.	13.				14.		
		15.			16.			
17.	18.			19.		20.	21.	
22.			23.		24.		25.	26.
	27.							
28.			29.				30.	

Across

1. $[(18 + 2) \times 4] + 19$
5. $\frac{14 + 3}{9 + 8} \times 66$
6. $500{,}000{,}000 \div 9$ (Omit the decimal part of the number.)
8. The middle two digits of the product of $2 \times 2 \times 2 \times 2 \times 3 \times 37$
12. $\frac{(41 - 29) \times (63 - 23)}{95 \div 19}$
15. The sum of the digits in the product of $13 \times 9004 \times 77$
16. The sum of the first seven odd numbers
17. March 30, 1974: A "freeze model" in a department store posed motionless for 5 hours, 32 minutes. How many minutes in all did he pose?
23. $\frac{13{,}334 - 6566}{8 \times 9} \times \frac{992 \div 4}{17 + 14}$
27. $(10{,}000{,}000 \div 81) \times 100$ (nearest whole number)
28. Take the current year, subtract 17, multiply by 2, add 6, divide by 4, add 21, subtract $\frac{1}{2}$ of the current year.
29. The first two digits of the product of 150×12

Down

3. $(34 \times 53) - (137 \times 2)$
4. Asphalt roof shingles are sold in bundles of 27. Three bundles cover 100 ft^2. How many bundles are needed for 11,700 ft^2 of roofing in a housing development?
5. $\frac{88{,}984 \times 493}{17 \times 392}$
7. $\frac{51 \times 153 \times 62 \times 57}{17 \times 969 \times 31}$
11. $\frac{(62 \times 21) + 23}{25} - 11$
14. The number of different combinations of U.S. coins into which a dollar can be changed. (This number is a palindrome, and the sum of its digits is 13.)
17. Remainder of $1816 \div 35$
18. $(343 \times 5 + 242) \times 2$
19. $4 \times [(37 \times 24) - 1]$
21. The number halfway between 7741 and 8611
26. $\frac{119 \times 207 \times 415}{23 \times 747}$

ACTIVITY SET 3.4

DIVIDING WITH BASE-TEN PIECES

Purpose To use base-ten number pieces to illustrate two physical interpretations of division and the long-division algorithm

Materials Base-ten pieces from the Manipulative Kit

Activity In this activity set we will begin by using base-ten pieces to look at two different ways to approach division. The *measurement* or *subtractive* approach to division is illustrated by answering the question "How many piles of 3 cards can you make from a total of 39 cards?"

The *sharing* or *partitive* approach to division is illustrated by answering the question "If 39 cards are separated into 3 piles, how many are in each pile?"

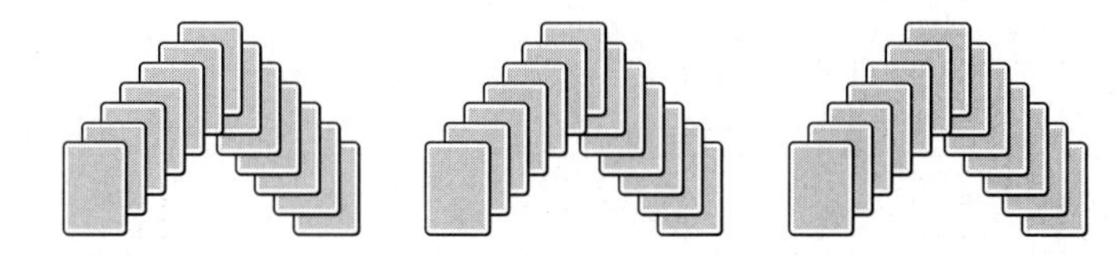

Each question can be answered by computing $39 \div 3$. The answer to both questions is 13. With cards, however, the first answer would be obtained by repeatedly subtracting groups of 3 cards and counting the number of groups, the second by dealing the cards into 3 piles and counting the number of cards in each group.

Base-ten pieces also will be used to illustrate the inverse relationship between multiplication and division and to provide a concrete interpretation of the long-division algorithm.

$36 \div 3 = 12$
(12 in each of 3 equal groups)
Sharing

1. Using base-ten number pieces, we can determine the quotient $36 \div 3$ by simple counting procedures. With the *sharing* method, 36 is shared equally among 3 groups. Each group has 1 long and 2 units.

   The *measurement* approach to division asks how many groups of 3 can be formed from 36. In this case, all the longs must be traded for units so that groups of 3 can be formed.

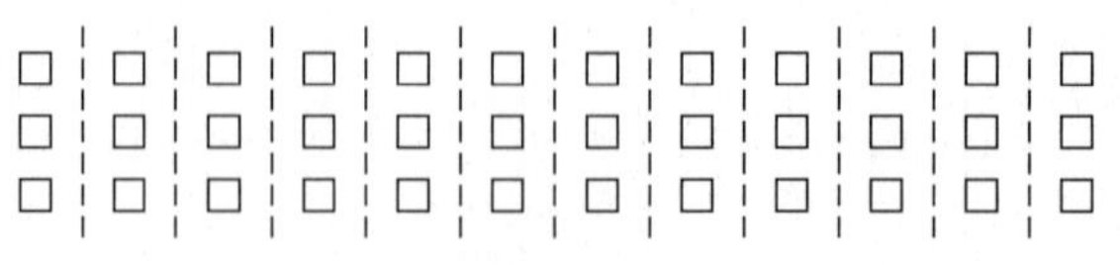

$36 \div 3 = 12$
(12 groups of 3)
Measurement

a. Represent 24 as a minimal collection of base-ten pieces. Then, regrouping when necessary, determine 24 ÷ 6 using the measurement interpretation of division. Record your results with sketches of base-ten pieces.

★ **b.** Represent 132 as a minimal collection of base-ten pieces. Determine 132 ÷ 12 using the sharing interpretation of division. Record your results with sketches of base-ten pieces.

2. Many times division does not come out evenly, so there is a remainder. When we divide 20 by 3 using the measurement approach, there are 6 groups of 3 units and 2 remaining units. Dividing 20 by 3 using the sharing method, we have 3 groups of 6 with 2 remaining.

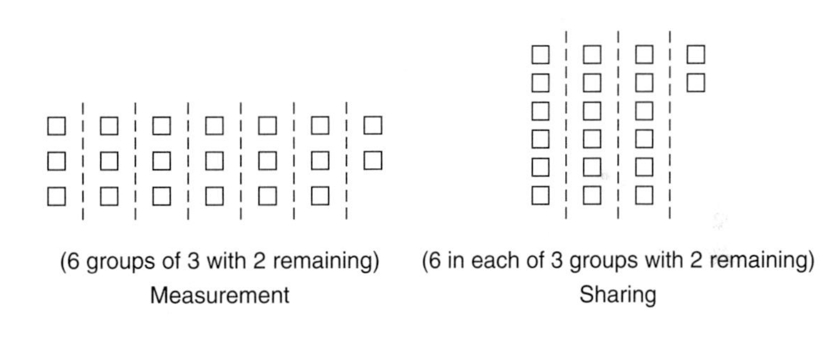

Use base-ten pieces to determine each of the following quotients and record your results in a diagram.

a. 57 ÷ 4 (sharing approach)

★ **b.** 114 ÷ 12 (measurement approach)

3. When 36 was divided by 3 in activity 1, the answer 12 was obtained in two different ways. Notice that with either approach the base-ten number pieces can be pushed together to form a rectangular array of 36 squares with one dimension of 3. So *division can be thought of as finding the missing dimension of a rectangle when the total number of unit squares and one dimension are known.*

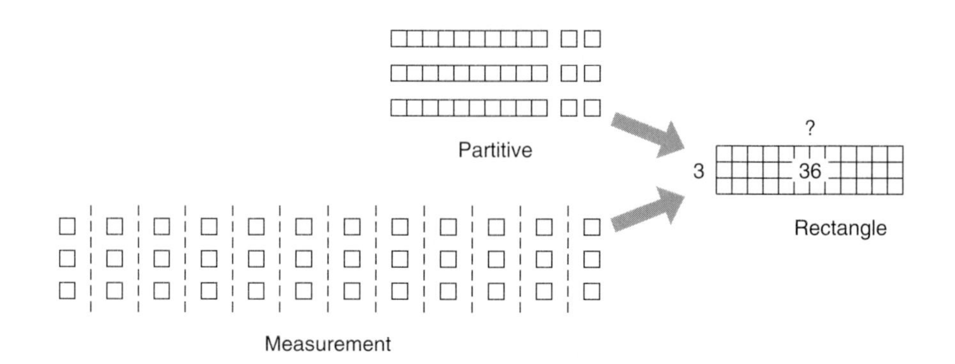

In each of the following activities, use the specified base-ten pieces to construct a rectangle with the given dimension. You may have to trade pieces for other pieces of equal value. Sketch a diagram of your result, and record the missing dimension of the rectangle.

a. Start with 9 longs and 6 units, and build a rectangle that has one dimension of 4 units.

★ **b.** Start with 2 flats, 7 longs, and 3 units, and build a rectangle that has one dimension of 13 units.

c. Start with 4 flats, 6 longs, and 2 units, and build a rectangle that has one dimension of 21 units.

4. To divide 143 by 11 using the base-ten number pieces, form the minimal collection of base-ten number pieces for 143. Then draw a line segment of length 11 units to represent one dimension of the rectangle. Finally, distribute the base pieces representing 143 (regrouping as necessary) until a rectangle has been formed. In this example we can see that the other dimension is 13, so $143 \div 11 = 13$.

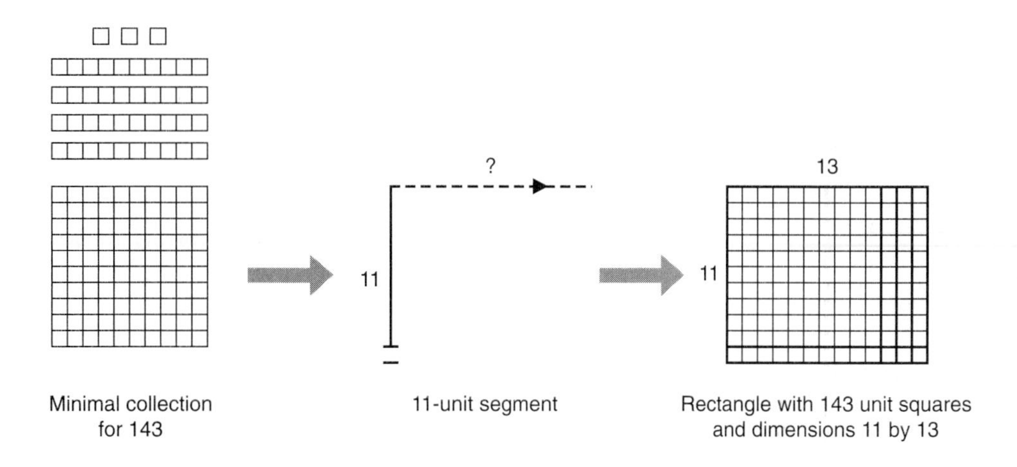

In a similar manner, use your base-ten number pieces to determine each of the following quotients. Sometimes you may need to trade pieces. Draw a sketch of your results. (*Note:* If there is a remainder, you will get a rectangle with units left over.)

a. $221 \div 17$

★ **b.** $529 \div 23$

c. $397 \div 34$

5. The paper-and-pencil algorithm for long division can be visualized using base-ten number pieces and the rectangle model. Suppose the goal is to compute $477 \div 14$. Notice that the usual symbol for division, $14\overline{)477}$, suggests two sides of a rectangle, where the length of one side is 14, the total number of units is 477, and the length of the other side (the quotient) is unknown. The following steps and full-page diagram show how the rectangle model for division is related to the division algorithm.

Step 1: First select the base-ten number pieces to represent 477.
Step 2: Draw a line segment of 14 units to represent one dimension of the rectangle. Begin distributing the base-ten number pieces representing 477 to form a rectangle of width 14.
Step 3: Three sections of width 10 can be incorporated into the rectangle. This takes 4 flats and 2 longs and leaves 5 longs and 7 units, which is not enough to complete another section of 10 (1 flat had to be traded for 10 longs to build this much).
Step 4: Four sections of width 1 can be added onto the rectangle (1 long was traded for 10 units). This uses 5 longs and 6 units and leaves 1 unit remaining—not enough to make the rectangle any wider unless the remaining unit is cut into pieces.
Step 5: The desired dimension of the rectangle (the quotient) is 34, and there is a remainder of 1.

Use your base-ten pieces to compute $255 \div 11$. Make a sketch of your number piece rectangle and describe how it relates to the standard division algorithm shown below.

$$\begin{array}{r} 23 \\ 11\overline{)255} \\ \underline{22} \\ 35 \\ \underline{33} \\ 2 \end{array}$$

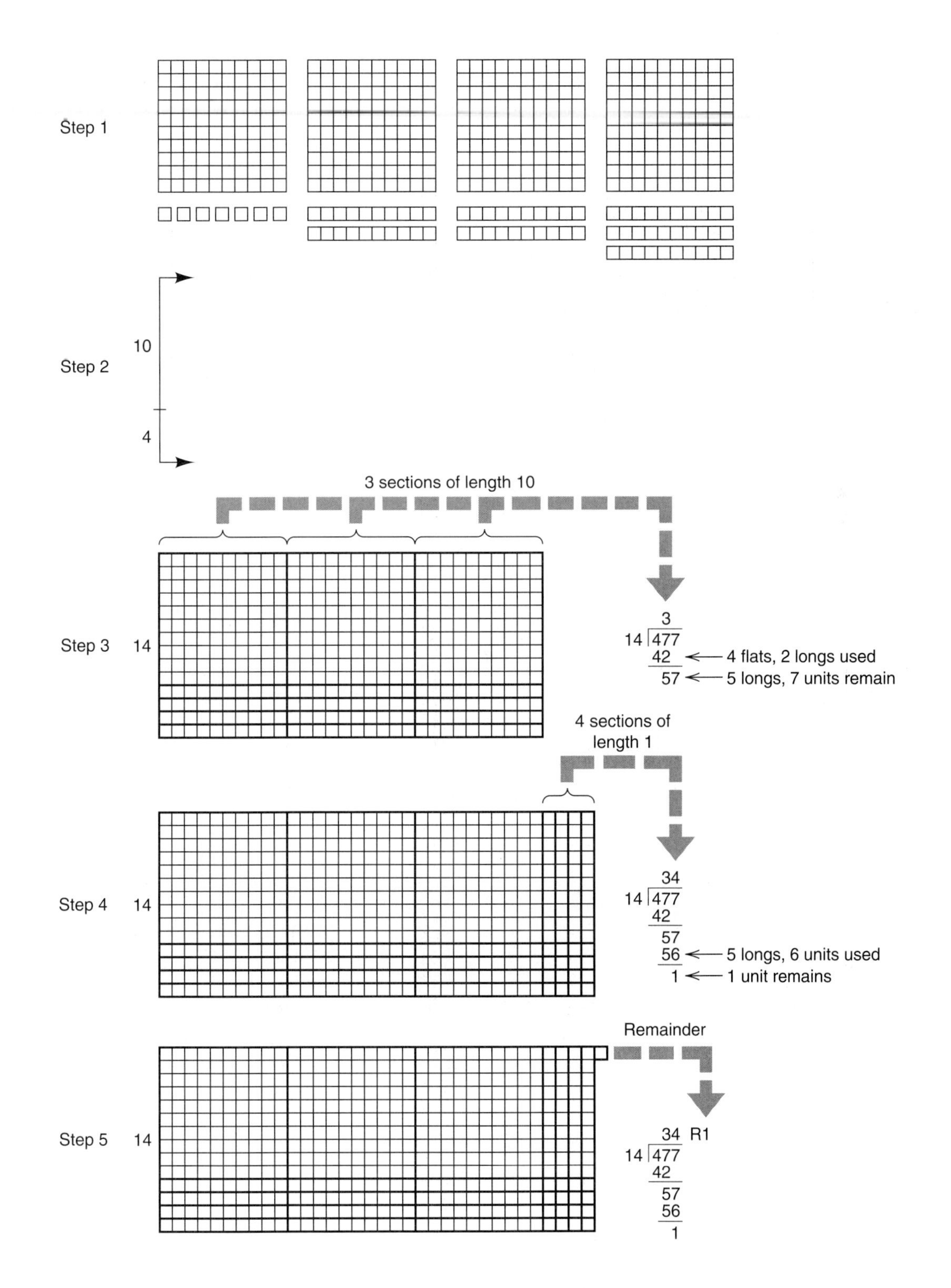
Step 1
Step 2
10
4
3 sections of length 10
Step 3
14
3
14 | 477
42 ← 4 flats, 2 longs used
57 ← 5 longs, 7 units remain
4 sections of length 1
Step 4
14
34
14 | 477
42
57
56 ← 5 longs, 6 units used
1 ← 1 unit remains
Remainder
Step 5
14
34 R1
14 | 477
42
57
56
1

6. Compute the following quotients using your base-ten pieces and the rectangle model illustrated in activity five. Record you results by making a sketch of each final number piece rectangle and showing how it relates to the standard division algorithm.

a. $144 \div 6$

$6\overline{)144}$

b. $134 \div 3$

$3\overline{)134}$

c. $587 \div 23$

$23\overline{)587}$

JUST FOR FUN

Here are several games and number tricks to try using a calculator.

ARABIAN NIGHTS MYSTERY

Select any three-digit number, such as 837, and enter it on your calculator twice to form a six-digit number (in this case, 837837). Then carry out the following steps: divide this number by 11, then divide the result by 7, and finally divide the result by 13. You may be surprised by the outcome. This trick has been called the Arabian nights mystery because it can be explained by using the number 1001, in reference to the book *Tales from the Thousand and One Nights.* Try this trick for some other three-digit numbers. Explain why it works.

KEYBOARD GAME (2 players)

The first player begins the game by entering a whole number on a calculator. The players then take turns subtracting single-digit numbers, with the following restriction: each player must subtract a nonzero digit, using a button adjacent to the one used by the preceding player. For example, if a player subtracts 6, the next player must subtract 9, 8, 5, 2, or 3. When a player's turn results in a number less than 0, that player loses the game.

MAGIC FORMULAS

Usually if someone asks you to select a number on which to perform some operations, you will choose an easy one. With the aid of a calculator, however, you do not have to be so careful about the number you select. For example, try a three- or four-digit number in the following formula.

- Select any number (remember this number, as it will be needed later), add 221, multiply by 2652, subtract 1326, divide by 663, subtract 870, divide by 4, and subtract the number you started with. Your result will be 3.

Most magic formulas such as the preceding one require that the original number be subtracted in the computation. There is no such requirement in the following formula. This trick, however, will work only on calculators with an eight-digit display and no "hidden digits."

- Select a three-digit number, divide by 9, divide by 10,000,000, multiply by 1,000,000, add 300, divide by 10,000,000, divide by 100, and multiply by 10,000,000. Your result will be 3.

PUZZLER

What is the minimum number of coins that you need to carry to pay the exact amount for any item costing from 1 to 99 cents?

IDEAS FOR THE ELEMENTARY CLASSROOM

SUGGESTED CLASSROOM ACTIVITY: VISUALIZING BASIC OPERATIONS

This activity encourages students to visualize and discuss whole number operations in nonsymbolic ways. It can be done with individuals, small groups, or the entire class, at all grade levels. Distribute about 24 tiles to each student. Write $24 \div 6$ on the chalkboard or overhead, and ask the students to make up a story about that expression, representing the story with their tiles. Encourage creativity and individuality as you walk around the classroom and acknowledge the representations. Then ask the students to share their stories and tile representations with the class. If the two representations below do not occur among the many representations, you might give them a story (to represent with the tile) so that both the *sharing* and *measurement* interpretations of division are represented.

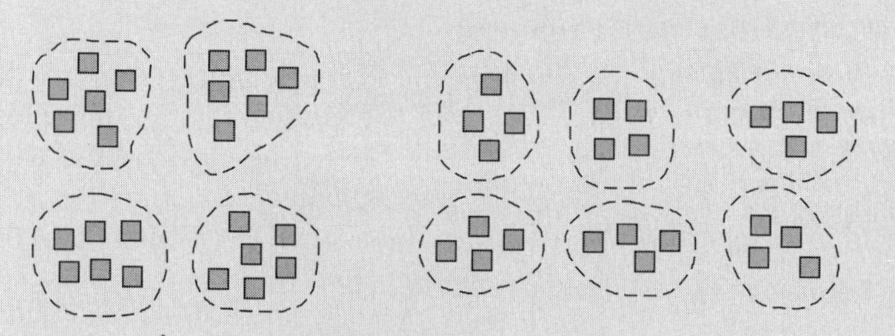

Do similar activities for expressions such as 3×5, $7 + 6$, and $13 - 7$. You may be surprised by the many different interpretations.[2]

Readings for More Classroom Ideas

Bennett, A., E. Maier, and L. Ted Nelson. "Modeling Whole Numbers," in *Math and the Mind's Eye* (Unit II). Salem, OR: The Math Learning Center, Box 3226, 1988.

Bohan, Harry, and P. H. Shawaker. "Using Manipulatives Effectively: A Drive Down Rounding Road." *Arithmetic Teacher* 41 (January 1994): 246–248.

Burton, Grace M., et al. *Number Sense and Operations,* in the *Curriculum and Evaluation Standards for School Mathematics Addenda Series: Grades K–6.* Reston, VA: National Council of Teachers of Mathematics, 1993.

Englert, Gail, and R. Sinicrope. "Making Connections with Two-Digit Multiplication." *Arithmetic Teacher* 41 (April 1994): 446–448.

Gluck, D. "Helping Students Understand Place Value." *Arithmetic Teacher* 38 (March 1991): 10–13.

Graeber, Anna O. "Research into Practice: Misconceptions About Multiplication and Division." *Arithmetic Teacher* 40 (March 1993): 408–411.

Joslyn, Ruth E. "Using Concrete Models to Teach Large-Number Concepts." *Arithmetic Teacher* 38 (November 1990): 6–9.

[2]See A. Bennett, E. Maier, and L. Ted Nelson, "Basic Operations," in *Math and the Mind's Eye* (Unit II, Activity I) (Salem, OR: The Math Learning Center, Box 3226, 1988).

Rathmell, Edward C., and Larry Leutzinger. "Implementing the Standards: Number Representations and Relationships." *Arithmetic Teacher* 38 (March 1991): 20–23.

Reys, Barbara J., et al. *Developing Number Sense,* in the *Curriculum and Evaluation Standards for School Mathematics Addenda Series: Grades 5–8.* Reston, VA: National Council of Teachers of Mathematics, 1991.

Ross, Rita, and Ray Kurtz. "Making Manipulatives Work: A Strategy for Success." *Arithmetic Teacher* 40 (January 1993): 254–257.

Sutton, John T., and Tonya Urbatsch. "Transition Boards: A Good Idea Made Better." *Arithmetic Teacher* 38 (January 1991): 4–8.

Thompson, Frances. "Two-Digit Addition and Subtraction: What Works?" *Arithmetic Teacher* 38 (January 1991): 10–13.

Van Erp, Jos W. M. "The Power of Five: An Alternative Model." *Arithmetic Teacher* 38 (April 1991): 48–53.

Zepp, Raymond A. "Numbers and Codes in Ancient Peru: The Quipi." *Arithmetic Teacher* 39 (May 1992): 42–44.

CHAPTER 4

NUMBER THEORY

Number theory offers many rich opportunities for explorations that are interesting, enjoyable, and useful.

. . . Challenging but accessible problems from number theory can be easily formulated and explored by students. For example, building rectangular arrays with a set of tile can stimulate questions about divisibility and prime, composite, square, even, and odd numbers . . . .[1]

ACTIVITY SET 4.1

MODELS FOR EVEN NUMBERS, ODD NUMBERS, FACTORS, AND PRIMES

Purpose To use models to provide visual images of basic concepts of number theory

Materials Colored tiles from the Manipulative Kit

Activity Some problems in number theory are simple enough for children to understand yet are unsolvable by mathematicians. Maybe that is why this branch of mathematics has intrigued so many people, novices and professionals alike, for over 2000 years. For example, is it true that every *even number* greater than 4 can be expressed as the sum of two *odd prime numbers*? It is true for the first few even numbers:

$6 = 3 + 3 \quad 8 = 3 + 5 \quad 10 = 5 + 5 \quad 12 = 5 + 7 \quad 14 = 7 + 7 \quad 16 = 3 + 13$

However, mathematicians have not been able to prove it is true for all even numbers greater than 4.

The ideas of odd, even, factors, and primes are basic concepts of number theory. In this activity set, these ideas will be given geometric form to show that visual images can be associated with them.

[1]*Curriculum and Evaluation Standards for School Mathematics* (Reston, VA: National Council of Teachers of Mathematics, 1989), 91–93.

Even and Odd Numbers

1. The nonzero whole numbers can be represented geometrically in many different ways. Here the first five consecutive numbers are represented as a sequence of tile figures. Extend the sequence by drawing the figures for the numbers 6, 7, and 13.

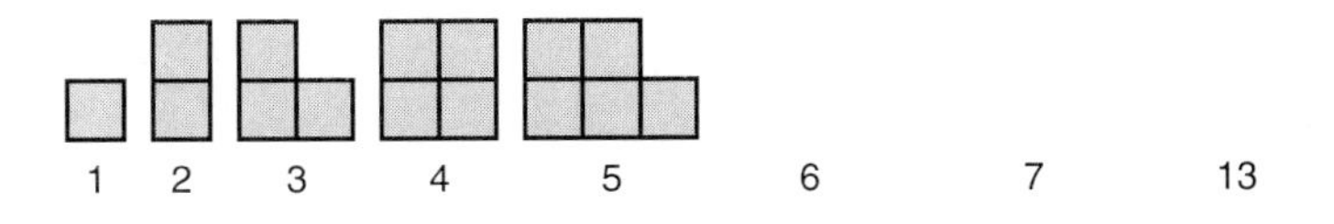

Describe, in your own words, how the figures for the even counting numbers differ from those for the odd counting numbers in the sequence above.

2. Here is a sequence for the first four *even numbers*.

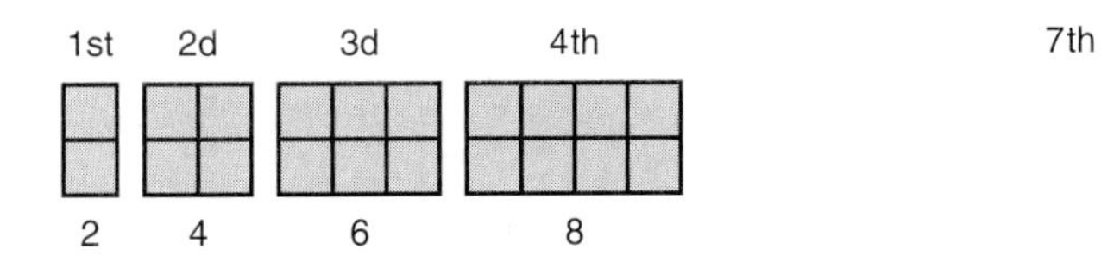

a. Draw the figure for the seventh even number. What is the seventh even number?

Seventh even number ________

★ **b.** Describe in words what the figure for the 125th even number would look like. What is the 125th even number?

125th even number ________

c. Describe in words what the figure for the nth even number would look like. Write a mathematical expression for the nth even number.

nth even number ________

3. This sequence represents the first four *odd numbers.*

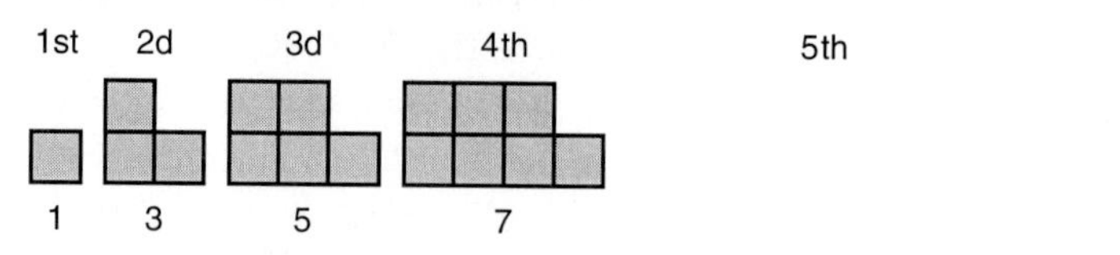

a. Draw the figures for the fifth and eighth odd numbers above.

★ **b.** Describe in words how to draw the figure representing the 15th odd number. What is the 15th odd number?

15th odd number ________

c. Determine the 50th, 100th, and nth odd numbers by imagining what their figures would look like, based on the preceding sequence.

50th odd number ________ 100th odd number ________ nth odd number ________

★ **d.** When the odd numbers, beginning with 1, are arranged in consecutive order, 7 is in the fourth position and 11 is in the sixth position. In what position is 79? 117? (*Hint:* Think about the figures for these numbers. You may wish to draw a rough sketch of the figures.)

79 is in the ________ position. 117 is in the ________ position.

4. The following diagram illustrates that when an even number is added to an odd number, the sum is an odd number. Determine the oddness or evenness of each of the sums and differences below by drawing similar diagrams.

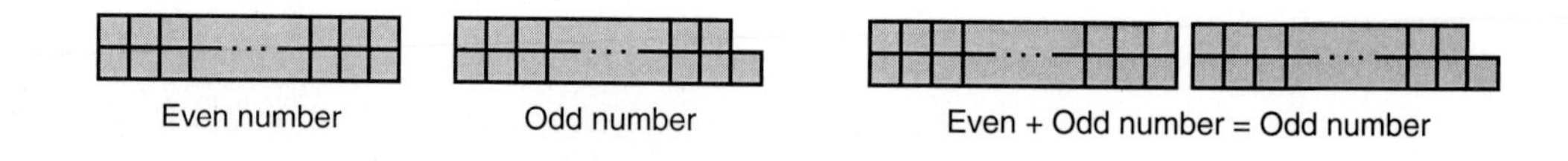

a. The sum of any two even numbers

★ **b.** The sum of any two odd numbers

c. The sum of any two consecutive whole numbers

d. The sum of any three consecutive whole numbers

★ **e.** The difference of any two odd numbers

f. The sum of any three odd numbers and two even numbers

5. The figures in the sequence for consecutive odd numbers can be rearranged as follows:

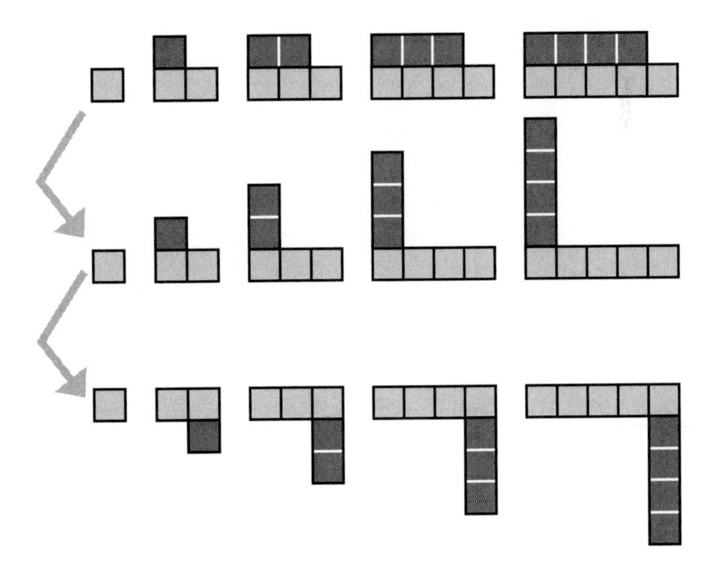

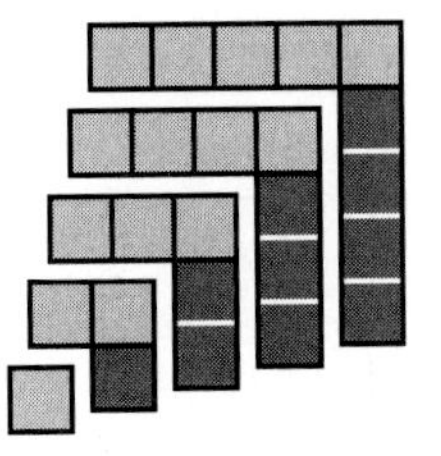

a. The L-shaped figures for the first five consecutive odd numbers can be pushed together to form a square. What does this tell you about the sum of the first five consecutive odd numbers?

★ **b.** Visualize the L-shaped figures for the first 10 consecutive odd numbers. What size square can be formed from these figures? What is the sum of the first 10 consecutive odd numbers?

c. Consider the sum of all odd numbers from 1 to 79.

$$1 + 3 + 5 + 7 + \cdots + 77 + 79$$

How many numbers are there? Determine the sum of the consecutive odd numbers from 1 to 79 by visualizing L-shaped figures forming a square.

Factors and Primes

6. All possible rectangular arrays that can be constructed with exactly 4 tiles, exactly 7 tiles, and exactly 12 tiles are diagrammed below. Use the tiles from your Manipulative Kit to form all possible rectangular arrays that can be constructed for the remaining numbers from 1 to 12 (remember a square is a rectangle). Record them in the spaces provided below.

Rectangles with 1 tile

Rectangles with 2 tiles

Rectangles with 3 tiles

Rectangles with 4 tiles

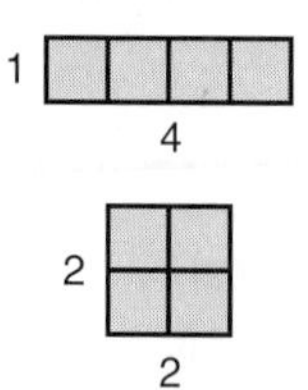

Rectangles with 5 tiles

Rectangles with 6 tiles

Rectangles with 7 tiles

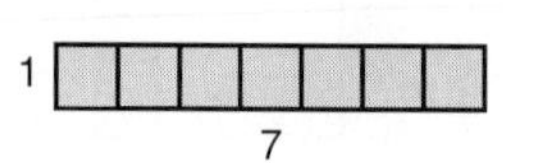

Rectangles with 8 tiles

Rectangles with 9 tiles

Rectangles with 10 tiles

Rectangles with 11 tiles

Rectangles with 12 tiles

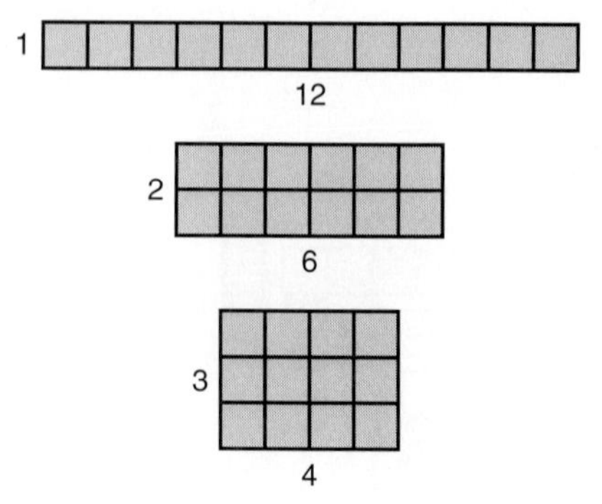

7. The dimensions of a rectangular array are factors of the number the array represents. For example, the number 12 has a 1×12 array, a 2×6 array, and a 3×4 array, and the factors of 12 are 1, 2, 3, 4, 6, and 12. Use your results from activity 6 to complete the following table by listing all the factors of each number and the total number of factors. Then extend the table to number through 20.

NUMBER	FACTORS	NUMBER OF FACTORS
1	______	______
2	______	______
3	______	______
4	1, 2, 4	3
5	______	______
6	______	______
7	1, 7	2
8	______	______
9	______	______
10	______	______
11	______	______
12	1, 2, 3, 4, 6, 12	6
13	______	______
14	______	______
15	______	______
16	______	______
17	______	______
18	______	______
19	______	______
20	______	______

8. Numbers with more than one rectangle, such as 12 and 4, are called *composite numbers*. Except for the number 1, all numbers that have exactly one rectangle are called *prime numbers*.

a. Write a sentence or two to explain how you can describe prime and composite numbers in terms of *numbers of factors*.

★ **b.** How does the number 1 differ from the prime and composite numbers?

c. Is there another number, like 1, that is neither prime nor composite? Explain why or why not in terms of rectangles.

d. Some numbers are called *square numbers* or *perfect squares*. Look at the sketches in activity 6. Which numbers from 1 to 12 do you think could be called square numbers? State your reason.

★ e. Examine the rectangles representing numbers that have an odd number of factors. What conclusions can you draw about numbers that have an odd number of factors?

9. The numbers listed in your chart have 1, 2, 3, 4, 5, or 6 factors.

a. Identify another number that has exactly 5 factors. Are there other numbers with exactly 5 factors? Explain.

b. Find two numbers which have more rectangles than the number 12. What are the factors of each of these numbers?

10. The numbers 6, 10, 14, and 15 each have exactly 4 factors. Looking at the rectangles for those numbers, determine what special characteristics the numbers with exactly 4 factors possess. List a few more numbers with exactly 4 factors, and explain how you could generate even more.

PUZZLER

Change the coins from the arrangement shown in the top row to the one shown in the bottom row by moving one pair of coins at a time. On each move, you may slide a pair of adjacent coins to a new position in the row without interchanging them.

JUST FOR FUN

NUMBER CHART PRIMES AND MULTIPLES

A seventh-grade student, Keith, chose to do a math project based on identifying prime numbers with the sieve of Eratosthenes.[2] Keith's twist to the sieve added extra information. He used a number chart like the one here and put a triangle around 1 because it is neither prime nor composite. Then he circled 2 because it was prime and then put a black dot above each multiple of 2. Next he moved from 2 to the next undotted number, 3, circled 3 and put a blue dot to the right of each multiple of 3.

The chart below has been started using Keith's method. The multiples of 2 have a black dot above them. Circle 3 and put a blue dot to the right of each multiple of 3. Circle 5 and put a green dot beneath all multiples of 5. Circle 7 and put a red dot to the left of all multiples of 7. Finally, circle all the remaining numbers which have no colored dots in their squares.

1. How can we be sure that all the circled numbers are prime?
2. How can you use the chart to find all numbers which are multiples of 3 and 5?
3. Keith's chart actually included all numbers up to 1000. How many different colors did he use?

1	2	3	•4	5	•6	7	•8	9	•10
11	•12	13	•14	15	•16	17	•18	19	•20
21	•22	23	•24	25	•26	27	•28	29	•30
31	•32	33	•34	35	•36	37	•38	39	•40
41	•42	43	•44	45	•46	47	•48	49	•50
51	•52	53	•54	55	•56	57	•58	59	•60
61	•62	63	•64	65	•66	67	•68	69	•70
71	•72	73	•74	75	•76	77	•78	79	•80
81	•82	83	•84	85	•86	87	•88	89	•90
91	•92	93	•94	95	•96	97	•98	99	•100
101	•102	103	•104	105	•106	107	•108	109	•110
111	•112	113	•114	115	•116	117	•118	119	•120

△ neither prime nor composite

◯ prime numbers

black dot	multiple of 2
blue dot	multiple of 3
green dot	multiple of 5
red dot	multiple of 7

[2] C. L. Bradford, "Keith's Secret Discovery of the Sieve of Eratosthenes," *Arithmetic Teacher* 21 (March 1974): 239–241.

ACTIVITY SET 4.2

MODELS FOR GREATEST COMMON FACTOR AND LEAST COMMON MULTIPLE

Purpose To use a linear model to illustrate the concepts of greatest common factor and least common multiple and show how they are related

Materials No supplementary materials are needed.

Activity Greatest common factor (GCF) and least common multiple (LCM) are important concepts that occur frequently in mathematics. The GCF of two numbers is usually introduced by listing all the factors of two numbers, identifying the *common factors,* and then choosing the *greatest* of the common factors.

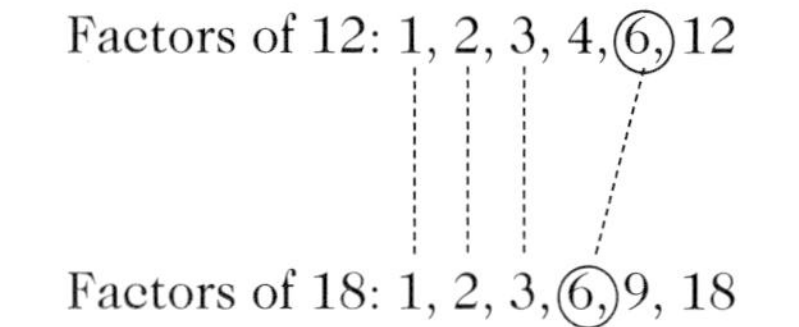

The LCM of two numbers is often introduced by listing multiples of each number, identifying *common multiples,* and choosing the *least* of the common multiples.

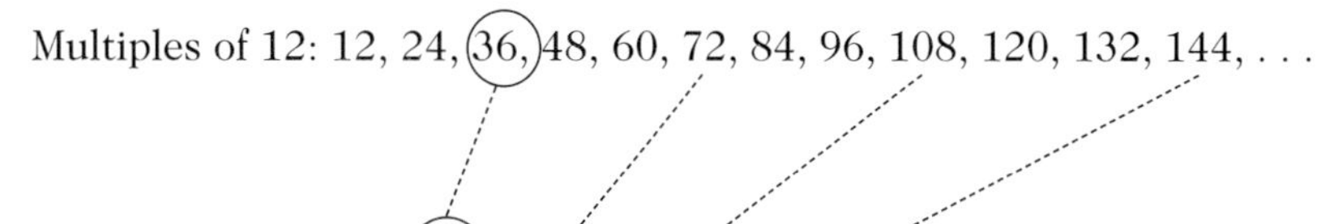

In this activity set, the concepts of GCF and LCM will be visually represented by rods. The GCF will be viewed as the greatest common length into which two (or more) rods can be cut.

The LCM will be viewed as the shortest common length into which two (or more) rods will fit. This will be determined by placing copies of each rod end to end.

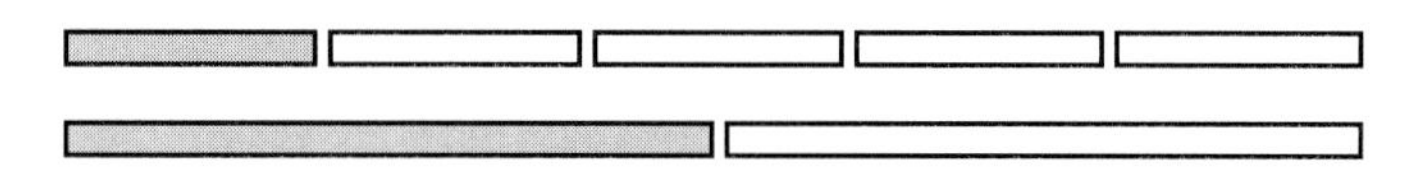

Greatest Common Factor

1. Here are rods of length 36 units and 54 units. Both rods can be cut evenly into pieces with a common length of 6 units.

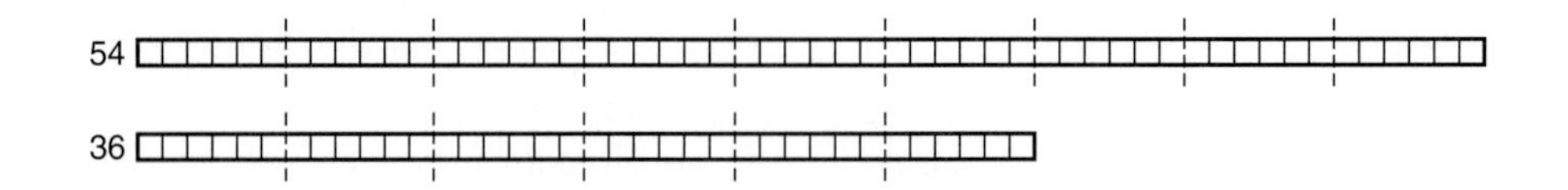

a. There are five other ways to cut both rods evenly into pieces of common length. Mark those on the following five pairs of rods.

★ **b.** Because rods of length 36 and 54 can both be cut evenly into pieces of length 6, 6 is called a *common factor* of 36 and 54. List the other common factors of 36 and 54.

Common factors of 36 and 54: 6, ______, ______, ______, ______, ______

c. Circle the *greatest common factor*[3] of 36 and 54. The greatest common factor of 36 and 54 is abbreviated as GCF(36, 54).

2. Determine the greatest common factor of each pair of numbers below by indicating how you would cut the rods into common pieces of greatest length. Record your answer next to the diagram.

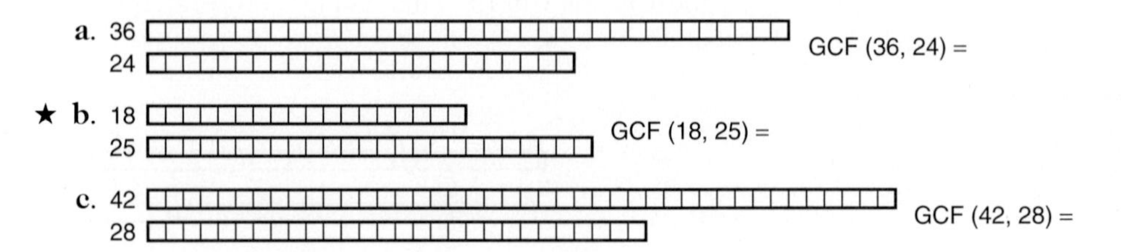

[3]Also commonly called the *greatest common divisor* or GCD.

3. a. Determine the greatest common factor of the numbers 20 and 12 by indicating how you would cut the two rods below into pieces of greatest common length.

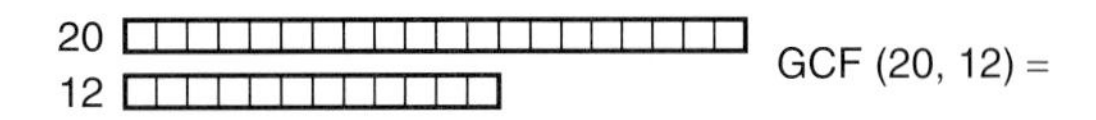

b. The amount by which one rod exceeds the other is represented by the *difference rod.* Determine the GCF of the difference rod and the shorter rod using the diagram below.

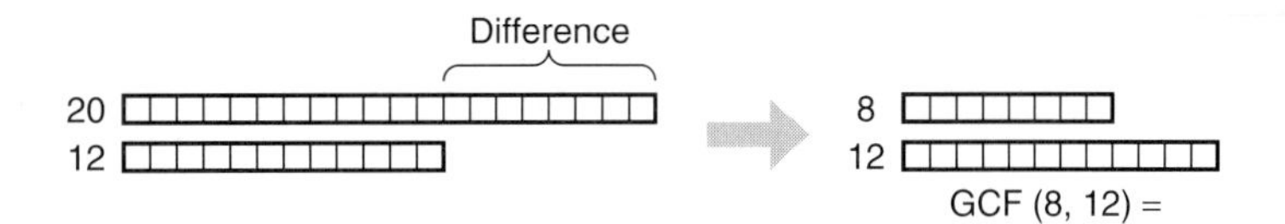

4. For each pair of rods shown below, determine the GCF of the shorter rod and the difference rod. Indicate how you would cut the rods on the diagrams below.

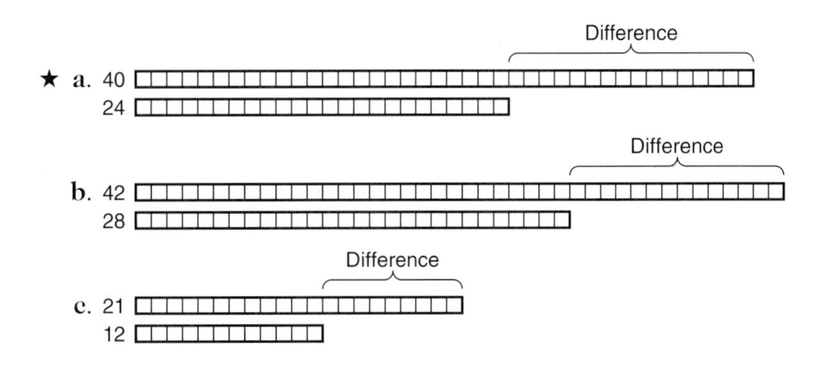

★ **5.** Why do you think that the GCF of two numbers is equal to the GCF of their difference and the smaller number? Use a diagram with rods to explain.

6. The GCF of two numbers can be determined by computing the difference of the two numbers and then finding the GCF of the difference and the smaller of the two numbers. If the GCF of the smaller numbers is not apparent, this method of taking differences can be continued. For example,

$$\begin{aligned} GCF(78, 60) &= GCF(18, 60) = GCF(18, 42) = GCF(18, 24) \\ &= GCF(18, 6) = GCF(12, 6) = GCF(6, 6) = 6 \end{aligned}$$

and

$$\text{GCF}(198, 126) = \text{GCF}(72, 126) = \text{GCF}(72, 54) = \text{GCF}(18, 54)$$
$$= \text{GCF}(18, 36) = \text{GCF}(18, 18) = 18$$

Find the GCF of the following pairs of numbers.

a. 144, 27

★ b. 280, 168

c. 714, 420

★ d. 306, 187

Least Common Multiple

7. In the following diagram, the numbers 3 and 5 are represented by rods. When the rods of length 3 are arranged end to end alongside a similar arrangement of rods of length 5, the distances at which the ends evenly match are *common multiples* of 3 and 5 (15, 30, 45, etc.). The least distance at which they match, 15, is the *least common multiple* of 3 and 5. This least common multiple is written LCM(3, 5) = 15.

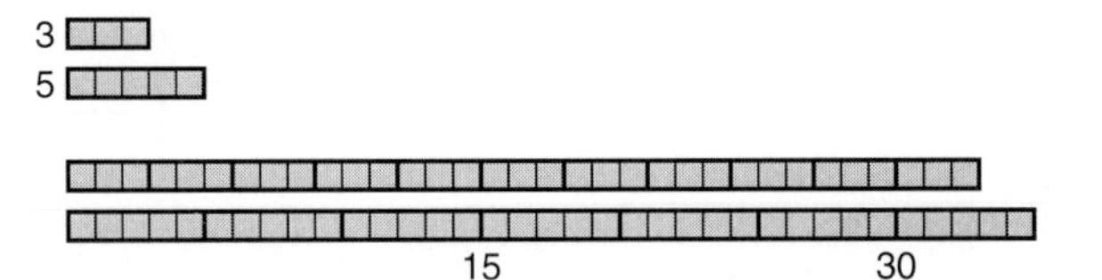

Find the LCM of each of the following pairs of numbers by drawing the minimum number of end-to-end rods of each length needed to make both rows the same length.

a. 8
12

★ b. 14
21

c. 5
7

★ d. 8
10

8. There is a relationship between the GCF and the LCM of two numbers. The first figure below shows the numbers 6 and 15; GCF(6, 15) = 3 is indicated by marks on the rods. The second figure shows that LCM(6, 15) = 30.

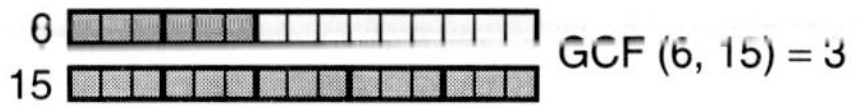

6
15
GCF (6, 15) = 3

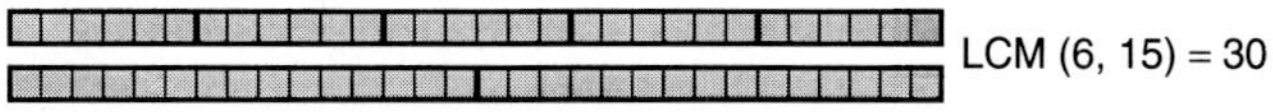

LCM (6, 15) = 30

a. GCF(6, 15) = 3 and 3 divides the 6-unit rod into 2 parts and the 15-unit rod into 5 parts. Notice that 2 rods of length 15 or 5 rods of length 6 equal 30, the LCM(6, 15). Use the rod diagrams in activity 7 to study the relationship between the GCF and LCM as you complete the following table.

	A	B	GCF (A, B)	LCM (A, B)
(1)	6	15	3	30
(2)	8	12	______	______
★ (3)	14	21	______	______
(4)	5	7	______	______
★ (5)	8	10	______	______

b. Based on your observations from part a, write a brief set of directions for finding the LCM of two numbers once you have determined the GCD.

9. For each of the following pairs of numbers, first compute the GCF of the pair and then use that information to compute the LCM.

★ **a.** 9, 15

GCF(9, 15) =

LCM(9, 15) =

b. 8, 18

GCF(8, 18) =

LCM(8, 18) =

★ **c.** 14, 35

GCF(14, 35) =

LCM(14, 35) =

d. 35, 42

GCF(35, 42) =

LCM(35, 42) =

JUST FOR FUN

STAR POLYGONS

Star polygons are often constructed to provide decorative and artistic patterns. The star polygon pictured here was formed from colored string and 16 equally spaced tacks on a piece of plywood. In the following activities, these patterns are analyzed by using the concepts of factor, multiple, greatest common factor, and least common multiple.[4]

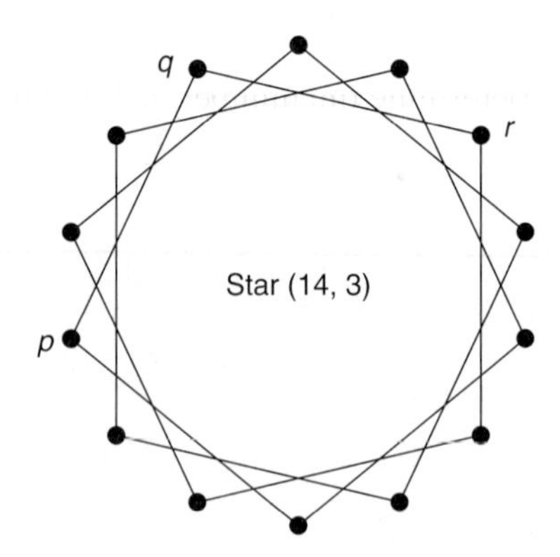

Star polygons can be constructed by taking steps of a given size around a circle of points. Star (14, 3) was constructed by beginning at point p and taking a step of 3 spaces to point q. Three spaces from q is point r. Through this process we eventually come back to point p, after having hit all 14 points. The resulting figure is a star polygon.

In general, for whole numbers n and s, star (n, s) will denote a star polygon with n points and steps of s, provided that $s < n$. In the special cases where $s = 1$ or $s = n - 1$, the resulting figure will be a polygon.

★ **1.** Sketch the following star polygons by beginning at point x and taking steps of s in a clockwise direction. Will the same star polygons be obtained if the steps are taken in a counterclockwise direction?

[4]A. B. Bennett, Jr., "Star Patterns," *Arithmetic Teacher* 25 (January 1978), 12–14.

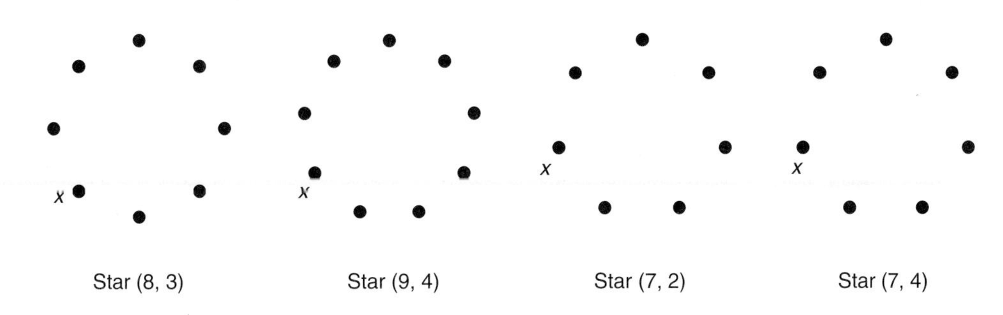

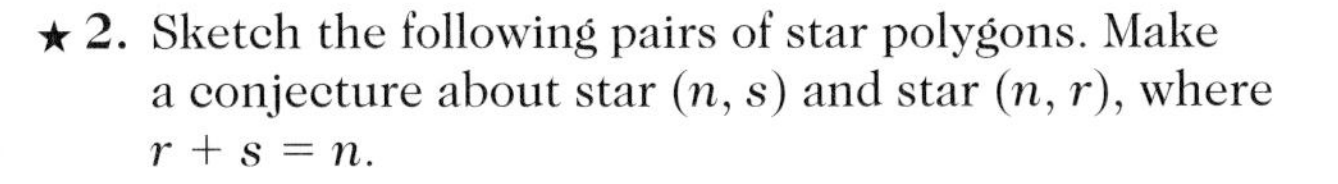

★ **2.** Sketch the following pairs of star polygons. Make a conjecture about star (n, s) and star (n, r), where $r + s = n$.

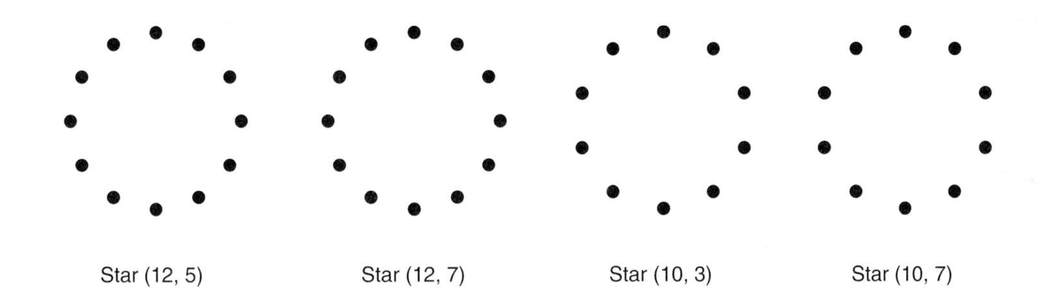

★ **3.** The star polygons in activities 1 and 2 can each be completed by beginning at any point and drawing one continuous path. In each of those examples, the path returns to the starting point after hitting all points. For star (15, 3), however, the path closes after hitting only 5 points. To complete this star, 3 different paths are needed. Determine the number of different paths for each of the following star polygons.

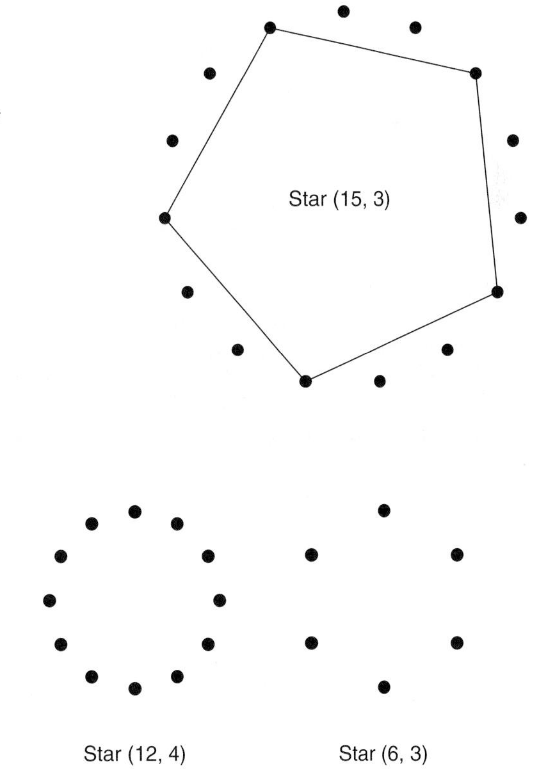

Star (5, 1) Star (10, 4) Star (12, 4) Star (6, 3)

★ **4.** What conditions must n and s satisfy in order for star (n, s) to be formed with 1 continuous path? Use your conjecture to sketch the 15-point star polygons that can be formed by 1 continuous path. There are 4 of these star polygons. Star (15, 1), which is congruent to star (15, 14), is one of them. Determine and sketch the remaining three star polygons, and write the number of steps for each.

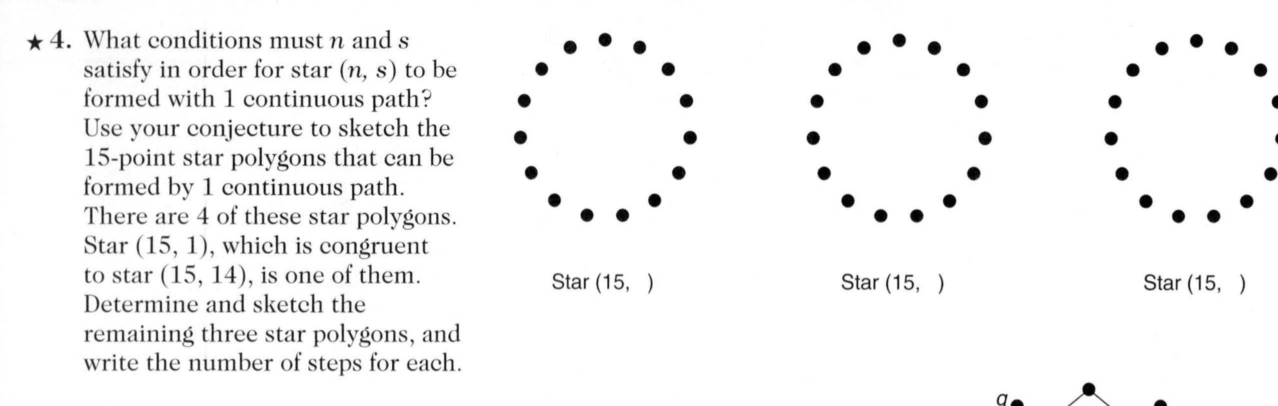

★ **5.** Star (14, 3) was constructed by beginning at point p and taking steps of 3 to points q, r, s, t, and so forth. After point t, the next step of 3 completes 1 *orbit* of the circle (once around the circle) and starts us on the second orbit. It takes 3 orbits before the path returns to its starting point. For the following star polygons, how many orbits are needed before a given path will return to its starting point?

q r p s t
Star (14, 3)

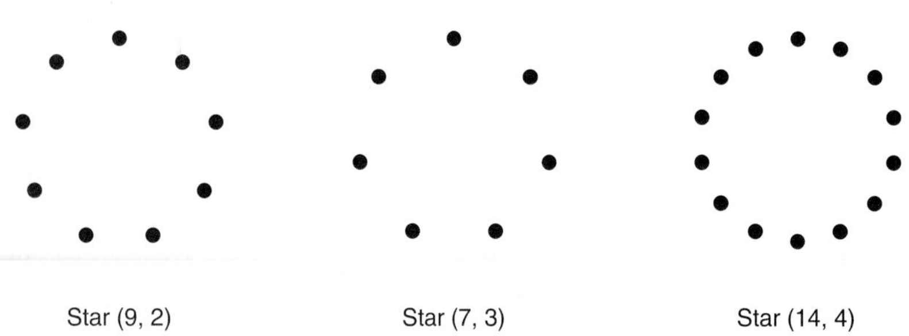

Star (9, 2) Star (7, 3) Star (14, 4)

6. Star (7, 3) was drawn by starting at point 1 and connecting the points in the following order: 1, 4, 7, 3, 6, 2, 5, and 1. Three orbits are needed to complete this star. The number of orbits is just the number of times we need to go around the 7 points of the circle before the steps of 3 bring us back to the beginning point. That is, after taking steps of 3 for 21 spaces, we arrive back at the beginning point. The 21 spaces represent the least common multiple of 7 and 3.

4 3 5 2 6 1 7
Star (7, 3)

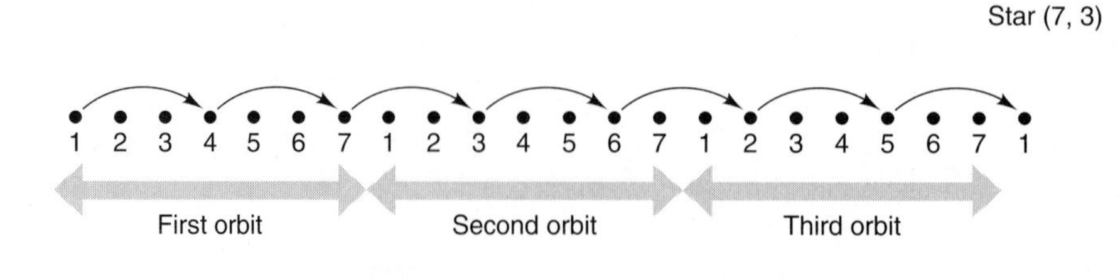

a. Explain how the concept of least common multiple can be used to determine the number of orbits for 1 path of star (15, 6). Check your answer.

★ **b.** Make a conjecture about the number of orbits needed to complete 1 path for star (n, s).

7. Create a star polygon. Any number of points may be used; they do not have to lie on a circle. Star polygons can be made by drawing with colored pencils (different colors for steps of different sizes) or by using colored thread or yarn. With a needle and thread, they can be stitched on posterboard. Several star polygons can be formed on the same set of points. Three star polygons have been stitched from 52 holes on the black posterboard pictured here. One star polygon is made from orange thread with steps of 2, another from green thread with steps of 10, and a third from yellow thread with steps of 15.

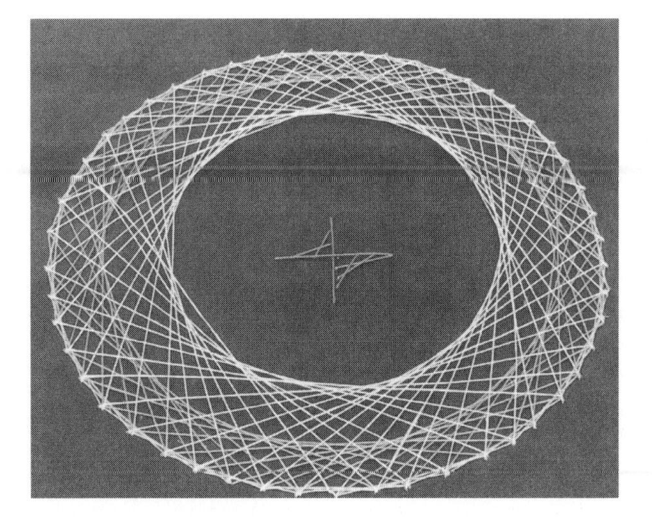

a. The star polygon with steps of 15 is the only one that can be constructed by 1 continuous path. Why?

★ **b.** How many orbits are needed to complete star (52, 15)?

PUZZLER

From a group of 10 people, can you form more committees of 2 or more committees of 8?

Ideas for the Elementary Classroom

SUGGESTED CLASSROOM ACTIVITY: ODD-EVEN CLASS MODELS

Make several large demonstration (generic) odd and even cards, like those shown on the next page, out of cardstock. Vary the lengths of the cards you make. These cards have no particular number value and can be used in a variety of ways as a physical model for odd and even numbers.

1. Students can use the cards to illustrate their answers to questions like "Is the sum of three odd numbers odd or even?" (See Activity Set 4.1, problem 4.)
2. Students may use the cards to make up questions of their own about odd and even numbers.

continued

(concluded)

3. You may use the cards as models for problem-solving situations you invent: "I'm thinking of two consecutive whole numbers whose sum is 77. What can you tell me about these numbers?"[5]

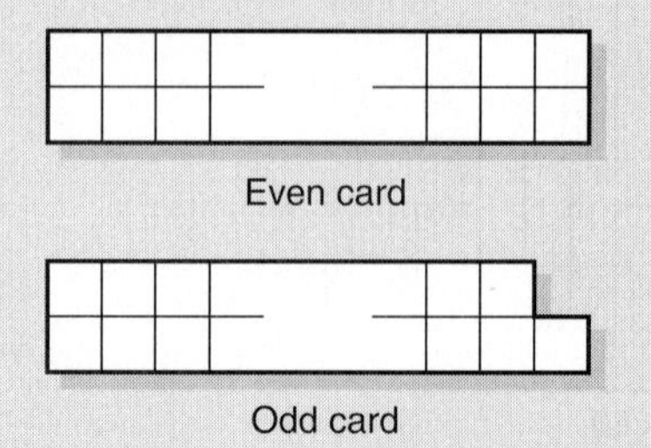

Even card

Odd card

[5]See A. Bennett, E. Maier, and L. Ted Nelson, "Odd and Even Numbers," in *Math and the Mind's Eye* (Unit II, Activity II), (Salem, OR: The Math Learning Center, Box 3226, 1988).

Readings for More Classroom Ideas

Beattie, Ian D. "Building Understanding with Blocks." *Arithmetic Teacher* 34 (October 1986): 5–11.

Bennett, A., E. Maier, and L. Ted Nelson. "Visualizing Number Concepts," in *Math and the Mind's Eye* (Unit ME2). Salem, OR: The Math Learning Center, Box 3226, 1988.

Cavanaugh, W. "The Spirograph and the Greatest Common Factor." *The Mathematics Teacher* 68 (February 1975): 162–163.

Fitzgerald, William M., and Jane Boyd. "Teacher to Teacher: A Number Line with Character." *Arithmetic Teacher* 41 (March 1994): 368–369.

Fitzgerald, William, Glenda Lappan, Elizabeth Phillips, and Mary Winter. *Factors and Multiples.* Middle Grades Mathematics Project. Menlo Park, CA: Addison-Wesley Publishing Company, 1986.

Newton, Clarie M., and S. Turkel. "Integrating Arithmetic and Geometry with Numbered Points on a Circle." *Arithmetic Teacher* 36 (January 1989): 28–31.

Robold, Alice I. "Patterns in Multiples." *Arithmetic Teacher* 29 (April 1982): 21–23.

Rockwell, C. "Another 'Sieve' for Prime Numbers." *Arithmetic Teacher* 20 (November 1973): 603–605.

Schaefer, Sister M. G. "Motivational Activities in Elementary Mathematics." *Arithmetic Teacher* 28 (May 1981): 17–18.

CHAPTER 5

INTEGERS AND FRACTIONS

Fraction symbols, such as $\frac{1}{4}$ and $\frac{3}{2}$, should be introduced only after children have developed the concepts and oral language necessary for symbols to be meaningful and should be carefully connected to both the models and oral language.[1]

ACTIVITY SET 5.1

BLACK AND RED TILE MODEL FOR INTEGERS

Purpose To use black and red tiles to provide a model for adding, subtracting, multiplying, and dividing integers

Materials Black and red tiles from the Manipulative Kit

Activity The *integers* (sometimes called *positive and negative numbers* or *signed numbers*) are the numbers . . . $^-6$, $^-5$, $^-4$, $^-3$, $^-2$, $^-1$, 0, $^+1$, $^+2$, $^+3$, $^+4$, $^+5$, $^+6$. . . . Two thousand years ago, the Chinese dealt with positive and negative numbers by using black and red rods. We will use a similar model, black and red tiles, to illustrate addition, subtraction, multiplication, and division of integers. Black tiles will represent positive numbers and red tiles will represent negative numbers.

In this model every collection of red and black tiles has a *net value*. The *net value* of a collection is the number of red or black tiles remaining after all possible red and black tiles have been matched. As the following figures show, an excess of red tiles will be designated by a negative integer and an excess of black tiles by a positive integer. The + and − signs are usually written as superscripts so as not to be confused with adding and subtracting.

[1]*Curriculum and Evaluation Standards for School Mathematics* (Reston, VA: National Council of Teachers of Mathematics, 1989), 58.

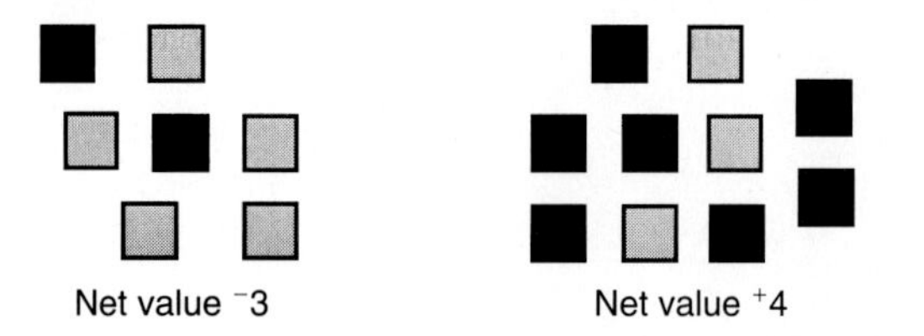

Using this model every integer can be represented in many ways. For example, the following three collections of tiles each have a net value of $^{-}2$ because when pairs of red and black are matched there is an excess of 2 red tiles in each collection.

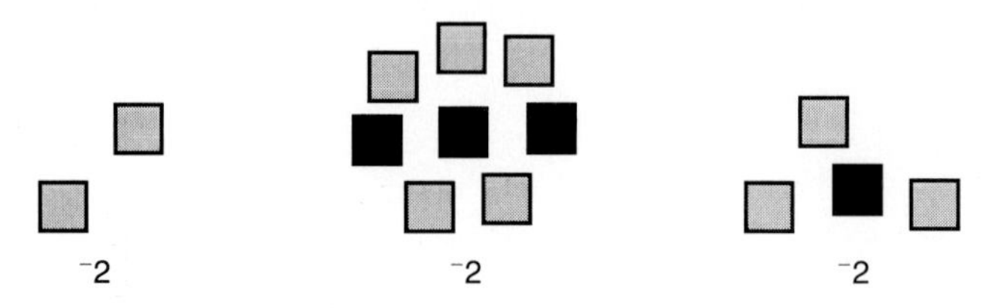

If there are the same number of black tiles as there are red tiles in a collection, the collection has *net value 0.* In this case, the positive integer for the black tiles and the negative integer for the red tiles are called *opposites* of each other.

---

1. Take a small handful of red and black tiles (from the Manipulative Kit) and drop them on a flat surface. Record the information about your collection in part a of the following table. Repeat the above directions three times and record the information in parts b, c, and d.

	TOTAL NUMBER OF TILES	NUMBER OF RED TILES	NUMBER OF BLACK TILES	NET VALUE
a.	______	______	______	______
b.	______	______	______	______
c.	______	______	______	______
d.	______	______	______	______

2. The following table contains partial results from collections of red and black tile that have been dropped. Use your red and black tiles to determine the missing table entries.

	TOTAL NUMBER OF TILES	NUMBER OF RED TILES	NUMBER OF BLACK TILES	NET VALUE
★ **a.**	______	8	29	______
b.	42	26	______	______
★ **c.**	______	______	14	$^{+}8$
d.	17	______	______	$^{+}5$
e.	20	______	______	$^{-}4$
★ **f.**	______	______	11	$^{-}2$

Addition

3. The following two collections represent the integers $^{-}3$ and $^{+}2$, respectively. Form each collection with your tiles. Combine the two collections to determine the sum, $^{-}3 + ^{+}2$. What integer represents the sum of the two collections?

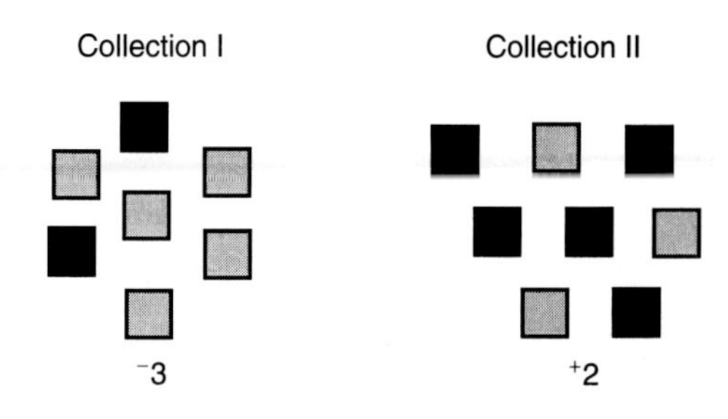

a. Represent $^{-}3$ with the fewest number of tiles. Do the same for $^{+}2$. Sketch these collections and show that, when combined, they represent the same sum as in the preceding figure.

b. Use your red and black tiles to determine these sums. Draw a sketch to indicate how you got your answer.

$^{+}7 + ^{-}3 =$

$^{-}4 + ^{-}3 =$

$^{-}8 + ^{+}2 =$

★ c. Try to visualize black and red tiles as you determine these sums.

$^{+}50 + ^{-}37 =$ $^{-}34 + ^{-}25 =$ $^{-}132 + ^{+}70 =$

4. Write directions that will enable the reader to do the following. (You may wish to use the black and red tile terminology in your directions.)

   a. Add any two negative integers.

   ★ b. Add a positive integer and a negative integer.

Subtraction

5. The collections A and B show one way to compute $^{+}5 - {}^{+}8$ with black and red tiles. It is not possible to take 8 black tiles from collection A, but collection A can be changed to collection B by adding 3 black tiles and 3 red tiles. Collection B still represents 5, but now 8 black tiles can be removed to determine that $^{+}5 - {}^{+}8 = {}^{-}3$.

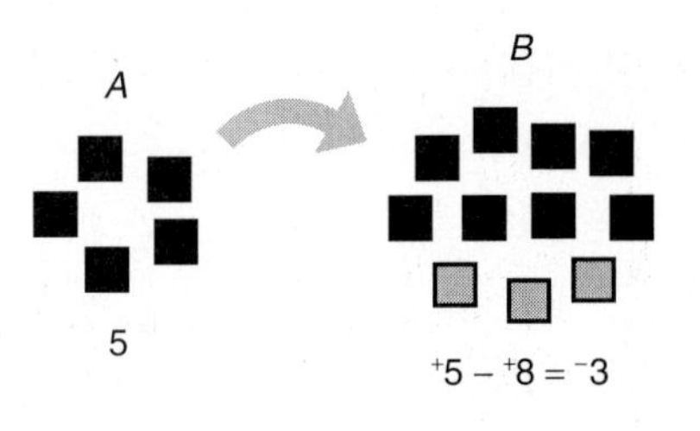

Use your tiles to form the following collections. Then change each collection to one that represents the same number but has enough tiles so that the appropriate number of tiles can be taken away. In each example, sketch the new collection. Use the results to complete the equations.

   a. Change this collection so that 3 black tiles can be taken away.

   $^{-}4$ $^{-}4$

   $^{-}4 - {}^{+}3 =$

   ★ b. Change this collection so that 2 red tiles can be taken away.

   $^{+}5$ $^{+}5$

   $^{+}5 - {}^{-}2 =$

   c. Change this collection so that 5 black tiles can be taken away.

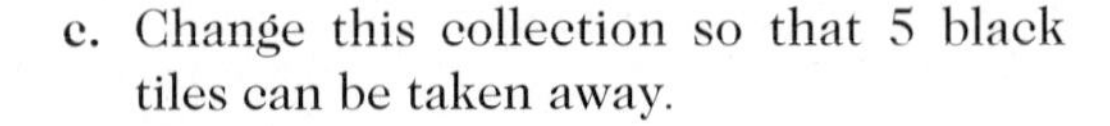

   $^{-}3$ $^{-}3$

   $^{-}3 - {}^{+}5 =$

★ 6. Explain how parts a, b, and c of activity 5 can be used to show that subtraction can be performed by adding opposites. That is, taking away a given number of one color is the same as adding the same number of the opposite color. Symbolically this is written $a - b = a + {}^{-}b$. (*Hint:* Reconstruct the sets in parts a, b, and c with your black and red tiles and watch what happens when you bring in extra tiles and then subtract.)

Multiplication[2]

7. The red and black tiles can be used to illustrate a model for multiplication if we agree that in the product $n \times s$, n tells the number of times we *put in* (+) or *take out* (−) s red or black tiles. For example, $^{-}2 \times {}^{+}3$ means that 2 times we take out 3 black tiles. This changes set *A*, which represents 0, to set *B*, which represents $^{-}6$. This shows that $^{-}2 \times {}^{+}3 = {}^{-}6$.

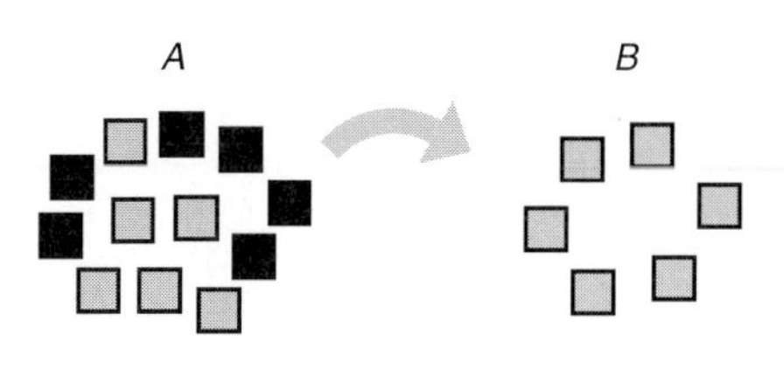

For each of the following activities, start with a collection of 8 red and 8 black tiles. Make the described changes to the collection. Sketch the resulting collection and record its value. Then write the complete number equation that has been illustrated.

a. Three times, put in 2 red tiles.

★ b. Two times, take out 4 red tiles.

c. Three times, take out 2 black tiles.

8. Use your tiles to show how you would illustrate $^{-}3 \times {}^{-}4$ with red and black tiles. Illustrate your results by drawing a sketch. Explain how you determined the number of red and black tiles you used in the initial collection.

[2]This model for multiplication can be described as the mail-delivery model. Bringing mail to you is a positive action and taking mail from you is a negative action. The mail contains bills for \$1 each (debts) or checks for \$1 each (assets). The mail carrier can bring (positive) or take away (negative), bills (negative) or checks (positive). Then, for example, $^{-}2 \times {}^{+}4$ signifies that 2 times the mail carrier takes away four \$1 checks, and your net worth has decreased by \$8.

Division

9. Review the measurement and sharing approaches to whole number division discussed in activity set 3.4.

   a. Form a collection of 12 red tiles and use the sharing approach to determine the quotient $^{-}12 \div {}^{+}3$. Draw a sketch of your tiles to indicate how you obtained your answer.

   b. Form a collection of 12 red tiles and use the measurement approach to determine the quotient $^{-}12 \div {}^{-}3$. Draw a sketch of your tiles to indicate how you obtained your result.

   ★ c. Explain why neither the sharing nor measurement approach can be used to illustrate $^{+}12 \div {}^{-}3$.

10. In whole number arithmetic we know that $6 \div 2 = 3$ because $2 \times 3 = 6$; that is, division can be defined in terms of multiplication. This same approach can be used for integer multiplication.

    a. Show that this approach gives the same answers as the methods in 9a and 9b above.

    b. What answer does it give for 9c?

    ★ c. Use any approach you wish to determine the following:

    $^{+}91 \div {}^{-}7 =$ $\quad$ $^{-}4 \div {}^{-}4 =$ $\quad$ $^{-}1001 \div {}^{+}11 =$

    $^{-}221 \div {}^{-}17 =$ $\quad$ $0 \div {}^{-}9 =$

JUST FOR FUN

The following four games for negative numbers require two dice, one white and one red. The white die represents positive numbers, and the red die represents negative numbers.

GAME 1 (Addition—2 to 4 players)

Each player begins the game with a marker on the zero point of the number line. On a turn, the player rolls the dice and adds the positive number from the white die to the negative number from the red die. This sum determines the amount of movement on the number line. If it is positive, the marker moves to the right, and if it is negative, the marker moves to the left. If the sum is zero, the player rolls again. The first player to reach either $^{-}10$ or 10 wins the game. If this game is played at first by moving the marker separately for each die—to the right for the white die and to the left for the red die—it will quickly lead to the rule for adding positive and negative numbers.

GAME 2 (Addition and subtraction—2 to 3 players)

Each player selects one of the three cards shown below. On a player's turn, the dice are rolled and the two numbers are either added or subtracted. Suppose, for example, that you rolled $^{+}4$ and $^{-}6$; you could add $^{+}4 + {}^{-}6 = {}^{-}2$, subtract $^{+}4 - {}^{-}6 = 10$, or subtract $^{-}6 - {}^{+}4 = {}^{-}10$. The answer is then marked on the player's card. The first player to complete a row, column, or diagonal wins the game.

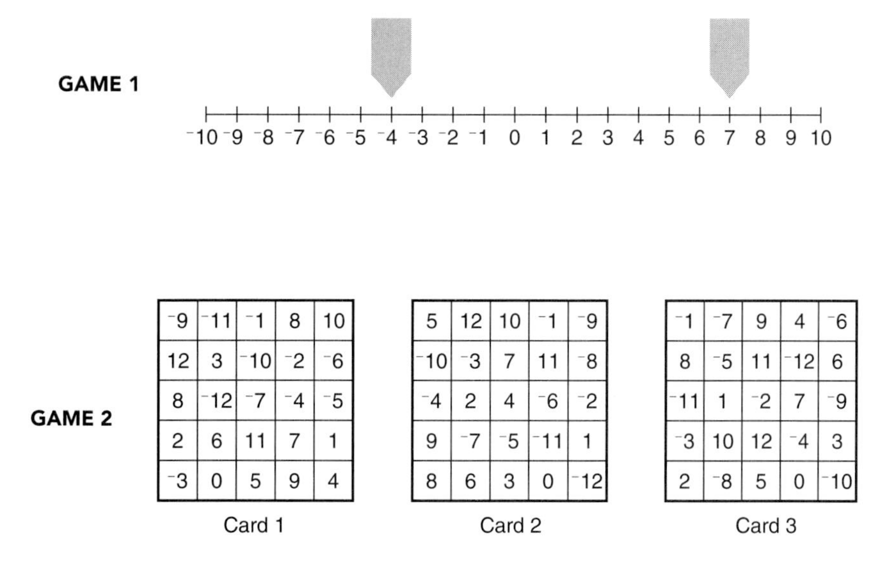

⁻9	⁻11	⁻1	8	10
12	3	⁻10	⁻2	⁻6
8	⁻12	⁻7	⁻4	⁻5
2	6	11	7	1
⁻3	0	5	9	4

Card 1

5	12	10	⁻1	⁻9
⁻10	⁻3	7	11	⁻8
⁻4	2	4	⁻6	⁻2
9	⁻7	⁻5	⁻11	1
8	6	3	0	⁻12

Card 2

⁻1	⁻7	9	4	⁻6
8	⁻5	11	⁻12	6
⁻11	1	⁻2	7	⁻9
⁻3	10	12	⁻4	3
2	⁻8	5	0	⁻10

Card 3

GAME 3 (Multiplication and inequality—2 to 4 players)

On a turn, the player rolls the dice and multiplies the positive number from the white die by the negative number from the red die. This operation is repeated a second time, and the player selects the greater of the two products. In this example, each player has had four turns. The greater product in each case has been circled. Player B is ahead at this point with the greater score: a total of $^{-}12$. After 10 rounds, the player with the greater total wins the game.

Chance Option: If a player obtains two undesirable numbers, such as $^{-}25$ and $^{-}15$, he or she may use the *chance option*, which involves rolling the dice a third time. If the third product is greater than the other two—say, $^{-}12$—the player may select this number. If the third product is less than or equal to either of the other two—say, $^{-}18$—then the smallest of the three numbers ($^{-}25$ in this example) must be kept as the player's score.

GAME 4 (All four basic operations—1 to 4 players)

Roll the dice four times, and record the four negative numbers from the red die and the four positive numbers from the white die. Using each number no more than once, write these eight numbers in the blanks of the equations under Round 1 so that you complete as many equations as possible. Repeat this activity for Round 2 and Round 3. You receive 1 point for each equation. If you complete all four equations for a given round, you receive 4 points plus 2 bonus points. The total number of points for the three rounds is your score.

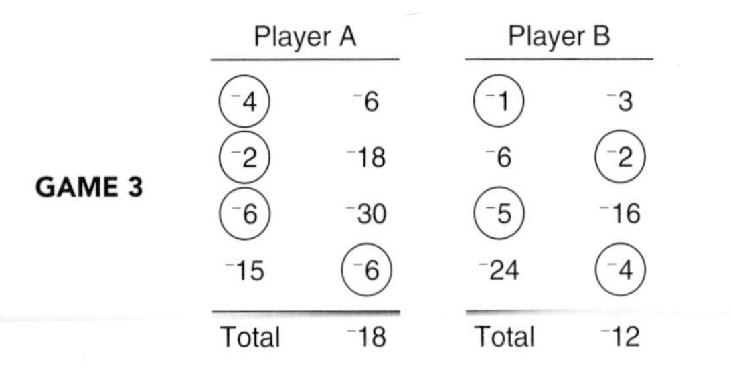

GAME 3

Player A		Player B	
$^{-}4$ (circled)	$^{-}6$	$^{-}1$ (circled)	$^{-}3$
$^{-}2$ (circled)	$^{-}18$	$^{-}6$	$^{-}2$ (circled)
$^{-}6$ (circled)	$^{-}30$	$^{-}5$ (circled)	$^{-}16$
$^{-}15$	$^{-}6$ (circled)	$^{-}24$	$^{-}4$ (circled)
Total	$^{-}18$	Total	$^{-}12$

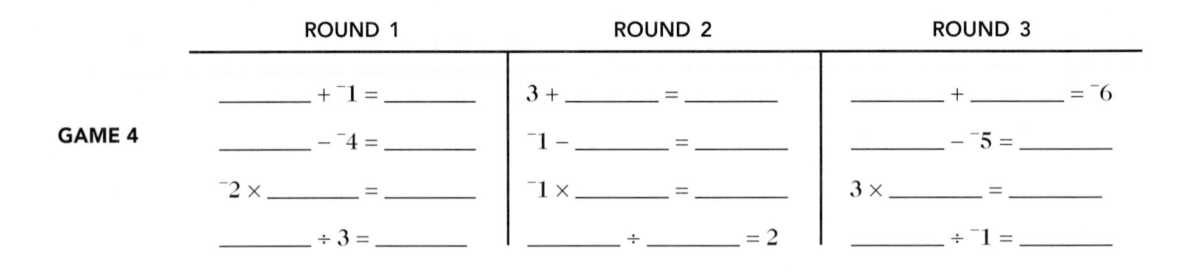

GAME 4

ROUND 1	ROUND 2	ROUND 3
____ + $^{-}1$ = ____	3 + ____ = ____	____ + ____ = $^{-}6$
____ − $^{-}4$ = ____	$^{-}1$ − ____ = ____	____ − $^{-}5$ = ____
$^{-}2$ × ____ = ____	$^{-}1$ × ____ = ____	3 × ____ = ____
____ ÷ 3 = ____	____ ÷ ____ = 2	____ ÷ $^{-}1$ = ____

ACTIVITY SET 5.2

FRACTION BAR MODEL FOR EQUALITY AND INEQUALITY

Purpose To use Fraction Bars to provide a visual model for the *part-to-whole* and *division* concepts of fractions, fraction equality and inequality, and fraction rounding and estimation[3]

Materials Fraction Bars from the Manipulative Kit

Activity The set of Fraction Bars consists of 32 bars divided into 2, 3, 4, 6, or 12 equal parts. The bars are colored (halves—green, thirds—yellow, fourths—blue, sixths—red, and twelfths—orange) so that the user can easily distinguish bars with different numbers of parts. Various fractions are named by shading parts of the bars. For example, the bar to the left represents $\frac{5}{6}$ (five-sixths). The top number in the fraction, the *numerator*, tells the number of shaded parts, and the bottom number, the *denominator*, tells the number of equal parts into which the bar is divided. This model illustrates the *part-to-whole* concept of fractions. It emphasizes the unit and models fractions in relation to this unit. The *division* concept of fractions is introduced in activity 10.

$\frac{5}{6}$

A Fraction Bar with all of its parts shaded is called a whole bar and its fraction equals 1. A bar with no parts shaded is called a zero bar and its fraction equals zero.

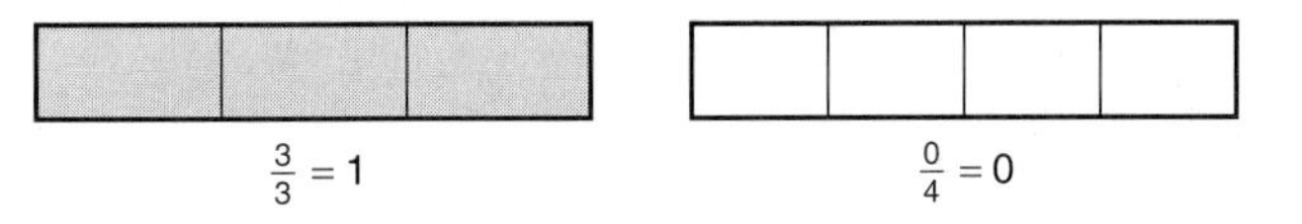

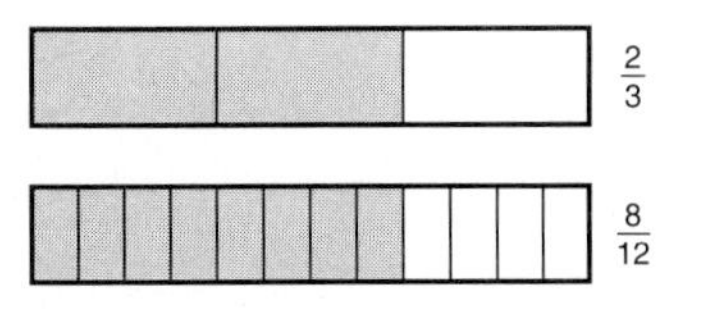

1. The fractions for the bars in the figure to the left are equal because both bars have the same amount of shading. In the set of 32 Fraction Bars, there are three bars whose fractions equal $\frac{2}{3}$. Sort your set of bars into piles so that the bars with the same shaded amount are in the same pile.

★ a. There are five bars whose fractions are equal to $\frac{0}{4}$. How many bars represent the other fractions listed here?

FRACTION	NUMBER OF BARS WITH EQUAL FRACTIONS
$\frac{0}{4}$	5
$\frac{1}{2}$	_____
$\frac{2}{3}$	_____
$\frac{6}{6}$	_____
$\frac{1}{4}$	_____

[3]Fraction Bars is a registered trademark of Scott Resources, Inc.

b. Complete this table by writing in the sixths, fourths, thirds, and halves that are equal to the twelfths. (*Note:* Many of the squares will be blank.)

Twelfths	$\frac{0}{12}$	$\frac{1}{12}$	$\frac{2}{12}$	$\frac{3}{12}$	$\frac{4}{12}$	$\frac{5}{12}$	$\frac{6}{12}$	$\frac{7}{12}$	$\frac{8}{12}$	$\frac{9}{12}$	$\frac{10}{12}$	$\frac{11}{12}$	$\frac{12}{12}$
Sixths													
Fourths													
Thirds													
Halves													

2. The $\frac{1}{3}$ bar has more shading than the $\frac{1}{4}$ bar, so $\frac{1}{3}$ is greater than $\frac{1}{4}$. This is writen $\frac{1}{3} > \frac{1}{4}$ (or $\frac{1}{4} < \frac{1}{3}$).

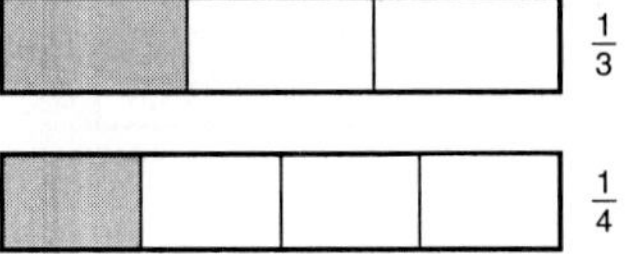

★ **a.** Select the following Fraction Bars from your set of bars:

$$\frac{5}{6}, \frac{1}{4}, \frac{11}{12}, \frac{2}{3}, \frac{1}{6}, \frac{5}{12}, \frac{3}{4}, \frac{1}{3}, \frac{7}{12}, \frac{1}{2}, \text{ and } \frac{1}{12}$$

Place these bars in increasing order from the smallest shaded amount to the largest shaded amount and complete the following inequalities.

$$\_ < \frac{1}{6} < \_ < \_ < \_ < \frac{1}{2} < \_ < \_ < \_ < \_ < \frac{11}{12}$$

b. Each of the preceding fractions has a common denominator of 12. Rewrite these inequalities using only fractions with a denominator of 12.

$$\_ < \frac{2}{12} < \_ < \_ < \_ < \_ < \_ < \_ < \_ < \_ < \frac{11}{12}$$

3. By placing the five different types of Fraction Bars in a column, we can visualize many equalities and inequalities. For example, by comparing the vertical lines of these bars, we can see that

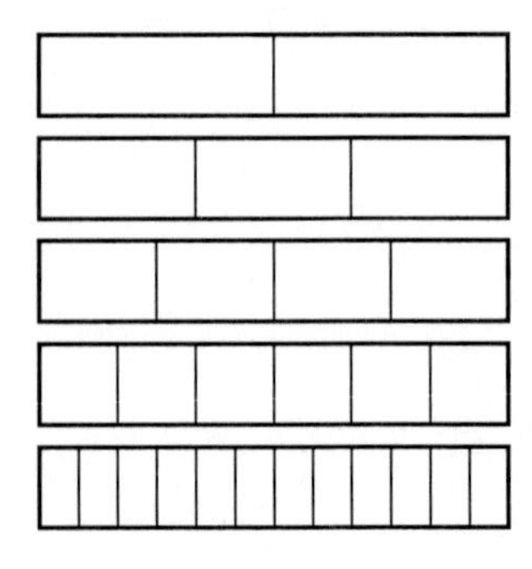

$$\frac{1}{2} = \frac{6}{12} \qquad \frac{2}{3} < \frac{3}{4} \qquad \frac{5}{6} = \frac{10}{12}$$

a. List 10 different equalities of pairs of fractions that can be illustrated by comparing the vertical lines of these bars.

★ **b.** List 10 different inequalities of pairs of fractions that can be illustrated by these bars.

4. Many inequalities for fractions can be determined mentally by visualizing the bars representing the fractions. For example, we can determine that $\frac{1}{5}$ is less than $\frac{1}{4}$, because each part of a bar with 5 equal parts is smaller than each part of a bar with 4 equal parts.

Mentally determine the correct inequality sign to put between each pair of fractions. Explain your reasoning in terms of Fraction Bars.

a. $\frac{1}{3}$ $\quad$ $\frac{1}{10}$

★ b. $\frac{5}{11}$ $\quad$ $\frac{4}{6}$

c. $\frac{2}{3}$ $\quad$ $\frac{9}{10}$

★ d. $\frac{1}{50}$ $\quad$ $\frac{1}{30}$

5. Rounding fractions to the nearest whole number can be facilitated by sketching or visualizing Fraction Bars. For example, on the bars representing $1\frac{2}{7}$, the shaded amount is closer to 1 whole bar than to 2 whole bars. So to the nearest whole number, $1\frac{2}{7}$ rounds to 1.

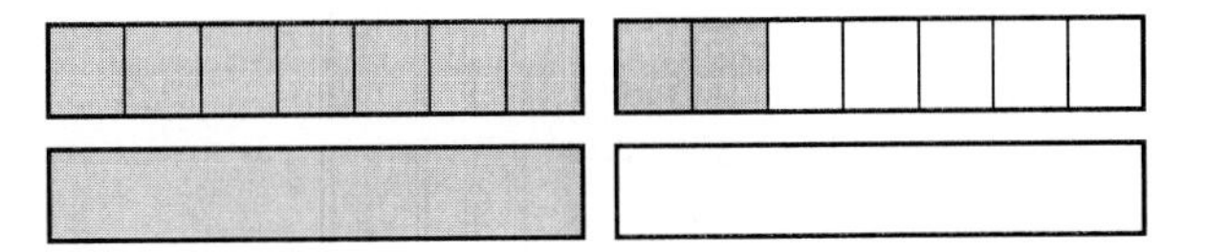

On the other hand, on the bars representing $1\frac{4}{7}$, the shaded amount is closer to 2 whole bars than to 1 whole bar. So to the nearest whole number, $1\frac{4}{7}$ rounds to 2.

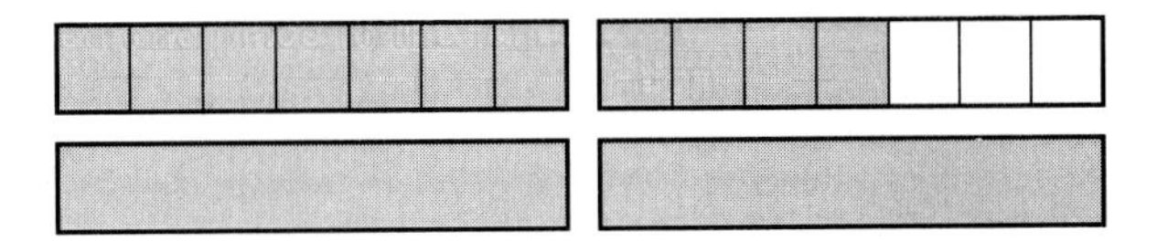

Round the following fractions to the nearest whole number. Draw a Fraction Bar sketch to justify your answer.

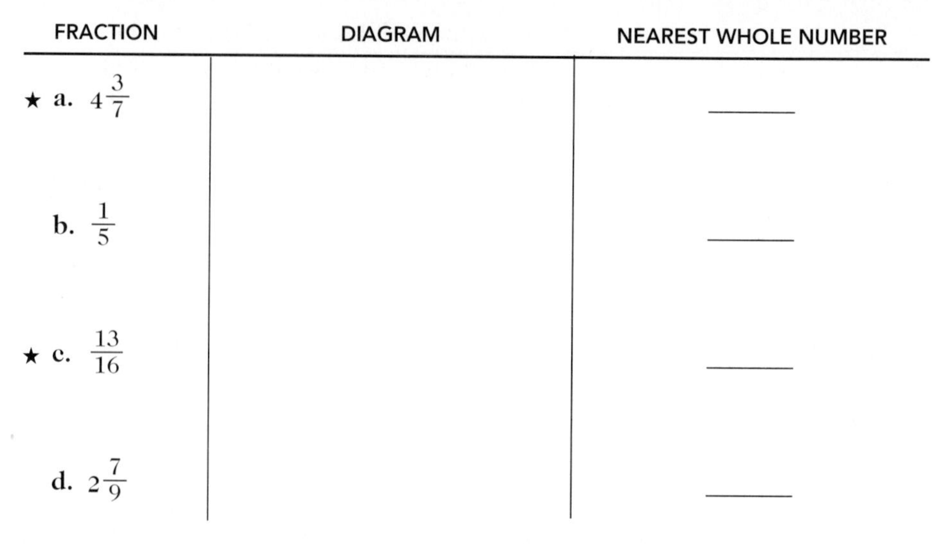

FRACTION	DIAGRAM	NEAREST WHOLE NUMBER
★ a. $4\frac{3}{7}$		______
b. $\frac{1}{5}$		______
★ c. $\frac{13}{16}$		______
d. $2\frac{7}{9}$		______

6. Each part of this $\frac{3}{4}$ bar has been split into 2 equal parts. There are now 8 parts, and 6 of these are shaded. Because this splitting has neither increased nor decreased the total shaded amount of the bar, both $\frac{3}{4}$ and $\frac{6}{8}$ are fractions for the same amount.

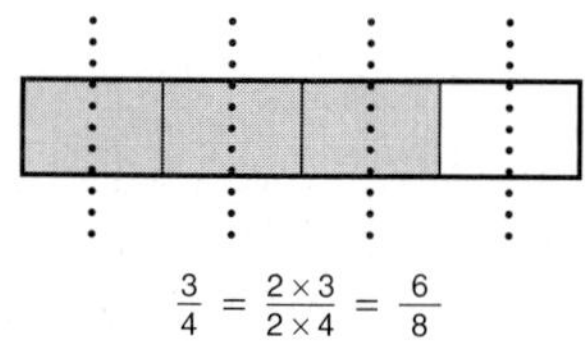

a. Split each part of these bars into 2 equal parts, and complete the equations.

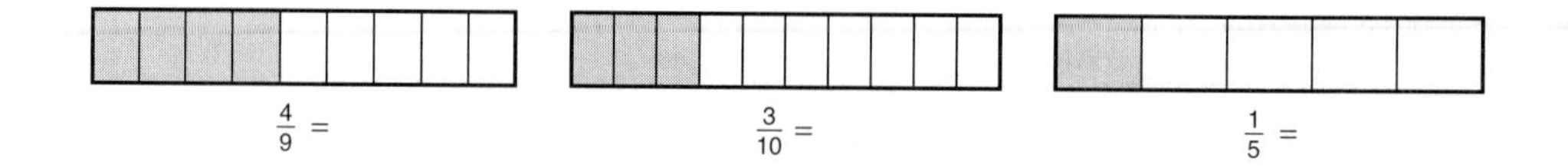

★ b. Split each part of these bars into 3 equal parts, and complete the equations.

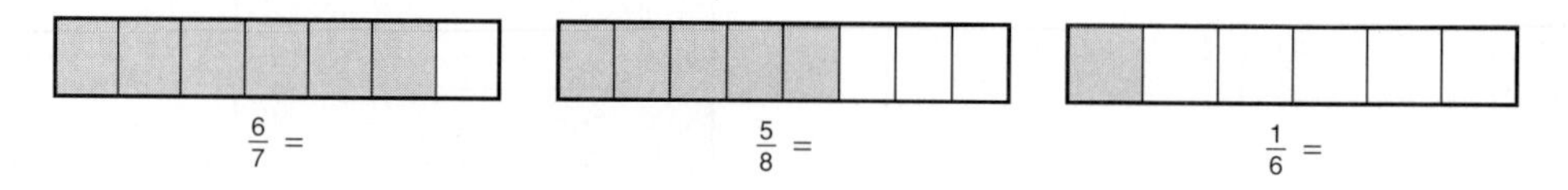

c. Split each part of these bars into 4 equal parts, and complete the equations.

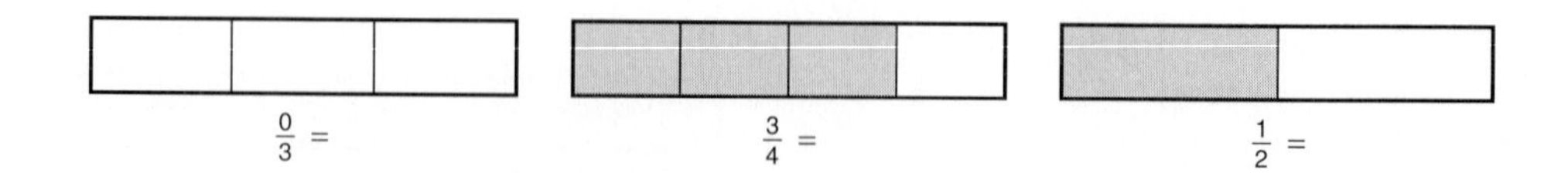

d. Suppose that each part of the Fraction Bar for $\frac{x}{y}$ is split into 3 equal parts. How many of the parts will be shaded? ______ How many parts will there be in all? ______ Write the fraction for this bar. ______

7. There are an infinite number of fractions that are equal to $\frac{2}{3}$. These fractions can be generated by equally splitting the parts of a $\frac{2}{3}$ bar.

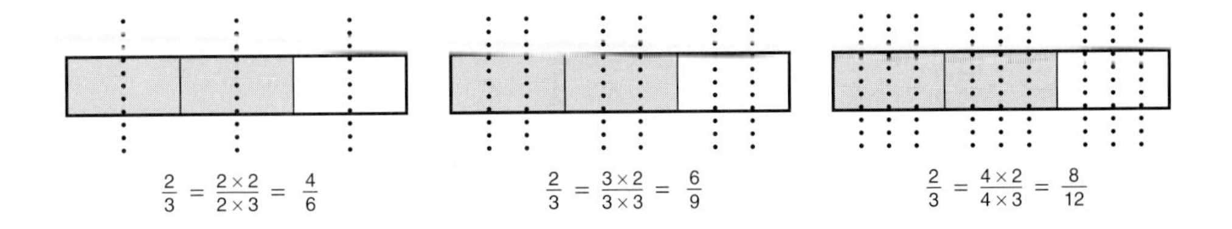

$$\frac{2}{3} = \frac{2 \times 2}{2 \times 3} = \frac{4}{6} \qquad \frac{2}{3} = \frac{3 \times 2}{3 \times 3} = \frac{6}{9} \qquad \frac{2}{3} = \frac{4 \times 2}{4 \times 3} = \frac{8}{12}$$

a. If each part of a $\frac{2}{3}$ bar is split into 5 equal parts, what equality of fractions does the bar represent?

★ b. If each part of a $\frac{2}{3}$ bar is split into 17 equal parts, what equality of fractions does the bar represent?

c. What effect does splitting each part of any Fraction Bar into 17 equal parts have on the numerator and denominator of the fraction for the bar?

★ d. Split each of the following bars to illustrate the given equality.

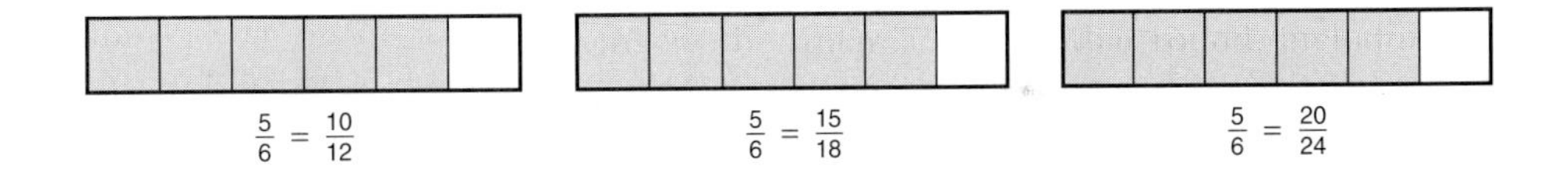

$$\frac{5}{6} = \frac{10}{12} \qquad \frac{5}{6} = \frac{15}{18} \qquad \frac{5}{6} = \frac{20}{24}$$

8. Each part of a $\frac{2}{3}$ bar is bigger than each part of a $\frac{3}{4}$ bar. If each part of the $\frac{2}{3}$ bar is split into 4 equal parts and each part of the $\frac{3}{4}$ bar is split into 3 equal parts, both bars will then have 12 parts of equal size. The new fractions for these bars will have a common denominator of 12.

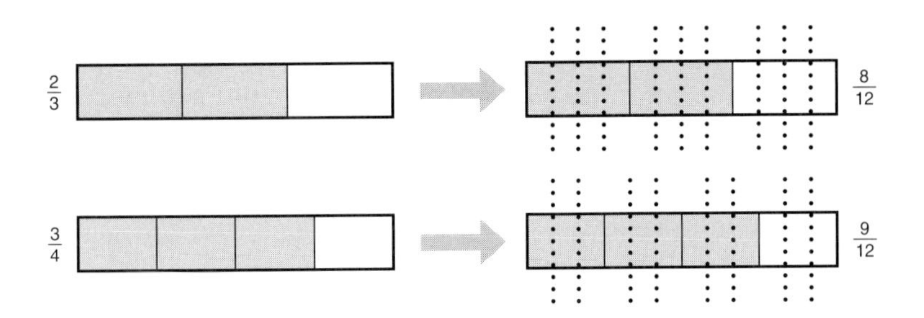

For each of the following pairs of bars, split the parts so that all the new parts for both bars are the same size. The corresponding fractions will have the same denominator. Complete the equations.

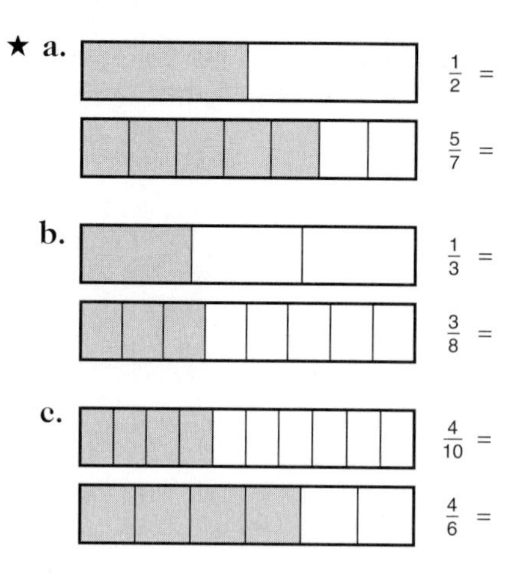

9. Sketch bars, at the left, for $\frac{3}{5}$ and $\frac{5}{7}$. Can you tell from your sketch which fraction is greater?

a. When both bars are split so that all parts have the same size, the two fractions for these bars will have the same denominators. Complete these equations so that both fractions have the same denominator.

$$\frac{3}{5} = \qquad\qquad \frac{5}{7} =$$

b. Which is greater, $\frac{3}{5}$ or $\frac{5}{7}$?

Replace the fractions in each of the following pairs with fractions having the same denominator. Circle the greater fraction in each pair (or state that they are equal).

c. $\frac{1}{3} =$ $\quad$ $\frac{2}{7} =$

★ d. $\frac{7}{9} =$ $\quad$ $\frac{8}{10} =$

e. $\frac{4}{6} =$ $\quad$ $\frac{6}{9} =$

★ f. $\frac{11}{15} =$ $\quad$ $\frac{8}{11} =$

10. The *division* concept of fractions relates quotients of whole numbers to fractions. For example, 3 whole bars are placed end to end to represent the number 3. If this 3 bar is divided into 4 equal parts, each part will be $\frac{3}{4}$ of a whole bar. This shows that $3 \div 4 = \frac{3}{4}$.

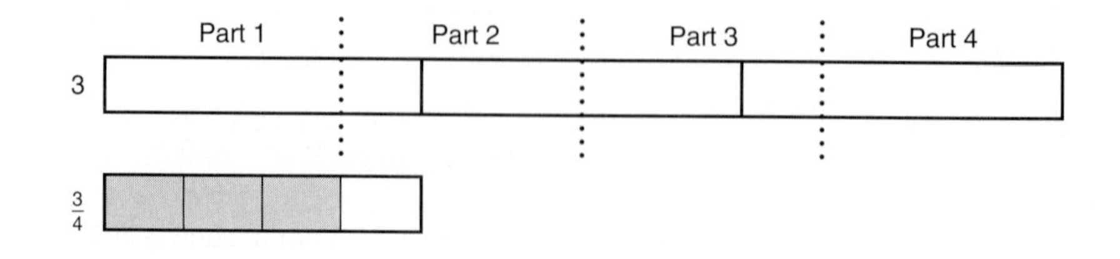

Illustrate each quotient of whole numbers by dividing the given bar into equal parts. Then complete the equation by writing a fraction or mixed number.[4]

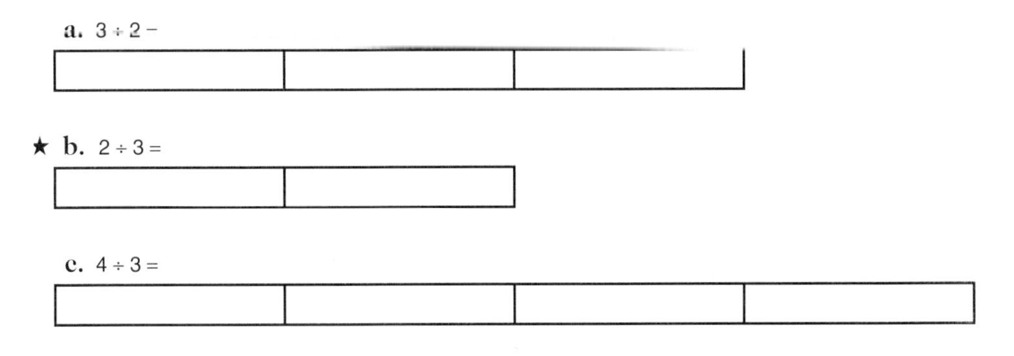

JUST FOR FUN

These two fraction games provide extra practice in comparing fractions. In FRIO[5] you compare unequal fractions, and in Fraction Bingo you search for equal fractions.

FRIO (2 to 4 players)

Each player is dealt five bars in a row face up. These bars should be left in the order in which they are dealt. The remaining bars are spread face down. Each player, in turn, takes a bar that is face down and uses it to replace any one of the five bars.

The object of the game is to get five bars in order from the *smallest* shaded amount to the *largest* or from the largest to the *smallest*. In this example the five bars will be in order if the $\frac{2}{6}$ bar is replaced by a whole bar and the $\frac{1}{4}$ bar is replaced by an $\frac{11}{12}$ bar. The first player to get five fraction bars in decreasing or increasing order wins the game.

FRACTION BINGO (2 to 4 players)

Each player selects a fraction bingo mat from those shown below. The deck of 32 bars is spread face down. Each player, in turn, takes a bar and circles the fraction or fractions on his or her mat that equal the fraction from the bar. The first player to circle four fractions in any row, column, or diagonal is the winner.

Strategy: If the bars are colored green, yellow, blue, red, and orange for halves, thirds, fourths, sixths, and twelfths, respectively, you can increase your chances of winning by selecting bars of the appropriate color.

$\frac{4}{6}$	$\frac{1}{6}$	$\frac{0}{4}$	$\frac{11}{12}$
$\frac{5}{12}$	$\frac{1}{4}$	$\frac{5}{6}$	$\frac{3}{6}$
$\frac{3}{4}$	$\frac{1}{12}$	$\frac{1}{3}$	$\frac{11}{12}$
$\frac{6}{12}$	$\frac{4}{4}$	$\frac{9}{12}$	$\frac{10}{12}$

Mat 1

$\frac{1}{3}$	$\frac{1}{4}$	$\frac{1}{6}$	$\frac{0}{3}$
$\frac{7}{12}$	$\frac{0}{6}$	$\frac{1}{12}$	$\frac{4}{6}$
$\frac{2}{3}$	$\frac{3}{4}$	$\frac{5}{6}$	$\frac{5}{12}$
$\frac{11}{12}$	$\frac{2}{2}$	$\frac{3}{12}$	$\frac{1}{2}$

Mat 2

$\frac{2}{3}$	$\frac{1}{4}$	$\frac{2}{6}$	$\frac{6}{12}$
$\frac{5}{6}$	$\frac{0}{3}$	$\frac{5}{12}$	$\frac{1}{6}$
$\frac{7}{12}$	$\frac{4}{4}$	$\frac{1}{2}$	$\frac{11}{12}$
$\frac{3}{4}$	$\frac{8}{12}$	$\frac{1}{12}$	$\frac{1}{3}$

Mat 3

$\frac{1}{4}$	$\frac{1}{6}$	$\frac{7}{12}$	$\frac{1}{2}$
$\frac{0}{3}$	$\frac{5}{12}$	$\frac{2}{3}$	$\frac{4}{4}$
$\frac{1}{12}$	$\frac{5}{6}$	$\frac{11}{12}$	$\frac{2}{12}$
$\frac{6}{6}$	$\frac{3}{4}$	$\frac{0}{12}$	$\frac{1}{3}$

Mat 4

PUZZLER

At a certain college, $\frac{1}{4}$ of the first-year women students are from homes where both parents are professionals. Of these, $\frac{3}{5}$ are interested in the same profession as one of their parents. The latter group is composed of 18 students. How many first-year women students attend the college?[6]

[4]If long strips of paper are available, the big bars such as the 2 bar (the length of 2 whole bars) and the 3 bar (the length of 3 whole bars) can be cut to actual length. A strip can then be folded and the folded amount compared to the fraction bars.

[5]R. Drizigacker, "FRIO, or FRactions In Order," *Arithmetic Teacher* 13 (December 1966), 684–685.

[6]From the "April Calendar" of *The Mathematics Teacher* 81 (April 1988), 278–279.

ACTIVITY SET 5.3

COMPUTING WITH FRACTION BARS

Purpose To use Fraction Bars to perform the basic operations of addition, subtraction, multiplication, and division in a visual and intuitive manner

Materials Fraction Bars from the Manipulative Kit

Activity Piaget has charted the cognitive development of preadolescents, and his research indicates that even at the age of 12, most children deal only with symbols that are closely tied to their perceptions. For example, the symbolic representation $\frac{1}{6} + \frac{2}{3} = \frac{5}{6}$ has meaning for most elementary school children only if they can relate it directly to concrete or pictorial representations.[7]

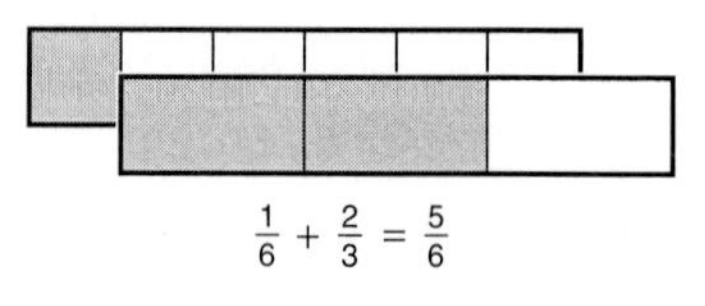

The following table documents the lack of conceptual understanding of fractions and fraction addition exhibited by 13-year-olds when they were asked to estimate the answer to $\frac{12}{13} + \frac{7}{8}$.[8] Over 50 percent of the students responded with the incorrect answers 19 and 21. What is a reasonable explanation of how students may have arrived at the incorrect answers 19 and 21?

RESPONSES	PERCENT RESPONDING, AGE 13
○ 1	7
● 2	24
○ 19	28
○ 21	27
○ I don't know	14

Students may learn fraction skills at a rote manipulation level, but when their memory fails or they encounter a nonstandard application, they have no conceptual basis to fall back on.

In this activity set, Fraction Bars will be used to develop visual images of fraction operations. These images will provide a conceptual basis for computations, estimations, and problem solving.

Addition and Subtraction

1. **Addition and Subtraction:** Write the missing fractions and the sum or difference of the fractions for each pair of bars. Determine the sum or difference by visual inspection of bars.

[7]M. J. Driscoll, "The Role of Manipulatives in Elementary School Mathematics," *Research within Reach: Elementary School Mathematics* (St. Louis, MO.: Cemrel Inc., 1983), 1.
[8]T. P. Carpenter, H. Kepner, M. K. Corbitt, M. M. Lindquist, and R. E. Reys, "Results and Implications of the Second NAEP Mathematics Assessment: Elementary School," *Arithmetic Teacher* 27 (April 1980), 10–12, 44–47.

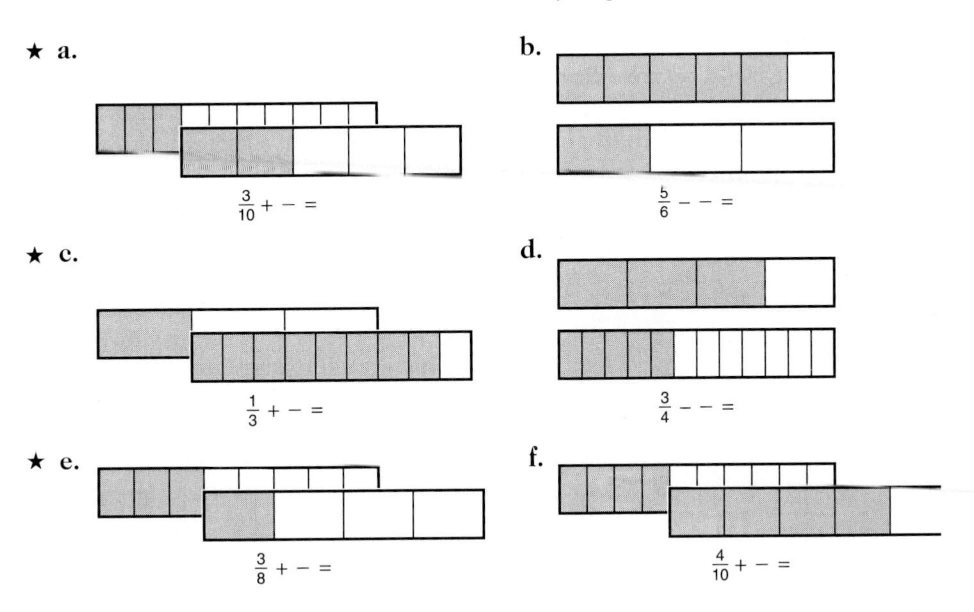

2. Select four pairs of fraction bars at random. Complete the following equations for the sum and difference of each pair of fractions. (Subtract the smaller from the larger if the fractions are unequal.)

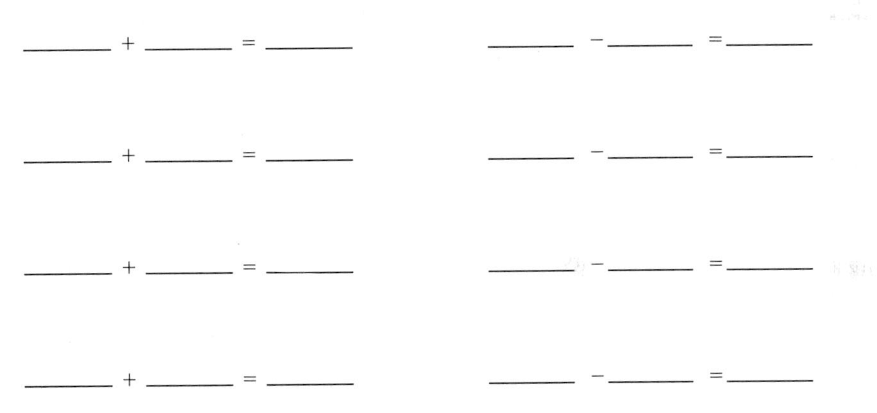

3. Obtaining Common Denominators: If each part of the $\frac{2}{3}$ bar is split into 4 equal parts and each part of the $\frac{1}{4}$ bar is split into 3 equal parts, both bars will have 12 parts of the same size. These new bars show that $\frac{2}{3} + \frac{1}{4} = \frac{11}{12}$ and $\frac{2}{3} - \frac{1}{4} = \frac{5}{12}$.

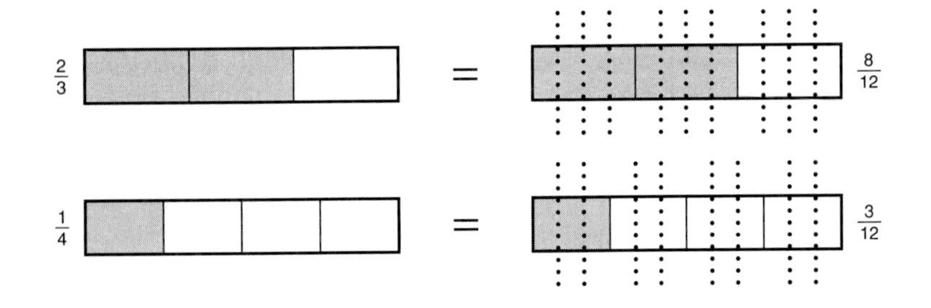

Split the parts of the following pairs of Fraction Bars so that each pair of bars has the same number of parts of the same size. Then write two fraction equalities under each pair. For example, in part a, $\frac{5}{6} = \frac{10}{12}$ and $\frac{1}{4} = \frac{3}{12}$.

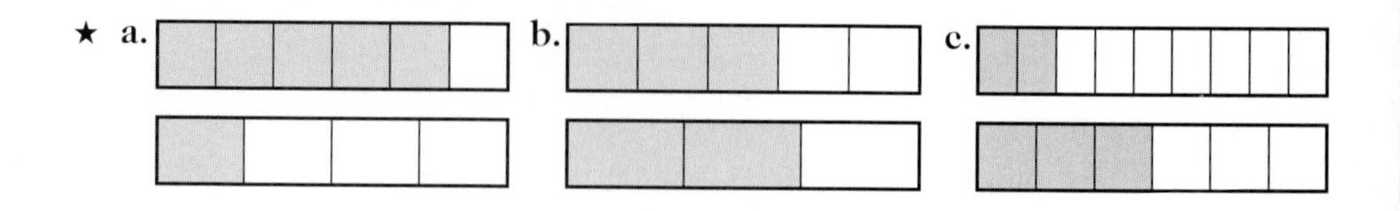

Use the bars from parts a, b, and c above to compute the following sums and differences.

★ d. $\frac{5}{6} + \frac{1}{4} =$ e. $\frac{3}{5} + \frac{2}{3} =$ f. $\frac{2}{9} + \frac{3}{6} =$

★ g. $\frac{5}{6} - \frac{1}{4} =$ h. $\frac{2}{3} - \frac{3}{5} =$ i. $\frac{3}{6} - \frac{2}{9} =$

4. Sketch Fraction Bars for each pair of fractions, and then split the parts of the bars to carry out the operation. Use your diagrams to explain how you reached your conclusions.

★ a. $\frac{5}{9} + \frac{1}{3}$

b. $\frac{1}{4} + \frac{3}{5}$

★ c. $\frac{2}{5} - \frac{1}{3}$

d. $\frac{1}{2} - \frac{2}{7}$

Multiplication

5. You can determine $\frac{1}{3} \times \frac{1}{6}$ by splitting the shaded part of a $\frac{1}{6}$ bar into 3 equal parts. One of these smaller parts is $\frac{1}{18}$ of a bar, because there are 18 of these parts in a whole bar.

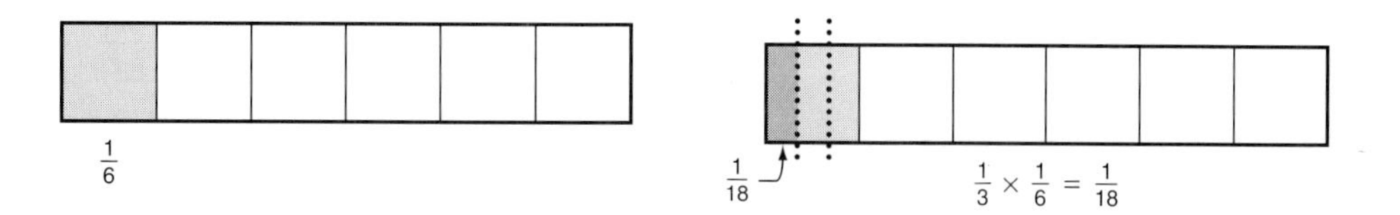

To determine $\frac{1}{3} \times \frac{4}{5}$, split each shaded part of the $\frac{4}{5}$ bar into 3 equal parts. One of these split parts is $\frac{1}{15}$ of the whole bar, and 4 of these split parts is $\frac{4}{15}$ of the whole bar.

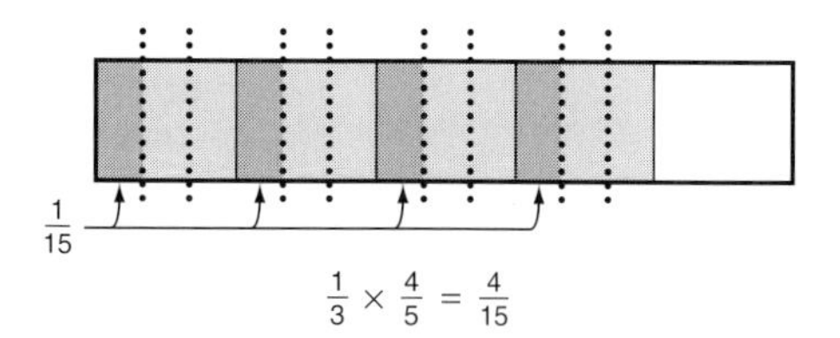

a. Split each shaded part of each bar into 2 equal parts. Use this result to complete the given equations.

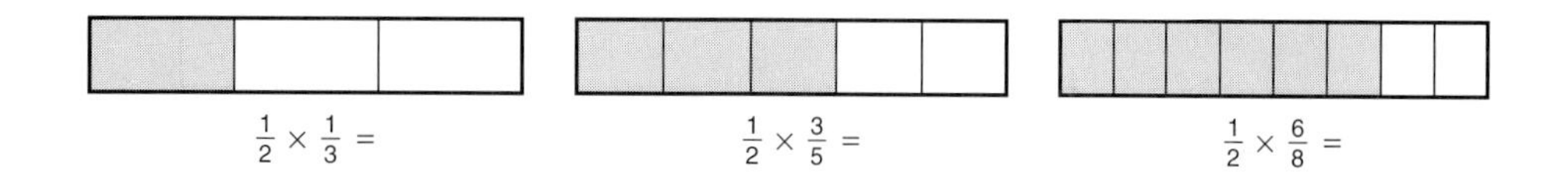

★ **b.** Split each shaded part of each bar into 3 equal parts, and complete the equations.

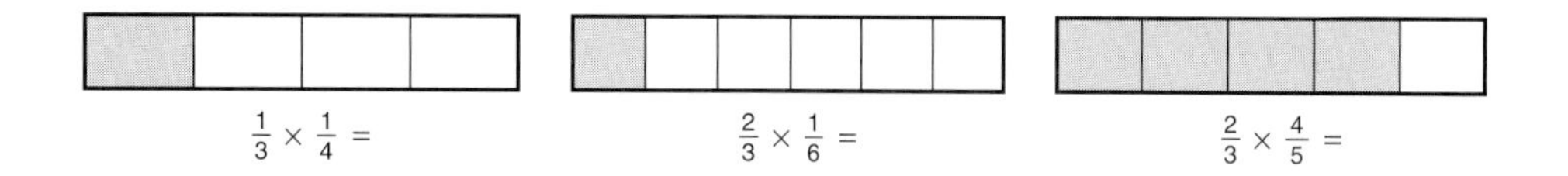

c. Split each shaded part of each bar into 4 equal parts, and complete the equations.

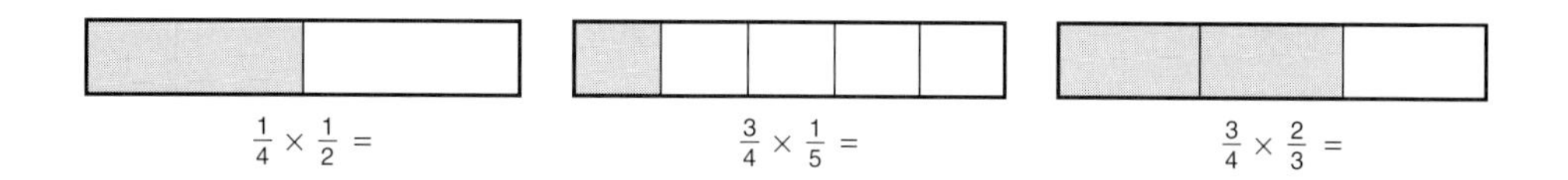

6. Sketch a Fraction Bar for $\frac{3}{4}$ and visually determine the product $\frac{2}{3} \times \frac{3}{4}$. Write a description of the procedure you used.

Division

7. Using the *measurement* approach to division (see activity set 3.4), we can write $\frac{1}{2} \div \frac{1}{6} = 3$, because the shaded portion of the $\frac{1}{6}$ bar can be measured off (or fits into) exactly 3 times on the shaded part of the $\frac{1}{2}$ bar.

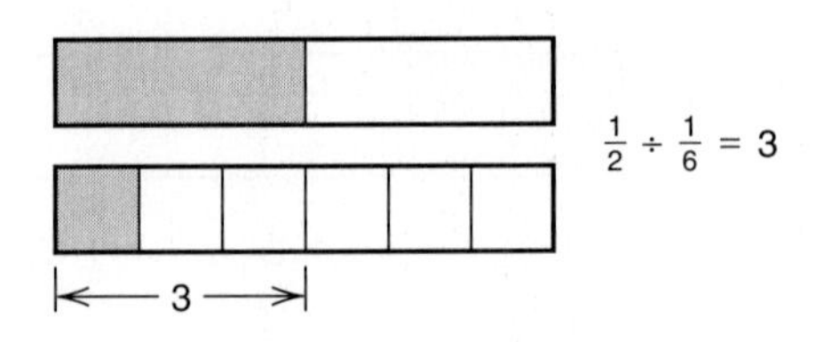

Similarly, $\frac{5}{6} \div \frac{1}{3} = 2\frac{1}{2}$, because the shaded portion of the $\frac{1}{3}$ bar can be measured off (or fits into) $2\frac{1}{2}$ times on the shaded part of the $\frac{5}{6}$ bar.

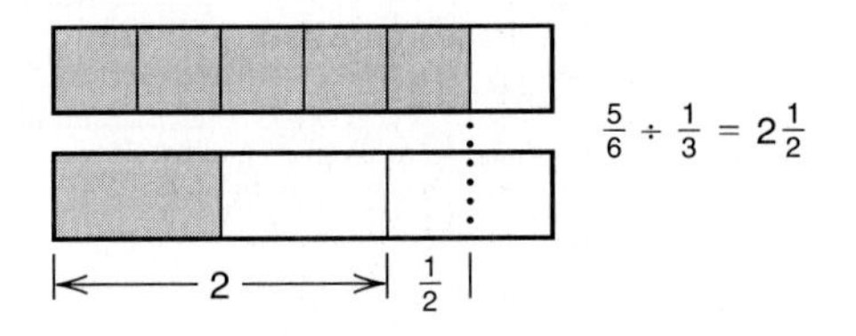

Use your Fraction Bars to determine the following quotients.

a. $\frac{3}{4} \div \frac{1}{2} =$ ★ b. $\frac{10}{12} \div \frac{1}{6} =$

c. $\frac{7}{12} \div \frac{1}{6} =$ ★ d. $1 \div \frac{2}{3} =$

e. $\frac{3}{2} \div \frac{3}{4} =$

Sketch Fraction Bars to determine the following quotients. Use your sketch to explain your reasoning.

★ f. $\frac{7}{8} \div \frac{1}{4} =$

g. $\frac{7}{10} \div \frac{1}{5} =$

Estimation

8. It is often helpful to draw a sketch or visualize a fraction model when you are trying to estimate with fractions. For example, to estimate the quotient $1\frac{5}{7} \div \frac{1}{8}$, you can compare Fraction Bars representing $1\frac{5}{7}$ to bars representing eighths. There are 8 eighths in 1 whole bar and about 6 eighths in the $\frac{5}{7}$ bar. So, $\frac{1}{8}$ can be measured off about 14 times, and $1\frac{5}{7} \div \frac{1}{8} \approx 14$.

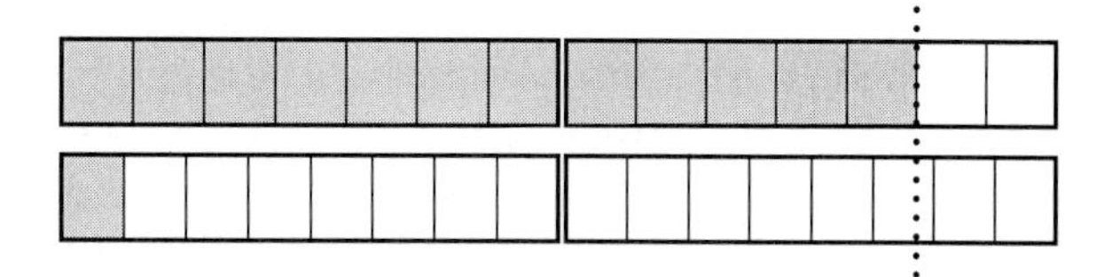

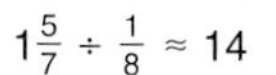

Use a visual approach to estimate the following fraction operations. Draw Fraction Bar sketches and explain how you arrived at your estimation.

a. $1\frac{3}{4} \div \frac{1}{3}$

★ b. $\frac{5}{8} - \frac{1}{6}$

c. $\frac{4}{5} + \frac{1}{2}$

★ d. $\frac{2}{3} \times \frac{5}{6}$

All Four Operations

9. ★ a. The following 10 fractions are from the Fraction Bars:

$$\frac{1}{6}, \frac{2}{4}, \frac{5}{12}, \frac{2}{6}, \frac{0}{3}, \frac{3}{4}, \frac{3}{12}, \frac{2}{3}, \frac{1}{2}, \frac{1}{4}$$

Using each fraction only once, try placing these fractions in the 10 blanks to form four equations.

— + — = —   3 × — = —

— − — = —   — ÷ — = 2

b. Spread your bars face down and select any 10 of them. Using the fraction from each bar only once, complete as many of these four equations as possible. If you cannot form all four equations, continue to select bars, one at a time, until all the equations can be completed.

c. Solitaire: The activity in part b can be played like a solitaire game. See how many turns it takes you to complete the four equations by selecting only 10 bars on each turn.

JUST FOR FUN

These three games provide opportunities to perform operations on fractions.

FRACTION BAR BLACKJACK (Addition—2 to 4 players)

Spread the bars face down. The object is to select one or more bars so that the fraction or the sum of fractions is as close to 1 as possible, but not greater than 1. The first player selects bars one at a time, trying to get close to 1 without going over. (A player may wish to take only 1 bar.) Player 1 finishes his or her turn by saying "I'm holding." After every player in turn has finished, the players show their bars. The player who is closest to a sum of 1, but not over, wins the round.

Examples: Player 1 has a sum greater than 1 and is over. Player 2 has a greater sum than player 3 and wins the round.

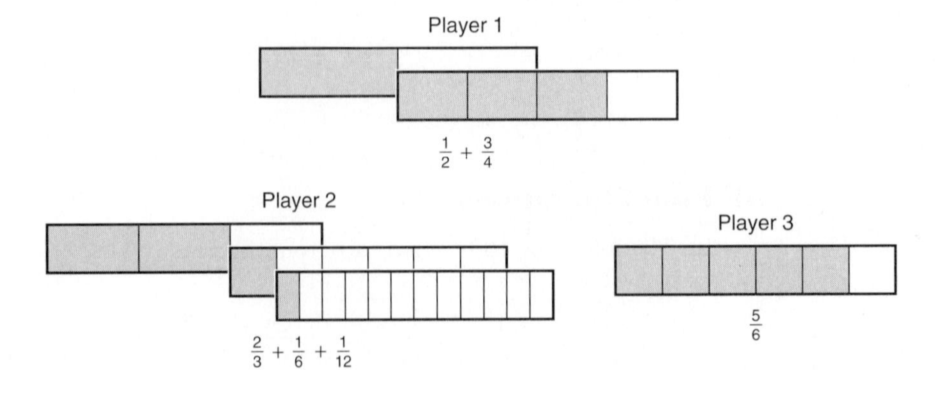

SOLITAIRE (Subtraction)

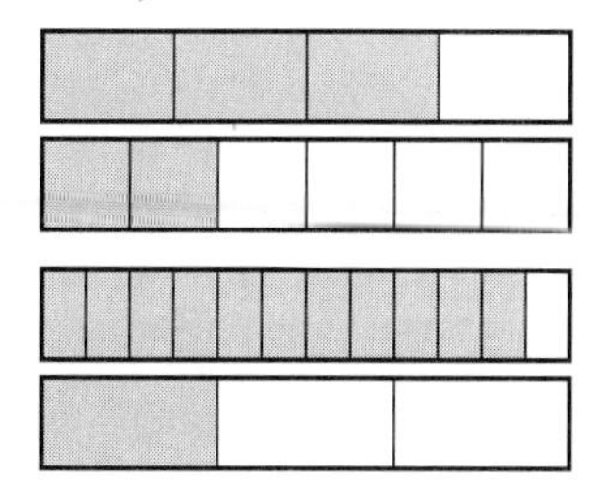

Spread the Fraction Bars face down. Turn over two bars and compare their shaded amounts. If the difference between the two fractions is less than $\frac{1}{2}$, you win the two bars. If not, you lose the bars. See how many bars you can win by playing through the deck. Will you win the top pair of bars shown here? the bottom pair?

GREATEST QUOTIENT (Division—2 to 4 players)

Remove the zero bars from your set of 32 bars and spread the remaining bars face down. Each player takes two bars. The object of the game is to get the greatest possible quotient by dividing one of the fractions by the other. The greatest *whole* number of times that one fraction divides into the other is the player's score. Each player has the option of taking another bar to improve his or her score or passing. If the player selects another bar, he or she must first discard one bar. The first player to score 21 points wins the game.

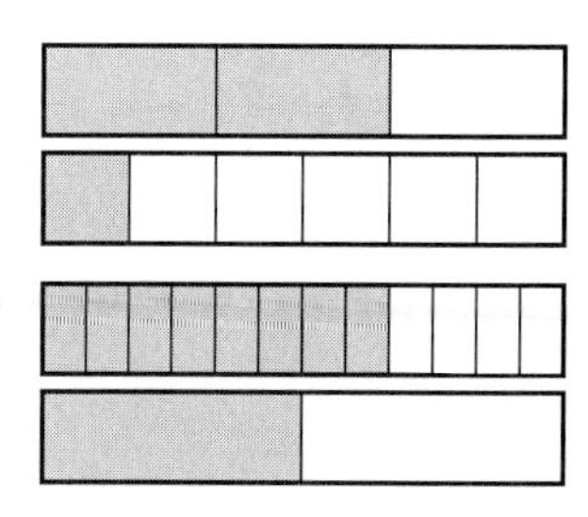

Examples: The whole number part of the quotient can be determined by comparing the shaded amounts of the bars. The score for the top two bars is 4, because the shaded amount of the $\frac{2}{3}$ bar is 4 times greater than the shaded amount of the $\frac{1}{6}$ bar. The score for the bottom two bars is 1, because the quotient is greater than 1 but less than 2.

IDEAS FOR THE ELEMENTARY CLASSROOM

SUGGESTED CLASSROOM ACTIVITY: INTEGER BALLOON

An intuitive understanding of a helium balloon is an effective model for beginning positive and negative number concepts. Students can make a balloon model (we'll call it UPORDOWN) like the one on the next page (with extra balloons and sandbags that can be attached or taken off) and place it near a vertical scale. It is also convenient to have a model for the overhead so that students can demonstrate their actions. Suppose UPORDOWN is in equilibrium with 10 balloons and 10 sandbags. When 1 balloon is attached, it goes up 1 level; when 2 sandbags are added, it goes down 2 levels; and so forth. Ask the students to demonstrate answers to questions like the following: What happens to UPORDOWN when 3 balloons are detached? 4 sandbags are taken off?

Students can use UPORDOWN to demonstrate sequences of operations as in "adding 4 sandbags and 3 balloons moves the big balloon 1 level down"; "adding 2 balloons and taking away 3 sandbags moves the big balloon 5 levels up." Depending on the backgrounds of your students, these examples might be summarized as

$$4S + 3B = 1 \text{ down} \qquad 2B - 3S = 5 \text{ up}$$

or, with + for B and "up" and minus for S and "down,"

$$^{-}4 + {}^{+}3 = {}^{-}1 \qquad {}^{+}2 - {}^{-}3 = {}^{+}5$$

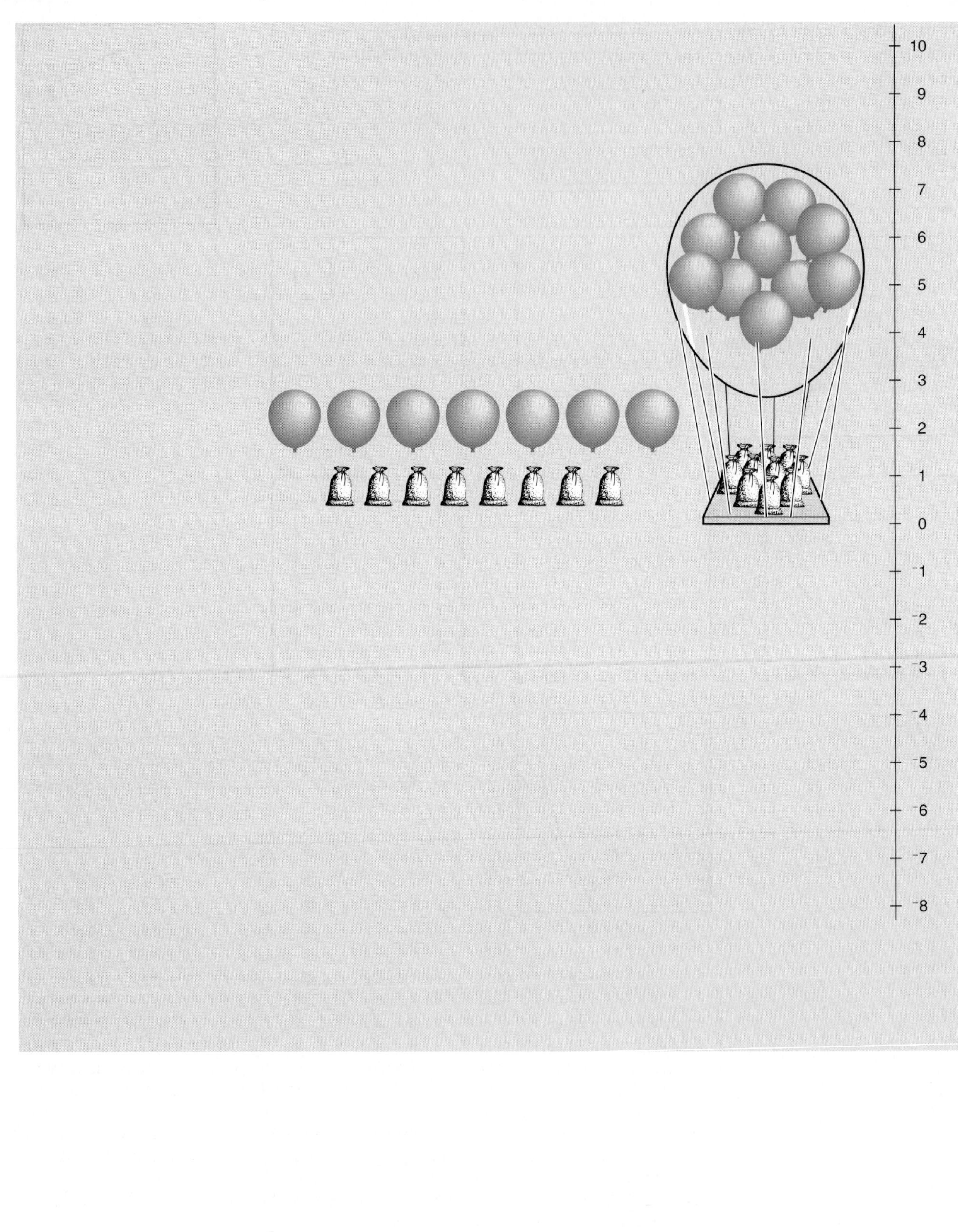
10
9
8
7
6
5
4
3
2
1
0
$^-1$
$^-2$
$^-3$
$^-4$
$^-5$
$^-6$
$^-7$
$^-8$

PUZZLER

Starting with a blank sheet of $8\frac{1}{2}$-by-11-inch paper, you can determine a length of $2\frac{1}{2}$ inches in one fold.

Then by folding the paper in half, you can determine a length of 3 inches. (*Note:* There are other ways to obtain a length of 3 inches.)

Circle all the remaining lengths in this list that you can obtain by folding the paper.

1 $1\frac{1}{2}$ 2 ($2\frac{1}{2}$) (3) $3\frac{1}{2}$ 4 $4\frac{1}{2}$ 5 $5\frac{1}{2}$ 6 $6\frac{1}{2}$
7 $7\frac{1}{2}$ 8 $8\frac{1}{2}$ 9 $9\frac{1}{2}$ 10 $10\frac{1}{2}$ 11

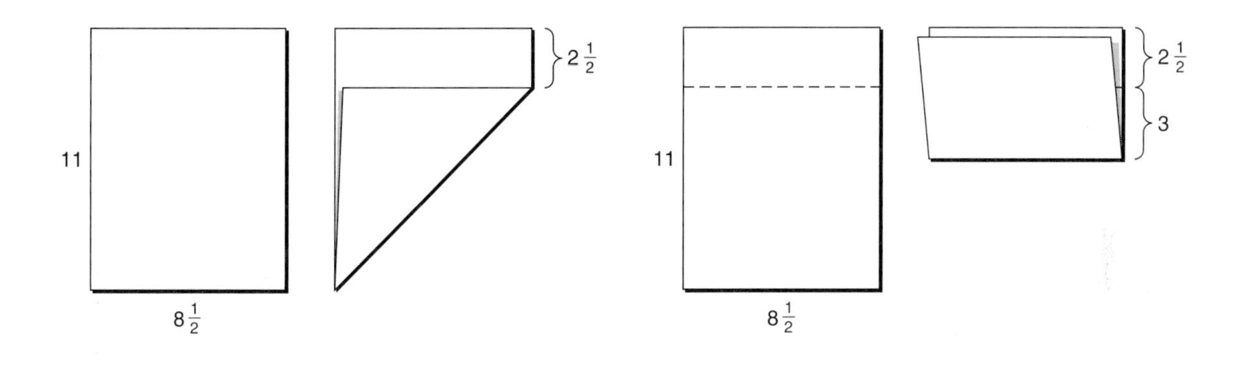

Readings for More Classroom Ideas

Bennett, A., and P. Davidson. *Fraction Bars Step-by-Step Teacher's Guide.* Fort Collins, CO: Scott Resources, Inc., 1982.

Bennett, A., E. Maier, and L. Ted Nelson. "Modeling Rationals" and "Modeling Integers," in *Math and the Mind's Eye* (Units ME4 and ME6). Salem, OR: The Math Learning Center, Box 3226, 1988.

Caldwell, Janet H. "Communicating About Fractions with Pattern Blocks." *Teaching Children Mathematics* 2 (November 1995): 156–161.

Cramer, Kathleen, and Nadine Bezuk. "Multiplication of Fractions: Teaching for Understanding." *Arithmetic Teacher* 39 (November 1991): 34–37.

Curcio, Frances R., Francine Sicklick, and Susan B. Turkel. "Divide and Conquer: Unit Strips to the Rescue." *Arithmetic Teacher* 35 (December 1987): 6–12.

Dirkes, M. Ann. "Draw to Understand." *Arithmetic Teacher* 39 (December 1991): 26–29.

Edge, Douglas, "Fractions and Panes." *Arithmetic Teacher* 34 (April 1987): 13–17.

Foreman, Linda, and Albert Bennett. "Linear Model for Fractions," in *Visual Mathematics* (Lessons 40–42). Salem, OR: The Math Learning Center, Box 3226, 1995.

Maher, C., R. Davis, and A. Alston. "Teachers Paying Attention to Students' Thinking." *Arithmetic Teacher* 39 (May 1992): 34–37.

Middleton, James A., Marja van den Heuvel-Panuizen, and Julia A. Shew. "Using Bar Representations as a Model for Connecting Concepts of Rational Number." *Mathematics Teaching in the Middle School* 3 (January 1998): 302–312.

Ott, J., D. Snook, and D. Gibson. "Understanding Partitive Division of Fractions." *Arithmetic Teacher* 39 (October 1991): 7–11.

Patterson, J. C. "Fourteen Different Strategies for Multiplication of Integers or Why $(-1)(-1) = 1$." *The Arithmetic Teacher* 19 (May 1972): 396–403.

Pothier, Yvonne, and Daiyo Sawada. "Partitioning: An Approach to Fractions." *Arithmetic Teacher* 38 (December 1990): 12–17.

Schultz, James E. "Area Models—Spanning the Mathematics of Grades 3–9." *Arithmetic Teacher* 39 (October 1991): 42–46.

Van de Walle, John, and C. S. Thompson. "Let's Do It: Fractions with Fraction Strips." *Arithmetic Teacher* 32 (December 1984): 4–9.

Warrington, Mary Ann, and Constance Kamii. "Multiplication with Fractions: A Piagetian, Constructivist Approach." *Mathematics Teaching in the Middle School* 3 (February 1998): 339–343.

Warrington, Mary Ann, and Constance Kamii. "Teaching Fractions: Fostering Children's Own Reasoning," in *Developing Mathematical Reasoning in Grades K–12,* 1999 Yearbook, edited by L. V. Stiff and R. R. Curcio, VA: National Council of Teachers in Mathematics, 1999, 82–92.

Zaweojewski, Judith S. "Ideas: Fractions as Areas." *Arithmetic Teacher* 34 (December 1986): 18–25.

CHAPTER 6

DECIMALS: RATIONAL AND IRRATIONAL

The approach to decimals should be similar to work with fractions, namely placing strong and continued emphasis on models and oral language and then connecting this work with symbols. . . . Exploring ideas of tenths and hundredths with models can include preliminary work with equivalent decimals.[1]

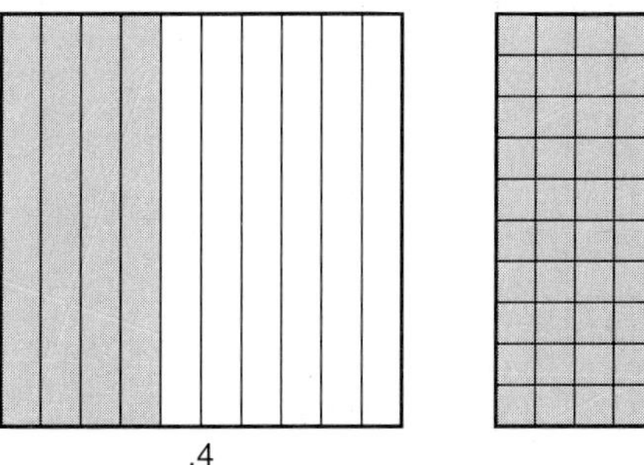

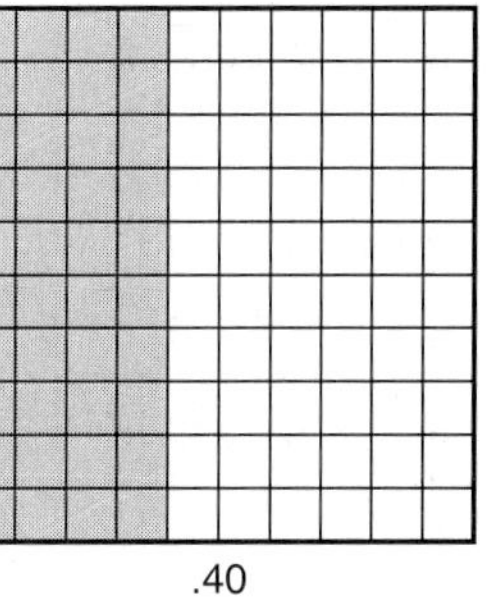

ACTIVITY SET 6.1

DECIMAL SQUARES MODEL

Purpose To use Decimal Squares[2] as a visual model for the part-to-whole concept of decimals and for illustrating decimal equality, inequality, place value, and estimation

Materials Decimal Squares are on Material Cards 23–26.

Activity The results of the second mathematics assessment of the National Assessment of Educational Progress (NAEP) have implications for the use of models in the teaching of decimal concepts.

[1]*Curriculum and Evaluation Standards for School Mathematics* (Reston, VA: National Council of Teachers of Mathematics, 1989), 59.
[2]Decimal Squares is a registered trademark of Scott Resources, Inc.

> In analyzing the computational errors made on the decimal exercises it is evident that much of the difficulty lies in a lack of conceptual understanding. . . . It is important that decimals be thought of as numbers and the ability to relate them to models should assist in understanding.[3]

The Decimal Squares model for decimals is an extension of the number piece model that was used for whole numbers in Chapter 3. The base-ten number piece model started with a unit and the pieces increased in size by a factor of 10 (10 units equal 1 long, 10 longs equal 1 flat, etc.). The Decimal Squares model starts with the unit, but the number pieces decrease in size by a factor of 10, as shown here. The number pieces representing decimals are individually too small to be handled, so they are indicated on the unit number piece by shading.

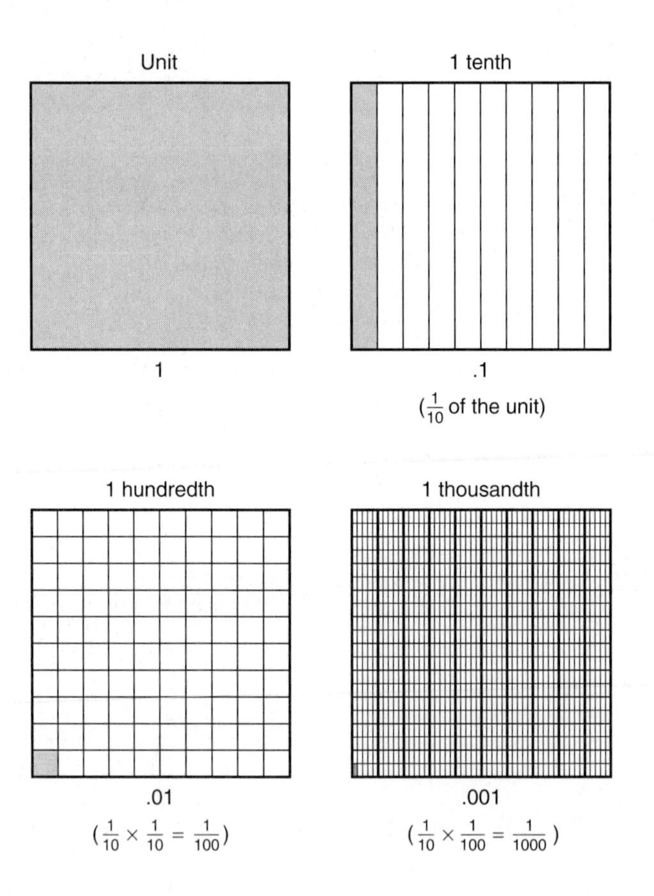

The decimals 3 tenths, 37 hundredths, and 379 thousandths can be illustrated as shown on the next page.

[3]T. P. Carpenter, M. K. Corbitt, H. S. Kepner, M. M. Lindquist, and R. E. Reys, "Decimals: Results and Implications from National Assessment," *Arithmetic Teacher* 28 (April 1981), 34–37.

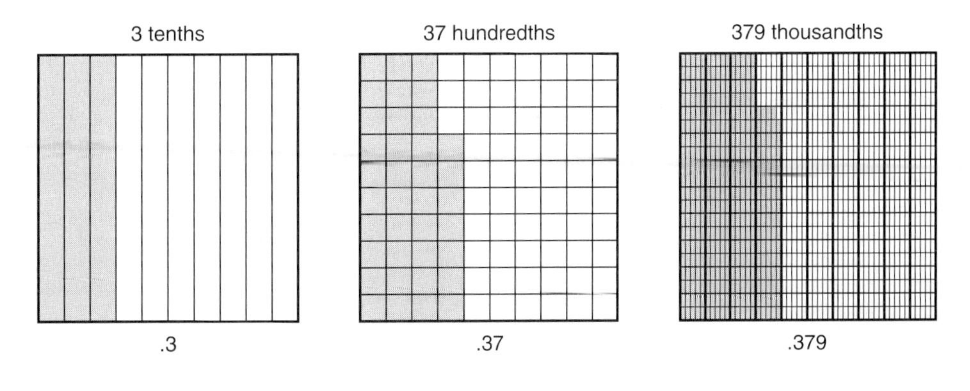

Part to Whole

1. Shade each square below so that the decimal tells how much of the square is shaded.

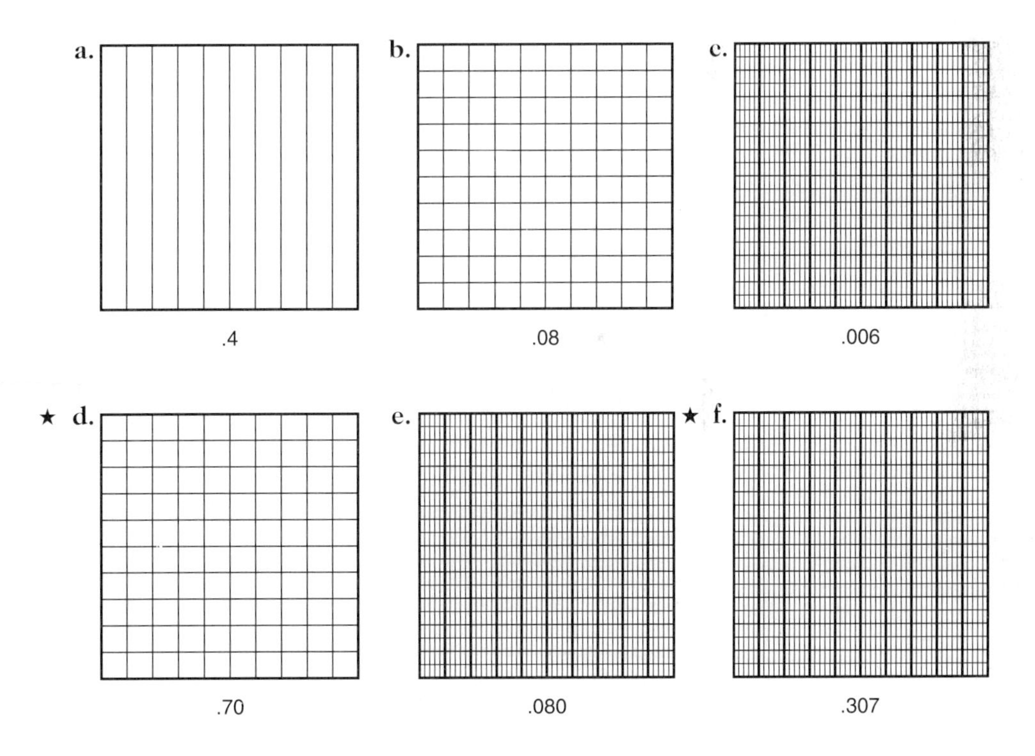

2. Below each square write the decimal that represents the shaded amount.

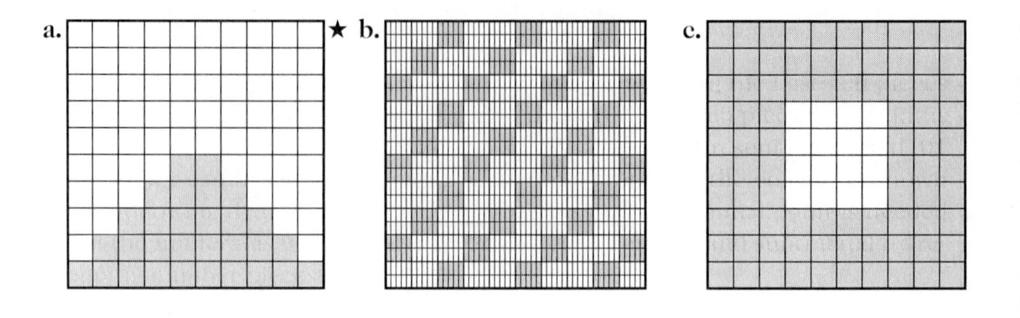

______ ______ ______

Write the decimal that represents each of the following squares:

d. 9 shaded parts out of 10 ______

e. 13 shaded parts out of 100 ______

★ f. 9 shaded parts out of 100 ______

g. 8 shaded parts out of 1000 ______

★ h. 90 shaded parts out of 1000 ______

Equality

3. The decimals for the following squares are equal because each square is the same size and has the same amount of shading. That is, 4 parts out of 10 is equal to 40 parts out of 100, and 40 parts out of 100 is equal to 400 parts out of 1000.

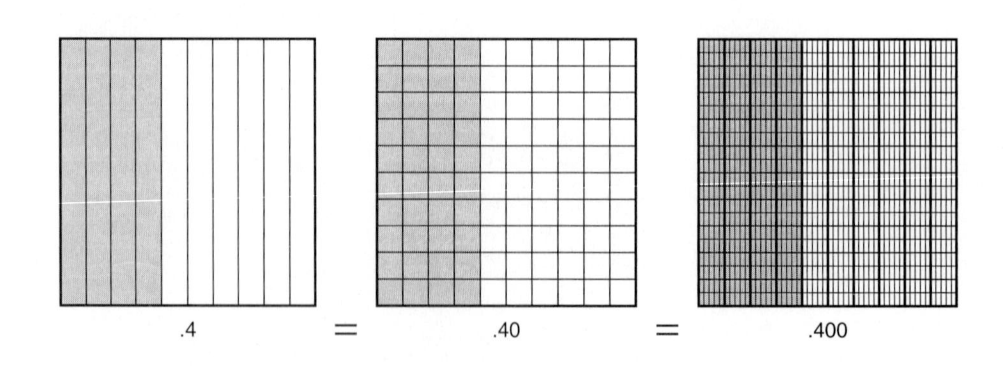

Fill in the boxes below each square to complete the statement and the equation.

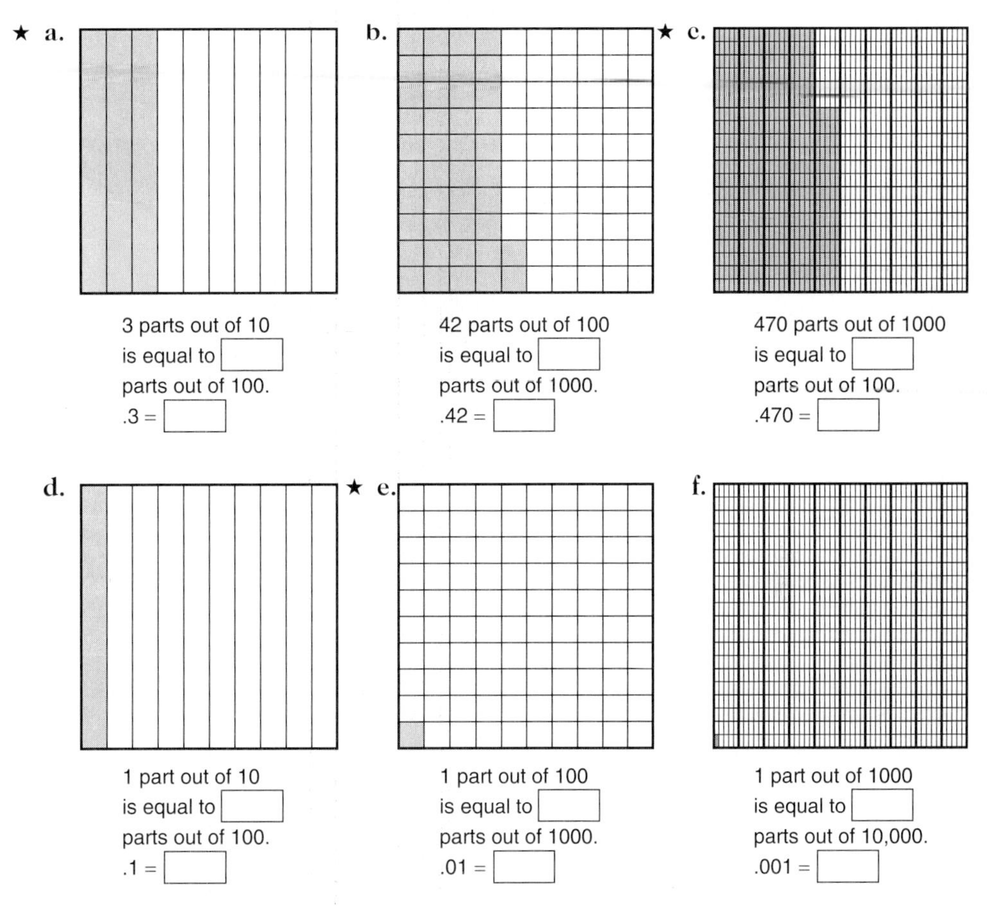

Place Value

4. The decimal .435 is represented by the shaded parts in both squares below. Figure a represents .435 the way we say its decimal name, "four hundred thirty-five thousandths," because 435 parts out of 1000 are shaded. In figure b, the encircled parts of the square show that .435 can also be thought of as 4 tenths, 3 hundredths, and 5 thousandths and that each digit has *place value* as recorded in the table.

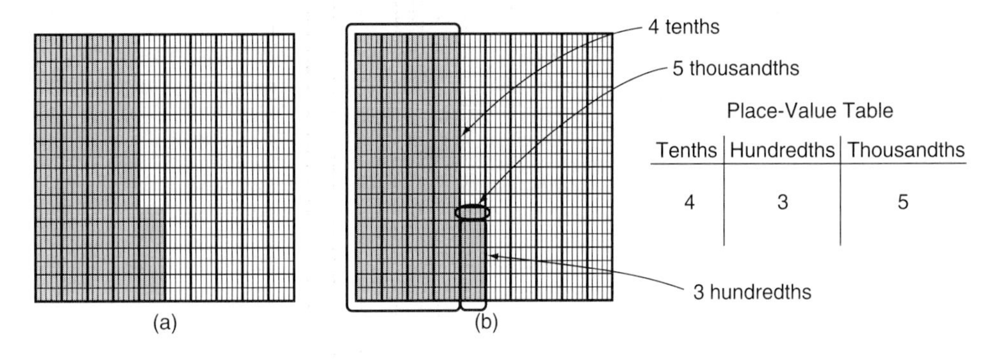

For each of the squares, circle the parts that show the decimal in place-value form. Write the digits for each decimal in the place-value table below the square.

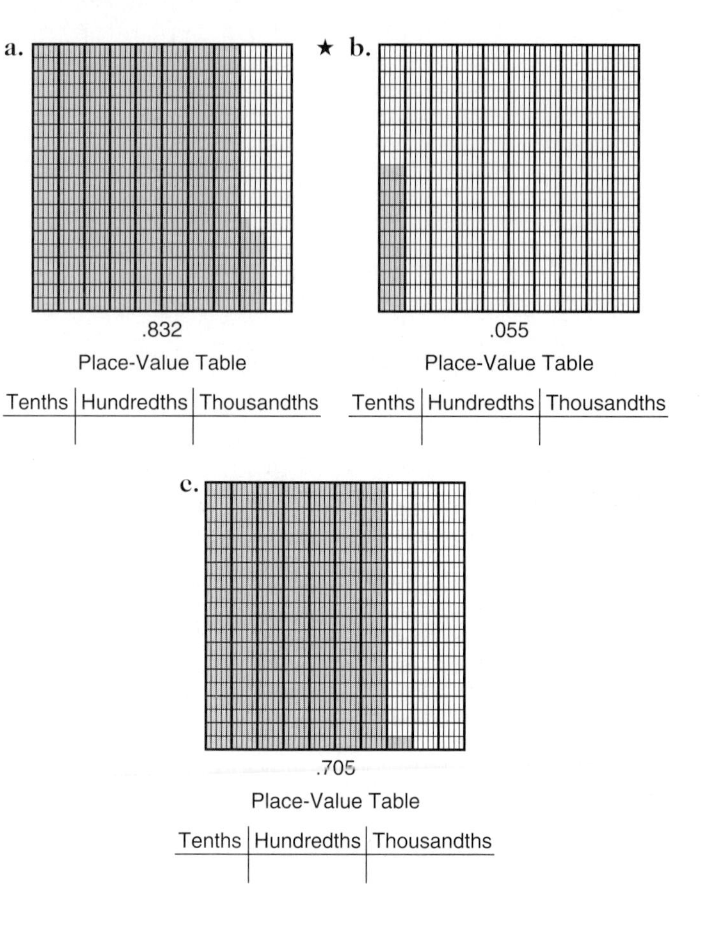

★ **d.** Explain why it is never necessary to use a digit greater than 9 in any column of a place-value table representing a decimal number.

5. When each part of a Decimal Square for hundredths is partitioned into 10 equal parts, there are 1000 parts. One of these parts is 1 thousandth of the whole square and represents the decimal .001.

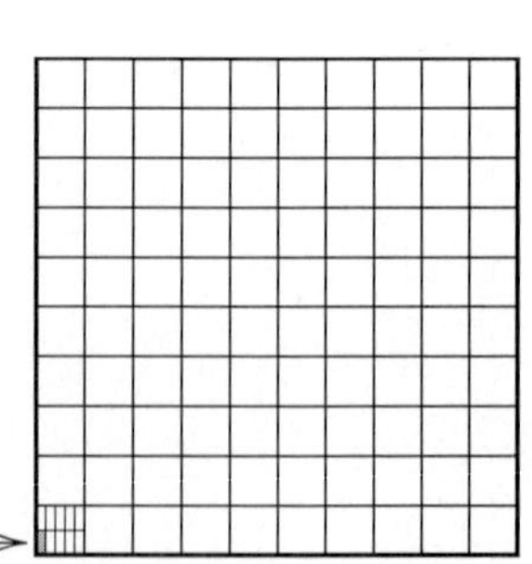

a. Suppose each part of a Decimal Square for thousandths is partitioned into 10 equal parts. What fraction of the whole square is 1 of these parts? What is the decimal for 1 of these small parts, and what is the name of the decimal?

★ **b.** Suppose each part of a Decimal Square for thousandths is partitioned into 100 equal parts. What fraction of the whole square is 1 of these parts? What is the decimal for 1 of these small parts, and what is the name of the decimal?

Inequality

6. The Decimal Squares for .62 and .610 show that .62 is greater than .610, because its square has the greater shaded amount. We can also determine that .62 > .610 by thinking in terms of place value. The digit 6 in .62 and .610 tells us that 6 tenths of both squares are shaded. That is, 6 full columns of both squares are shaded. Looking at the next column, we see that the square for .62 has 2 hundredths (or 20 thousandths) shaded, whereas the square for .610 has 1 hundredth (or 10 thousandths) shaded. So .62 > .610.

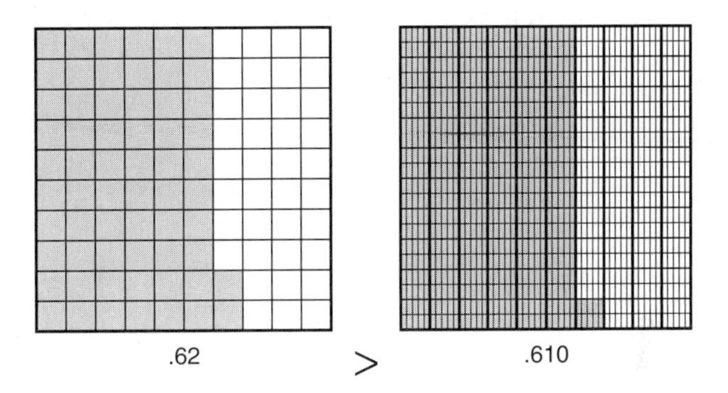

Determine an inequality for each pair of decimals, and explain the reason for the inequality by describing their Decimal Squares. For example, .4 > .27 because the square for .4 has 4 full columns shaded and the square for .27 has less than 3 full columns shaded.

a. .7 .43

★ b. .042 .04

c. .3 .285

d. Write the following decimals as equivalent decimals, with three decimal places. Circle the greatest and underline the least.

.3 .27 .298

★ e. Write the following decimals as equivalent decimals, with four decimal places. Circle the greatest and underline the least.

.042 .0047 .04

7. The following question is from a 1979 mathematics assessment test that was given to students entering universities.

Question: Which of the following numbers is the smallest?

a. .07 b. 1.003 c. .08 d. .075 e. .3

Less than 30 percent of 7100 students answered this question correctly.[4] Explain how to determine the correct answer by describing Decimal Squares for these decimals.

Approximation

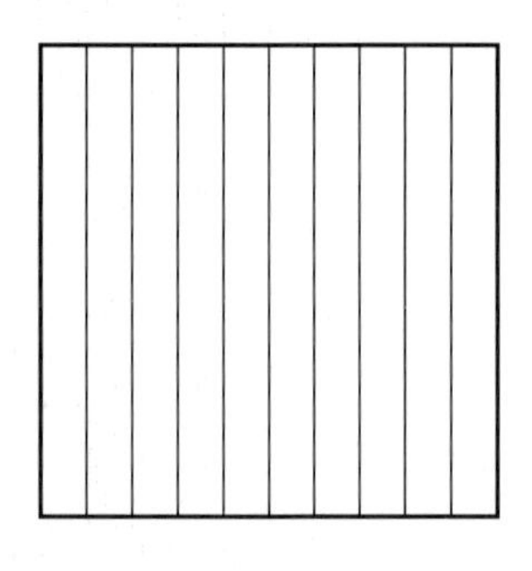

8. a. Using a Decimal Square for tenths, determine how many complete parts should be shaded to best approximate $\frac{1}{3}$ of the whole square. Shade these parts. Beneath the square, record the decimal that represents the shaded part.

b. Shade complete parts of each of these two Decimal Squares to best approximate $\frac{1}{3}$ of a whole square. Beneath each square, record the decimal that represents the shaded portion of the square.

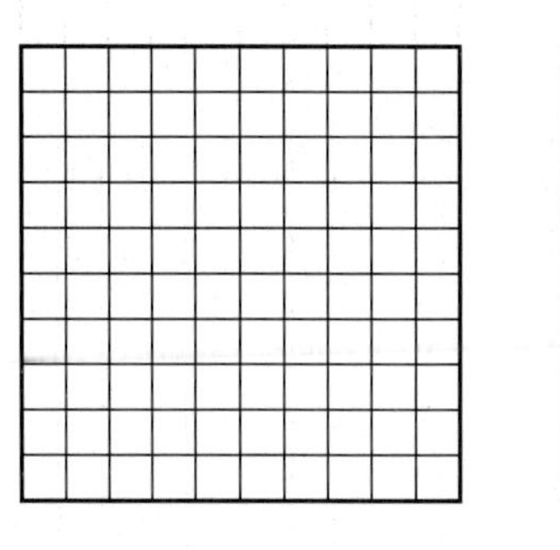

★ **c.** Shade complete parts of each of these Decimal Squares to best approximate $\frac{1}{6}$ of a whole square. Write the decimal that represents the shaded part of each square.

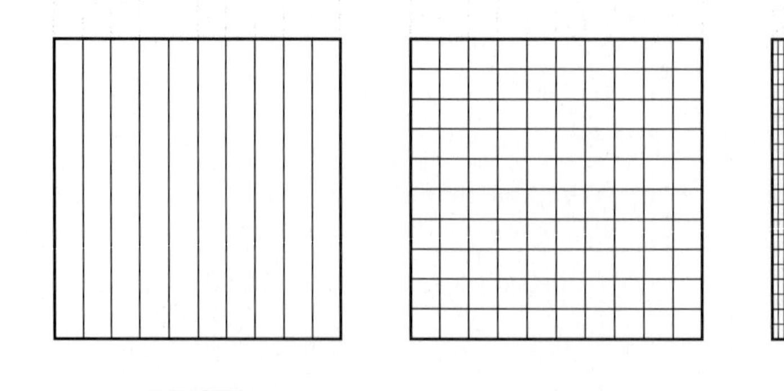

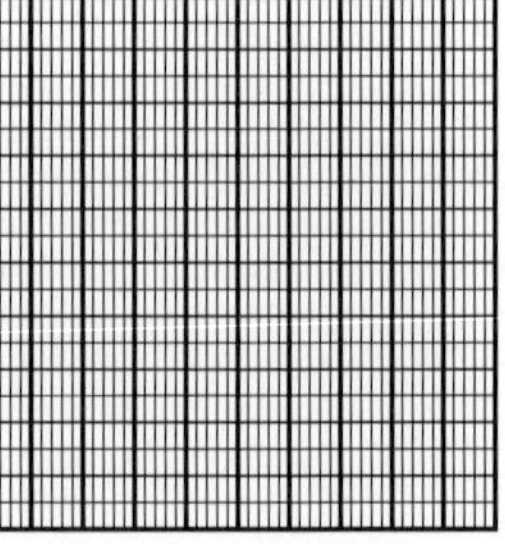

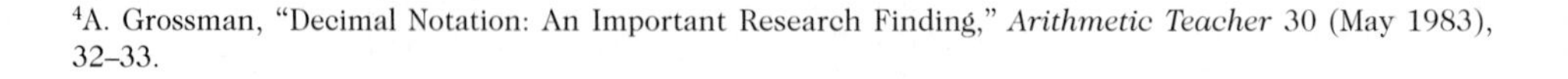

[4]A. Grossman, "Decimal Notation: An Important Research Finding," *Arithmetic Teacher* 30 (May 1983), 32–33.

JUST FOR FUN

In Decimal Bingo you practice identifying equal decimals. The Decimal Place-Value Game provides more insight on place values.

DECIMAL BINGO (2 to 4 players)

Each player selects one of the following four bingo mats with decimals. The deck of Decimal Squares should be spread face down. Each player in turn takes a square and circles the decimal or decimals on the bingo mat that equal the decimal for the square. The winner is the first player to circle four decimals in any row, column, or diagonal.

.30	.400	.1	.70
.7	.55	.3	.8
.10	.40	.85	.95
.8	.300	.05	.550

Mat 1

.60	.50	.3	.10
.1	.75	.600	.5
.450	.30	.15	.525
.500	.45	.425	.750

Mat 2

.70	.2	.50	.40
.5	.4	.450	.65
.20	.700	.375	.675
.400	.45	.475	.650

Mat 3

.75	.300	.60	.4
.600	.9	.30	.350
.40	.750	.325	.25
.35	.6	.575	.90

Mat 4

DECIMAL PLACE-VALUE GAME (2 to 4 players)

Players each draw a place-value table like the one next to the dice. Each player in turn rolls two dice until he or she obtains a sum less than 10. The player then records this number in one of the three columns in his or her place-value table. Only one number is written in one column on each turn, and the number cannot be moved to another column after it has been written down. After each player has had three turns, the player with the greatest three-place decimal is the winner. The first player to win five rounds wins the game.[5]

Variation: Use a place-value table with four columns, and roll the dice to obtain the digits in a four-place decimal.

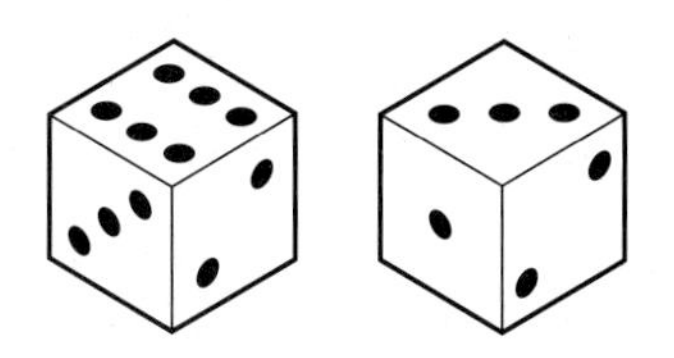

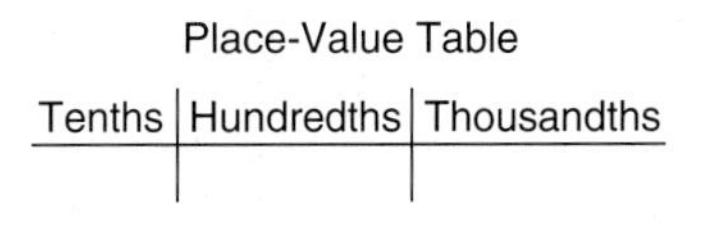

[5]Another way to generate numbers for this game is to write each digit from 0 to 9 on a separate card and then draw one card at random from this deck.

ACTIVITY SET 6.2

OPERATIONS WITH DECIMAL SQUARES

Purpose To use Decimal Squares to illustrate decimal addition, subtraction, multiplication, and division

Materials Decimal Squares are on Material Cards 23–26.

Activity It is almost impossible to compute on a calculator without getting decimal numbers. Because of our technology and the frequent occurrence of decimals, it has been suggested that decimal concepts be taught earlier in the elementary school curriculum.

> Calculators create whole new opportunities for ordering the curriculum and for integrating mathematics into science. No longer need teachers be constrained by the artificial restriction to numbers that children know how to employ in the paper-and-pencil algorithms of arithmetic. Decimals can be introduced much earlier since they arise naturally on the calculator. Real measurements from science experiments can be used in mathematics lessons because the calculator will be able to add or multiply the data even if the children have not yet learned how. They may learn first what addition and multiplication mean and when to use them, and only later how to perform these operations manually in all possible cases.[6]

Early introduction of the decimal concepts of addition, subtraction, multiplication, and division in the elementary school curriculum makes it imperative that children have good physical and visual models that embody these concepts. The goal is to develop concepts and a decimal number sense rather than paper-and-pencil manipulative skills. For example, a child may not know the traditional moving-the-decimal-point algorithm of computing .35 ÷ .05 but may "see" that the answer is 7 by using the Decimal Squares model—and then verify this result by using a calculator.

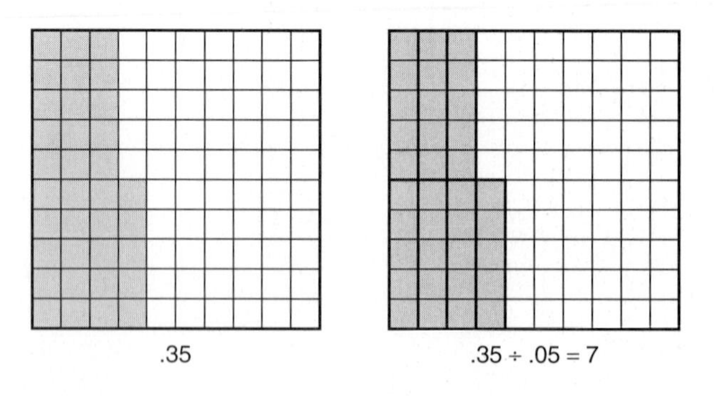

[6]National Research Council, *Everybody Counts: A Report to the Nation on the Future of Mathematics Education* (Washington, DC: The National Academy Press, 1989), 47–48.

Addition

1. Addition of decimals can be illustrated by determining the total shaded amount of two or more squares. Fill in the missing number, and compute the sum by counting the total number of shaded parts.

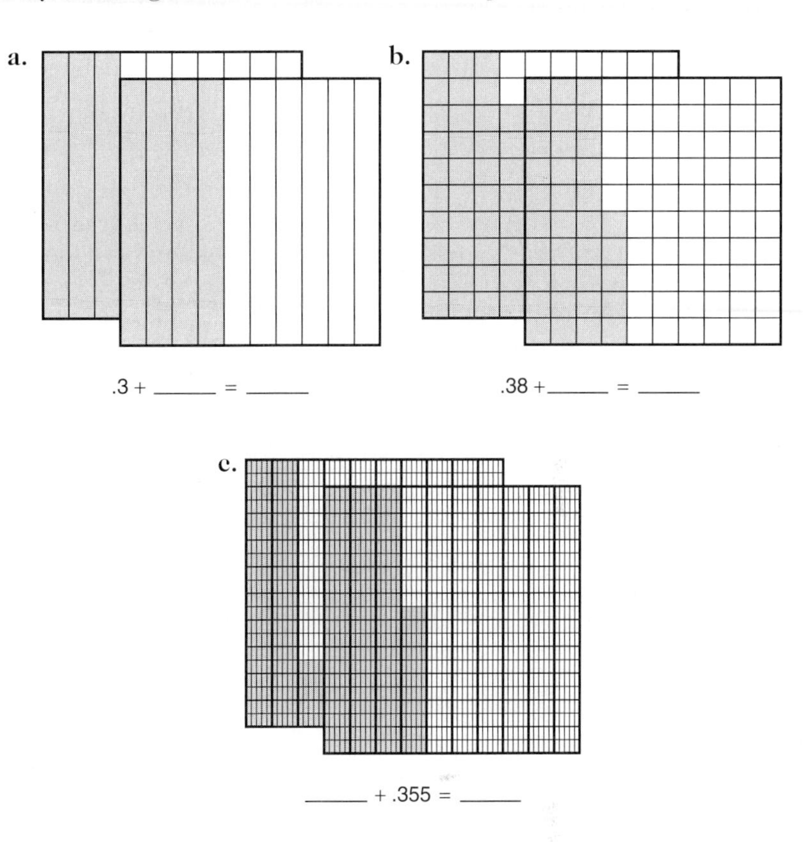

2. Shade each square to represent the decimal written beneath it. Add the decimals by determining the total shaded amount of the two squares. Explain how you arrived at your answer.

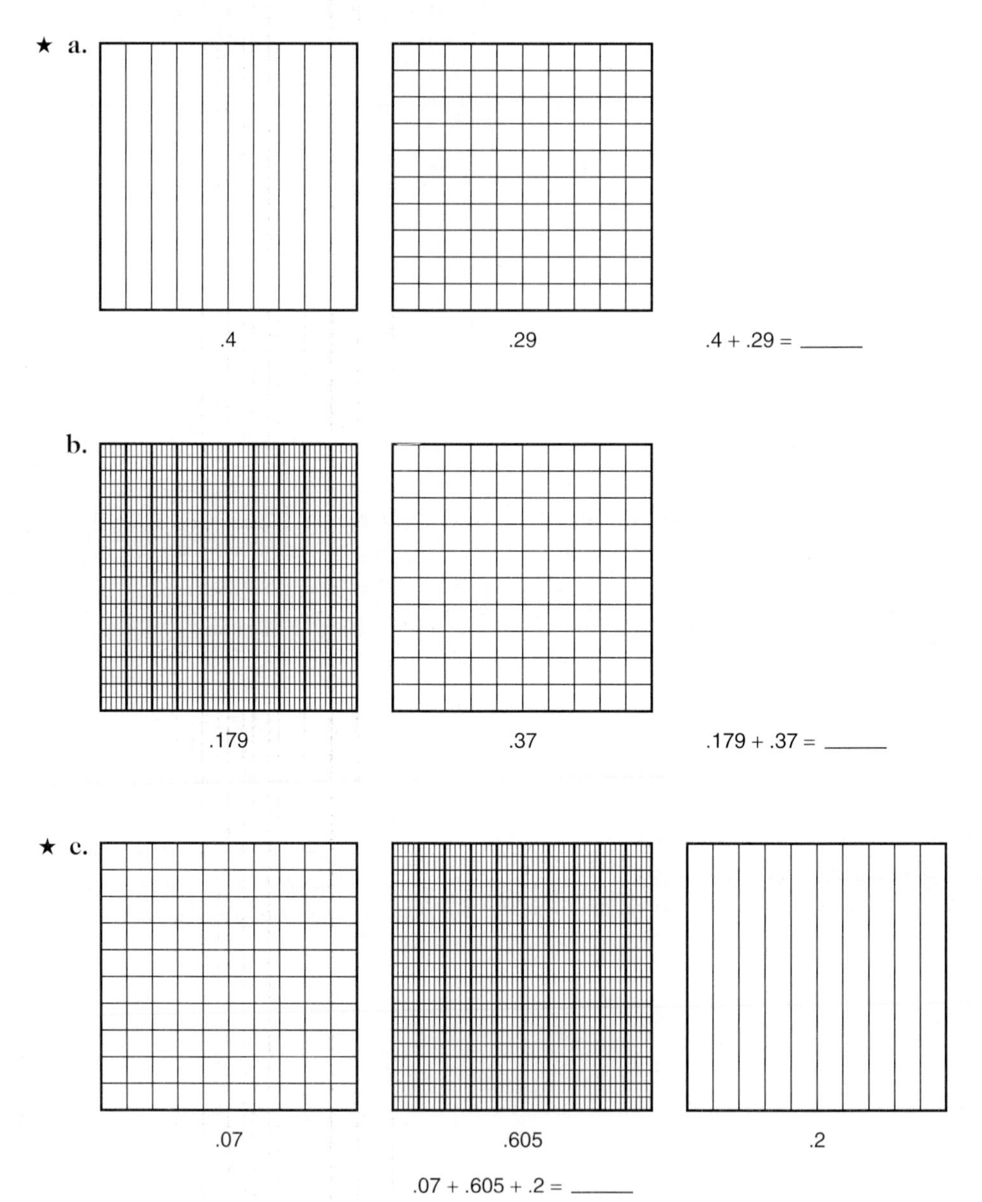

Subtraction

3. Subtraction of decimals can be illustrated with Decimal Squares by circling the amount to be taken away—*take-away concept for subtraction*—and counting the number of shaded parts that remain.

.47 − .15 = .32

For each of the following, circle the amount to be taken away and complete the equation.

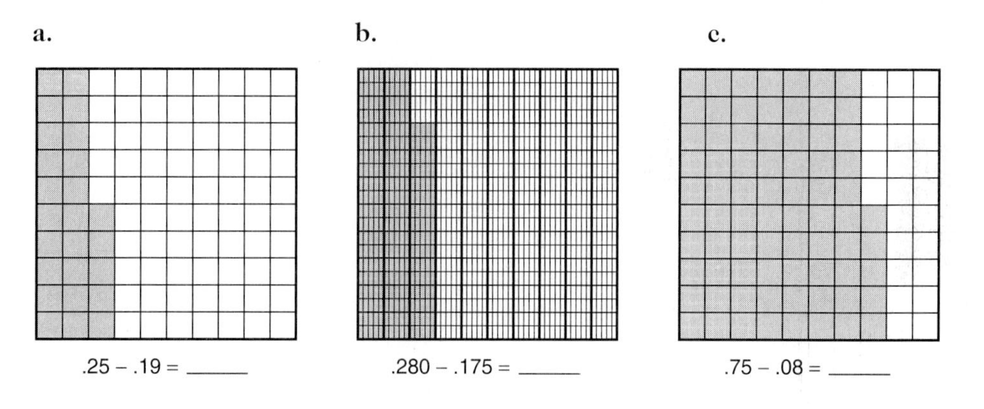

4. Shade each square to represent the decimal written beneath it. Determine the differences by *comparing* the shaded parts of the squares. In part a, for example, you may find the difference by determining what must be added to .07 to get .5. Briefly explain how you arrived at each answer.

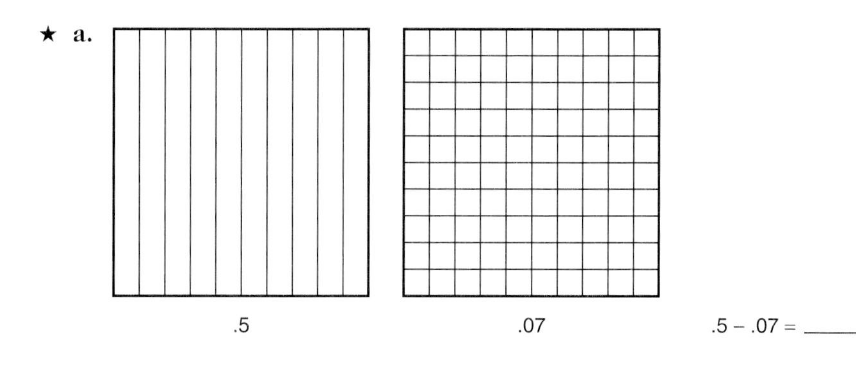

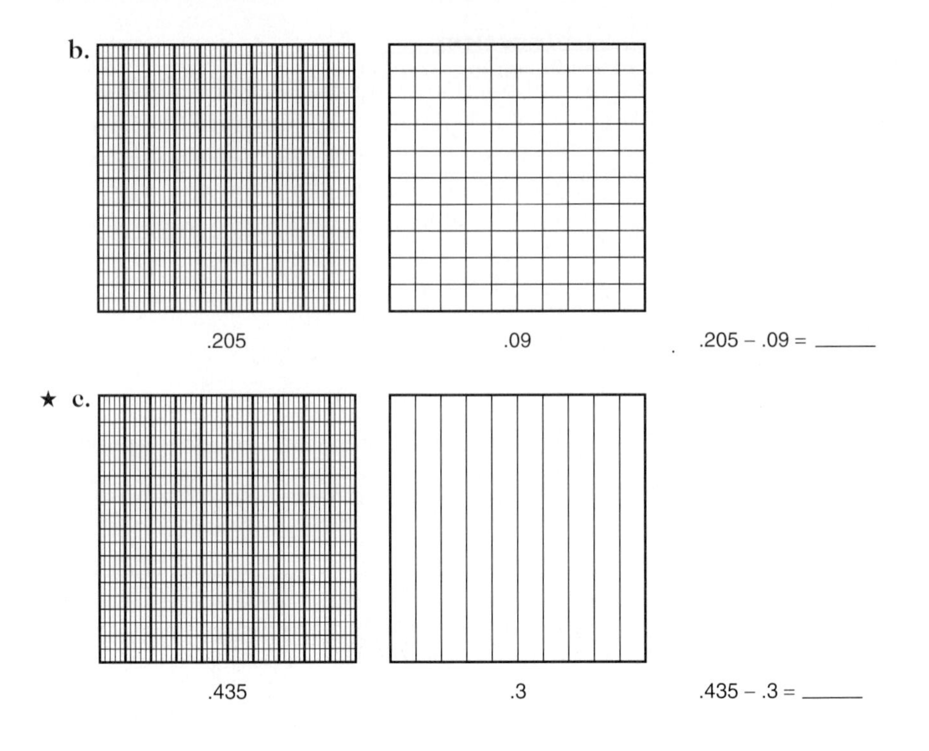

Multiplication

5. Here is the decimal square for .45. If there were 3 of these Decimal Squares, there would be a total of $3 \times 45 = 135$ shaded parts. Since there are 100 parts in each whole square, the total shaded amount would be 1 whole square and 35 parts out of 100.

.45

$3 \times .45 = 1.35$

In a similar manner, compute the following products and explain your answers in terms of the shaded parts of squares.

a. $6 \times .83 =$ ______

★ **b.** $4 \times .725 =$ ______

c. $10 \times .72 =$ ______

★ **d.** Explain, in terms of Decimal Squares, why multiplication by 10 results in a decimal number with the same digits but with the decimal point in a different place.

6. The product .3 × .2 means "take 3 tenths of 2 tenths." To do this, we split the shaded amount of the .2 square into 10 equal parts and take 3 of them. The result is 6 hundredths. This shows that .3 × .2 = .06.

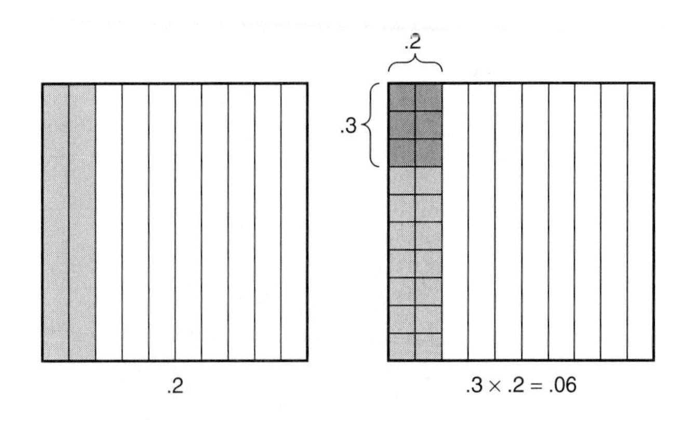

The shaded portion of the first figure below represents the decimal 1.2. When the shaded area representing 1.2 is split into 10 equal parts and 3 are taken, as in the second figure, the result is 3 tenths and 6 hundredths, or 36 hundredths.

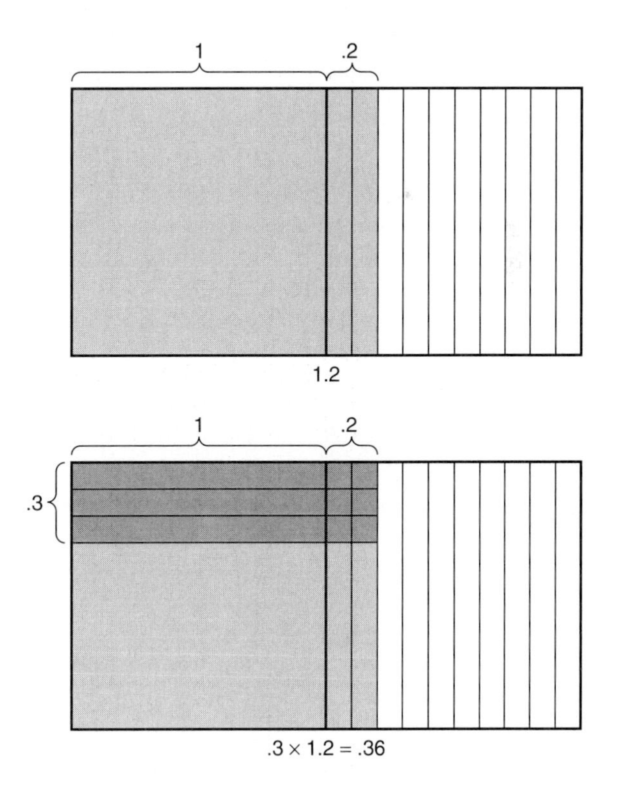

The product 1.3×1.2 can be thought of as 1×1.2 added to $.3 \times 1.2$ and represented as shown below. The shaded portion represents 1 unit, 5 tenths, and 6 hundredths. So $1.3 \times 1.2 = 1.56$.

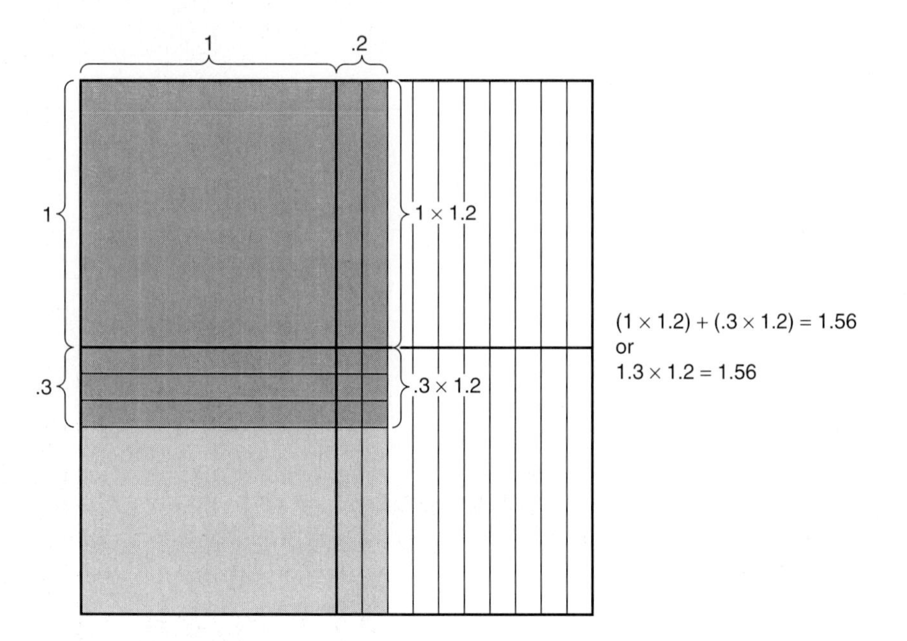

In a similar manner, sketch a Decimal Square diagram to illustrate each of the following. Determine the product from the diagram and explain your reasoning.

a. $.4 \times .3$

b. $.4 \times 1.3$

★ c. 1.4×1.3

d. 2.4×1.3

Division

7. The *sharing method of division* can be used to illustrate the division of a decimal by a whole number. In the following square, the shaded portion representing .60 has been divided into 3 equal parts to illustrate .60 ÷ 3. Since each of these parts has 20 shaded parts, .60 divided by 3 is .20.

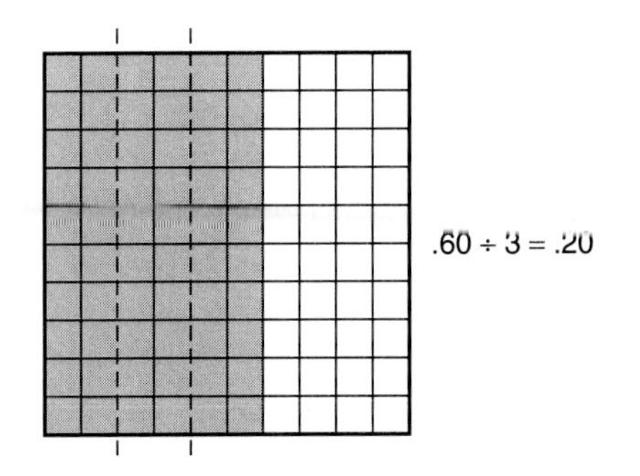

Divide each Decimal Square into parts, and use the result to determine the quotient.

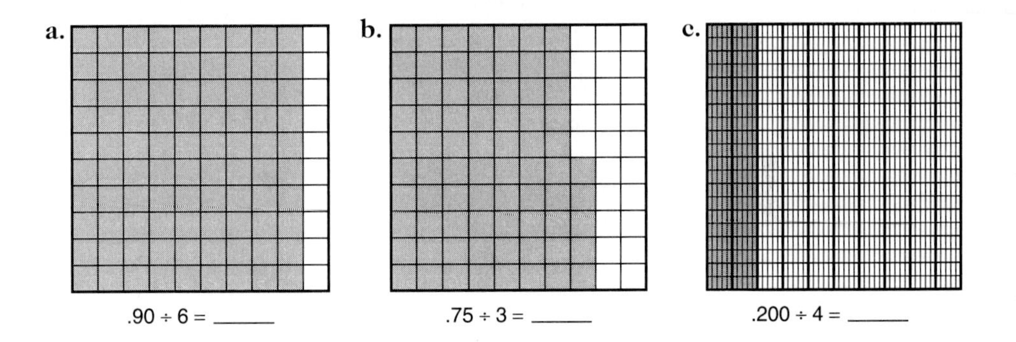

8. Use the following Decimal Squares to determine the indicated quotients. Record each quotient beneath the square, and shade that amount on the Decimal Square.

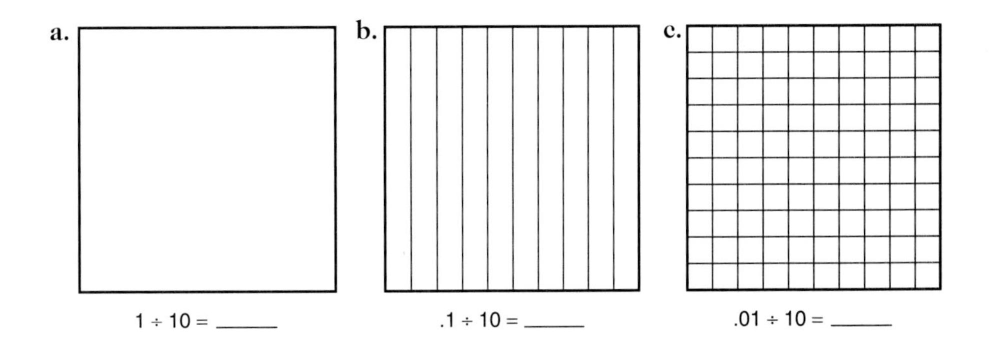

★ 9. In view of the preceding results, explain why dividing a decimal number, such as 2.87, by 10 results in a decimal with the same digits but the decimal point moved one place to the left.

10. The *measurement concept of division* involves repeatedly measuring off or subtracting one amount from another. It can be seen in the following square that 15 of the shaded parts can be measured off from 75 parts 5 times.

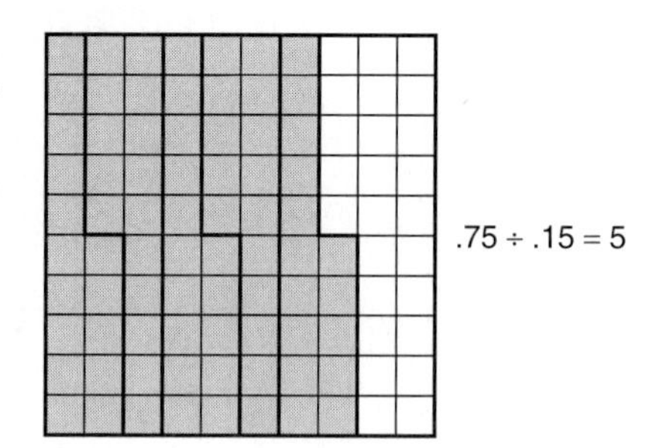

Draw lines on the following Decimal Squares to indicate how the measurement concept of division can be used to obtain the indicated quotient.

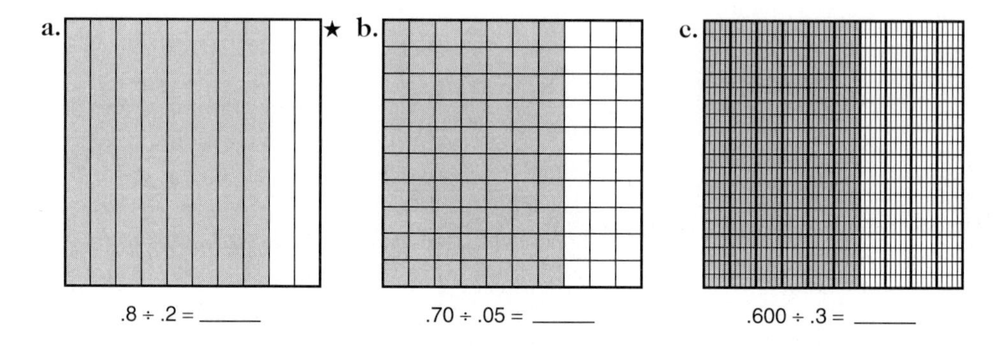

★ **d.** On the Second National Assessment of Educational Progress, students were asked to estimate 250 ÷ .5. Only 39 percent of the 17-year-olds correctly estimated this quotient; 47 percent ignored the decimal point, giving an answer of 50.[7] Use the measurement concept of division to explain why 250 ÷ .5 = 500.

All Four Operations

11. Use the Decimal Squares from Material Cards 23–26 for parts b and c of this activity.

a. The following 10 decimals represent Decimal Squares:

.3 .75 .8 .650 .15 .2 .35 .45 .400 .350

Complete these equations by placing the 10 decimals in the boxes.

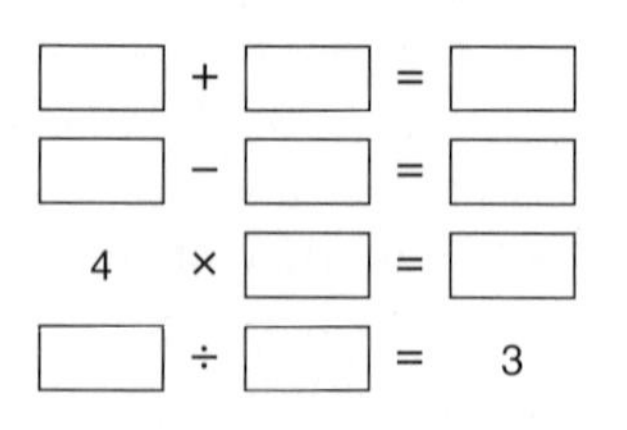

[7]National Assessment of Educational Progress, "Math Achievement Is Plus and Minus," *NAEP Newsletter* 12 (5) (October 1979), 2.

b. Spread the Decimal Squares face down and select 10 of them. List your 10 decimals. Using each of your decimals only once, complete as many of the four equations as possible. If you cannot complete all four equations, continue to select Decimal Squares one at a time (recording each decimal) until you can complete all the equations.

10 decimals selected: ______, ______, ______, ______, ______,
______, ______, ______, ______, ______

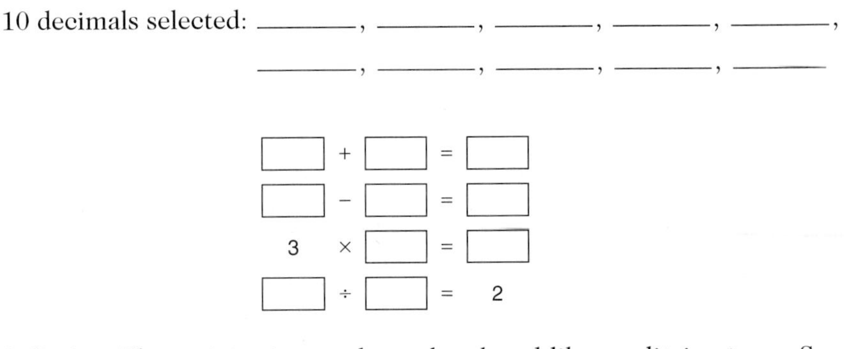

c. **Solitaire:** The activity in part b can be played like a solitaire game. See how many turns it takes you to complete the four equations by selecting only 10 Decimal Squares on each turn.

JUST FOR FUN

Build skills in performing operations on decimals with these games—addition in the first game, division in the second, and subtraction in the third.

DECIMAL SQUARES BLACKJACK (2 to 4 players)

The dealer shuffles the deck of Decimal Squares and deals one square face down to each player. The dealer also gets one square. The object of the game is to use one or more squares to get a decimal or a sum of two or more decimals that is less than or equal to 1, without going over. To get an additional square from the dealer, a player says "Hit me." When a player wants no more additional squares, he or she says "I'm holding." After every player has said "I'm holding," the players show their squares. The player whose sum is closest to 1, but not greater than 1, wins 1 point. The winner is the first player to win 5 points.

GREATEST QUOTIENT (2 to 4 players)

Spread the Decimal Squares face down. Each player chooses two squares and computes the quotient of one decimal divided by the other. The object is to get the greatest quotient. The player with the greatest quotient wins all the squares used in that round, including any discarded squares. If there is a tie, the squares are placed aside and the winner of the next round gets them. Play continues until the deck has been played through and no more rounds are possible. The winner is the player who has won the most squares.

Chance Option: A player can attempt to increase a quotient by choosing a new square. However, before choosing a new square, the player must discard one of the previously chosen squares.

GREATEST DIFFERENCE (2 to 4 players)

This game is similar to the greatest quotient game except that the player with the greatest difference wins the round.

ACTIVITY SET 6.3

A MODEL FOR INTRODUCING PERCENT

Purpose To use base-ten number pieces to introduce the concept of percent and to illustrate and solve various percent problems

Materials Base-ten pieces from the Manipulative Kit

Activity Percent is a common and widely used mathematical concept. From television to newspapers to shopping, the word *percent* enters our life almost daily, as noted in the following passage.

> Percentages are everywhere—in the food we eat ([a serving of] cornflakes contains 25% of our recommended daily allowance of vitamin A), in the clothes we wear (35% cotton), in our favorite television programs (watched by 55% of the television audience), and in the money we earn (27% tax bracket) and spend ($5\frac{1}{2}$% sales tax). In modern society, people who do not understand percentages will have difficulty reading articles and advertisements that appear in newspapers, determining if they are being cheated by financial institutions or retail establishments, or knowing if they are being misled by politicians.[8]

When something is divided into 100 equal parts, one part is called *one percent* and written 1%. The 10 by 10 grids from the base-ten number pieces are divided into 100 equal parts so the smallest square is 1 percent and the whole grid is 100 percent. These number pieces will be used to explore the concept of percent and solve various percent problems.

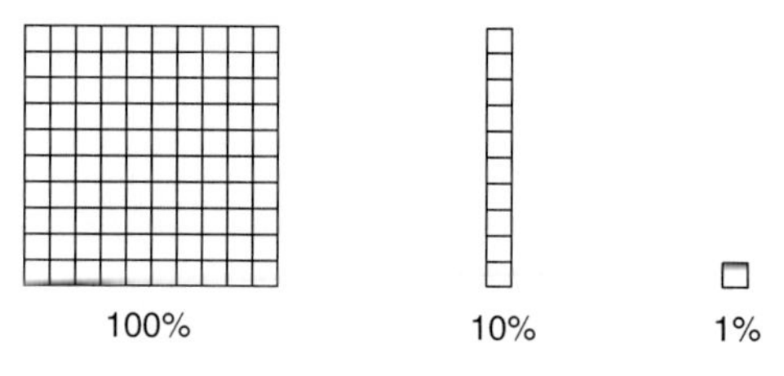

The classical three cases of percent are visualized using this model.

Percent and whole:	25% of 150 = ?
Part and whole:	?% of 150 = 90
Percent and part:	6% of ? = 24

At the end of this activity set, Fraction Bars are split into approximately 100 parts to provide percent and decimal estimations for fractions.

1. If the 10 by 10 grid represents 100%, determine what percent is represented by each of the following collections of number pieces.

[8]James H. Wiebe, "Manipulating Percentages," *The Mathematics Teacher* (January 1986), 23–26.

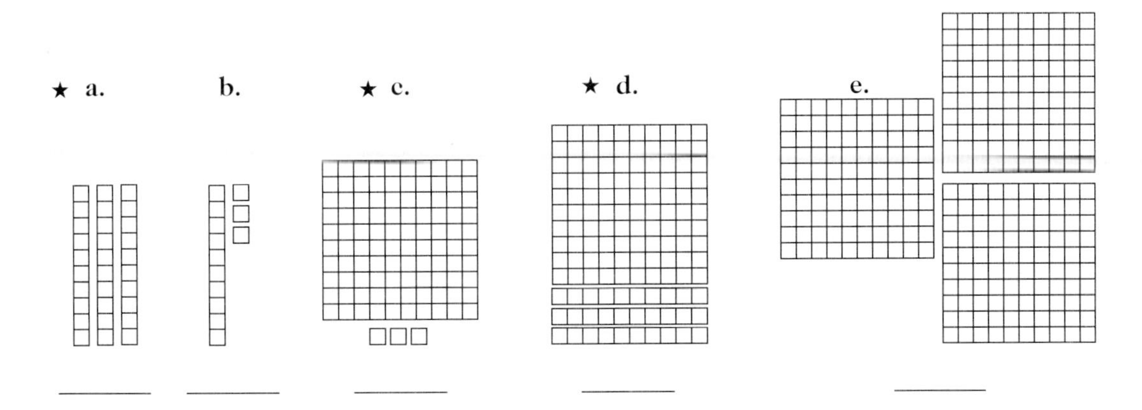

2. If the 10 by 10 grid represents 100%, use your pieces to form a collection to represent each of the percentages below. Draw a sketch to record your answers.

★ a. 23% b. 98% ★ c. 105%

d. 123% e. 202% ★ f. $\frac{1}{2}$%

Value 300

3. Suppose that a value of 300 is assigned to the 10 by 10 grid and that this value is spread evenly among all the small squares.

★ a. Describe an easy method for performing a mental calculation to find the value of this collection.

Using mental calculations determine the value of each of the following collections:

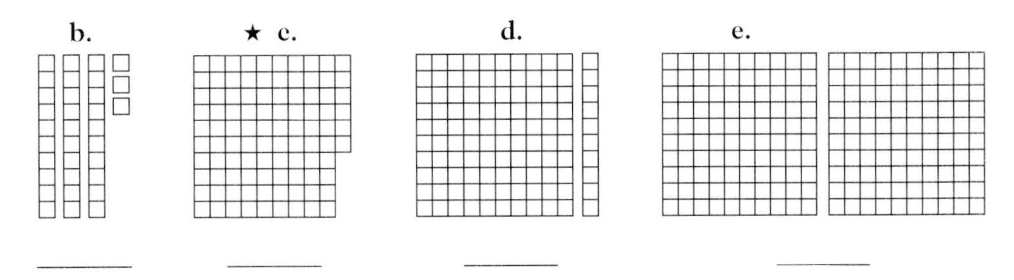

Value 150

4. Suppose that the 10 by 10 grid is assigned a value of 150 and that this value is spread evenly among all the small squares.

★ a. Form a collection of pieces that represents 25% of the 10 by 10 grid. Determine the value of this collection without doing paper-and-pencil computations. Explain your method.

Repeat part a by forming collections of pieces representing the following percentages and determining each of their values using mental computations. Explain your thinking in each case.

b. 75%

★ c. 7%

d. 111%

e. 210%

Value 450

5. Assign a value of 450 to the 10 by 10 grid.

★ a. Form a collection of pieces that has a value of 270. Determine what percent of the 10 by 10 grid the value 270 represents. Record a sketch of your collection and the percent it represents.

Repeat the directions of part a for collections having the following values:

b. 108

★ c. 441

d. 585

e. 909

6. ★ a. Form a collection of pieces that represents 6% of a 10 by 10 grid. If this collection has a value of 24, what is the value of the 10 by 10 grid? Draw a sketch and explain your thinking.

Repeat the directions of part a when

b. 40% has a value of 30

★ c. 55% has a value of 132

d. 125% has a value of 180.

Some Percent Statements

7. Many statements that deal with percent can be represented with the number piece model. Consider the statement "Last year the charity benefit raised $6720, which was 120% of its goal." 120% can be represented with these number pieces.

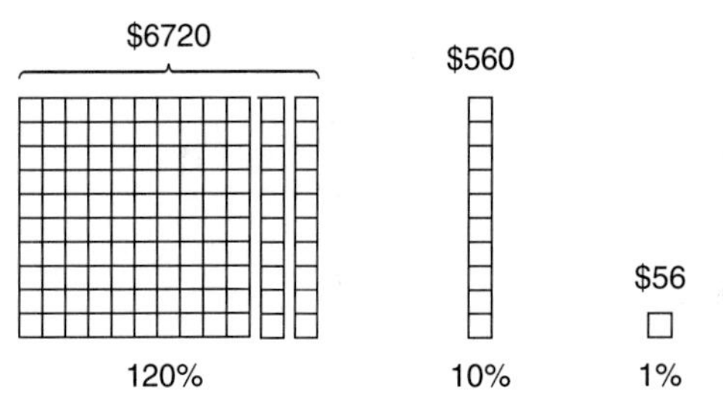

There are several observations that can be made by looking at the model:

(1) Because 120% has a value of $6720, one can see from the model that each long has a value of $560 ($6720 ÷ 12) or that each small square has a value of $56 (6720 ÷ 120).

(2) Last year's goal (100%) must have been $5600 (10 × $560 or 100 × $56).

(3) The goal was exceeded by $1120 ($6720 − $5600).

Model each of the following statements with your number pieces. Make a sketch of your model, record at least one *percent* observation, and explain how you arrived at it.

★ a. 65% of the firm's 320 employees were women.

★ **b.** 280 of the school's 800 students were absent.

c. I bought this sweater for $126 at the store's 30% off sale.

d. This year the charity benefit had a goal of $7000 and it raised 135% of its goal.

e. In a certain county, 57 of the schools have a student-to-teacher ratio that is greater than the recommendations for accreditation. The 57 schools are 38% of the number of schools in the county.

f. A new company entered the New York Stock Exchange in July and by December the price of each share of stock had risen (increased) 223% to $32.50 a share.

Fractions as Percents

8. Fractions with denominators that are factors of 100 can be readily expressed as percents. For example, if each part of a $\frac{3}{4}$ bar is split into 25 equal parts, the bar will have 75 parts shaded out of 100 equal parts. So $\frac{3}{4}$ equals 75%, or the decimal .75.

25 parts 25 parts 25 parts 25 parts

$\frac{3}{4} = \frac{75}{100} = 75\% = .75$

Imagine splitting each of the following bars into 100 equal parts. Indicate above each part of a bar the number of new parts after splitting, and then determine the percent and decimal equivalents for each fraction.

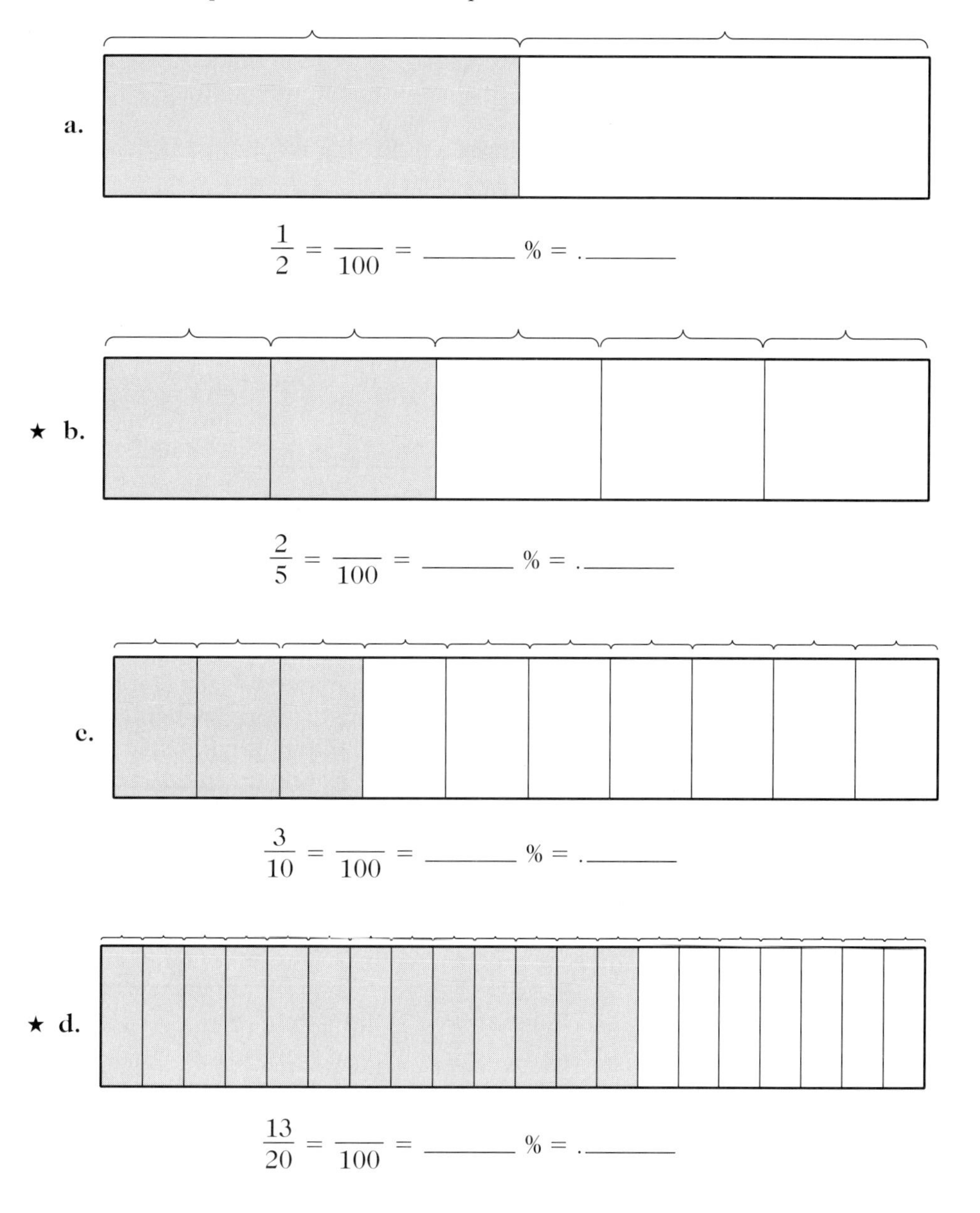

Fractions as Approximate Percents

9. To obtain an approximate percent for the fraction $\frac{1}{7}$, visualize a $\frac{1}{7}$ fraction bar. This bar has 1 part shaded out of 7 parts. If each part of the bar is split into 14 equal parts, the result will be 14 shaded parts out of 98 parts.

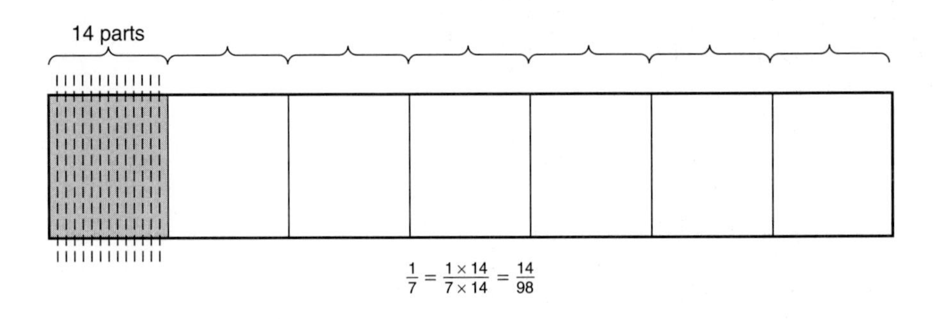

If each part of the $\frac{1}{7}$ bar is split in 15 equal parts, there will be 15 shaded parts out of 105 parts.

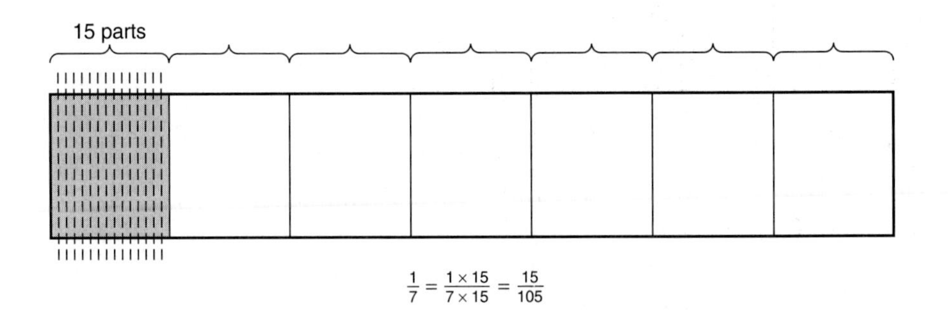

Because 105 is not as close to 100 as 98 is, we can say that $\frac{1}{7}$ is approximately 14%. Since 14% is $\frac{14}{100}$, the decimal .14 is also an approximation for $\frac{1}{7}$.

Determine a percent and decimal approximation for each of the following fractions by splitting the corresponding fraction bar into approximately 100 parts (or visualizing the splitting). Explain how you arrived at your percent for each fraction by indicating the number of parts due to splitting.

a. $\frac{1}{3} \approx \frac{\quad}{100} =$ _______% = ._______

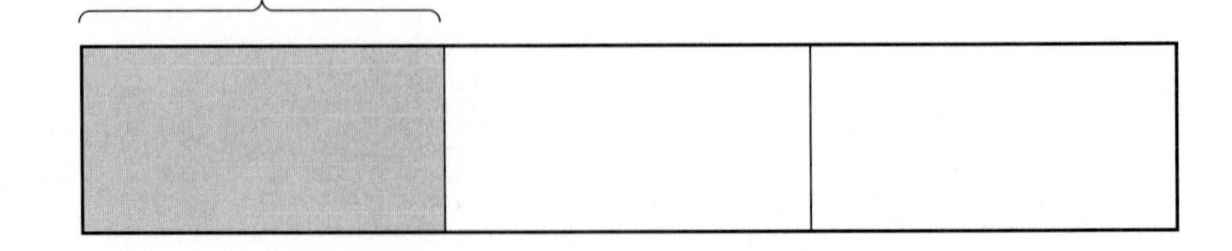

★ **b.** $\frac{5}{6} \approx \frac{\quad}{100} = \underline{\qquad}\% = .\underline{\qquad}$

c. $\frac{2}{9} \approx \frac{\quad}{100} = \underline{\qquad}\% = .\underline{\qquad}$

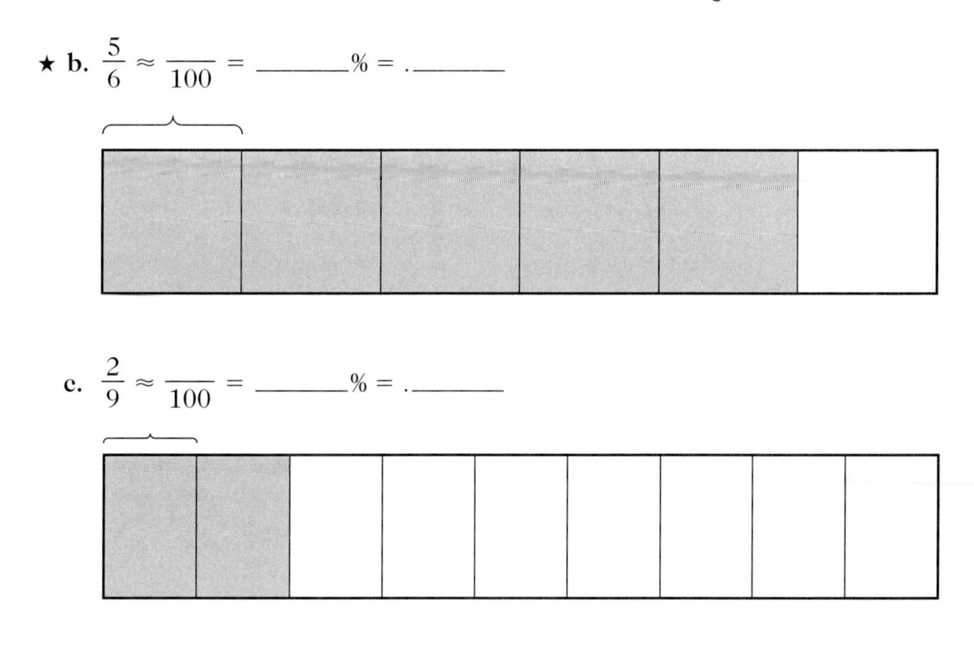

JUST FOR FUN

GAME OF INTEREST (2 to 4 players)

This game requires the Color Tiles and the 1-to-12 Spinner from the Manipulative Kit, and the game mat on Material Card 40.

Each player chooses markers of one color and players spin to see who will play first (high number). In turns, players spin to obtain a percent from 1% to 10%. (If the spinner ends on the numbers 11 or 12 spin again.) The player spinning then mentally computes that percent of one of the following amounts, $50, $100, $200, $300, $400, $500, or $600, and places a marker on the number in the table corresponding to the correct percent of that amount (the *interest*). The game continues until one player, the winner, gets 4 adjacent markers in a row either vertically, horizontally, or diagonally.

Example: Spin 7%; choose to take 7% of $400 which is $28; place a color marker on the number 28 on the gameboard.

Interest Gameboard

$50 $100 $200 $300 $400 $500 $600

.5	1	1.5	2	2.5	3	3.5	4	4.5	5
1	2	3	4	5	6	7	8	9	10
2	4	6	8	10	12	14	16	18	20
3	6	9	12	15	18	21	24	27	30
4	8	12	16	20	24	28	32	36	40
5	10	15	20	25	30	35	40	45	50
6	12	18	24	30	36	42	48	54	60

1-to-12 Spinner

PUZZLER

Which rectangle has the greater percent of its area shaded?

ACTIVITY SET 6.4

IRRATIONAL NUMBERS ON THE GEOBOARD

Purpose To use geoboards and/or dot paper to represent figures whose sides and perimeters have lengths that are irrational numbers

Materials A geoboard and rubber bands (optional). There is a template for making a rectangular geoboard on Material Card 27.

Activity When we used the rod model for greatest common factors (Activity Set 4.2), we saw that any two rods with whole number lengths could be cut evenly into rods of equal common length. In some cases the lengths of the rods may be relatively prime, like 3 and 8, but the rods can still be cut evenly into pieces with common length 1.

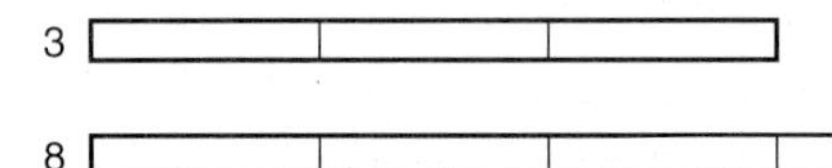

If two rods do not have whole number lengths but do have rational number lengths, can they still be cut evenly into rods of equal common length? Again the answer is yes. Suppose, for example, the rods have lengths $3\frac{1}{3}$ and $5\frac{1}{2}$. Because these numbers are equal to $3\frac{2}{6}$ and $5\frac{3}{6}$, the rods can both be cut evenly into pieces of length $\frac{1}{6}$.

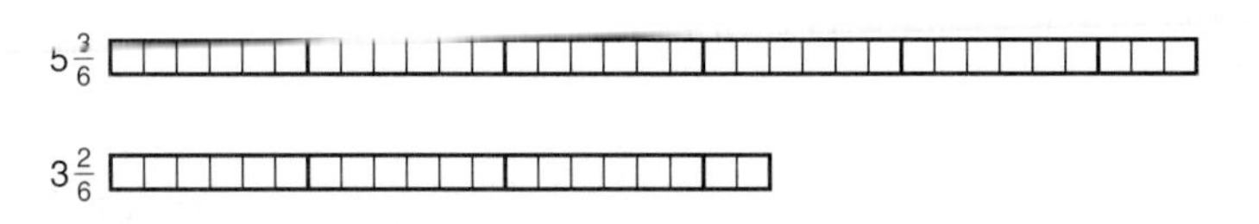

Does it then follow that any two lengths can be evenly cut into pieces of equal common length? About 2300 years ago, mathematicians would have answered yes. They thought that any two lengths are *commensurable*—that is, have a common unit of measure. This unit might be very small, such as a millionth or a billionth of an inch, but surely there would always be such a length. When the Greek mathematician Hippasus (470 B.C.) discovered that the side and the diagonal of a square are *incommensurable*—that is, do not have a common unit of measure—he caused such a crisis in mathematics that, reportedly, he was taken to sea and never returned. His discovery ultimately led to new types of numbers, called *irrational numbers,* which are represented by infinite nonrepeating decimals.

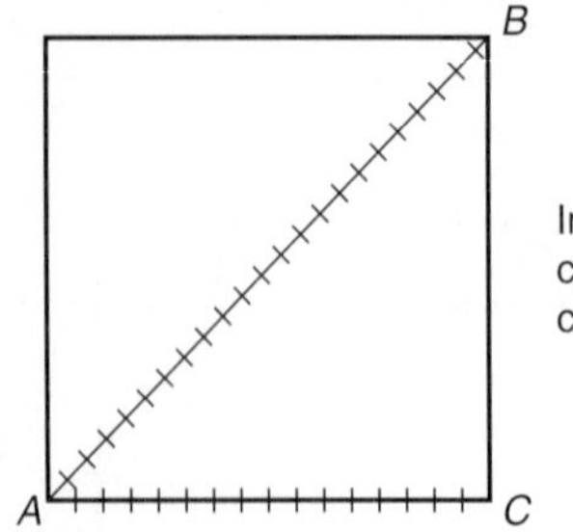

In a square, diagonal $\overline{AB}$ and side $\overline{AC}$ cannot be cut into pieces of equal common length.

1. Region X represents 1 unit of area on the geoboard. Region Y can be subdivided into 6 unit squares and 4 halves so that it has an area of 8 square units. If this figure is formed on the geoboard, it may be helpful to use rubber bands to subdivide it into unit squares and halves. Region Z is half of a rectangle of area 6 square units, so Z has an area of 3 square units.

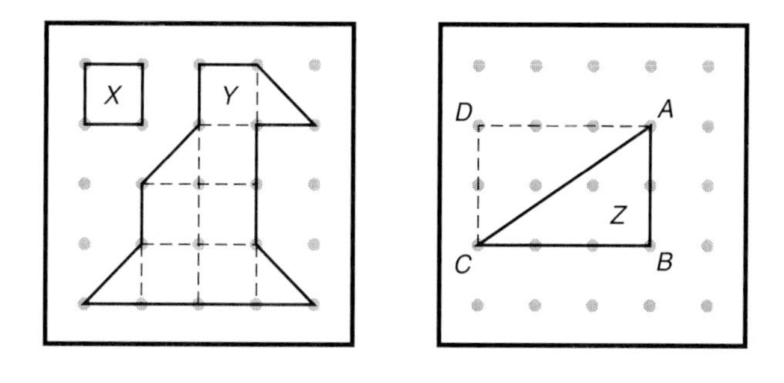

Determine the area of each of the following geoboard regions. Record the areas inside the figures.

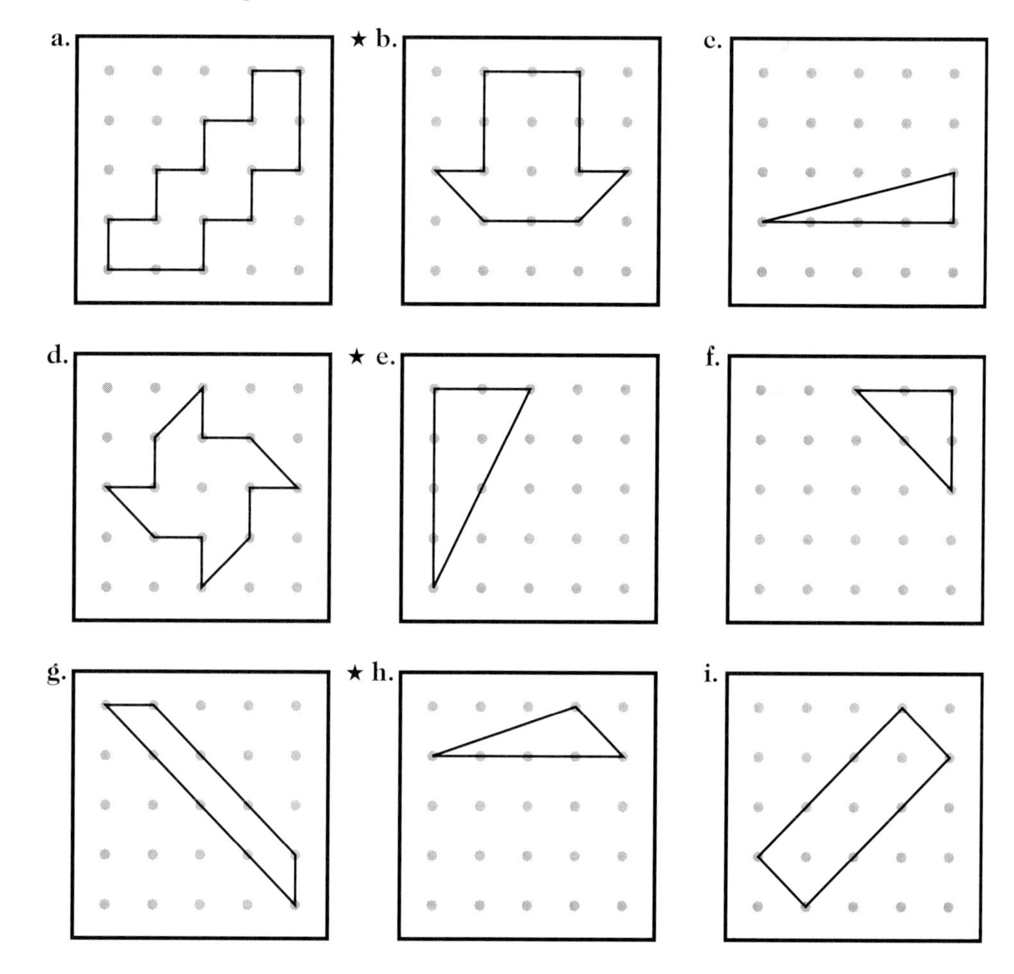

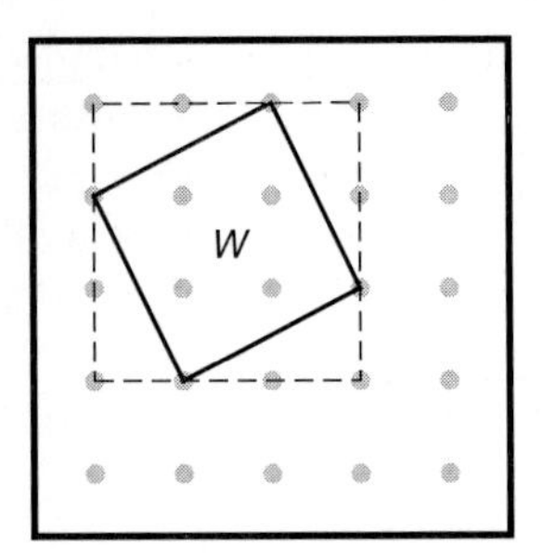

2. One way to determine the area of square W is to enclose it inside the larger square that has an area of 9 square units and then subtract the area of the triangular corners. Each of the triangular corners has an area of 1 square unit. So the area of square W is $9 - 4 = 5$ square units.

Here are seven other squares that can be constructed on the geoboard. Determine the area of each square. Record the areas inside the squares.

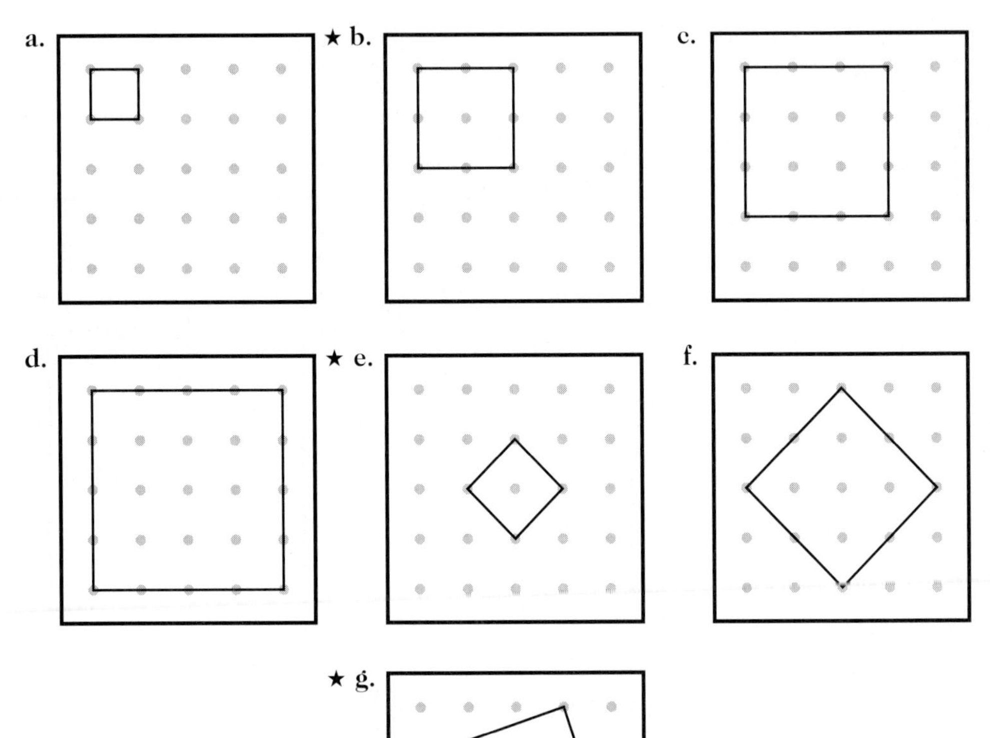

3. For any square, the length of a side times itself gives the area of the square.

$$\text{side length} \times \text{side length} = \text{area}$$

Square W in activity 2 has an area of 5 square units. The length of a side of square W is $\sqrt{5}$ because by definition $\sqrt{5}$ is *the number that, when multiplied times itself, gives 5.* That is, $\sqrt{5} \times \sqrt{5} = 5$. One of the squares in activity 2 has an area of 9, and the length of a side is $\sqrt{9}$. Notice that $\sqrt{9} = 3$ because $3 \times 3 = 9$.

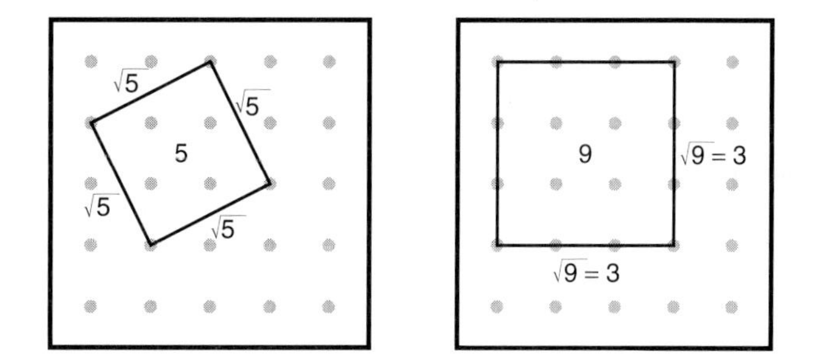

Finish recording the areas (from least to greatest) and the corresponding side lengths for the eight squares in activity 2.

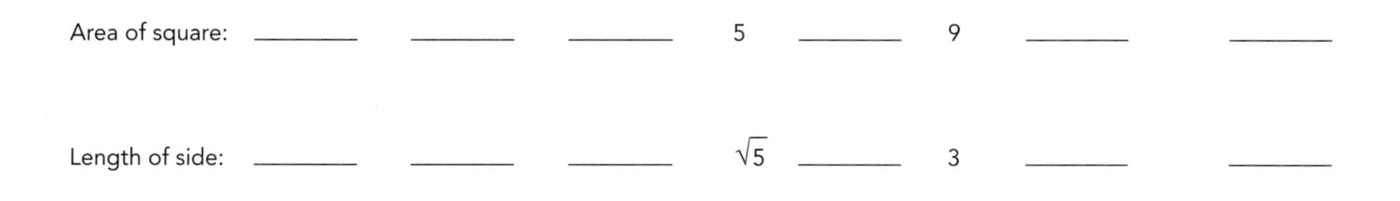

Area of square: ______ ______ ______ 5 ______ 9 ______ ______
Length of side: ______ ______ ______ √5 ______ 3 ______ ______

★ **4.** Determine the lengths of the three line segments in the next figure by first drawing a square on each segment and finding the area of the square. The corners of a sheet of paper can be used to draw the sides and locate the corners of the square. The corners of each square will be on dots. (The dotted lines form a square on segment a.)

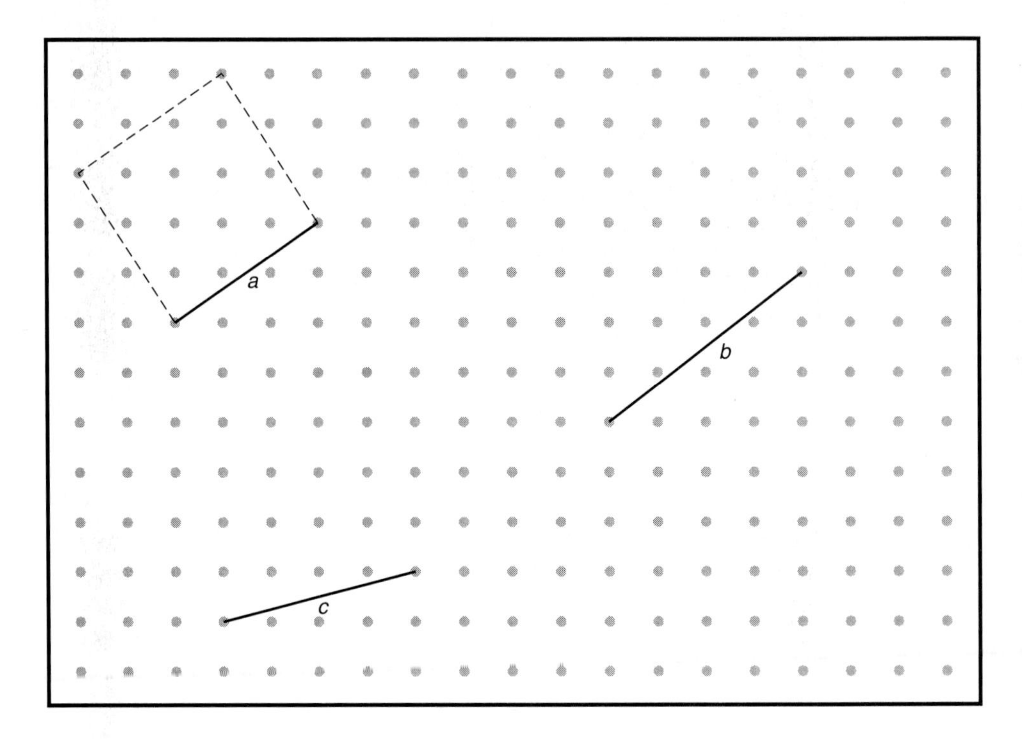

5. Calculator Activity: If n is a positive integer and not a perfect square (that is, n is not 0, 1, 4, 9, 16, 25, 36, . . .), then $\sqrt{n}$ is an *irrational number.* For example, $\sqrt{5}$ is an irrational number. When written as decimals, irrational numbers have infinite and nonrepeating decimal expansions.

a. Use the $\sqrt{\ }$ key on your calculator to obtain an approximate value for $\sqrt{5}$. Record the decimal from your calculator display.

b. Multiply the calculator's decimal for $\sqrt{5}$ times itself and record the result. Did you get 5? If not, why do you think you didn't?

c. Explain the advantages of labeling the length of a segment with a symbol like $\sqrt{5}$ rather than writing out the digits for $\sqrt{5}$ that appear on a calculator display.

6. Each of the following figures has three squares surrounding a right triangle (a right triangle has one angle of 90°). For each figure, determine the area of each square and complete the table.

	SQUARE A	SQUARE B	SQUARE C
a. Figure 1	______	______	______
★ b. Figure 2	______	______	______
c. Figure 3	______	______	______
★ d. Figure 4	______	______	______

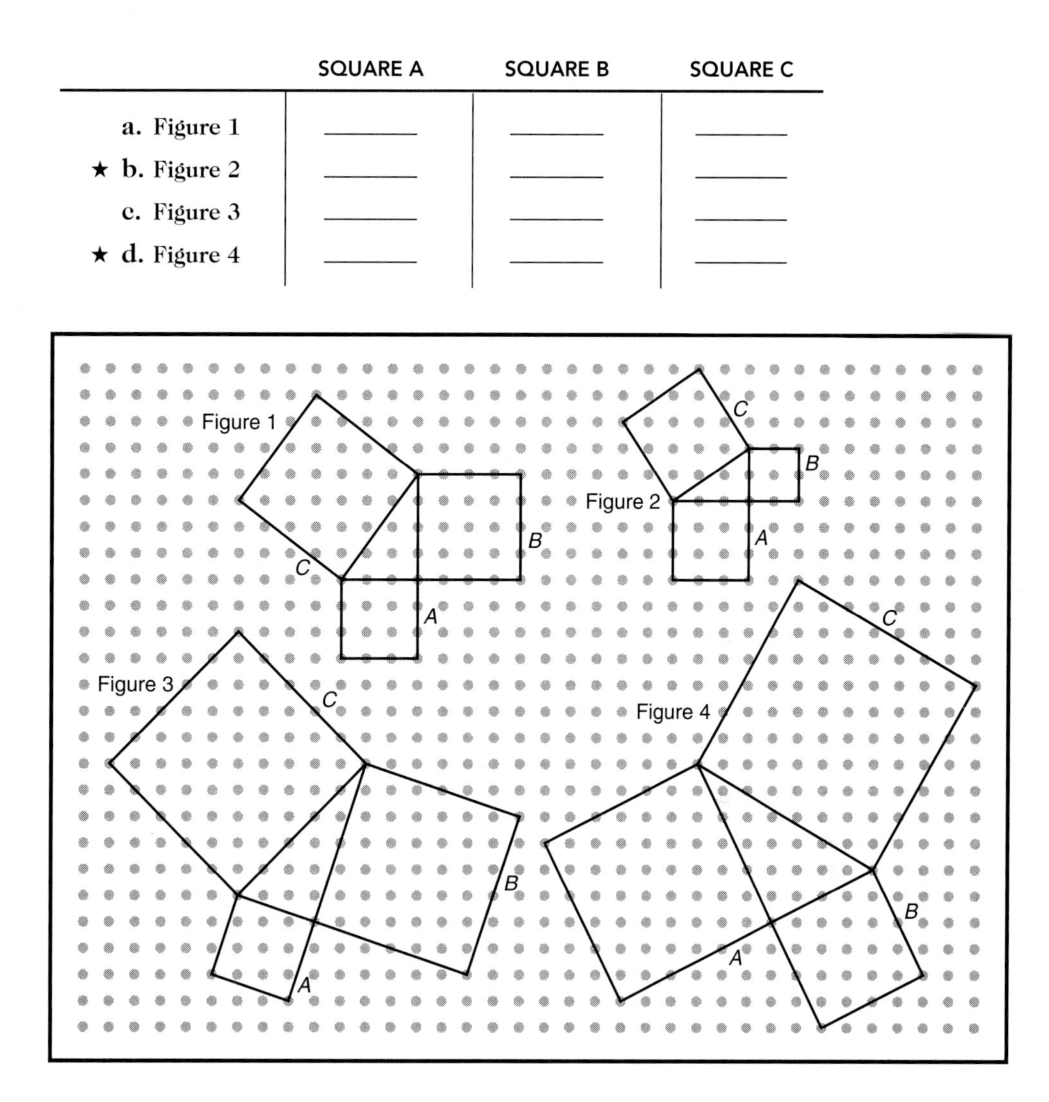

7. For every right triangle, the same relationship holds between the area of the square on the longest side (the hypotenuse) and the area of the squares on the other two sides (the legs).

   a. Use the data from activity 6 to make a conjecture, in writing, about this relationship.

   b. Use this relationship to find the area of the square on the hypotenuse for each of the following figures. Record your results in the table, and then use them to determine the length of the hypotenuse.

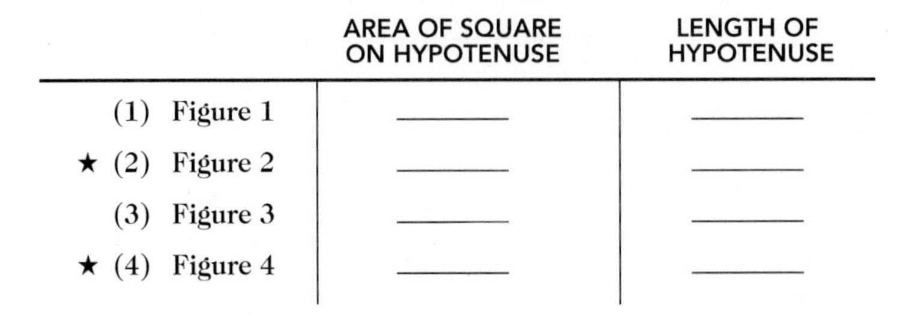

	AREA OF SQUARE ON HYPOTENUSE	LENGTH OF HYPOTENUSE
(1) Figure 1	______	______
★ (2) Figure 2	______	______
(3) Figure 3	______	______
★ (4) Figure 4	______	______

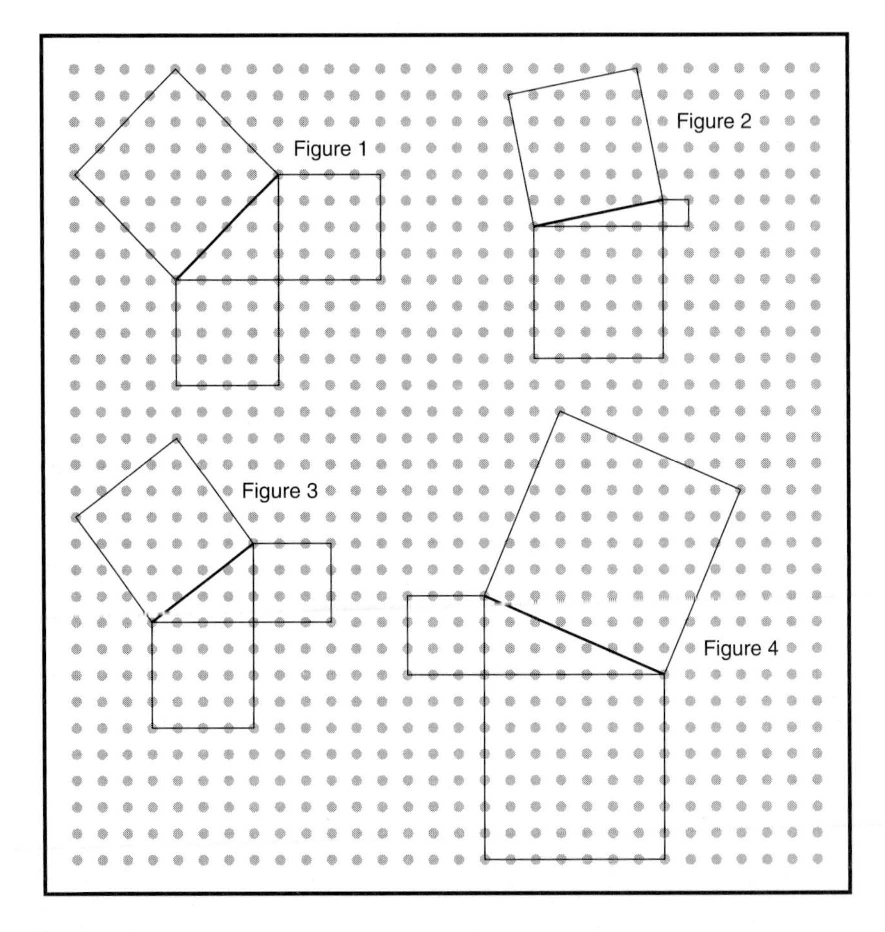

JUST FOR FUN

GOLDEN RECTANGLES

Some rectangles are more aesthetically pleasing than others. One in particular, the *golden rectangle,* has been the favorite of architects, sculptors, and artists for over 2000 years. The Greeks of the fifth century B.C. were fond of this rectangle. The front of the Parthenon in Athens fits into a golden rectangle.

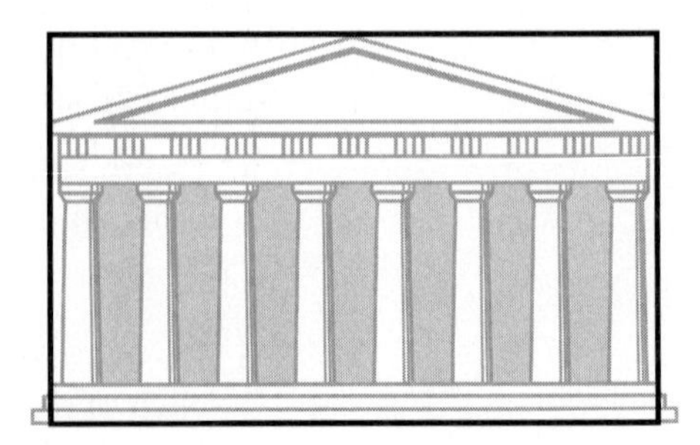

In the nineteenth century, the German psychologist Gustav Fechner found that most people unconsciously favor the golden rectangle. Which one of the following rectangles do you feel has the most pleasing dimensions?

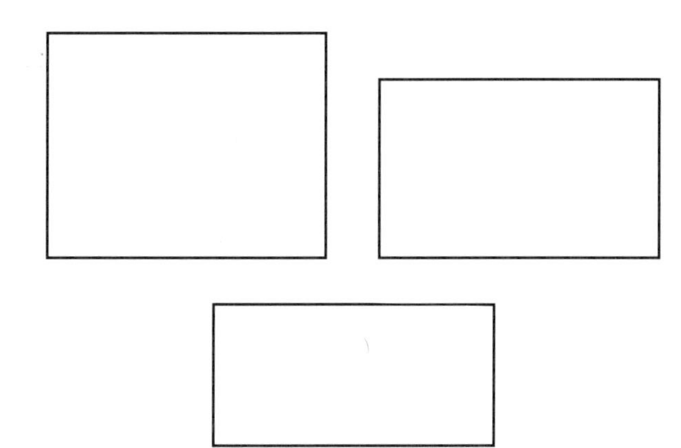

The length of a golden rectangle divided by its width is an irrational number that, when rounded to three decimal places, is 1.618. Measure the preceding rectangles to determine which is closest to a golden rectangle. Measure some other rectangles, such as greeting cards, credit cards, business cards, index cards, pictures, and mirrors, to see if you can find a golden rectangle. (A ratio of 1.6 is close enough.) Check the ratio of the length of the front of the Parthenon to its width.

A golden rectangle can be constructed from a square such as *ABCD* by placing a compass on the midpoint of side $\overline{AB}$ and swinging an arc *DF*. Then *FBCE* is a golden rectangle.

Make a golden rectangle, and check the ratio of its sides. If the sides of the square have length 2, then $\overline{BF} = 1 + \sqrt{5}$. Use your calculator to determine $(1 + \sqrt{5}) \div 2$ (the ratio of the sides of the rectangle) to three decimal places. Cut out your rectangle and fold it in half along its longer side. Is the result another golden rectangle?

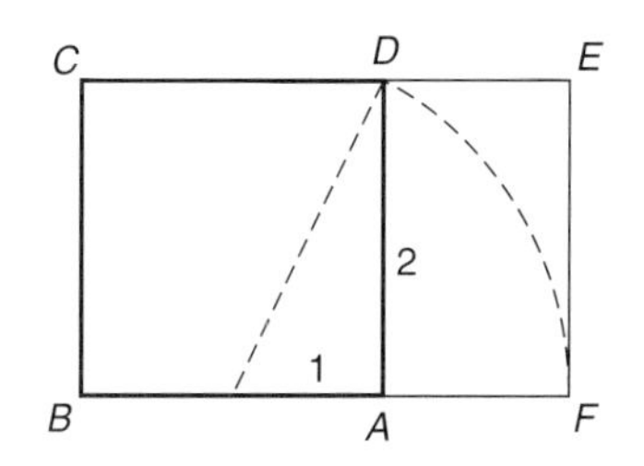

The ratio of the sides of a golden rectangle is called the *golden mean* or *golden ratio.*[9] It is surprising that this number (approximately 1.618) is also associated with the Fibonacci numbers: 1, 1, 2, 3, 5, 8, 13, ... The ratios of successive terms in this sequence ($\frac{1}{1}, \frac{2}{1}, \frac{3}{2}, \frac{5}{3}, \frac{8}{5}, \ldots$) get closer and closer to the golden ratio. The ratio $\frac{8}{5}$, for example, is 1.6. Use your calculator to determine how far you must go in the Fibonacci sequence before the ratio of numbers is approximately 1.618.

PUZZLER

These nine plants are in a square fenced garden. Construct two more square fences so that *each* plant is contained inside its own fence.

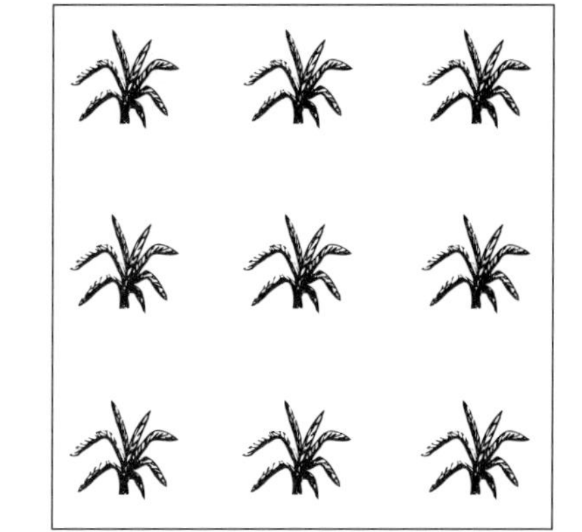

[9]Several occurrences of the golden ratio are described in C. F. Linn, *The Golden Mean* (New York: Doubleday, 1974).

IDEAS FOR THE ELEMENTARY CLASSROOM

SUGGESTED CLASSROOM ACTIVITY: BASE-TEN DECIMAL MODEL

Another physical model for decimals can be made by using the base-ten pieces from the Manipulative Kit. If the largest of the three base-ten pieces in the kit is designated as 1 (the unit), then the two smaller pieces represent .1 ($\frac{1}{10}$) and .01 ($\frac{1}{100}$). So, for example, the decimal 2.37 is represented by the collection shown in the following figure. The students can discuss why a decimal point is needed when the numeral is written. Also, decimals can be added and subtracted by regrouping number pieces.

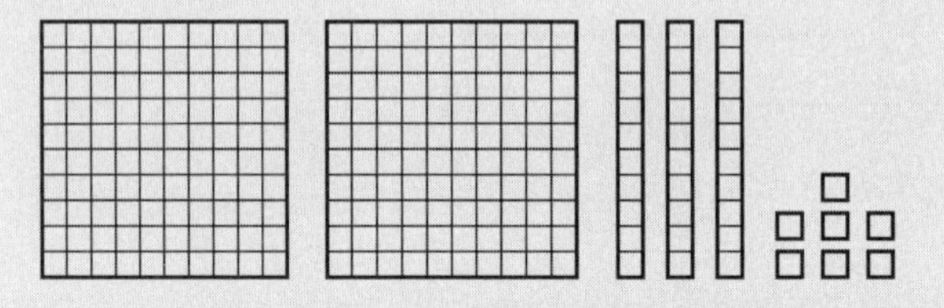

For classroom demonstration, a number piece representing 10 can be made by taping 10 units in a strip (or cutting a piece of cardstock 10 cm by 100 cm). A number piece representing 100 could be made with a square piece of paper or cardstock measuring 1 m on each edge. This model provides a nice way to visualize the relative sizes of numbers.[10]

Readings for More Classroom Ideas

Bennett, A., E. Maier, and L. Ted Nelson. "A Conceptual Model for Solving Percent Problems." *Mathematics Teaching in the Middle School* 1 (April 1994): 20–25.

Bennett, A., E. Maier, and L. Ted Nelson. "Modeling Percentages and Ratios," in *Math and the Mind's Eye* (Unit ME7). Salem, OR: The Math Learning Center, Box 3226, 1991.

Bennett, A., E. Maier, and L. Ted Nelson. "Modeling Rationals," in *Math and the Mind's Eye* (Unit ME4). Salem, OR: The Math Learning Center, Box 3226, 1988.

Caldwell, Janet H. "Communicating About Fractions with Pattern Blocks." *Teaching Children Mathematics* 2 (November 1995): 156–161.

Carraher, Terezinha N., and A. D. Schliemann. "Research into Practice: Using Money to Teach About the Decimal System." *Arithmetic Teacher* 36 (December 1988): 42–43.

Curcio, Frances R., et al. *Understanding Rational Numbers and Proportions, in the Curriculum and Evaluation Standards for School Mathematics Addenda Series: Grades 5–8.* Reston, VA: National Council of Teachers of Mathematics, 1994.

Fennell, Francis. "The Newspaper: A Source for Applications in Mathematics." *Arithmetic Teacher* 30 (October 1982): 22–26.

Haubner, Mary Ann. "Percents: Developing Meaning Through Models." *Arithmetic Teacher* 40 (December 1992): 232–234.

Quintero, Ana H. "Helping Children Understand Ratios." *Arithmetic Teacher* 34 (May 1987): 17–21.

Sowder, Judith. "Place Value as the Key to Teaching Decimal Operations." *Teaching Children Mathematics* 3 (April 1997): 448–453.

[10]For more information about this model for decimals, see A. Bennett, E. Maier, and L. Ted Nelson, "Modeling Rationals" (Unit IV, Activities 6–9), *Math and the Mind's Eye* (Salem, Ore.: The Math Learning Center, Box 3226, 1988).

CHAPTER 7

STATISTICS

In grades K–4, students begin to explore basic ideas of statistics by gathering data appropriate to their grade level, organizing them in charts or graphs, and reading information from displays of data. These concepts should be expanded in the middle grades. . . . Instruction in statistics should focus on the active involvement of students in the entire process: formulating key questions; collecting and organizing data; representing the data using graphs, tables, frequency distributions, and summary statistics; analyzing the data; making conjectures; and communicating information in a convincing way.[1]

ACTIVITY SET 7.1

RANDOMNESS, SAMPLING, AND SIMULATION IN STATISTICS

Purpose To clarify the idea of a random event and apply that idea to sampling and simulation in statistics

Materials Green and red tiles from the Manipulative Kit, the table of random digits from Material Card 8, and containers such as boxes and bags for selecting samples

© 1974 United Feature Syndicate, Inc.

[1]*Curriculum and Evaluation Standards for School Mathematics* (Reston, VA: National Council of Teachers of Mathematics, 1989), 105.

Activity One of the main objectives of statistics is to make decisions or predictions about a population (students in schools, voters in a country, lightbulbs, fish, and so forth) based on information taken from a sample of the population. A poorly chosen sample will not truly reflect the nature of the population. An important requirement is that the sample be randomly chosen. *Randomly chosen* means that every member of the population is equally likely to be selected.

Once a sample has been chosen, it is useful to determine a single number that describes the data. An *average* is one type of value or measure that represents a whole set of data. There are three averages commonly used to describe a set of numbers: mean, median, and mode. Let's use these averages to describe the following set of numbers:

$$2, 3, 4, 6, 8, 11, 12, 12, 12, 18$$

The *mean,* which is commonly referred to as the average, is the sum of the values divided by the total number of values. The sum of the 10 numbers is 88, so the mean is 8.8 ($88 \div 10$). The *median* is the middle value when all the values are arranged in order. The median for these values is 9.5, which is halfway between the two middle values 8 and 11. The *mode* is the value that occurs most frequently—in this case, the mode is 12.

Simulations are processes that help answer questions about a real problem; they are experiments that closely resemble the real situation. These experiments usually involve random samples.

In this activity set, we will explore the notion of random numbers, experimentally select random samples, and use the idea of randomness to simulate a real situation.

1. **Random Digits:** Three methods for generating a list of random digits are described below.

 a. *Container Selection Method:* One way to obtain random digits is to write each digit, 0 through 9, on a chip; place all 10 chips in a container; shake the container to thoroughly mix the chips; and then draw one chip, without looking, from the container. If every chip in the container has an equal chance of being selected, then the one you draw will be a *random digit.* By replacing that chip and repeating the process many times, you can form a list of random digits. Write the digits on chips or slips of paper, put them in a container, and draw them out to generate a list of 100 random digits. Record five digits on each of the 20 blanks below.

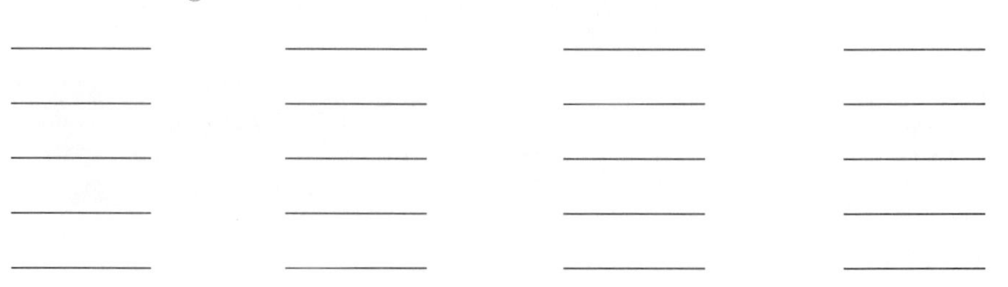

b. *Calculator Method:* Here is a method for generating a list of random digits using a calculator.

	Example
(1) Choose an arbitrary five-digit number:	72194
(2) Square that number:	$(72194)^2 = 52\underline{11973}636$
(3) Select 5 middle digits of the result, and square that number:	$(11973)^2 = 14\underline{33527}29$
(4) Repeat the process:	$(33527)^2 = 11\underline{24059}729$

Each string of middle digits—11973, 33527, and 24059—is a group of random digits. Select your own arbitrary five-digit number and generate 100 random digits using the calculator process. Record five digits on each of the blanks below.

________	________	________	________
________	________	________	________
________	________	________	________
________	________	________	________
________	________	________	________

c. *"Top-of-the-Head" Method:* Can a list of random digits be produced by writing numbers down as quickly as possible, as they pop into your head? Try this method of generating 100 digits. Record the digits below in groups of five.

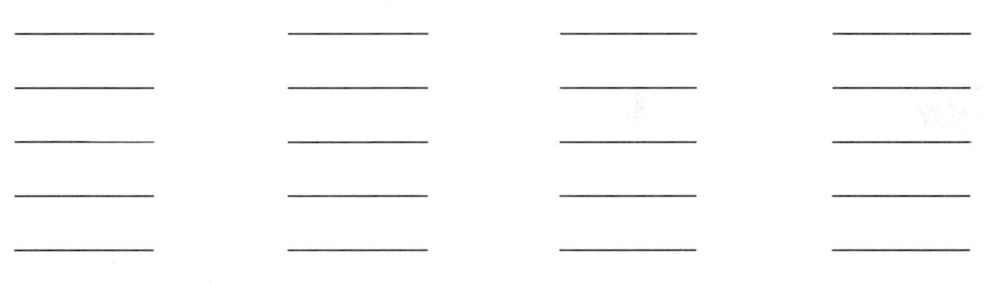

2. For each list of digits that you generated in activity 1, determine, by tallying, the number of times (frequency) each digit occurred in that list.

Container Selection Method			Calculator Method			Top-of-the-Head Method		
DIGIT	TALLY	FREQUENCY	DIGIT	TALLY	FREQUENCY	DIGIT	TALLY	FREQUENCY
0	_____	_____	0	_____	_____	0	_____	_____
1	_____	_____	1	_____	_____	1	_____	_____
2	_____	_____	2	_____	_____	2	_____	_____
3	_____	_____	3	_____	_____	3	_____	_____
4	_____	_____	4	_____	_____	4	_____	_____
5	_____	_____	5	_____	_____	5	_____	_____
6	_____	_____	6	_____	_____	6	_____	_____
7	_____	_____	7	_____	_____	7	_____	_____
8	_____	_____	8	_____	_____	8	_____	_____
9	_____	_____	9	_____	_____	9	_____	_____
	Total	100		Total	100		Total	100

a. If you had to determine some commonsense ways to judge whether or not a sequence of digits was indeed random, what are some factors you would consider?

b. If each digit in your list has a frequency from 7 to 13, then you can have confidence that you have a random sequence of digits.[2] Compare the three methods of generating random digits. How would you rate the top-of-the-head method? Was the top-of-the-head method biased in favor of any particular digit?

3. **Sampling and Predictions:** The objective of sampling is to make predictions about a large population. Samples are usually relatively small. For example, some national television ratings are based on a sample of only 1100 people. The following activity is designed to show how a sample can represent the given population.

Have a classmate put any one of the following three populations of tiles in a container, without your knowing the numbers of red and green tiles. *Don't look at the contents of the container until you have completed all parts of this activity.*

[2]Such limits as these (7 and 13) for determining if a distribution is uniform can be obtained from a statistical test called the chi-square test.

POPULATION 1	POPULATION 2	POPULATION 3
25 red tiles 5 green tiles	20 red tiles 10 green tiles	10 red tiles 20 green tiles

Draw a tile and check off its color in the following table. Return the tile to the container and repeat this activity. After every five draws, make a prediction as to which population you think is in the container. Continue the experiment until you have selected 25 times.

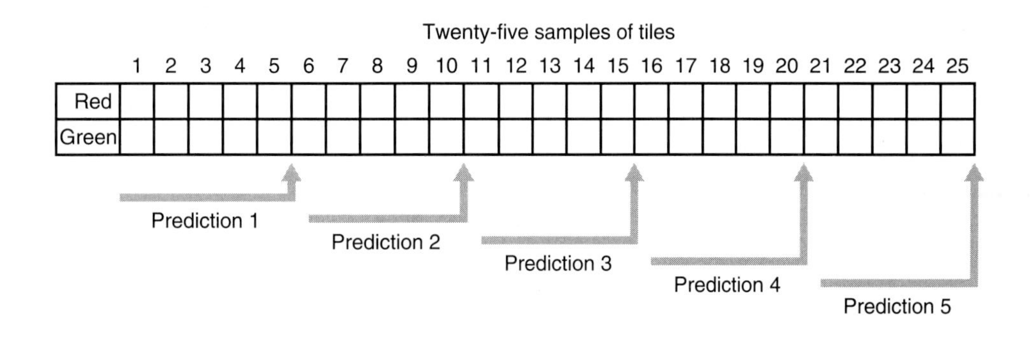

a. Compute the percentage of red tiles in each of the following groups.

______	______	______	______	______
First 5	First 10	First 15	First 20	All 25

b. As the size of the sample increases, the likelihood that it represents the population becomes greater. Based on your sample of 25 tiles, predict which of the three given populations you have sampled. Would you have selected the same population if your sample size had been only the first 5 tiles? The first 10? Now look at the contents of the container to check your prediction.

4. **Stratified Sampling:** Following are the income and regional living area for 18 households. Write the information for each household on a separate piece of paper, and place the pieces of paper in a container.

Lake Property	Lake Property	Lake Property	Suburban	Suburban	Suburban	Suburban	Suburban	Suburban
$210,000	$180,000	$140,000	$70,000	$60,000	$44,000	$40,000	$35,000	$26,000
City	City	City	City	City	City	City	City	City
$40,000	$35,000	$24,000	$22,000	$20,000	$20,000	$18,000	$17,000	$15,000

a. The average of all 18 incomes is $56,444.44. Select 6 pieces of paper at random from the container. Compute the average (mean) income of your sample and record it here. By how much does it differ from the average of all 18 incomes?

★ **b.** There are many different samples of 6 incomes that you could have selected. Suppose, for example, your sample contained the 6 greatest incomes. What would be the average of these 6 incomes? What would be the average of the 6 smallest incomes?

c. The average of the 6 greatest incomes and the average of the 6 smallest incomes differ significantly from the average of all 18 incomes. By using a technique called *stratified sampling,* you can avoid getting an unrepresentative sample. Separate the total population of 18 incomes by living area into 3 different containers. Notice that the lake property, suburban, and city salaries represent $\frac{1}{6}$, $\frac{1}{3}$, and $\frac{1}{2}$ of the population, respectively. To obtain a proportional *stratified sample* of 6 incomes, select 1 lake property income ($\frac{1}{6}$ of 6), 2 suburban incomes ($\frac{1}{3}$ of 6), and 3 city incomes ($\frac{1}{2}$ of 6) from their individual containers. Compute the average of these 6 incomes. By how much does it differ from the average of all 18 incomes?

★ **d.** If the stratified sampling procedure in part c is used, what is the greatest possible average that can be obtained for 6 incomes?

What is the smallest possible average that can be obtained?

Compare these averages to those in part b and discuss the advantages of stratified sampling.

5. Simulation: People with type O blood are called universal donors because their blood can be used in transfusions to people with different blood types. About 45 percent of the population has type O blood. If blood donors arrive at the blood bank with random blood types, how many donors, on the average, would it take to obtain five donors with type O blood? Make a conjecture.

a. There are 100 pairs of digits from 00 to 99 and the pairs (00, 01, 02, . . . , 44) represent 45 percent of them. Without looking, put your pencil point on the page of random digits on Material Card 8. Beginning with the digit closest to the pencil point, start listing consecutive pairs of digits from the table of random digits until your list includes exactly five pairs from the list (00, 01, 02, . . . , 44). If a pair of digits represents a donor, how many donors would have to arrive at the blood bank to obtain five donors with type O blood?

b. The procedure you followed in part a is a *simulation*—a way to answer a problem by running an experiment which resembles the problem.[3] Assuming that blood donors who go to the blood bank have random blood types and that 45 percent have type O, your experiment simulated the number of donors needed to obtain five donors with type O blood. Run the same experiment a total of 10 times and record your results here.

	TALLIES		
EXPERIMENT	DIGIT PAIRS FROM 00 TO 44	DIGIT PAIRS FROM 45 TO 99	NUMBER OF DONORS NEEDED TO OBTAIN 5 OF BLOOD TYPE O
1			
2			
3			
4			
5			
6			
7			
8			
9			
10			

c. The mean of the numbers of donors in the 10 experiments is the *average number* for these experiments. Based on your experimental results, what is the average number of donors the blood bank would have to receive to obtain five donors with type O blood?

[3]An excellent source for simulation activities is M. Gnanadesikan, R. Scheaffer, and J. Swift, *The Art and Techniques of Simulation* (Palo Alto, CA: Dale Seymour Publications, 1987).

JUST FOR FUN

SIMULATED RACING GAME (2 players)

In this race, one player flips a coin and the other rolls a die. At the start of play, each player rolls the die, and the high roller gets to use the die in the game. The race track is 12 steps long, and the racers move simultaneously. That is, the coin is flipped and the die is rolled at the same time. The coin tosser can move 1 step for heads and 2 steps for tails. The die roller can move the number of steps that comes up on the die. The race continues until 1 person lands on the last step. However, *the winner must land on the last step by exact count.* It is possible for this race to end in a tie. Find a person to race against and complete several races using a die and a coin.

1. Does this appear to be a fair race? That is, do the wins seem to be evenly distributed between the die roller and the coin tosser? You may wish to have a larger sample of completed races on which to base your conclusions. Use the table of random numbers on Material Card 8 to simulate die rolling and coin flipping as follows:
   - *Coin flipping*—Put your pencil down on an arbitrary point in the table. If the closest digit is even, call the outcome heads; if it is odd, call the outcome tails. For the succeeding flip, move to the next digit in the table and do the same. Repeat this process until the game is complete.
   - *Die tossing*—Put your pencil down on an arbitrary point in the table. If the closest digit is a 1, 2, 3, 4, 5, or 6, let it represent your first roll of the die. If that digit is a 0, 7, 8, or 9, move to the next digit in the table that is not a 0, 7, 8, or 9 and let it represent your roll. For the succeeding roll, move to the next digit in the table and do the same. Repeat this process until the game is complete.
2. What happens when the length of the race (number of steps) is shortened? For example, will a two-step race be won more frequently by one player than the other? Try races of different lengths. Can you draw some conclusion about the fairness of the race and its length?

	START	1	2	3	4	5	6	7	8	9	10	11	FINISH
Coin flipper	*X*	—	—	—	—	—	—	—	—	—	—	—	—
Die roller	*X*	—	—	—	—	—	—	—	—	—	—	—	—

PUZZLER

Assume the probability of a baby being a boy is $\frac{1}{2}$. Which of the following sequences of births is more likely to occur in a six-child family?

GBBGBG or GGGGBG

ACTIVITY SET 7.2

SCATTER PLOTS: LOOKING FOR RELATIONSHIPS

Purpose To look for relationships or trends between two sets of data

Materials A tape measure or ruler and string

Activity Scatter plots provide a visual means of looking for a relationship between two variables. As an example, one middle school student with two very tall parents wanted to see if, in general, taller people married taller people. She randomly selected students in her school and recorded the heights of their parents. Then for each set of parents she placed a point on the following graph that pairs the father's height with the mother's. From this scatter plot she hoped to reach some conclusions. Do you think there is any association in parental heights based on her scatter plot?

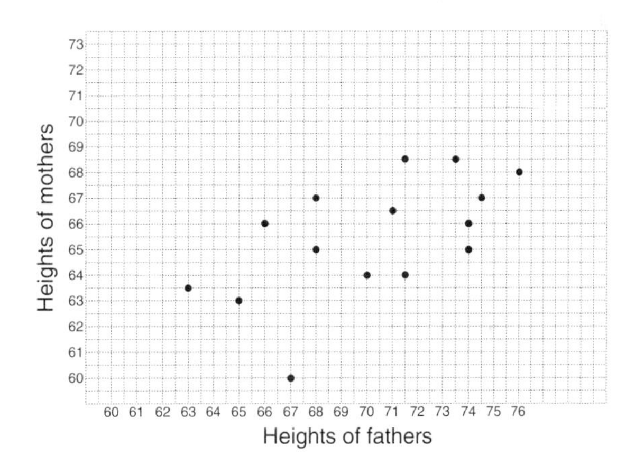

In this activity set you will collect data, form scatter plots, and look for relationships and trends.

1. Use the scatter plot above to answer the following questions.

★ a. What is the height of the shortest mother? The tallest mother?

b. What are the heights of the shortest and tallest fathers?

c. What is the difference in height between the shortest and tallest mother? The shortest and tallest father?

★ d. For how many couples was the woman taller than the man?

2. One method for determining a relationship or trend on a scatter plot is to draw a single straight line or curve that approximates the location of the points.

★ **a.** On the scatter plot at the beginning of this section, draw a single straight line through the set of points that you believe best represents the trend of the plotted points. This is called a *trend line.* (One strategy: Look for a directional pattern of the points from left to right and then draw a line in a similar direction so that approximately half the points are above the line and half below.)

b. Do you think there is a trend or relationship between the two sets of data? That is, from the data collected, do taller people seem to marry taller people? Justify your conclusion in a sentence or two.

★ **3. a.** Suppose that the middle school student's scatter plot had looked as follows. Draw a trend line that you believe best represents this data set. Write a sentence or two summarizing the relationship between heights of fathers and mothers for this data.

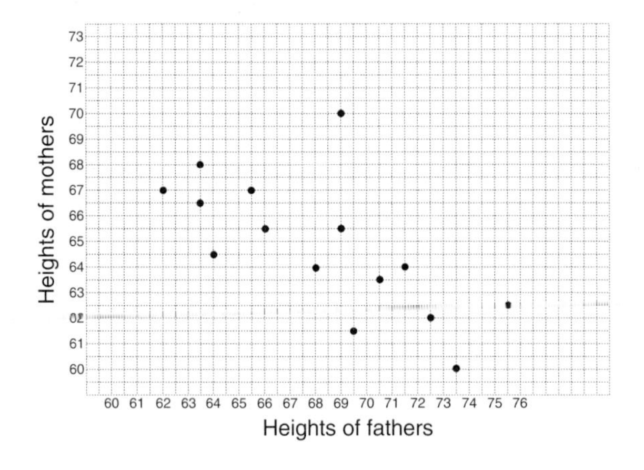

★ **b.** Now suppose that the student's data had looked like the scatter plot below. Draw a trend line for these points. Would you say there is a trend or relationship between the heights of the fathers and mothers for this data? Explain your reasoning.

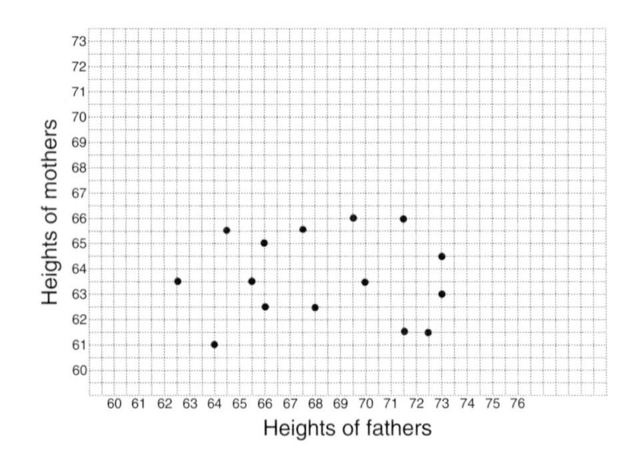

c. The relationship between two sets of data can be described as positive, negative, or no relationship, depending on the slope of the trend line. Classify the relationships for the scatter plots in activities 2 and 3 above as positive, negative, or no relationship. Describe two other sets of data that you believe will result in a scatter plot with a negative relationship.

4. Form a scatter plot for each of the following sets of data and sketch a trend line. Determine if there is a positive, negative, or no relationship.

a. This table contains data on one aspect of child development—the time required to hop a given distance. The age of each child is rounded to the nearest half year. Use your trend line to predict the hopping time for an average 7.5-year-old.

Age (years)	5	5	5.5	5.5	6	6	6.5	7	7	7.5	8	8	8.5	8.5	9	9	9.5	10	11
Time (seconds) to hop 50 feet	10.8	10.8	10.5	9.0	8.4	7.5	9.0	7.1	6.7		7.5	6.3	7.5	6.8	6.7	6.3	6.3	4.8	4.4

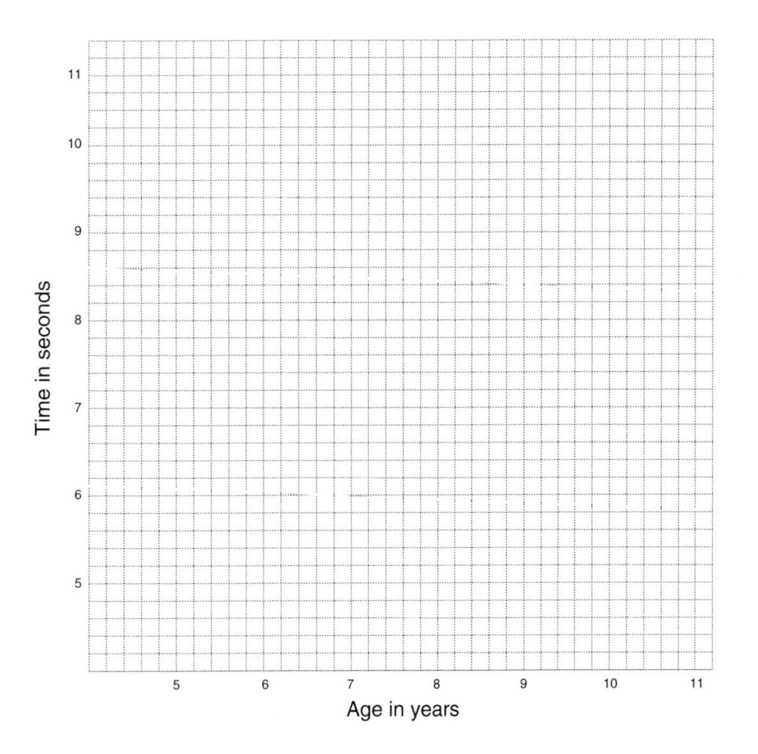

b. The following table was recorded by a forester who did a sample cutting of oak trees and recorded their diameters (in inches) and age (by counting rings). Use your trend line to predict the age of a tree that is 5.5 inches in diameter and the diameter of a tree that is 40 years old.

AGE (YEARS)	DIAMETER (INCHES)
10	2.0
8	1.0
22	5.8
30	6.0
18	4.6
13	3.5
38	7.0
38	5.0
25	6.5
8	3.0
16	4.5
28	6.0
34	6.5
29	4.5
20	5.5
4	0.8
33	8.0
23	4.7
14	2.5
35	7.0
30	7.0
12	4.9
8	2.0
5	0.8
10	3.5

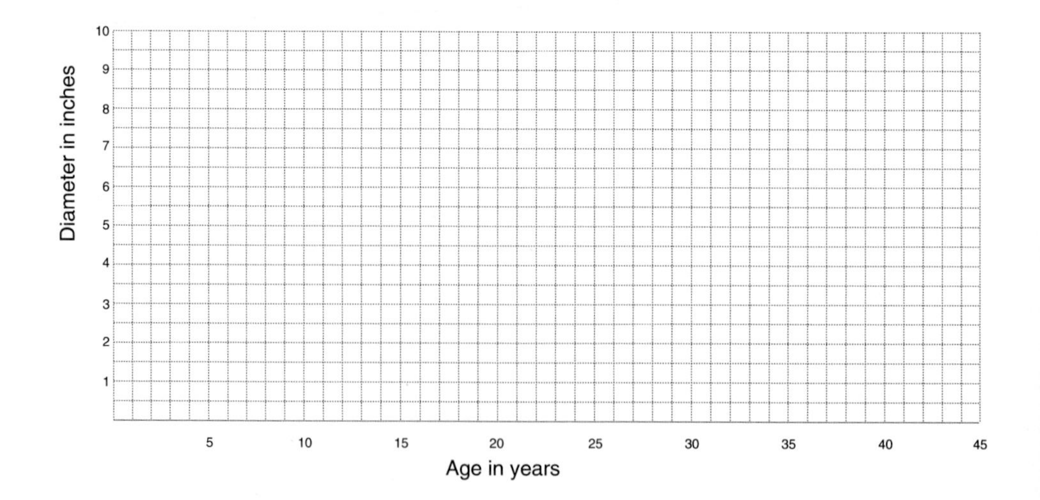

5. Collect the following information from a sample of at least 20 people: height, arm span, and foot length. Select a body measurement of your choice for the last row of the table and collect the data for it. (Arm span means the distance from the tip of one index finger to the other when the arms are outstretched in opposite directions and parallel to the floor.)

Data Record	1	2	3	4	5	6	7	8	9	10	11	12	13	14	15	16	17	18	19	20
Height																				
Arm span																				
Foot length																				

On grid I below, form a scatter plot that compares heights and arm spans. Put a scale on the horizontal axis for heights and a scale on the vertical axis for arm spans. The range of each scale will depend on the range of the measures you obtained. Draw a trend line for the scatter plot. Then write a summary of your observations and conclusions that includes your judgment of whether or not there is a relationship between the two sets of measurements.

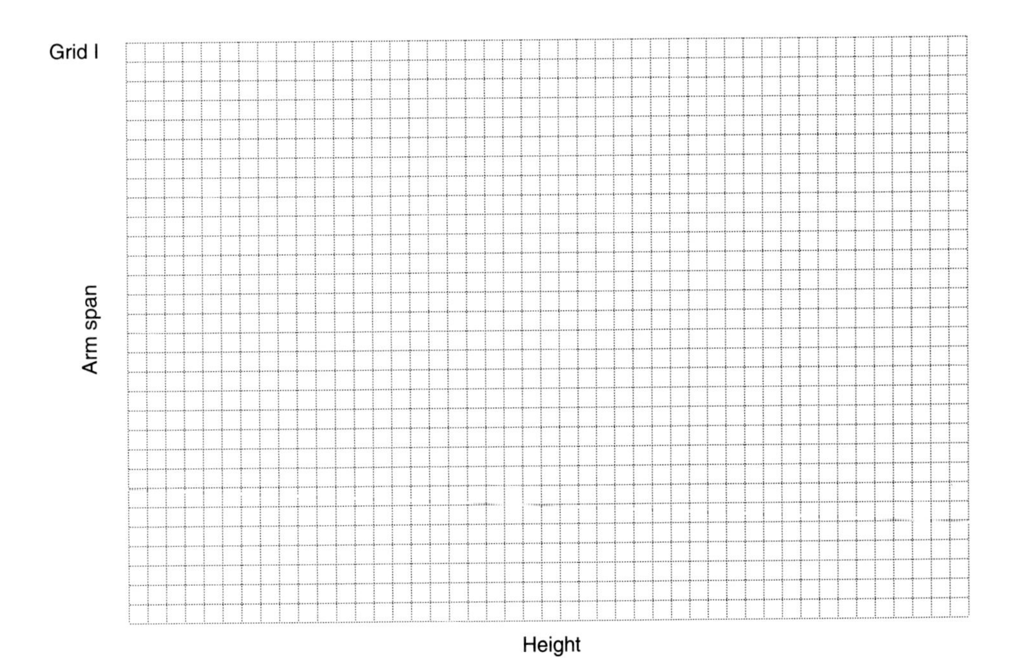

On grid II repeat the process but put the foot measurements on the vertical axis. Is there a relationship between these two sets of measures? Explain.

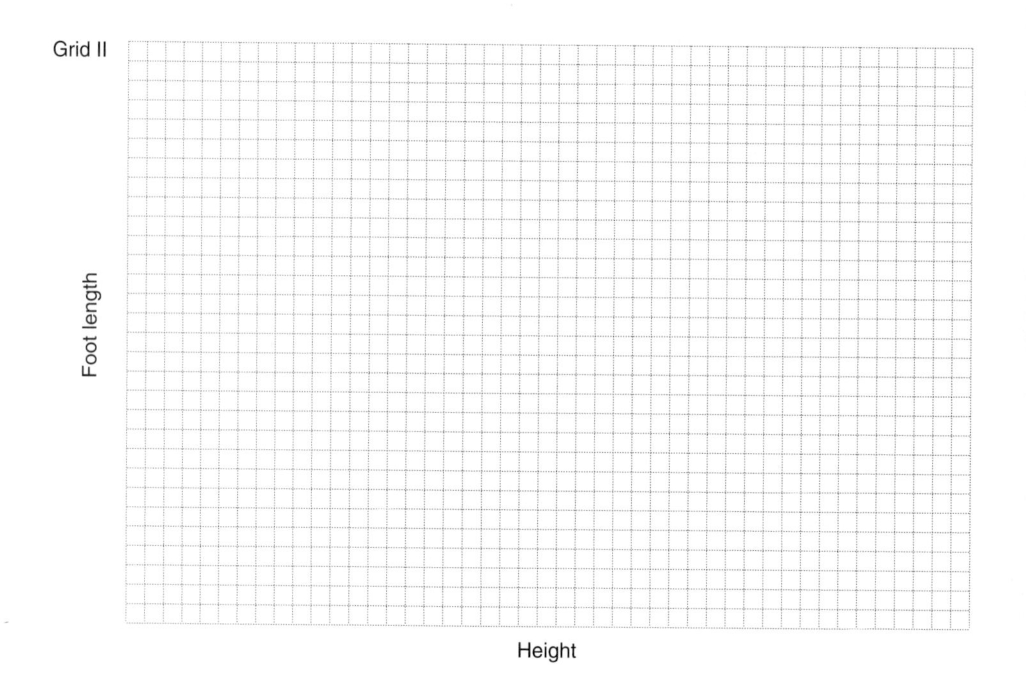

On grid III construct a scatter plot that compares the data in the table for the measure of your choice to one of the other three measures. Draw a trend line. Does there appear to be a relationship between these measurements? Explain.

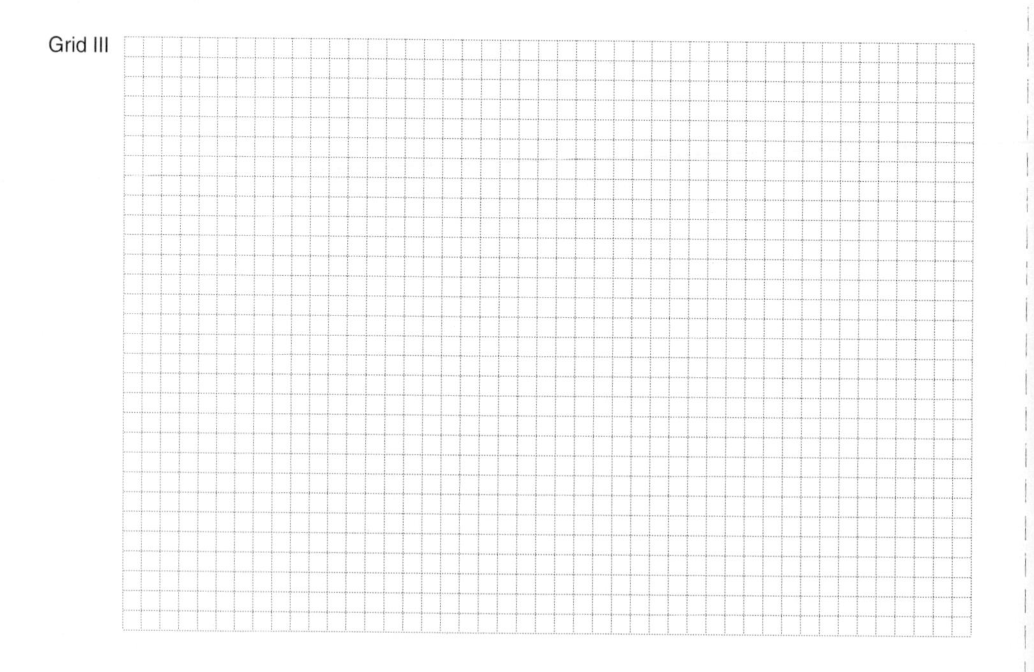

JUST FOR FUN

PAGE GUESSING

Does performance improve with practice? In this activity you will see if you can improve your page location performance with practice.

Find a book with several hundred pages and note the total number of pages it contains. Arbitrarily pick a page number in that book and record it as your target page. Before opening the book, estimate where you think that target page is located and then try to open the book as close to that page as possible. In the following table record the page number opened to and the number of pages you were off target (the difference between the target number and the page you opened). Close the book and try again. After five trials, graph your results on the grid provided below. Then, plot what you think the next few points on the graph would be if you were to continue. Continue your experiment a few more times to check your predictions. Repeat with another book to see if you improve with practice.

Target Page Number ________

Trial number	1	2	3	4	5	6	7	8	9	10
Page opened	____	____	____	____	____	____	____	____	____	____
Number of pages off target	____	____	____	____	____	____	____	____	____	____

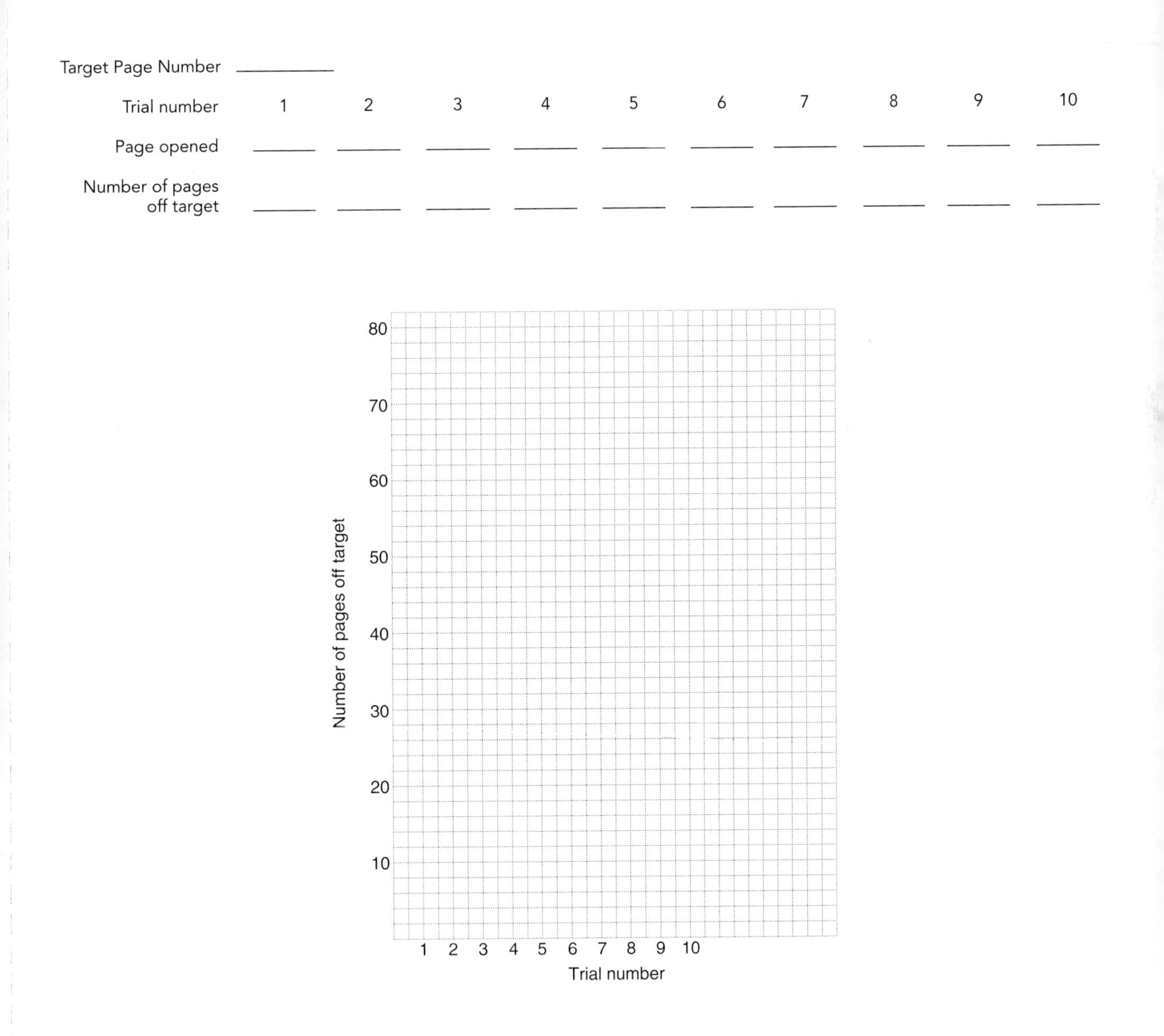

ACTIVITY SET 7.3

STATISTICAL DISTRIBUTIONS: OBSERVATIONS AND APPLICATIONS

Purpose To construct and apply frequency distributions

Materials The table of random digits on Material Card 8

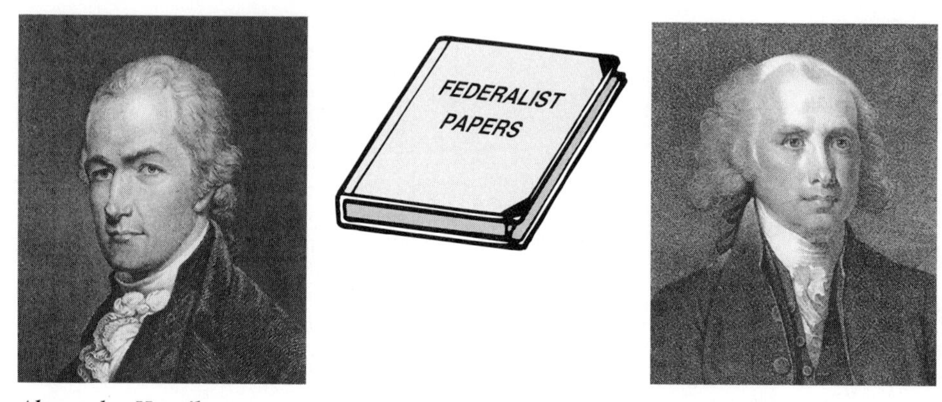

Alexander Hamilton *James Madison*

Activity Once measurements from a population sample have been collected, they are arranged by grouping similar measurements together in a table, chart, or graph. A *frequency distribution* is a tabulation of values by categories or intervals. An interesting illustration of the use of frequency distributions was in the resolution of the controversy over authorship of the *Federalist Papers.*[4] This is a collection of 85 political papers written in the eighteenth century by Alexander Hamilton, John Jay, and James Madison. There is general agreement on the authorship of all but 12 of these papers. In order to determine who should be given credit for the 12 disputed papers, frequency distributions were compiled for certain filler words such as *by, to, of,* and *on.*

Forty-eight of Hamilton's papers were analyzed to see how many times he used the word *by* in every 1000 words. In 18 out of 48 papers, he used *by* an average of 7 times. This is indicated by the longest bar of the top distribution in figure a. In others of his papers he used *by* at different rates, as indicated by the other bars of that distribution. A distribution of Madison's usage of *by* in 50 of his papers is shown in the middle distribution of figure a. Notice that Madison's distribution looks like the distribution of *by* in the disputed papers shown in the bottom of figure a. A similar analysis for the word *to* is shown in figure b. These and similar distributions for other filler words support the contention that Madison was the author of the disputed papers.

[4]F. Mosteller et al., *Statistics: A Guide to the Unknown* (New York: Holden-Day, 1972), 164–175.

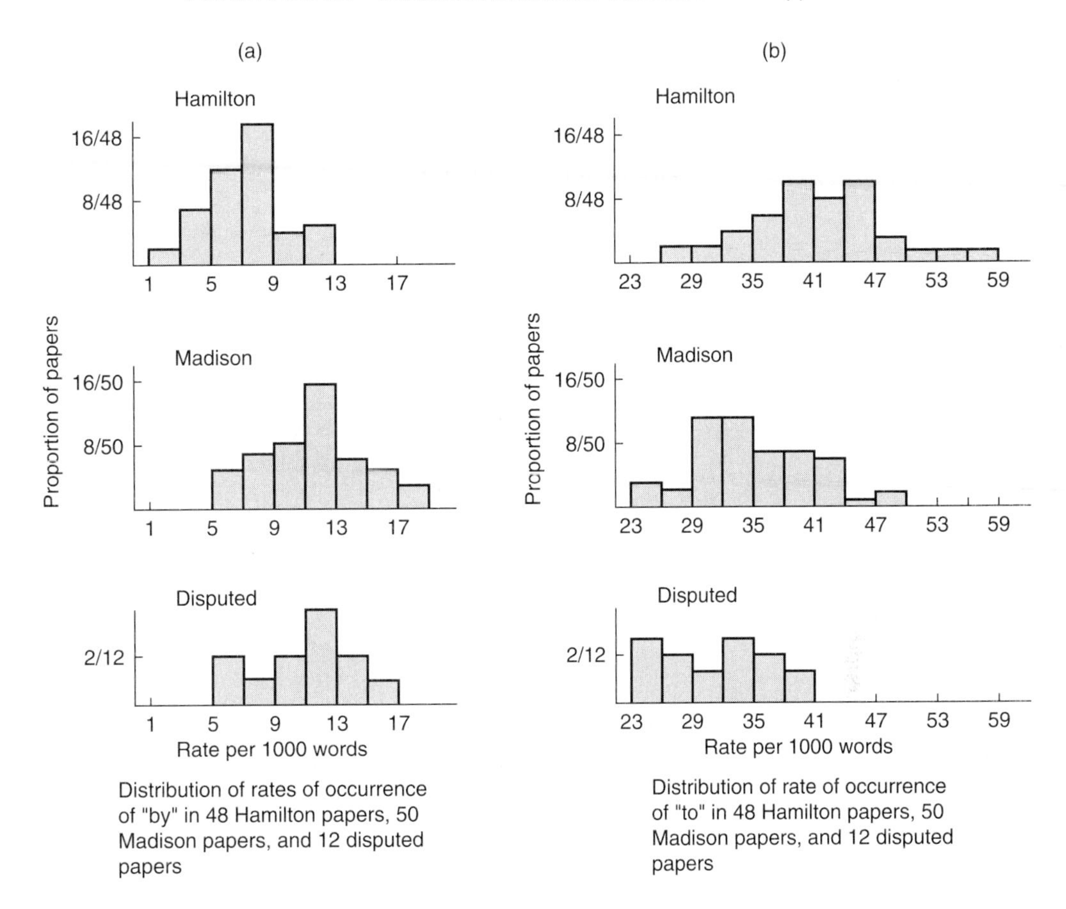

Distribution of rates of occurrence of "by" in 48 Hamilton papers, 50 Madison papers, and 12 disputed papers

Distribution of rate of occurrence of "to" in 48 Hamilton papers, 50 Madison papers, and 12 disputed papers

Several types of frequency distributions commonly occur. If each category or interval of the distribution has approximately the same number of values, the distribution is called a *uniform distribution*. For example, the following bar graph shows the frequency distribution for the outcomes of rolling a die 3000 times. The frequency of each outcome varies between approximately 480 and 520.

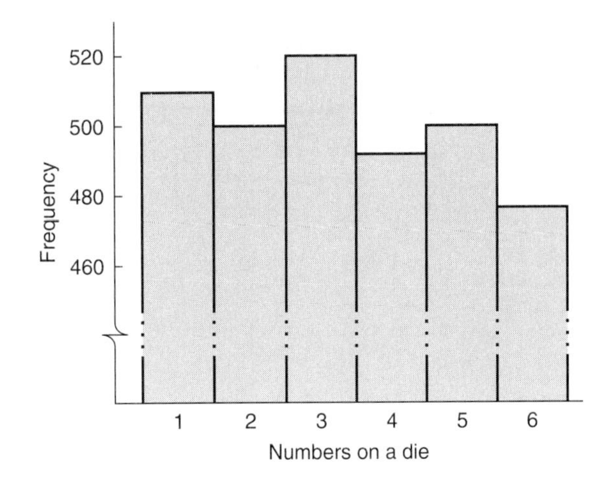

One of the most useful distributions in statistics is the *normal distribution.* This distribution often occurs in the physical, social, and biological sciences. The graph of a normal distribution is the familiar bell-shaped curve. This curve is symmetric about the mean, with a gradual decrease at both ends. The heights of a large sample of adult men will be *normally distributed,* as shown in the following graph.

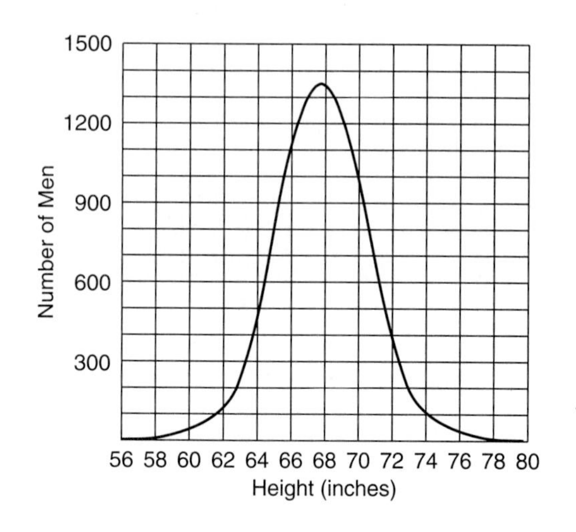

1. When four coins are tossed, there are six ways to get exactly two heads. Think of the coins as being a penny, nickel, dime, and quarter. The penny and dime may be heads and the nickel and quarter tails; the penny and nickel heads and the dime and quarter tails; and so forth. Fill in the remaining four ways of getting exactly two heads. Then fill in all other possible ways the coins can come up.

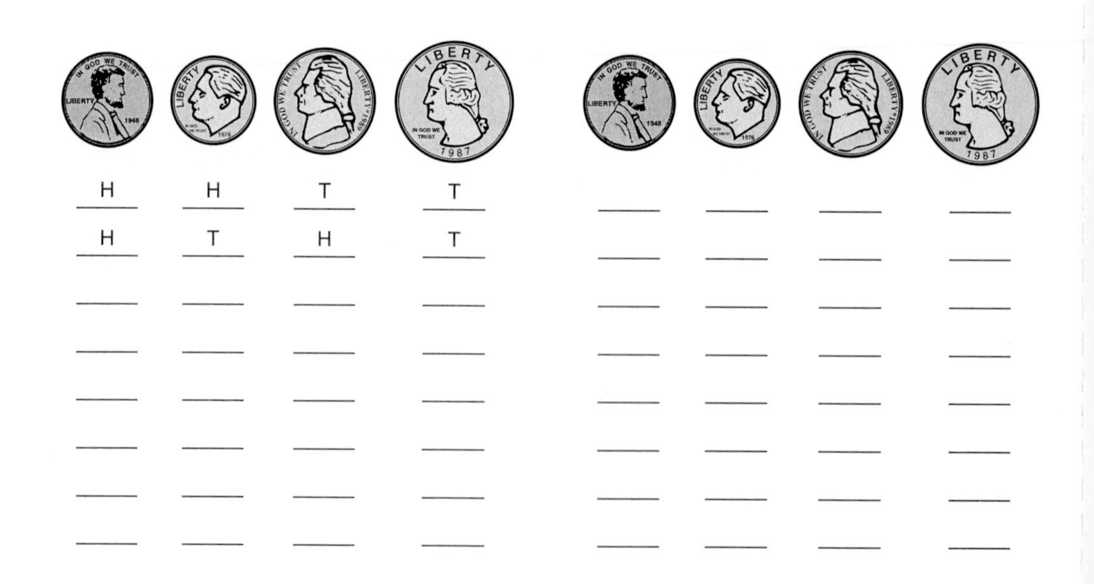

This graph shows the number of different outcomes for tosses of four coins at a time. The frequency of 6 above 2 heads corresponds to the six different combinations of two heads. There are frequencies of 1 above 0 heads and 4 heads, because these two events can occur only if the tosses result in all tails or all heads. This distribution looks like a normal curve. As the number of coins being tossed increases, the distribution of outcomes theoretically gets close to a normal curve.

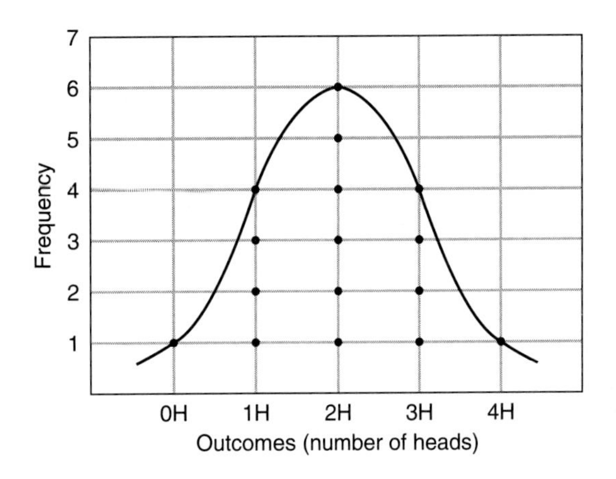

a. Use the table of random digits on Material Card 8 to simulate the tossing of six coins. Pick an arbitrary starting point in the table and, counting even digits as heads and odd digits as tails, record the number of heads in each block of six digits. Simulate 64 tosses of six coins, and record the number of heads for each toss in the table below. Each blank in the table will contain a digit from 0 to 6, depending on the number of heads counted in the toss.

1	2	3	4	5	6	7	8	9	10	11	12	13
14	15	16	17	18	19	20	21	22	23	24	25	26
27	28	29	30	31	32	33	34	35	36	37	38	39
40	41	42	43	44	45	46	47	48	49	50	51	52
53	54	55	56	57	58	59	60	61	62	63	64	

b. Plot the frequency of each outcome on this grid. For example, the number of times you got two heads should be plotted above 2H on the graph. Theoretically, about 78 percent of the outcomes should be 2, 3, or 4 heads. What percent of your outcomes are 2, 3, or 4 heads?

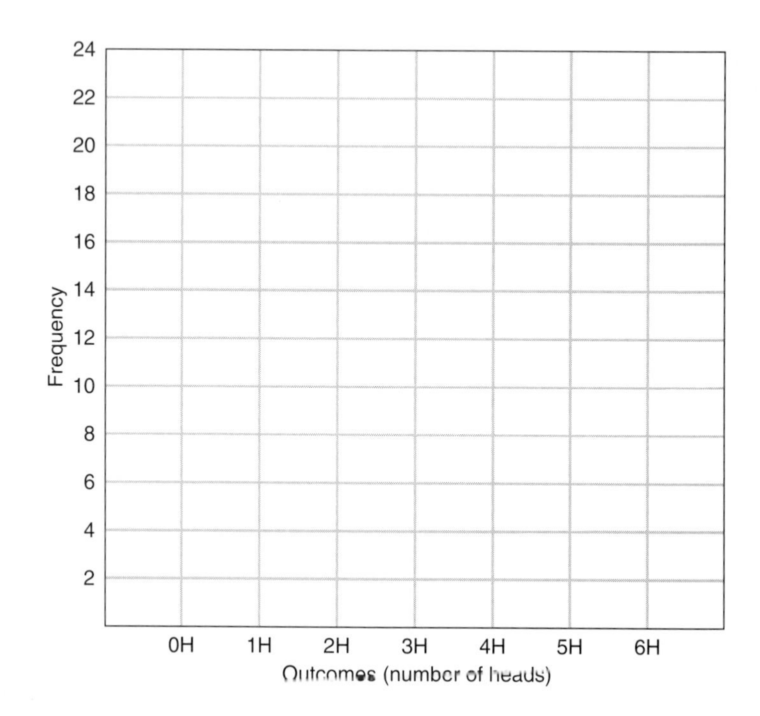

★ **c.** Suppose that an outcome of heads represents the birth of a girl and an outcome of tails represents that of a boy. In addition, suppose that each of the coin tosses in part a represents a family that is planning to have six children. Explain how you would use the data to predict approximately what percent of those families will have *at least* three girls.

2. When measurement such as weight, range of eyesight, or IQ are compiled for a large number of people, they are usually normally distributed. The following list contains the pulse rates of 60 students. Use these data to complete the bar graph below. Do the pulse rates appear to you to be close to a normal distribution?

Pulse rates (beats per minute)/Frequency

Pulse	53	56	57	58	61	62	63	64	65	66	67	68	69	70	71	72	73	74	75	76	77	78	79	80	81	82	84	86	89	91	92
Freq.	1	2	1	1	1	3	1	1	3	1	2	2	2	6	1	4	3	4	2	3	1	1	2	3	1	1	2	2	1	1	1

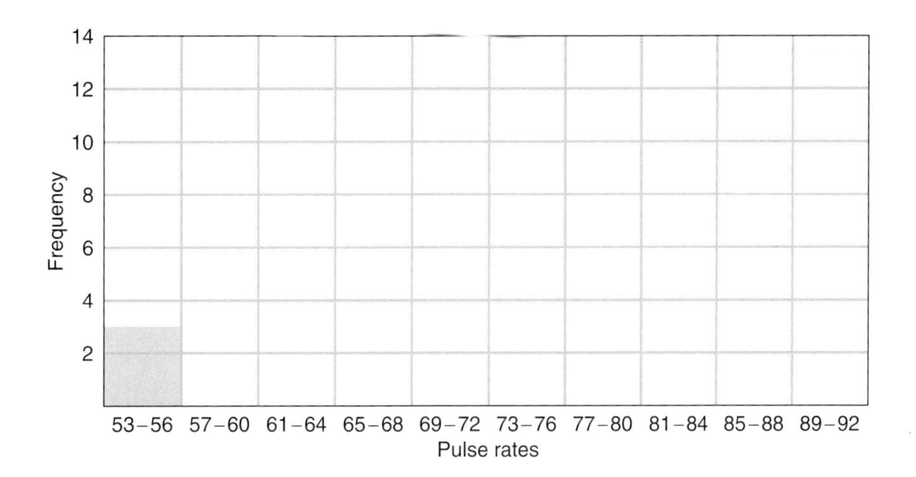

★ **a.** One property of a normal distribution is that 68 percent of the measurements are within 1 standard deviation on either side of the mean and 95 percent are within 2 standard deviations on either side of the mean, as illustrated below. The set of pulse rates has a mean of 72 and a standard deviation of 8.7. Does the distribution of pulse rates satisfy this property? Record your findings.

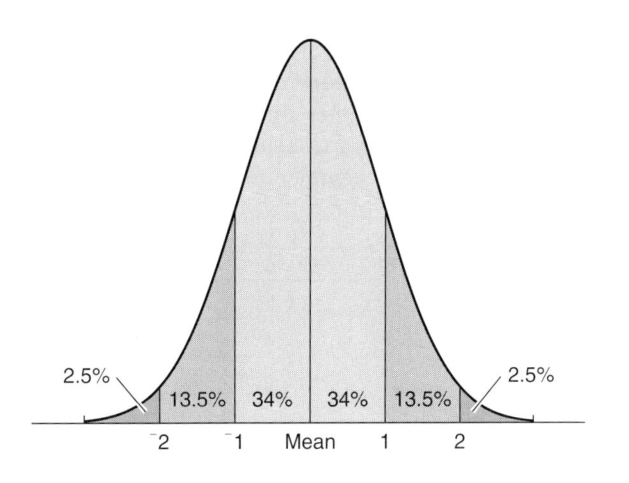

b. Even smaller numbers of measurements may tend to be normally distributed. Draw a bar graph of your classmates' pulse rates taken over a 10-second interval. One property of approximately normal curves is that the mean, median, and mode are approximately equal. Determine the mean, median, and mode for the pulse rates of your class. Are they approximately equal?

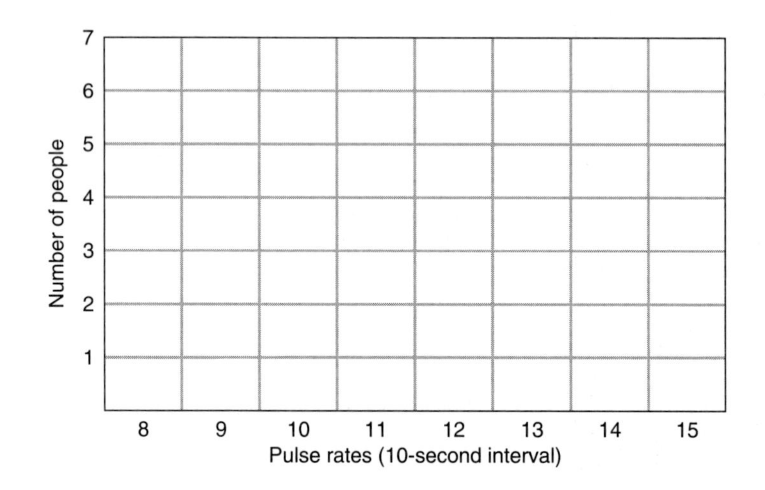

3. Some people are not aware of the uneven distribution of letters which occurs in words. One such person may have been Christofer Sholes, the inventor of the typewriter. He gave the left hand 56 percent of all the strokes, and he gave the two most agile fingers on the right hand two of the least often used letters of the alphabet, *j* and *k*. This table shows the frequencies, in terms of percentages of occurrence, of letters in large samples.

E	12.3%	S	6.6%	U	3.1%	B	1.6%	J	0.1%
T	9.6%	R	6.0%	P	2.3%	G	1.6%	Z	0.1%
A	8.1%	H	5.1%	F	2.3%	V	0.9%		
O	7.9%	L	4.0%	M	2.2%	K	0.5%		
N	7.2%	D	3.7%	W	2.0%	Q	0.2%		
I	7.2%	C	3.2%	Y	1.9%	X	0.2%		

★ **a.** There are 350 letters in the preceding paragraph. Count the number of times that *e, s, c, w,* and *k* occur and compute their percentages. Compare your answer with the percentages that have been computed for large samples.

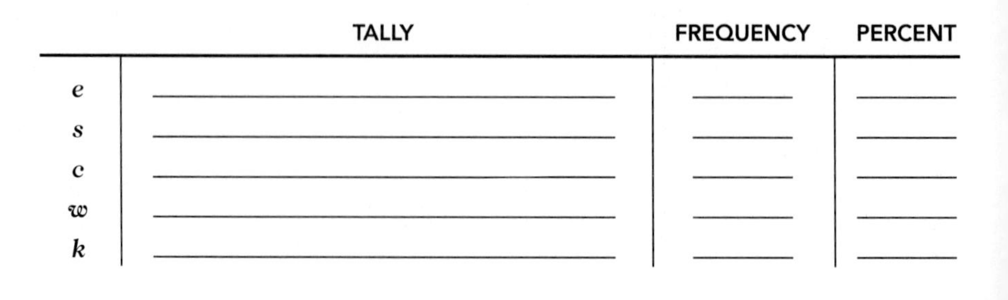

	TALLY	FREQUENCY	PERCENT
e	______	______	______
s	______	______	______
c	______	______	______
w	______	______	______
k	______	______	______

b. The most frequently occurring letter in the English language is *e*. In 1939 a 267-page novel entitled *Gadsby* was released by the Wetzel Publishing Company. The novel was known not for its literary merit, but rather for one distinctive feature: Not 1 of its 50,000 words contained the letter *e*. Try to compose a sentence of at least ten words that does not use the letter *e*.

★ **c.** In Morse code, letters are represented by dots and dashes so that they can be sent by electrical impulses. A dot consumes 1 time unit, a dash consumes 3 time units, and 1 time unit is needed for each space between symbols. In 1938 Samuel Morse assigned the symbols with the shortest time intervals to the letters that occurred most frequently, according to his sample of 12,000 words from a Philadelphia newspaper. Although Morse's code assignment is more efficient than it would have been if he had assigned letters haphazardly, it can still be improved. Devise a new and more efficient assignment of Morse's code symbols for the nine letters in this table. How many fewer time units do you require for these nine letters than Samuel Morse did?

CODE SYMBOL	NUMBER OF TIME UNITS	LETTERS	REASSIGNED CODE SYMBOL
•	1	e	________
—	3	t	________
• —	5	a	________
— — —	11	o	________
— •	5	n	________
••	3	i	________
•••	5	s	________
• — •	7	r	________
••••	7	h	________

4. It is surprising to learn that the leading digits of numbers *do not occur with the same frequency*—that is, are not evenly distributed.[5] Find a partner to help you tally the first digit in the list of metropolitan populations that follows and try this experiment. For each number in the table beginning with a 1, 2, 3, or 4, you win; for each number beginning with a 5, 6, 7, 8, or 9, your partner wins. Record your tallies in the table on page 175 and determine the total for each digit to see who wins.

[5]An explanation of this phenomenon can be found in W. Weaver, *Lady Luck* (New York: Doubleday, 1963), 270–277.

POPULATIONS OF 166 SELECTED METROPOLITAN AREAS

Philadelphia to Atlantic City	
Cecil	82,522
Chester	421,686
New Castle	482,807
Bucks	587,942
Gloucester	247,897
Burlington	420,323
Montgomery	719,718
Atlantic	238,047
Cape May	98,069
Cumberland	140,341
Camden	505,204
Salem	64,912
Delaware	542,593
Philadelphia	1,436,287
Phoenix-Mesa	
Maricopa	2,784,075
Pinal	146,929
Pittsburgh	
Butler	170,785
Westmoreland	372,103
Washington	205,566
Fayette	144,847
Beaver	184,406
Allegheny	1,268,446
Portland-Salem	
Clark	326,943
Washington	399,697
Yamhill	82,085
Polk	61,560
Clackamas	334,732
Columbia	44,416
Marion	268,541
Multnomah	631,082
Providence-Warwick-Pawtucket	
Washington	120,649
Bristol	49,114
Kent	161,811
Providence	574,038
Raleigh-Durham-Chapel Hill	
Wake	570,615
Johnston	106,582
Franklin	44,743
Orange	110,116
Chatham	45,406
Durham	202,411
Richmond-Petersburg	
Powhatan	21,950
Hanover	81,975
Goochland	17,823
New Kent	13,052
Chesterfield	245,915
Henrico	246,052
Charles City	7,092
Dinwiddie	24,657
Prince George	30,135
Colonial Heights	16,955
Hopewell	22,529
Richmond	194,173
Petersburg	34,724
Rochester	
Wayne	94,977
Orleans	44,518
Livingston	66,000
Ontario	99,662
Genesee	60,654
Monroe	716,072
Sacramento-Yolo	
Placer	229,259
El Dorado	158,502
Sacramento	1,144,202
Yolo	153,849
St. Louis	
St. Charles	272,353
Lincoln	36,556
Warren	24,600
Monroe	26,586
Jefferson	195,675
Franklin	91,763
Clinton	35,591
Jersey	21,373
Madison	259,351
St. Louis County	998,696
St. Clair	261,941
City of St. Louis	339,316
Salt Lake City-Ogden	
Davis	233,013
Salt Lake	850,667
Weber	184,065
San Antonio	
Comal	73,391
Wilson	31,423
Guadalupe	80,472
Bexar	1,353,052
San Diego	
San Diego	2,780,592
San Francisco Bay Area	
Contra Costa	918,200
Sonoma	433,304
Solano	377,415
Santa Clara	1,641,215
San Mateo	700,765
Napa	119,288
Alameda	1,400,322
Santa Cruz	242,994
San Francisco	745,774
Marin	236,770
Sarasota-Bradenton	
Manatee	239,682
Sarasota	303,400
Scranton-Wilkes-Barre-Hazleton	
Wyoming	29,149
Columbia	64,120
Luzerne	313,767
Lackawanna	208,455
Seattle Area	
Snohomish	587,783
Thurston	202,255
Kitsap	232,623
Island	70,319
Pierce	676,505
King	1,654,876
Springfield, Mass.	
Hampshire	149,384
Hampden	439,609
Stockton-Lodi	550,445
San Joaquin	550,445
Syracuse	
Madison	71,069
Oswego	124,006
Cayuga	81,264
Onondaga	458,301
Tampa-St. Petersburg-Clearwater	
Hernando	127,227
Pasco	325,824
Hillsborough	925,277
Pinellas	878,231
Toledo	
Fulton	41,895
Wood	119,498
Lucas	448,542
Tucson	
Pima	790,755
Tulsa	
Rogers	68,128
Wagoner	55,259
Creek	67,142
Tulsa	543,539
Osage	42,838
Washington-Baltimore	
Loudoun	143,940
Spotsylvania	83,692
Stafford	87,055
Calvert	71,877
Manassas Park	8,711
King George	17,236
Manassas	35,336
Howard	236,388
Frederick	186,777
Carroll	149,697
Prince William	259,827
Berkeley	70,970
Culpeper	33,083
Harford	214,668
Queen Anne's	39,672
Charles	117,963
Warren	30,126
Jefferson	41,368
Fredericksburg	21,686
Fairfax County	929,239
Anne Arundel	476,060
Fauquier	54,109
Montgomery	840,879
Prince George's	777,811
Alexandria	118,300
Clarke	12,779
Falls Church	10,042
Washington County	127,352
Baltimore County	721,874
City of Fairfax	20,697
Arlington	177,275
City of Baltimore	645,593
Washington	523,124
West Palm Beach-Boca Raton	
Palm Beach	1,032,625
Wichita	
Butler	61,932
Sedgwick	448,050
Harvey	34,361
Youngstown-Warren	
Columbiana	111,521
Trumbull	225,066
Mahoning	255,165

FIRST DIGIT OF NUMBER	1	2	3	4	5	6	7	8	9
Tallies									
Totals									

a. Would you have won the game if you had used the digits 1, 2, and 3 only? Use the results from the table to make one or more conjectures regarding the occurrence of the leading digits of large numbers from sets of data.

b. The populations in the list of metropolitan areas are four-, five-, six-, and seven-digit numbers. The least number is 7,092 and the greatest number is 2,784,075. In the table below, all numbers from 7,092 to 2,784,075 have been separated into six intervals. There are 2,908 numbers (9,999 − 7,092 + 1) in the first interval and all have leading digits of 7, 8, and 9. The second interval has numbers with leading digits 1, 2, 3, 4. Determine the number of numbers in each of the remaining intervals and whether or not you win with leading digits 1, 2, 3, and 4.

INTERVAL	NUMBER OF NUMBERS	WIN OR LOSE
7,092 to 9,999	2,908	LOSE
10,000 to 49,999	______	______
50,000 to 99,999	______	______
100,000 to 499,999	______	______
500,000 to 999,999	______	______
1,000,000 to 2,784,075	______	______

★ **c.** Considering *all* numbers from 7,092 to 2,784,075, how do the number of numbers with leading digits 1, 2, 3, and 4 compare to those with leading digits of 5, 6, 7, 8, and 9? Describe how this result can explain why it is likely that the person choosing digits 1, 2, 3, and 4 will win using the list of populations of metropolitan areas.

JUST FOR FUN

CRYPTANALYSIS

Cryptanalysis is the analysis and deciphering of cryptograms, codes, and other secret writings. Successful cryptanalysis has been important in gaining many military victories and in preventing crime and espionage. One of the earliest cryptographic systems known was used by Julius Caesar and is called the Caesar cipher. Enciphering a message using this system involves replacing each letter of the alphabet by the letter that is 3 letters beyond it: A is replaced by D, B by E, C by F, and so forth. Use the Caesar cipher to decipher the statement on this scroll.

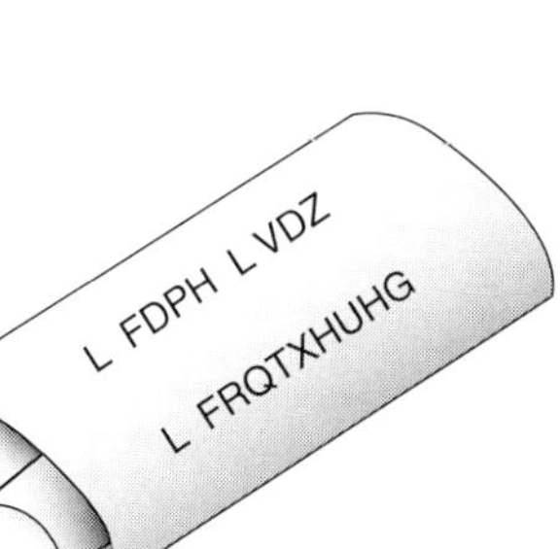

It is not known why Caesar selected 3 as the number of places to shift the alphabet. Any whole number greater than 1 can be used.

One way of improving cryptograms is to eliminate word lengths. The message on the next page is written in groups of 5 letters. It was enciphered by shifting each letter of the alphabet the same number of letters to the right.

OITQT MWAPW EMLUM VWNAK QMVKM BPIBE MQOPQ VOIVL UMIAC

ZQVOI VMEWZ BPEPQ TMVME BWVKW VDQVK MLITI ZOMXZ WXWZB

QWVWN BPMUB PIBEM QOPQV OIVLU MIACZ QVOIZ MBPMW VTGQV

DMABQ OIBQW VABPI BIZME WZBPE PQTMK PIZTM AAQVO MZ

One method of determining the amount of shift is to make a frequency distribution of the letters in the message and compare it to a frequency distribution of letters in common use. The following bar graph shows the percentage of occurrence of each letter (rounded to the nearest whole number) in a large sample of unencrypted words. There are some important patterns. Among the high-frequency letters, we see that three (A, E, and I) are 4 letters apart; two (N and O) are next to each other; and three (R, S, and T) are side by side. Among the low-frequency letters, two (J and K) occur as a pair, and six (U, V, W, X, Y, and Z) are side by side.

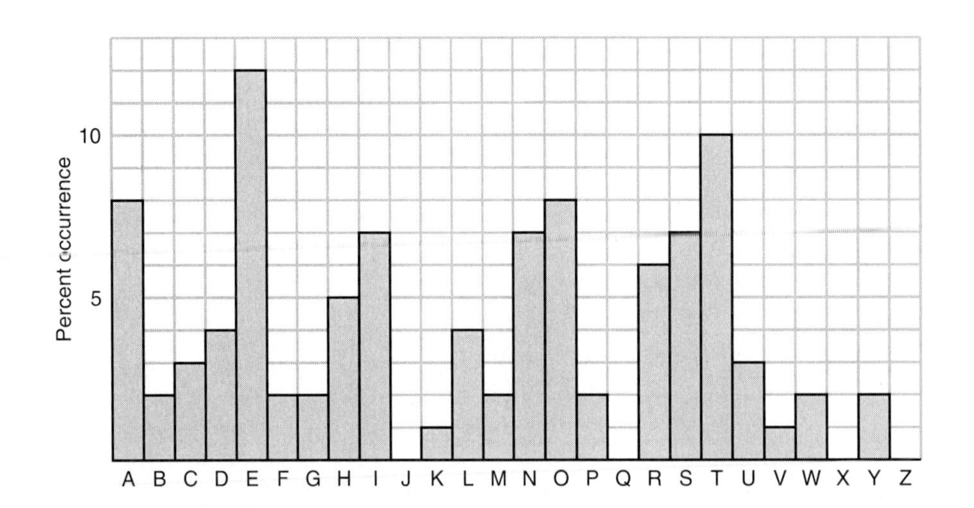

★ Make a frequency distribution of letters in the coded message. Predict the amount of shift by comparing this distribution with the one above, and by looking for patterns of high- and low-frequency letters. Try your prediction by deciphering a few words. When you have found the correct amount of shift, decipher the message.[6]

[6]For further techniques in cryptography, see A. Sinkov, *Elementary Cryptanalysis* (Washington, DC: Mathematical Association of America, 1966).

IDEAS FOR THE ELEMENTARY CLASSROOM

SUGGESTED CLASSROOM ACTIVITY: STUDENT-CENTERED DATA COLLECTION

Collecting, organizing, displaying, describing and interpreting of data can be done at all grade levels. At early grade levels, children can form living graphs as they stand in rows according to number of pets, month of birth, blouse or shirt color, and so forth. The "living graph" can be replaced by a similar display on a chart, with names in place of the children. Many mathematical terms arise in a discussion of the results, such as *more, less, fewer, greatest, how many, average,* and *frequency.*

As students get older and develop greater curiosity about themselves, their peers, and their environment, the list of topics for statistical study becomes much longer. Issues such as favorite television program, number of hours spent watching TV each day, and favorite advertisement can be explored for an entire school rather than one classroom.[7] This can help students understand the need for random sampling techniques.

PUZZLER

Label all faces of two cubes so that when the cubes are rolled, the sum of the top two numbers will always be 0, 1, 2, or 3, and all of these sums are equally likely to occur.

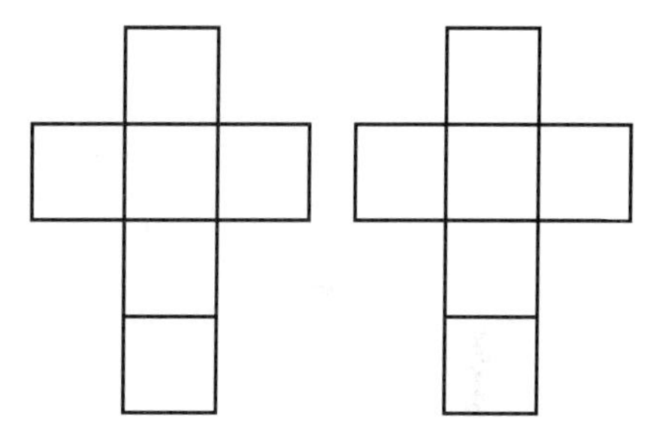

Readings for More Classroom Ideas

Bankard, Dianne, and Francis Fennell. "Ideas: Measurements, Surveys and Graphing." *Arithmetic Teacher* 39 (September 1991): 26–33.

Barson, Alan, and L. Barson. "Ideas: Collecting Data." *Arithmetic Teacher* 35 (March 1988): 19–24.

Brosnan, Patricia A. "Implementing Data Analysis in a Sixth-Grade Classroom." *Mathematics Teaching in the Middle School* 1 (January–February 1996): 622–628.

Bryan, Elizabeth H. "Exploring Data with Box Plots." *Mathematics Teacher* 81 (November 1988): 658–663.

Burrill, Gail, et al. *Data Analysis and Statistics Across the Curriculum,* in the *Curriculum and Evaluation Standards for School Mathematics Addenda Series.* Reston, VA: National Council of Teachers of Mathematics, 1992.

[7]An interesting example of a statistical study done by middle school students is described by David Pagni in "A Television Programming Challenge: A Cooperative Group Activity That Uses Mathematics," *Arithmetic Teacher* 36 (January 1989): 7–9.

Finlay, Ellen, and Ian Lowe. *Chance and Data: Exploring Real Data.* Mathematics Curriculum and Teaching Program. Carlton, Australia: Curriculum Corporation, 1993.

Friel, Susan N., and William T. O'Connor. "Sticks to the Roof of Your Mouth?" *Mathematics Teaching in the Middle School* 4 (March 1999): 404–411.

Greeley, Nansee, and Theresa R. Offerman. "Words, Words, Words, Ancient Communication." *Mathematics Teaching in the Middle School* 3 (February 1998): 358–362.

Hitch, Chris, and Georganna Armstrong. "Daily Activities for Data Analysis." *Arithmetic Teacher* 41 (January 1994): 242–245.

Hofstetter, Elaine B., and Laura A. Sgroi. "Data with Snap, Crackle, and Pop." *Mathematics Teaching in the Middle School* 1 (March–April 1996): 760–764.

Lindquist, Mary M., et al. *Making Sense of Data,* in the *Curriculum and Evaluation Standards for School Mathematics Addenda Series: Grades K–6.* Reston, VA: National Council of Teachers of Mathematics, 1992.

Lovitt, Charles, and Ian Lowe. *Chance and Data: Investigations,* vol. II. Mathematics Curriculum and Teaching Program. Carlton, Australia: Curriculum Corporation, 1993.

O'Keefe, James J. "The Human Scatterplot." *Mathematics Teaching in the Middle School* 3 (November–December 1997): 208–209.

Pagni, David L. "A Television Programming Challenge: A Cooperative Group Activity That Uses Mathematics." *Arithmetic Teacher* 36 (January 1989): 7–9.

Paull, Sandra. "Not Just An Average Unit." *Arithmetic Teacher* 38 (December 1990): 54–58.

Shaw, Jean M. "Let's Do It: Dealing with Data." *Arithmetic Teacher* 31 (May 1984): 9–15.

Shaw, Jean M. "Let's Do It: Making Graphs." *Arithmetic Teacher* 31 (January 1984): 7–11.

Speer, William R. "Exploring Random Numbers." *Teaching Children Mathematics* 3 (January 1997): 242–246.

Taylor, Judith V. "Young Children Deal with Data." *Teaching Children Mathematics* 4 (November 1997): 146–149.

Young, Sharon. "Ideas: Bicycle Projects." *Arithmetic Teacher* 38 (September 1990): 23–33.

Young, Sharon. "Ideas: Dinosaur Projects." *Arithmetic Teacher* 38 (December 1990): 23–33.

CHAPTER 8

PROBABILITY

Students should actively explore situations by experimenting and simulating probability models. . . . Probability is rich in interesting problems that can fascinate students and provide settings for developing or applying such concepts as ratios, fractions, percents, and decimals.[1]

ACTIVITY SET 8.1

PROBABILITY EXPERIMENTS

Purpose To determine probabilities experimentally and then analyze the experiments to determine the theoretical probabilities, when possible

Materials A penny, 10 identical tacks, game grids on Material Cards 9 and 10, and the table of random digits on Material Card 8

Activity

> The word *stochastic,* strange sounding but currently very popular in scientific circles, means random, chancy, chaotic. It is pronounced "stoh-kastic." The stochastization of the world (forgive this tongue-twister) means the adoption of a point of view wherein randomness or chance or probability is perceived as real, objective and fundamental aspect of the world. . . .
>
> Of the digits that crowd our daily papers many have a stochastic basis. We read about the percentage of families in New York City that are childless, the average number of cars owned by four-person families in Orlando, Florida, the probability that a certain transplant operation will succeed. We read about the odds that Nick the Greek is offering on a certain horse race, a market survey that estimated the monthly gross of a fast-food store in a certain location, and the fact that in a certain insurance pool favorable experience has lowered the rate by $.82 per thousand per month. It is implied that a certain attitude is to be engendered by these disclosures, that a course of action should be set in motion. If it is reported that the English scores of Nebraskan tenth-graders are such and such while that of Iowan

[1] *Curriculum and Evaluation Standards for School Mathematics* (Reston, VA: National Council of Teachers of Mathematics, 1989), 109.

tenth-graders are this and that, then presumably someone believes that something ought to be done about it.

The stochastization of the world so permeates our thinking and our behavior that it can be said to be one of the characteristic features of modern life. . . . [2]

The importance of stochastization (probability) in the education experience of youth has become widely recognized. The *Curriculum and Evaluation Standards for School Mathematics*[3] recommends that probability activities be included in all grades, kindergarten through 12th, and that a spirit of investigation, exploration, and discussion permeate this study. Lack of probabilistic thinking in the school curriculum appears to result in a lack of "probability sense" in students, and even misconceptions about probability.

In this activity set, we will consider both *experimental probabilities,* which are determined by observing the outcomes of experiments, and *theoretical probabilities,* which are determined by mathematical calculations. The theoretical probability of rolling a 4 on a die is $\frac{1}{6}$ because there are 6 faces on a die, each of which is equally likely to turn up, and only one of these faces has a 4. The experimental probability of obtaining a 4 is found by rolling a die many times and recording the results. Dividing the number of times a 4 comes up by the total number of rolls gives us the experimental probability.

1. Suppose someone wanted to bet that a tack would land with its point up when dropped on a hard surface. Would you accept the bet? Before you wager any money, try the following experiment. Take 10 identical tacks and drop them on a hard surface. Count the number of tacks that land point up. Do this experiment 10 times and record your results in the table on the next page.

[2]P. J. Davis and R. Hersh, *Descartes Dream* (New York: Harcourt Brace Jovanovich, 1986), 18–19.
[3]This document was written by the National Council of Teachers of Mathematics to provide guidelines for reform in school mathematics in the next decade.

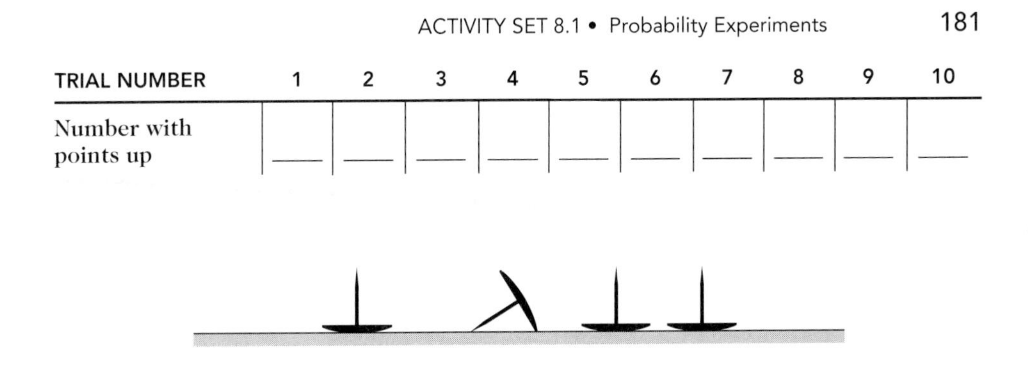

TRIAL NUMBER	1	2	3	4	5	6	7	8	9	10
Number with points up	___	___	___	___	___	___	___	___	___	___

a. Out of the 100 tacks you dropped, how many landed point up? On the basis of this experiment, explain why you would or would not accept the bet.

b. What is the experimental probability that this type of tack will land point up?

$$\frac{\text{Number with point up}}{\text{Total number dropped}} =$$

c. About how many tacks would you expect to land point up if you dropped 300?

d. What is the experimental probability that this type of tack will land point down? How is this probability related to the probability in part b?

2. Games of chance like the following are common at carnivals and fairs. A penny is spun on a square grid, and if it lands inside a square you win a prize. Compare the diameter of a penny to the width of the Two-Penny Grid (Material Card 9). Do you think there is a better chance that the penny will land inside a square or on a line? Make and record a conjecture before you begin the experiment described on the next page.

Conjecture:

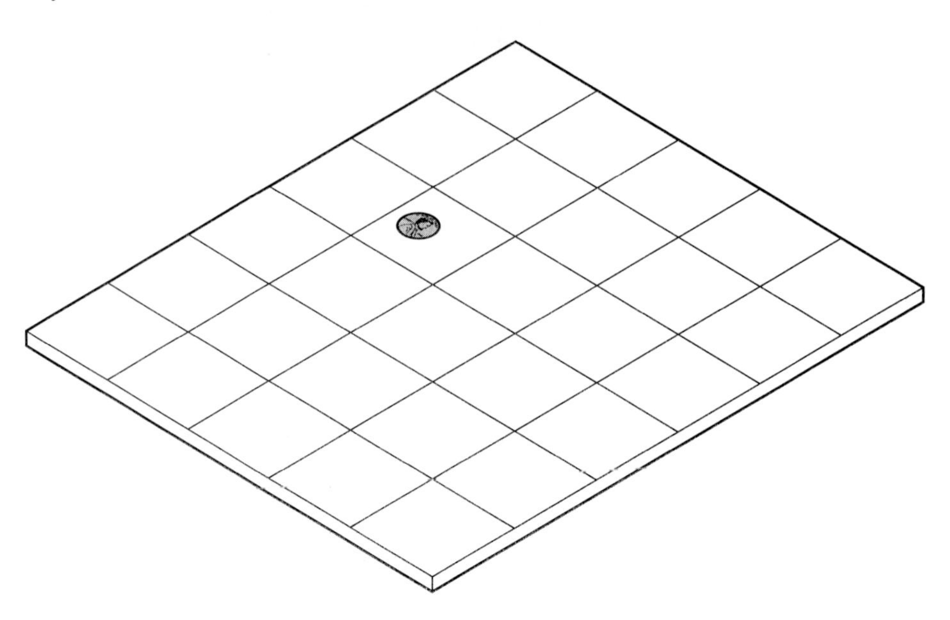

a. Spin a penny 24 times on the Two-Penny Grid and tally the results below.

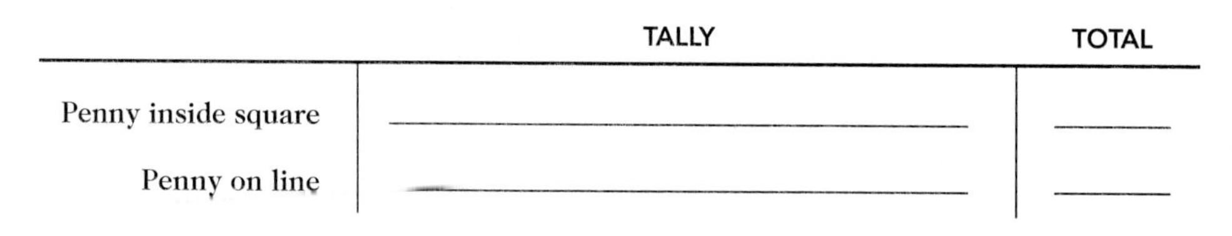

	TALLY	TOTAL
Penny inside square	____________________	______
Penny on line	____________________	______

b. Based on your experiment, compute and record the experimental probability that a penny will land inside a square.

c. The length of a side of a square is twice the diameter of a penny. Which of the points labeled A, B, and C will be winning points if they represent the center of a penny?

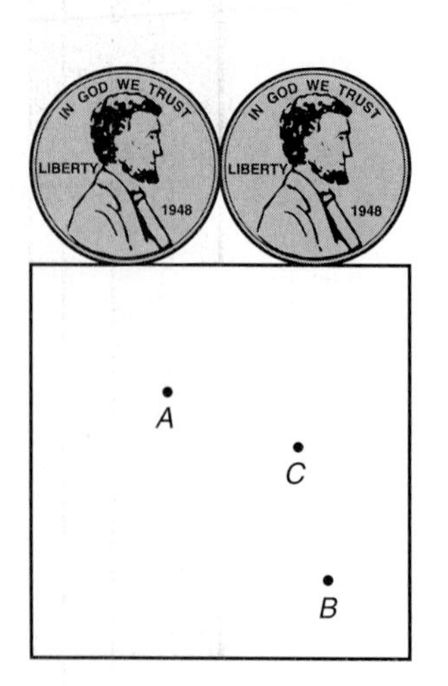

★ d. Shade the region of the square in which the center of every *winning* penny must fall. What fractional part of the square is the shaded region?

★ e. Compare the area of the shaded region to the area of the whole square to determine the theoretical probability that a penny will land inside a square. Is this a fair game, in the sense that there is a 50 percent chance of winning?

3. The sides of the squares on the Three-Penny Grid (Material Card 10) are three times the diameter of a penny. When a penny is spun on this grid, do you think the chances of its landing inside the lines are better than the chances of its landing on a line? Make and record a conjecture before you try the game.

Conjecture:

a. Spin a penny 24 times, tally your results, and compute the experimental probability of not landing on a line.

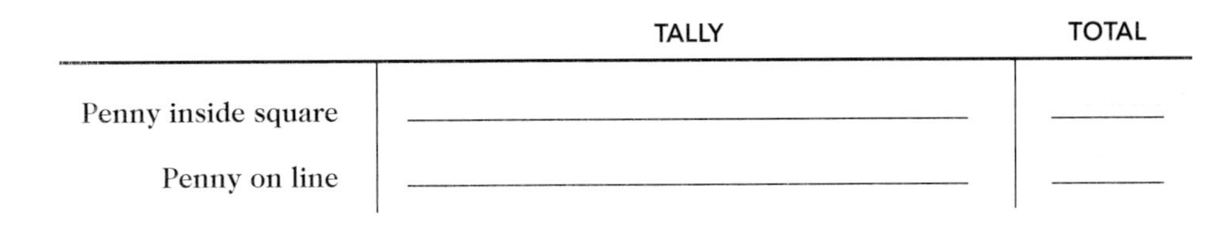

	TALLY	TOTAL
Penny inside square	______	______
Penny on line	______	______

★ **b.** Shade the region of the square in which the centers of winning pennies must fall. What is the theoretical probability that a penny will be a winning penny?

c. Explain how you would determine the theoretical probability that a spinning penny will not touch the lines on a Four-Penny Grid. Is this a fair game?

4. Cut out 3 cards of the same size. Label both sides of one with the letter A, both sides of the second with the letter B, one side of the third with A, and the other side of the third with B. Select a card at random and place it on the table. The card on the table will have either an A or a B facing up. What is the probability that the letter facing down will be different from the letter facing up? Make a conjecture and record it below. Then repeat this experiment 24 times and record the number of times that the letter facing down is different from the letter facing up in the table on the next page. Compute and record the experimental probability.

Conjecture:

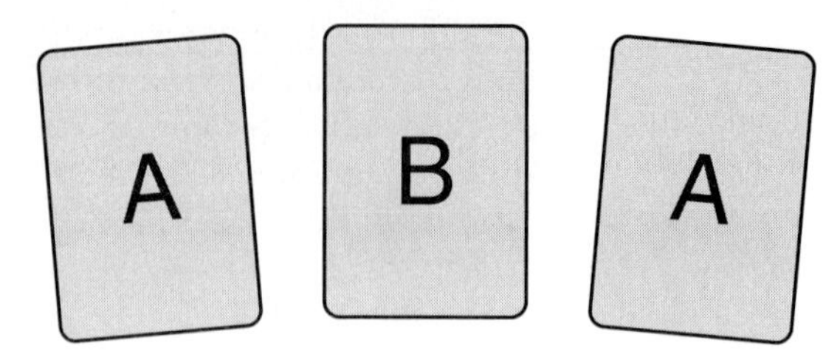

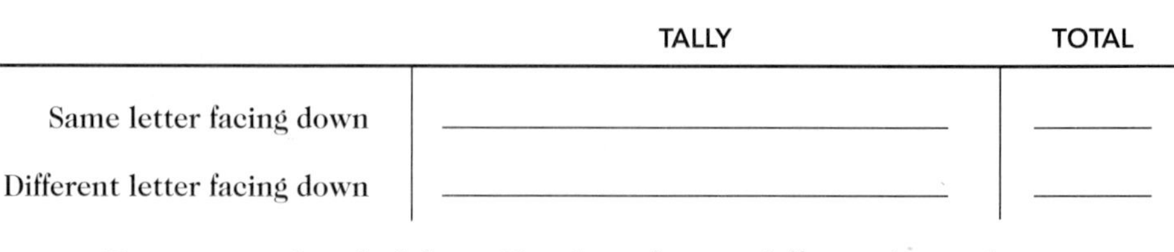

	TALLY	TOTAL
Same letter facing down	______	______
Different letter facing down	______	______

Experimental probability: $\dfrac{\text{Number of times different letter down}}{\text{Total number of random draws}} =$

Once a card has been selected and it is clear which letter is facing up, people often feel that there is a 50-50 chance that the letter facing down will be different. This is a false assumption. Notice that the 3 cards have 6 faces altogether. To distinguish between the two sides of the A-A card, let's refer to one side as A_1 and the other as A_2. Similarly, the two sides of the B-B card will be referred to as B_1 and B_2. We will continue to refer to the sides of the A-B card as A and B. When the first card is selected, 1 of 6 sides is facing up: A_1, A_2, B_1, B_2, A, B. For each of these 6 sides, record the letter of the side that would be facing down. Then answer the following questions.

Outcome	1	2	3	4	5	6
Top face	A_1	A_2	B_1	B_2	A	B
Bottom face	______	______	______	______	______	______

★ **a.** In how many of these outcomes is the top face of the card different from the bottom face? (Remember, the faces are labeled with only As and Bs.)

b. What is the probability of selecting a card on which the letter on the top face is different from the letter on the bottom face?

★ **c.** Suppose that a card has been selected and an A is face up. List the 3 outcomes from the table for this event.

d. What is the probability that the letter on the bottom face of the card in part c is a B?

★ **e.** Does knowing what letter is facing up help you determine the probability that the letter facing down is different? Explain.

5. Simulation: A student has 10 sweaters. If the student grabs a sweater at random each morning, what is the probability that the student will wear the same sweater more than once in a five-day week? Assign each sweater a digit from 0 through 9 and set up a simulation to determine the experimental probability. (*Hint:* Use 5 random digits selected from the table of random digits on Material Card 8 to represent 5 sweaters chosen at random in one week.)

JUST FOR FUN

RACETRACK GAME (2 or 3 players)

Each player chooses a track on the racetrack gameboard and places a marker on its starting square. In turn, the players roll two dice. If the sum of the numbers on the dice matches the number of the track on which the player's marker is located, the player may move 1 square toward the finish line. The first player whose marker reaches the last square on the track is the winner. Play the game a few times. Are you more likely to win on some tracks than on others?

Finish											
Start	2	3	4	5	6	7	8	9	10	11	12

Variation 1: Play the game as above, but with an individual gameboard and 2 markers for each player. A player's markers may be placed on different tracks or on the same track. Is there a way to place your markers so as to increase your chances of winning?

Variation 2: Use the sum of the numbers on three dice rather than two. Do some tracks have better winning records than others?

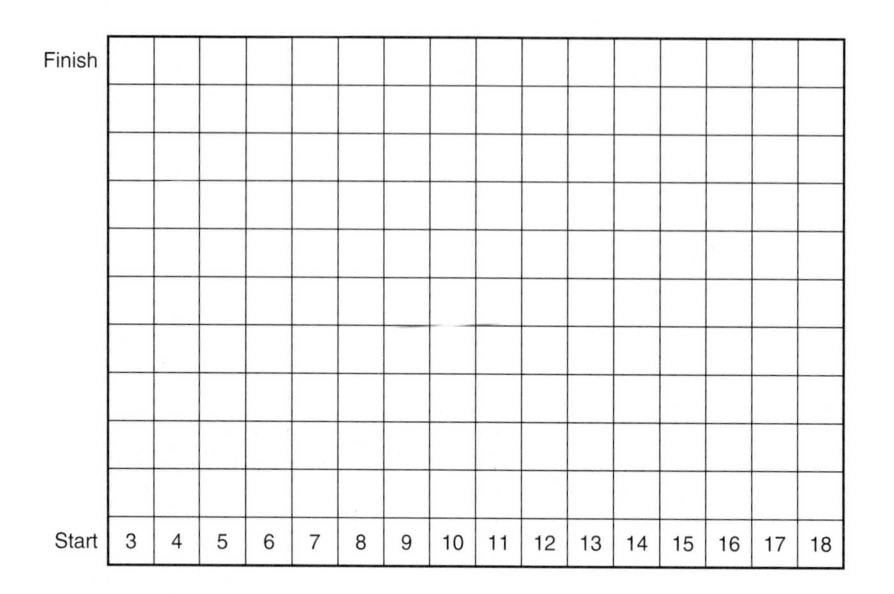

Variation 3: Use the product of the numbers on the dice instead of the sum. Will you need more tracks on the game mat? Will certain tracks give you a better chance of winning?

PUZZLER

Have a friend grasp four identical shoelaces (or pieces of string) in the middle so that four ends extend on each side of the hand. If you tie them in pairs on each side of your friend's fist, are your chances of having one big loop when the laces are released greater or less than 50 percent?

ACTIVITY SET 8.2

MULTISTAGE PROBABILITY EXPERIMENTS

Purpose To perform multistage probability experiments and then compare the resulting experimental probabilities with the corresponding theoretical probabilities

Materials Green and red tiles from the Manipulative Kit, four coins, a pair of dice, and the simulation spinners on Material Card 30

Activity It once happened in Monte Carlo that red came up on 32 consecutive spins of a roulette wheel. The probability that such a run will occur is about one out of 4 billion.

The probability of two or more events occurring is called *compound probability* or *multistage probability.* When one event does not influence or affect the outcome of another event, the two are called *independent events.* For example, because the second spin of a roulette wheel is not affected by the first spin, the two are considered independent events. When one event does influence the outcome of a second, the events are said to be *dependent events.*

Consider, for example, the probability of drawing 2 red tiles from a box containing 3 red tiles and 4 green tiles.

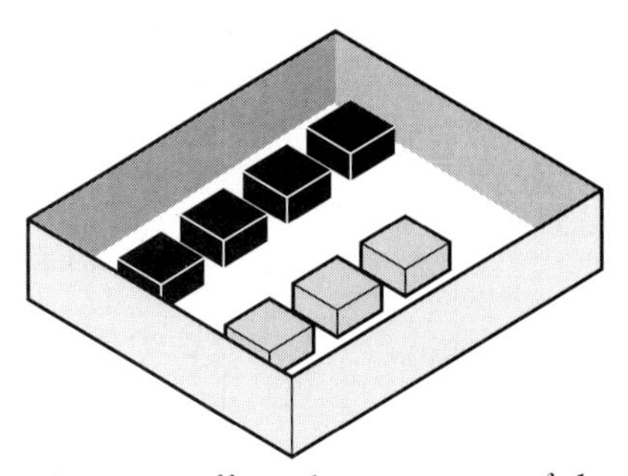

On the first draw, the probability of obtaining a red tile is $\frac{3}{7}$. If the first tile is replaced before the second draw, the probability of drawing another red tile is $\frac{3}{7}$, because the numbers of each color of tiles in the box are the same as they were on the first draw. These two drawings are *independent* events because the result of the first draw does not affect the outcome of the second draw. The probability of drawing 2 red tiles in this case is $\frac{3}{7} \times \frac{3}{7} = \frac{9}{49}$, as illustrated by the following probability tree.

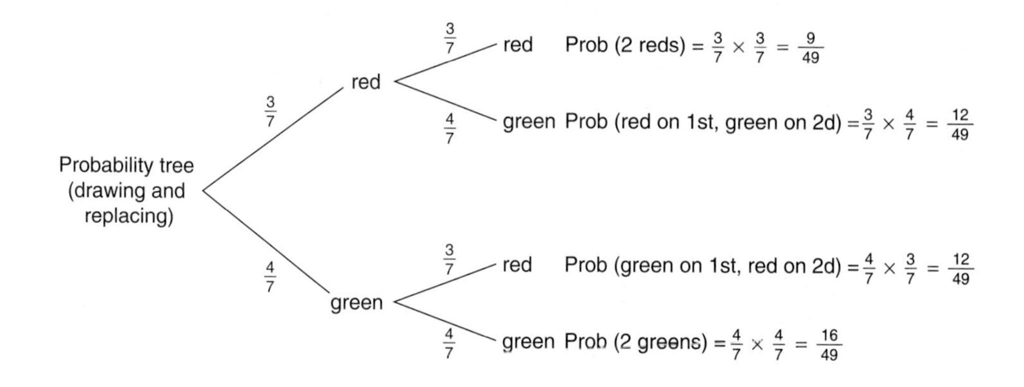

A completely different situation exists if the tile obtained on the first draw is not returned to the box before the second tile is drawn. In this case, the probability of drawing a red tile on the second draw depends on what happens on the first draw. That is, these events are *dependent.* For example, if a red tile is taken on the first draw, the probability of selecting a red tile on the second draw is $\frac{2}{6}$. In this case, the probability of drawing 2 red tiles is $\frac{3}{7} \times \frac{2}{6} = \frac{6}{42}$. The following probability tree illustrates this case.

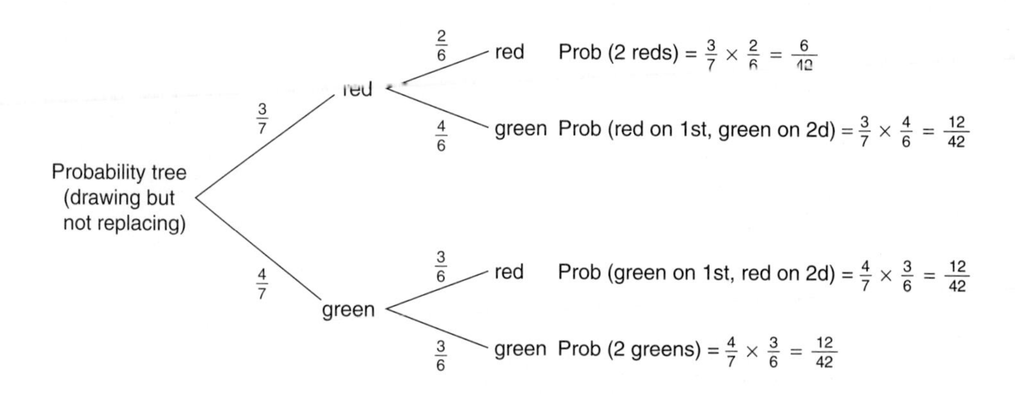

In the following activities, there are several experiments involving compound probabilities with both independent and dependent events. In each case, the experimental probability will be compared with the theoretical probability.

1. The following experiment has a result likely to be contrary to your intuition. Suppose that 3 red and 3 green tiles are placed in a container and that 2 are selected at random (both at once). What is the probability of getting 2 tiles of different colors? Make and record a conjecture about the probability, and then perform the experiment 40 times and tally your results on the next page.

   Conjecture:

	TALLY	TOTAL
Two tiles of different colors	______	______
Two tiles of the same color	______	______

a. Based on your results, what is the *experimental* probability of drawing the following?

2 tiles of different colors ______

2 tiles of the same color ______

★ b. One way to determine the *theoretical* probability is to systematically list all 15 possible pairs that could be drawn from the container. Complete the list of outcomes started below. (Notice that the tiles of each color have been numbered to preserve their identity.) Use the list to determine the theoretical probabilities by forming the following ratio. Compare these probabilities to the experimental probabilities from part a.

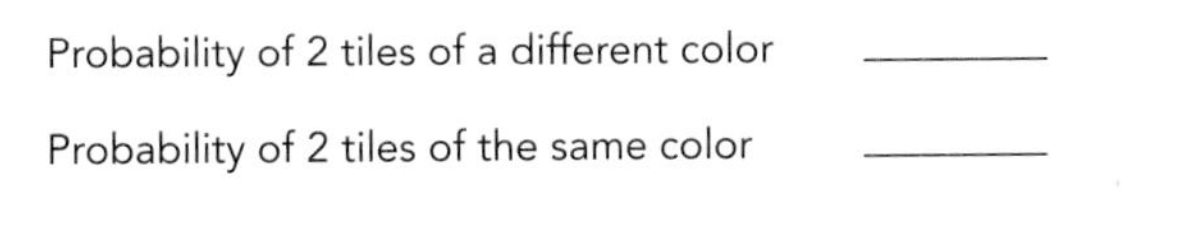

Probability of 2 tiles of a different color ______

Probability of 2 tiles of the same color ______

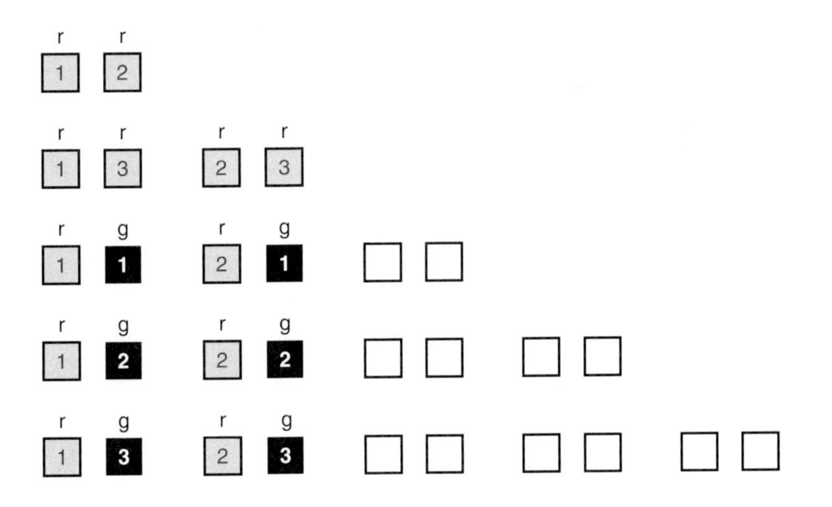

c. Drawing 2 tiles is the same as drawing 1 tile and then, without replacing it, drawing a second tile. The probability of drawing a red tile is $\frac{1}{2}$. The probability of then drawing a second red tile is $\frac{2}{5}$. The probability of drawing a red followed by a red is $\frac{1}{2} \times \frac{2}{5} = \frac{1}{5}$, as shown on the following probability tree. Determine the theoretical probabilities of the remaining three outcomes.

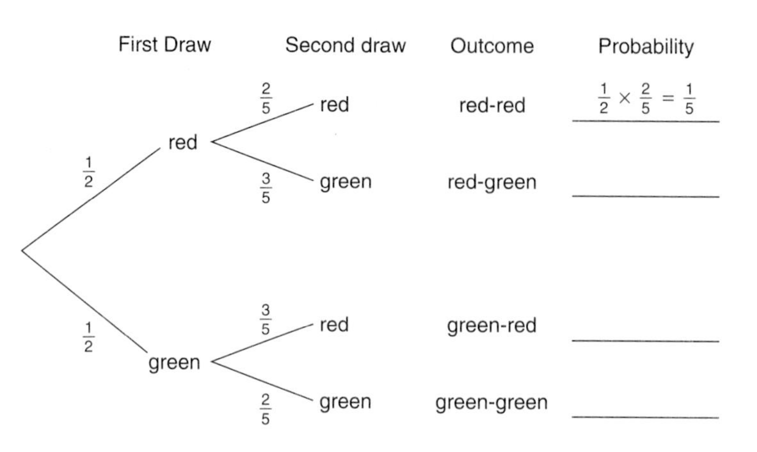

★ **d.** The theoretical probability of drawing 2 tiles of different colors can be obtained from the preceding tree diagram. Explain how this can be done and determine the probability. Compare this result to the theoretical probability determined in part b.

2. What is the probability that, on a toss of 4 coins, 2 will come up heads and 2 will come up tails? First make a conjecture about the probability and then perform the experiment by tossing 4 coins and recording the numbers of heads and tails in the table below. Do this experiment 32 times.
Conjecture:

	1	2	3	4	5	6	7	8	9	10	11	12	13	14	15	16	17	18	19	20	21	22	23	24	25	26	27	28	29	30	31	32
H																																
T																																

a. According to your table, what is the experimental probability of obtaining exactly 2 heads on a toss of 4 coins?

b. Tossing 4 coins at once is like tossing a single coin 4 times. Continue this tree diagram to show all possible outcomes when a coin is tossed 4 times in succession. There will be 16 possible outcomes. How many of these will have 2 heads and 2 tails?

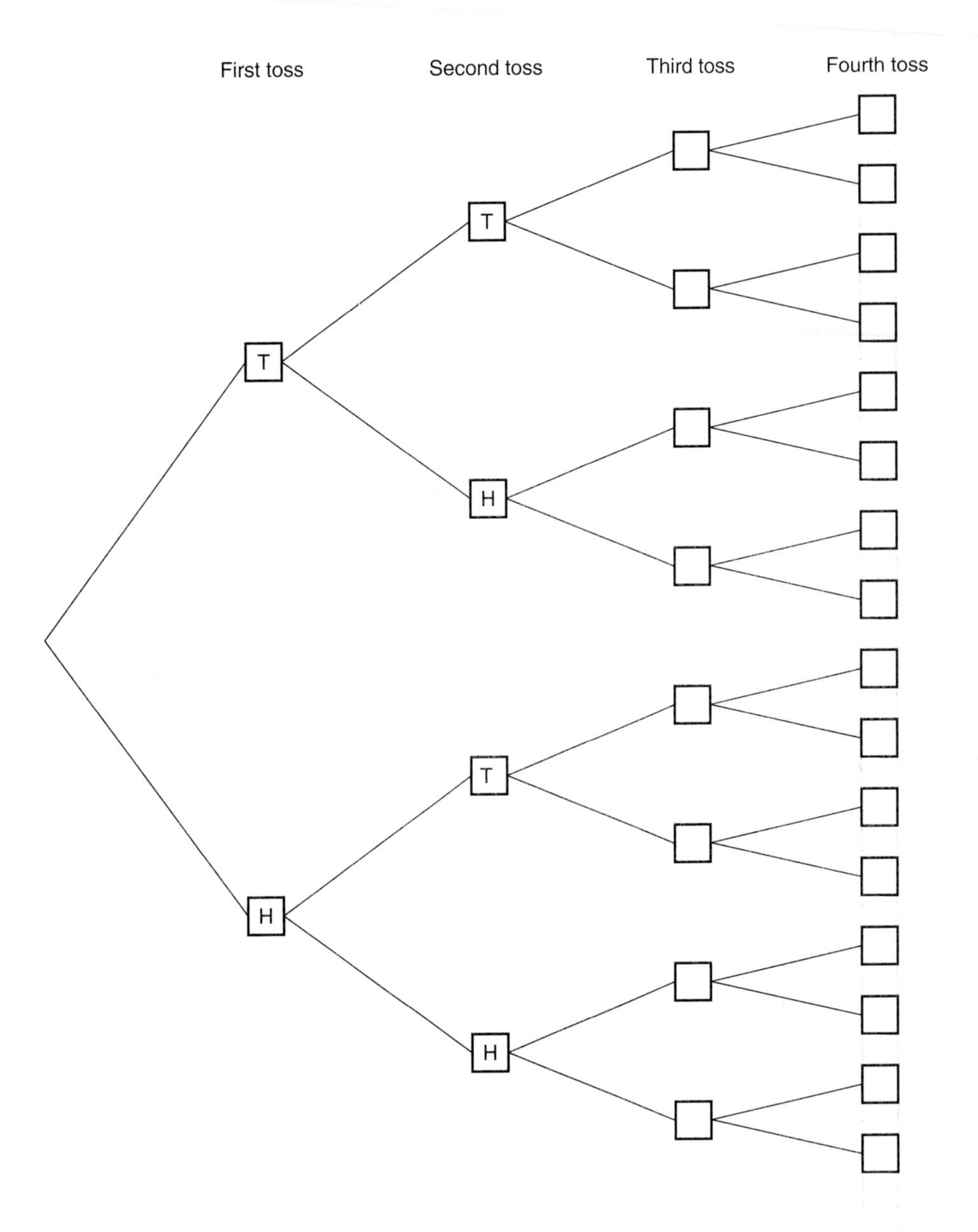

★ **c.** What is the theoretical probability of getting 2 heads and 2 tails on a toss of 4 coins?

3. The chances of succeeding in compound events are often deceiving. You might expect that if a box had 7 red tiles and 3 green tiles, your chances of drawing 2 red tiles at the same time would be fairly good. Try this experiment 30 times and record your results. In this experiment you can either select two tiles at a time or select one tile and then, without replacing it, select another.

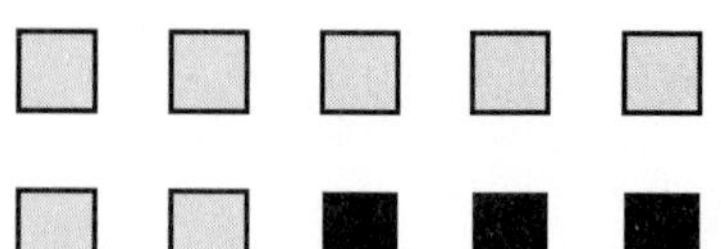

	TALLY	TOTAL
2 red tiles	__________	____
1 or more green tiles	__________	____

a. What is the experimental probability of drawing 2 red tiles?

b. Sketch a probability tree to determine the theoretical probability of drawing 2 red tiles.

4. Were you surprised in activity 3 that the chances of drawing 2 red tiles are so poor? Even if you replace the first tile before drawing the second, the chances of getting 2 red tiles are worse than you might expect. Carry out the experiment again, this time drawing the tiles separately and replacing the first tile each time. Repeat this procedure 30 times and record the results.

	TALLY	TOTAL
2 red tiles	__________	____
1 or more green tiles	__________	____

a. What is the experimental probability of drawing 2 red tiles in succession when the first tile is returned?

★ b. Compute the theoretical probability of getting 2 red tiles when the first tile is returned. Explain your method.

5. Do you think that at least one double 6 is likely to occur in 24 rolls of two dice? This is one of the seventeenth-century gambling questions that led to the development of probability. Toss two dice 24 times and record the results of each toss. (Record a Y for a double 6 and an N otherwise.)

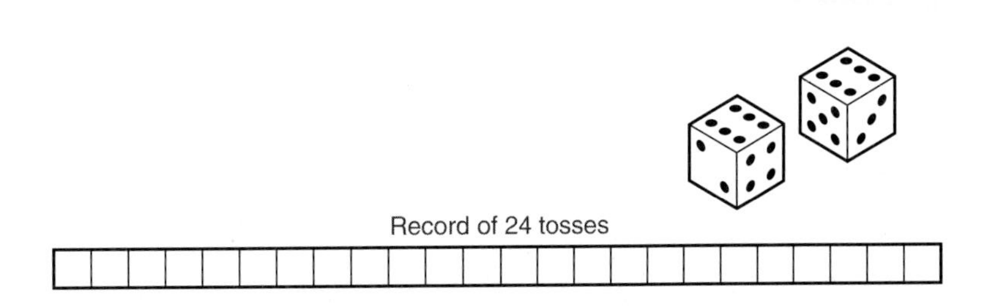

Did you roll a double 6? Even though this event may have occurred 1 or more times in your experiment, there is slightly less than a 50 percent chance of rolling one double 6 in 24 trials. The probability that an event will happen and the probability that it will not happen are called *complementary probabilities.* In calculating the following theoretical probabilities, you can make use of the fact that the sum of complementary probabilities is 1.

★ **a.** What is the probability of rolling a double 6 in 1 roll of the dice?

b. What is the probability of not rolling a double 6 in 1 roll of the dice?

★ **c.** What is the probability of not rolling a double 6 in 2 rolls of the dice?

d. Indicate how you would find the probability of not rolling a double 6 in 24 rolls of the dice. (Do not multiply your answer out.)

★ **e.** The product $\frac{35}{36} \times \frac{35}{36} \times \cdots \times \frac{35}{36}$, in which $\frac{35}{36}$ occurs 24 times, is approximately .51. What is the probability of rolling at least one double 6 in 24 rolls of the dice?

6. **Simulation:** A TV game show offers its contestants the opportunity to win a prize by choosing 1 of 3 doors. Behind 1 door is a valuable prize, but behind the other 2 doors is junk. After the contestant has chosen a door, the host opens 1 of the remaining doors, which has junk behind it, and asks if the contestant would like to *stick* with the initial choice or *switch* to the remaining unopened door. If you were the contestant, would you stick, switch, or, possibly, choose the options at random? You may be surprised to discover that your chances of winning vary depending on your strategy.

a. *Experimental Probability of Sticking:* Suppose you always stick with your first choice. To simulate the sticking strategy, perform the following experiment using the spinner from Material Card 30. Anchor a paper clip to the center of the circle with your pencil point, and then spin the paper clip to simulate choosing a door at random. Perform the experiment 50 times and tally your results in the table. Determine the experimental probability of winning the prize if you always stick with your first choice.

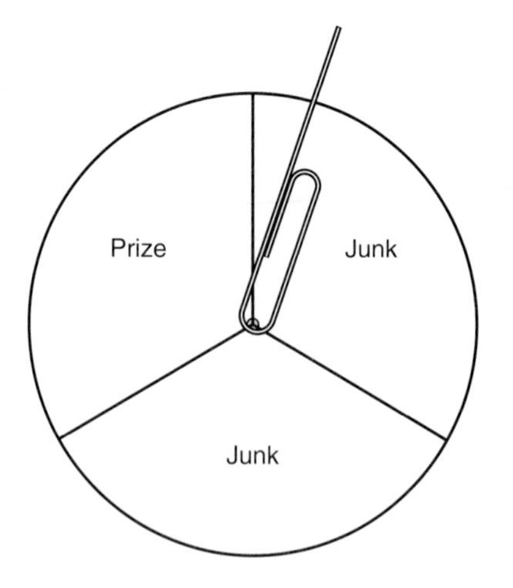

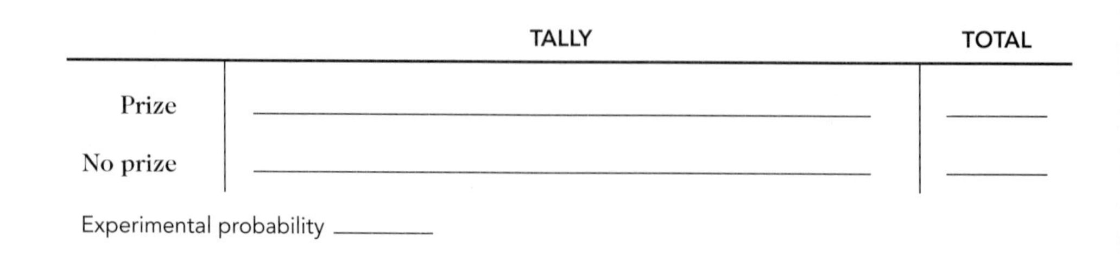

	TALLY	TOTAL
Prize	____________________	______
No prize	____________________	______

Experimental probability ________

b. *Experimental Probability of Switching:* No matter what door the contestant chooses, the host will always open a door with junk behind it and ask if the contestant wants to switch. Perform the following experiment to simulate a switching strategy. Choose the first door randomly, using the spinner as in part a. With the switch strategy, you win if the pointer stops on "junk," but you lose if the pointer stops on "prize." Explain why this happens. Then perform the experiment 50 times and compute the experimental probability of winning on a switch strategy.

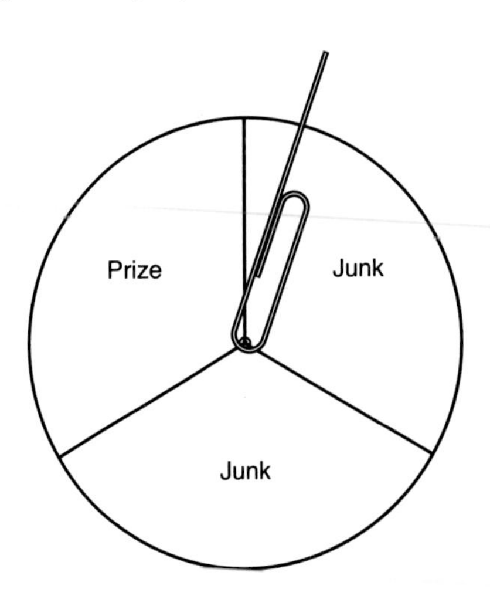

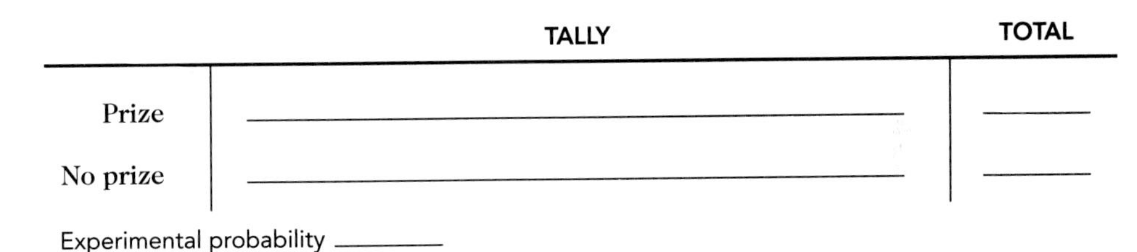

	TALLY	TOTAL
Prize	________________	______
No prize	________________	______

Experimental probability ________

c. *Experimental Probability of Randomly Sticking or Switching:* Maybe the best strategy would be to choose a door and then decide on a stick or switch strategy by chance. Perform the following experiment to simulate this strategy. Choose your first door randomly, using the spinner as above. Then use the second spinner (from Material Card 30) to decide whether you should stick or switch. What happens if your first spin selects "junk" and your second spin selects "switch"—do you win the prize or not? Then perform the experiment 50 times and compute the experimental probability of winning with this strategy.

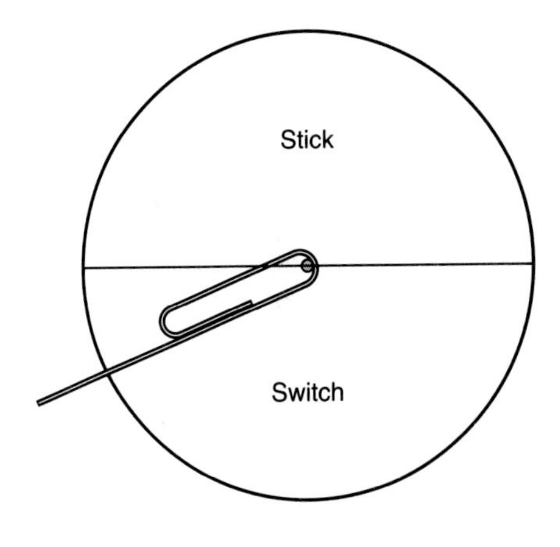

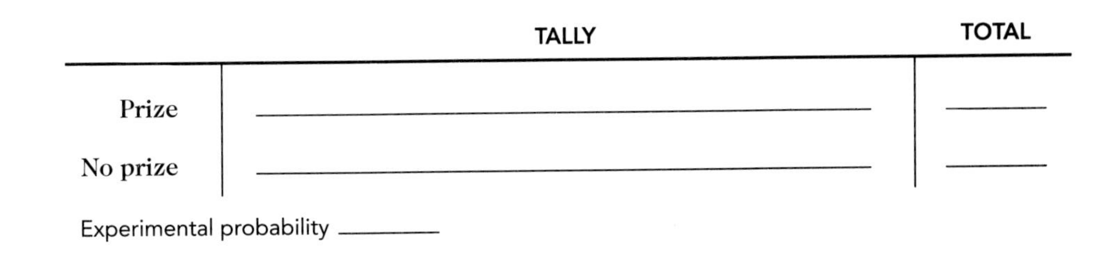

	TALLY	TOTAL
Prize	________________	______
No prize	________________	______

Experimental probability ________

★ **d.** The tree diagrams below represent the first stage of the game for the switch strategy and the stick strategy. Explain why we need show only one stage of the tree diagram to determine whether we have won or lost. For each strategy, determine the theoretical probability of winning the prize.

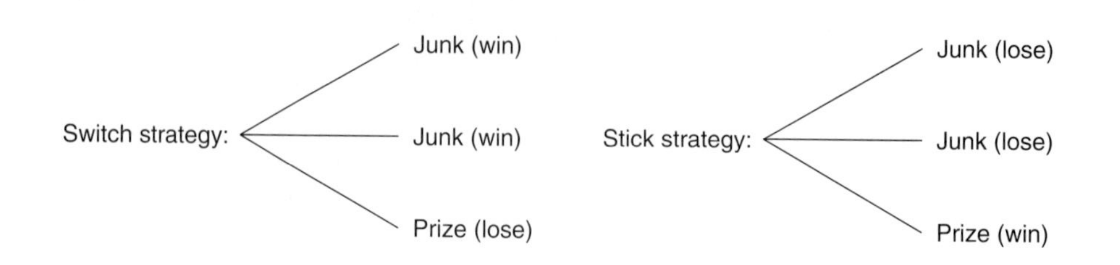

e. This two-stage tree diagram represents picking a door at random and then deciding to stick or switch at random. Notice that if you randomly choose a junk door and then randomly spin "switch," you switch choices when offered the opportunity and win the prize. Finish labeling the outcomes of the second stage and determine the theoretical probability of winning the prize with this strategy.

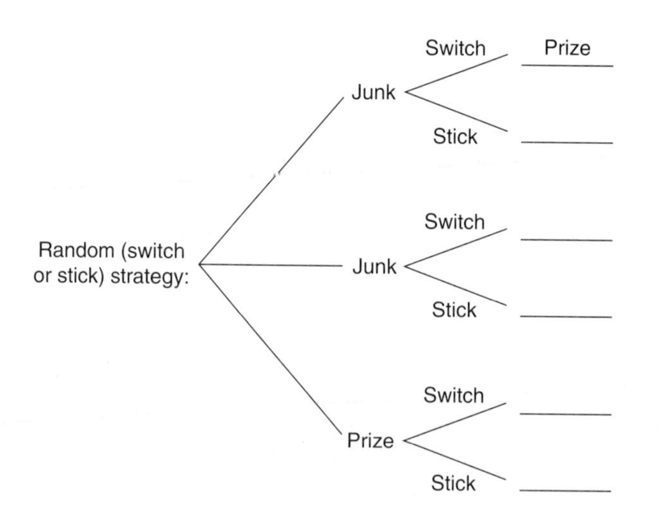

f. Suppose you were chosen to be a contestant for this game show. Write a short summary explaining what strategy you would choose and why you would choose it.

JUST FOR FUN

TRICK DICE

The four dice shown next have the following remarkable property: no matter which die your opponent selects, you can always select one of the remaining three dice so that the probability of winning is in your favor. (Winning means having a greater number of dots facing up.) Make a set of these dice, either by altering regular dice or by using the dice on Material Card 31.

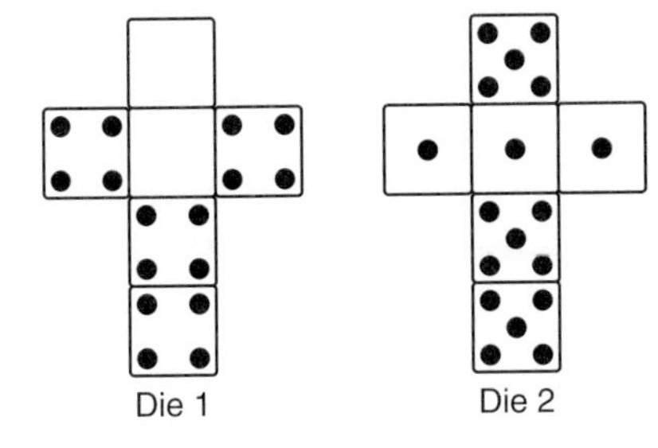

Die 1 Die 2

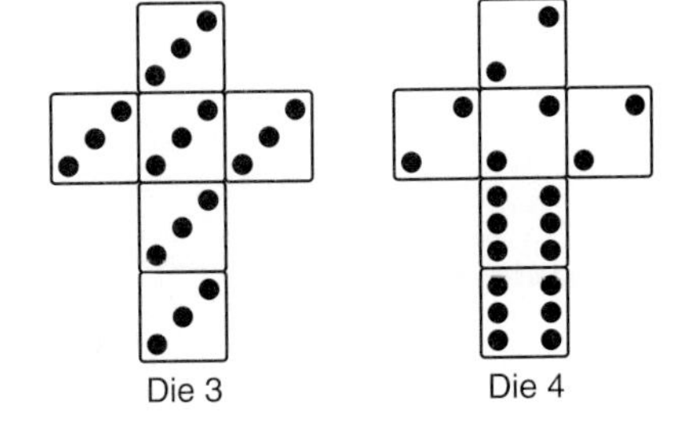

Die 3 Die 4

Try the following experiment with die 2 and die 4. Roll both dice 21 times and record the resulting numbers in the appropriate boxes of the following table. Then circle the numbers for which die 4 wins over die 2. Before beginning the experiment, try to predict which die has the better chance of winning.

Die 2																					
Die 4																					

1. For the winning die in the experiment above, compute the experimental probability of winning.

★ 2. Complete this table to show all 36 pairs of numbers that can result from the two dice. Circle the pairs in which die 4 has the greater number. What is the theoretical probability that die 4 will win over die 2? Compare this result to the experimental probability in problem 1.

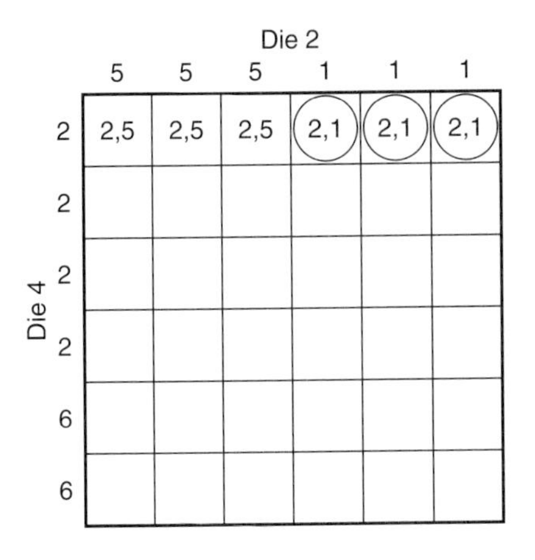

Die 4 \ Die 2	5	5	5	1	1	1
2	2,5	2,5	2,5	(2,1)	(2,1)	(2,1)
2						
2						
2						
6						
6						

★ 3. If a 2 is rolled on die 4 and a 1 is rolled on die 2, then die 4 wins. What is the probability of these events happening? (*Hint:* Multiply the 2 probabilities.)

★ 4. If a 6 is rolled on die 4, it wins over every number on die 2. What is the chance of rolling a 6 on die 4?

★ 5. The theoretical probability that die 4 will win over die 2 is the sum of the probabilities in problems 3 and 4. What is this probability? Compare this result to your answer in problem 2.

Similar approaches can be used to show that die 2 wins over die 1, die 1 wins over die 3, and die 3 wins over die 4. Surprisingly, the first die of each pair has a $\frac{2}{3}$ probability of winning over the second die. Try this experiment with some of your friends. Let them choose a die first, then you select one with a greater chance of winning.

IDEAS FOR THE ELEMENTARY CLASSROOM

SUGGESTED CLASSROOM ACTIVITY: RACETRACK PROBABILITY

The racetrack game (Just for Fun in Section 8.1) can be turned into a noncompetitive activity for young children involving addition, probability, and graphing. Post a chart like the one below at the front of the class. Each student, in turn, rolls a pair of dice (possibly large foam dice), adds the two top numbers, and marks the chart by putting an X in the square above the sum. The activity ends when one column becomes full of X's. The next time this activity is done, have the students each pick the number of the column they think will fill up first. Record their names beneath their chosen columns. Older students can engage in a similar activity with three dice. First have the students design the recording chart, deciding which sums can be obtained with three dice.

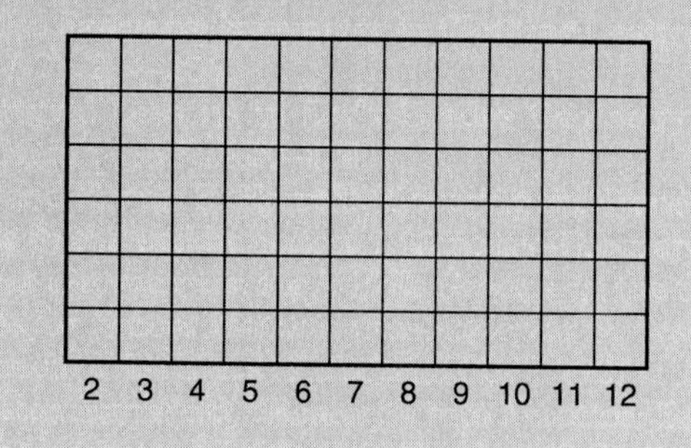

PUZZLER

If a string is cut at random into two pieces, what is the probability that one piece is less than or equal to half the length of the other?

Readings for More Classroom Ideas

Bright, George W. "Teaching Mathematics with Technology: Probability Simulations." *Arithmetic Teacher* 36 (May 1989): 16–18.

Bruni, James V., and H. J. Silverman. "Developing Concepts in Probability and Statistics—and Much More." *Arithmetic Teacher* 33 (February 1986): 34–37.

Dolan, Dan. "Implementing the Standards: Making Connections in Mathematics." *Arithmetic Teacher* 38 (February 1991): 57–60.

Freda, Andrew. "Roll the Dice—An Introduction to Probability." *Mathematics Teaching in the Middle School* 4 (October 1998): 85–89.

Horak, Virginia M., and W. J. Horak. "Let's Do It: Take a Chance." *Arithmetic Teacher* 30 (May 1983): 8–15.

Jones, Graham A., and Carol A. Thornton. *Data, Chance and Probability: Grades 1–3 Activity Book.* (Also grades 2–4 and 6–8 activity books) Lincolnshire, IL: Learning Resources, Inc., 1992.

Lovell, Robert. *Probability Activities.* Berkeley, CA: Key Curriculum Press, 1993.

Lovitt, Charles, and Ian Lowe. *Chance and Data: Investigations*, vol. I. Mathematics Curriculum and Teaching Program. Carlton, Australia: Curriculum Corporation, 1993.

Mason, Julia A., and Graham Jones. "The Lunch-Wheel Spin." *Arithmetic Teacher* 41 (March 1994): 404–408.

Phillips, Elizabeth, Glenda Lappan, Mary J. Winter, and William Fitzgerald. *Probability.* Middle Grades Mathematics Project. Menlo Park, CA: Addison-Wesley Publishing Company, 1986.

Shaughnessy, Michael, and M. J. Arcidiacono. "Visual Encounters with Chance," in *Math and the Mind's Eye* (Unit VIII). Salem, OR: The Math Learning Center, Box 3226, 1993.

Shaughnessy, Michael, and Tom Dick. "Monty's Dilemma: Should You Stick or Switch." *Mathematics Teacher* 84 (April 1991): 252–256.

Van Zoest, Laura R., and Rebecca K. Walker. "Racing to Understand Probability." *Mathematics Teaching in the Middle School* 3 (October 1997): 162–170.

Vissa, Jeanne M. "Probability and Combinations for Third Graders." *Arithmetic Teacher* 36 (December 1988): 33–37.

Wiest, Lynda R., and Robert J. Quinn. "Exploring Probability through an Evens-Odds Dice Game." *Mathematics Teaching in the Middle School* 4 (March 1999): 358–362.

Woodward, Ernest. "A Second-Grade Probability and Graphing Lesson." *Arithmetic Teacher* 30 (March 1983): 23–24.

Zawojewski, Judith S. *Dealing with Data and Chance*, in the *Curriculum and Evaluation Standards for School Mathematics Addenda Series: Grades 5–8.* Reston, VA: National Council of Teachers of Mathematics, 1991.

CHAPTER 9

GEOMETRIC FIGURES

As they [children] explore patterns and relationships with models, blocks, geoboards, and graph paper, they learn about the properties of shapes and sharpen their intuitions and awareness of spatial concepts. . . . Folding paper cutouts or using mirrors to investigate lines of symmetry are other ways for children to observe figures in a variety of positions, become aware of their important properties, and compare and contrast them.[1]

ACTIVITY SET 9.1

FIGURES ON RECTANGULAR AND CIRCULAR GEOBOARDS

Purpose To use rectangular geoboards to illustrate geometric figures and circular geoboards to study inscribed and central angles

Materials A rectangular geoboard, a circular geoboard, and rubber bands (optional). Templates for making rectangular and circular geoboards are on Material Cards 27 and 33, respectively. Recording paper for activity 1 is on Material Card 11, and the three-circle Venn diagram is on Material Card 5. A protractor is on Material Card 32.

Activity The geoboard is a popular physical model for illustrating geometric concepts and solving geometric investigations. The most familiar type of geoboard has a square shape, with 25 nails arranged in a 5 by 5 array. Rubber bands can be placed on the nails to form models for segments, angles, and polygons. For example, here is a 12-sided polygon.

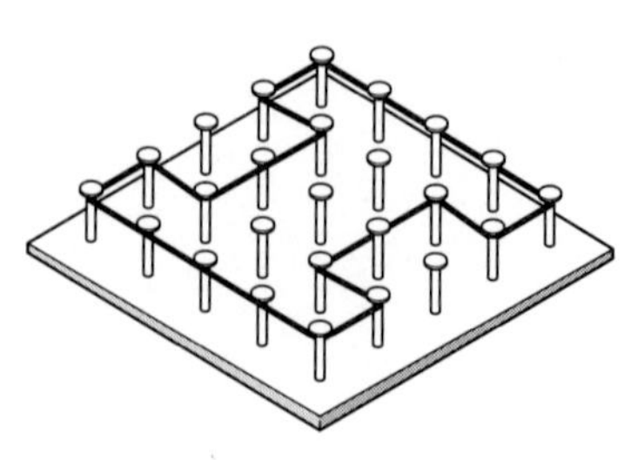

[1]*Curriculum and Evaluation Standards for School Mathematics* (Reston, VA: National Council of Teachers of Mathematics, 1989), 48–49.

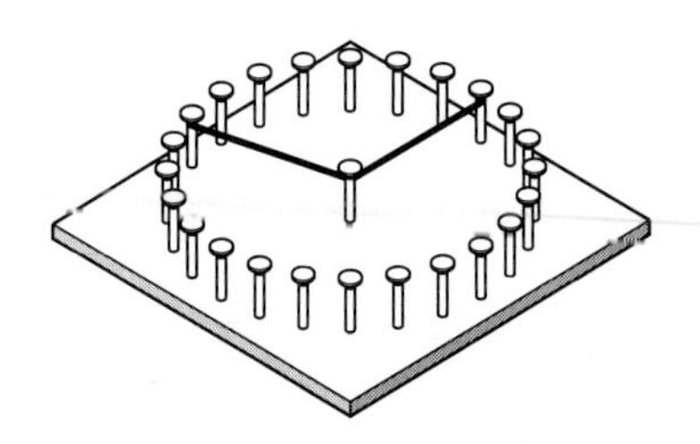

The circular geoboard is very helpful for developing angle concepts. The one shown below has a nail at the center and an outer circle of 24 nails. Many of the angles and polygons that can be formed on this geoboard cannot be made on the rectangular geoboard. The circular geoboard will be used to form central and inscribed angles and to establish an important relationship between them.

Since most geoboard activities can be carried out on dot paper by drawing figures, the use of geoboards is optional in the following activities. However, geoboards provide insights because of the ease with which figures can be shaped and reshaped during an investigation. They also encourage experimentation and creativity. If geoboards are not available, you may wish to make them from boards and finishing nails using the templates on the Material Cards.

Rectangular Geoboards

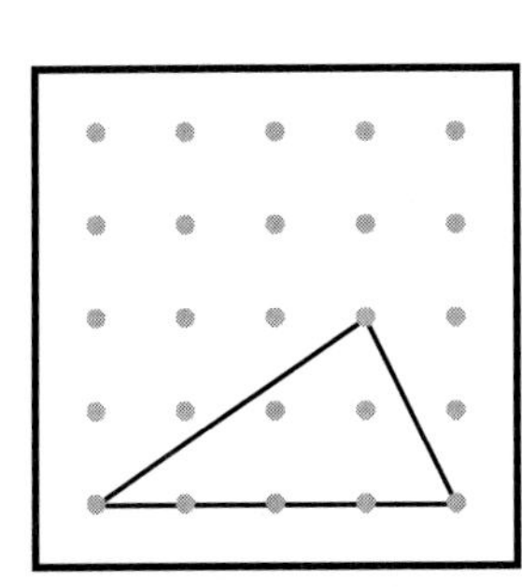

1. This figure shows a geoboard triangle with its base along the bottom row of pins. Determine all possible geoboard triangles that have the same base. Record your results by drawing each triangle on geoboard recording paper (Material Card 11).

a. Determine the area of each triangle using methods developed in Activity Set 6.4. Record the area inside the triangle.

★ **b.** Cut out each geoboard, that has a triangle drawn on it. (Cut around the entire geoboard, not just the triangles.) Group the triangles that have the same area. Describe the characteristics that triangles of equal area have in common.

★ **c.** The two triangles shown here are *congruent* because one can be cut out and placed on the other so that they coincide. Match all congruent triangles. Now, form the collection of all noncongruent triangles, by selecting only one to represent each group. How many noncongruent triangles are there?

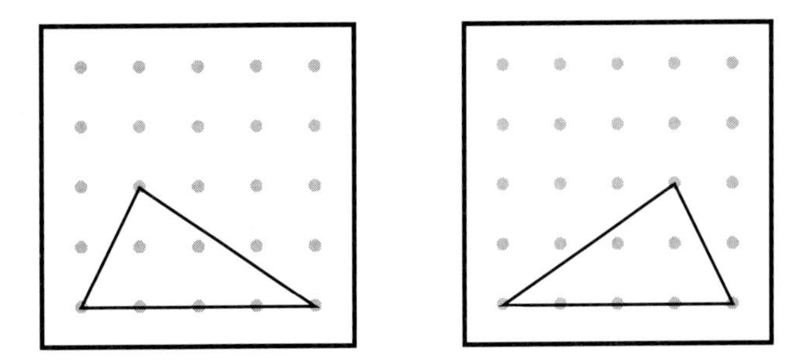

d. Label the three-circle Venn diagram, Material Card 5, as shown here. Sort the collection of noncongruent triangles from part c by placing them in the appropriate regions of the Venn diagram. Record the number of triangles in each region on the diagram shown here. (A scalene triangle has all sides of different length; an isosceles triangle has at least two sides of equal length; and a right triangle has one right angle.)

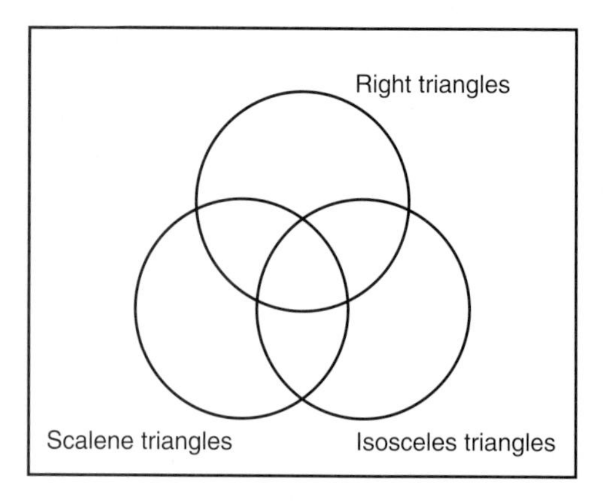

e. For the regions of the Venn diagram in part d which have no entries, can you draw triangles which could be placed in those regions? Explain how you arrived at your answer.

2. The first geoboard below shows three segments of different lengths. A geoboard segment has its endpoints on geoboard pins. Use your geoboard to determine how many geoboard segments of *different* lengths can be constructed. Record your solutions.

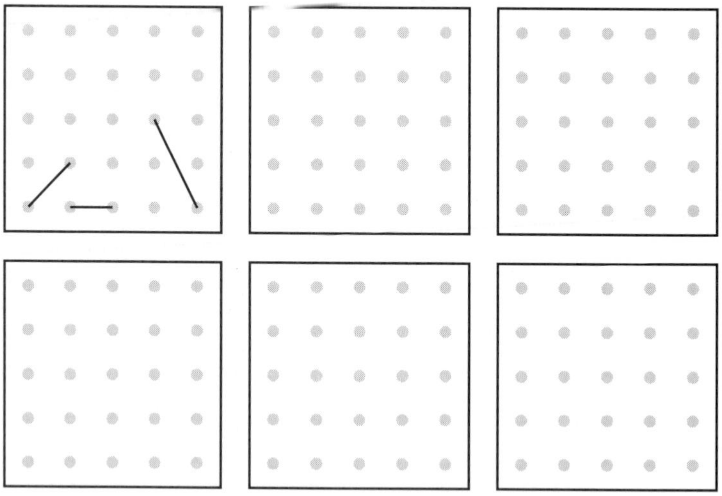

3. The geoboards following have a rubber band around the outer pins to form as large a square as possible. Additional rubber bands are used to divide the squares into two *congruent* halves. (Congruent figures have the same size and shape.)

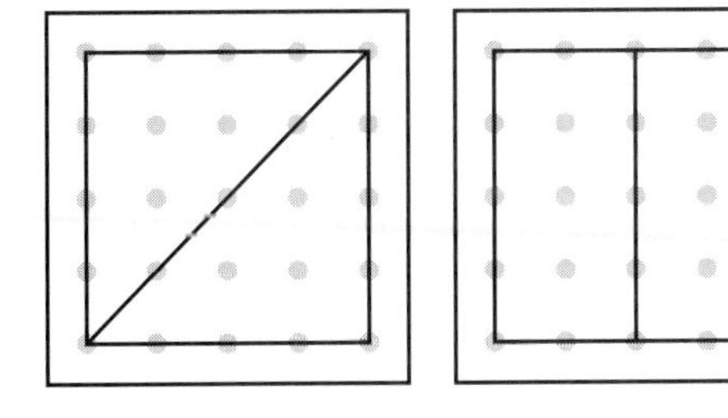
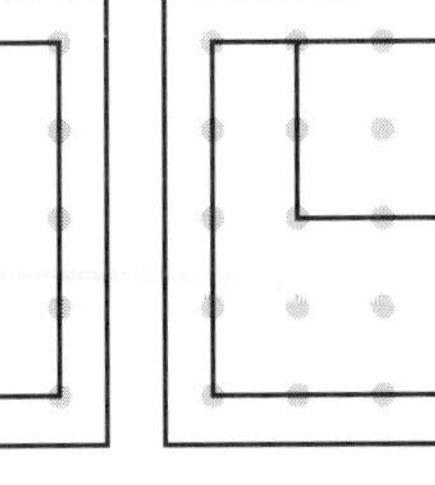
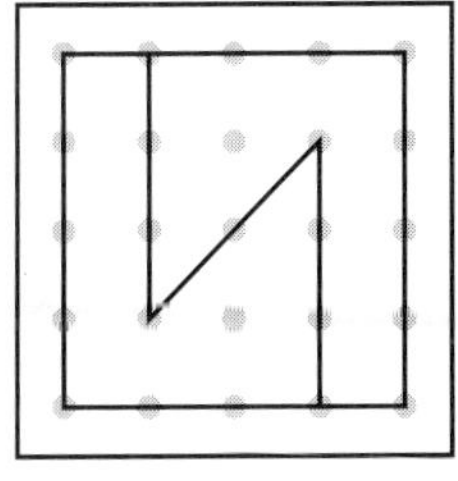

Find at least seven other ways to divide the square into congruent halves, and draw them below.

★ **4.** There are 16 noncongruent quadrilaterals that can be formed on a 3 by 3 geoboard. Sketch them here.

a. **b.** **c.** **d.** **e.** **f.**

g. **h.** **i.** **j.** **k.** **l.**

m. **n.** **o.** **p.**

Classify each of the above quadrilaterals by recording its letter following each description that applies to it. (A parallelogram has two pairs of opposite parallel sides, and a trapezoid has exactly one pair of parallel sides.)

Square ______________________

Rectangle, but not square ______________________

Parallelogram, but not rectangle ______________________

Trapezoid ______________________

Convex ______________________

Nonconvex ______________________

5. The first geoboard below shows a 21-sided polygon. Find polygons with more than 21 sides and draw them here.

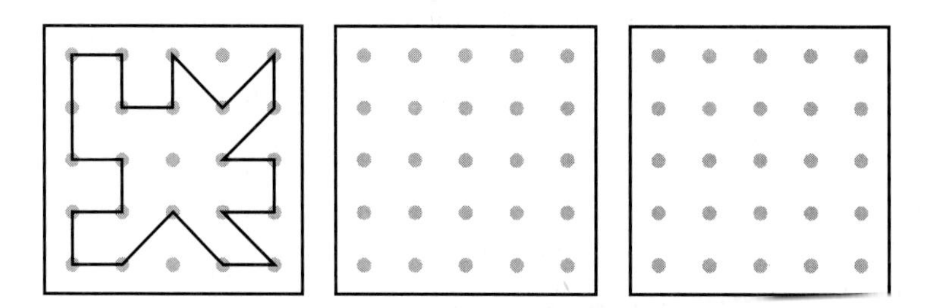

Circular Geoboards

6. Central Angles: An angle whose vertex is located at the center of the circle and whose sides intersect the circle is called a *central angle.* The portion of the circle that is cut off by the sides of the angle is called the *intercepted arc.* The central angle on this geoboard has a measure of 15 degrees (15°) because its intercepted arc is $\frac{1}{24}$ of a whole circle ($360 \div 24 = 15$).

★ **a.** Under each geoboard, write the number of degrees in the indicated central angle.

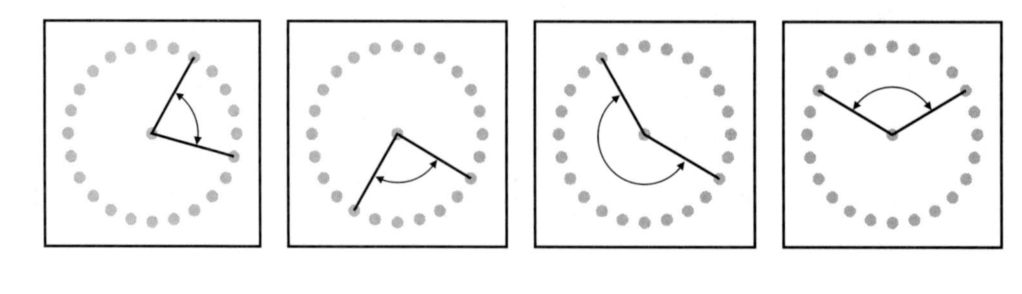

________ ________ ________ ________

★ **b.** On your geoboard form central angles that have the following angle measures. Then draw them here.

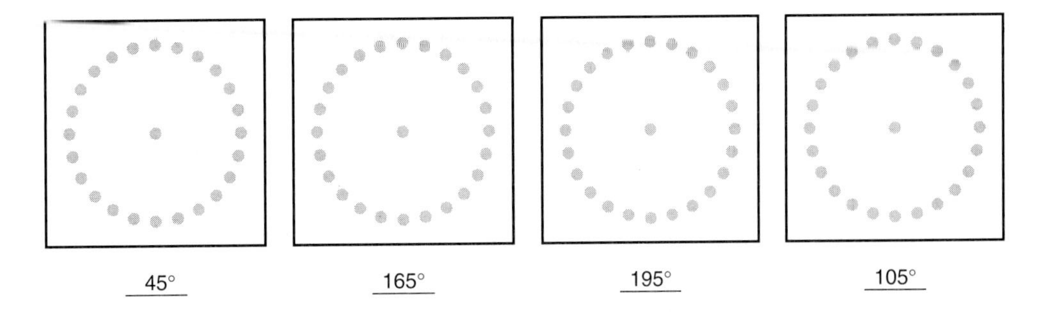

7. Inscribed Angles: An angle is *inscribed* in an arc of a circle when its vertex is on the arc and its sides contain endpoints of the arc.

★ **a.** Construct the following inscribed angles on your circular geoboard and measure them with your protractor. (There is a protractor on Material Card 32. If you do not have a circular geoboard, draw the angles lightly, in pencil, on the geoboard template on Material Card 33.)

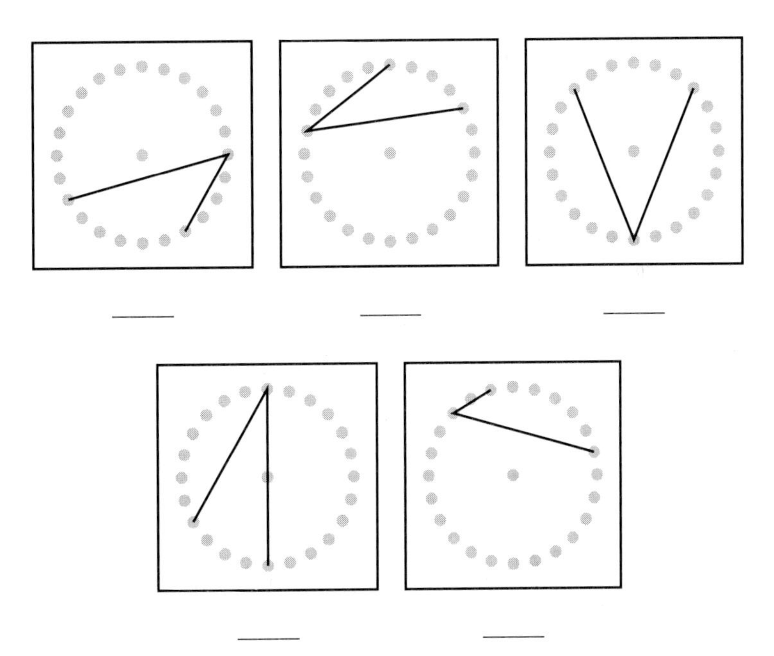

★ **b.** Look at the inscribed angles in part a that have the same angle measure. Make a conjecture about the *equality* of inscribed angles and the size of their corresponding intercepted arcs. Test your conjecture by drawing and measuring more inscribed angles. Record your conclusions.

Conjecture:

★ **c.** Angle PQR is inscribed in a semicircle (half a circle). Using your circular geoboard and protractor, construct and measure angle PQR and several other inscribed angles whose sides intersect the ends of a diameter. Record the angles and their measures below. Make a conjecture about such angles and record it below.

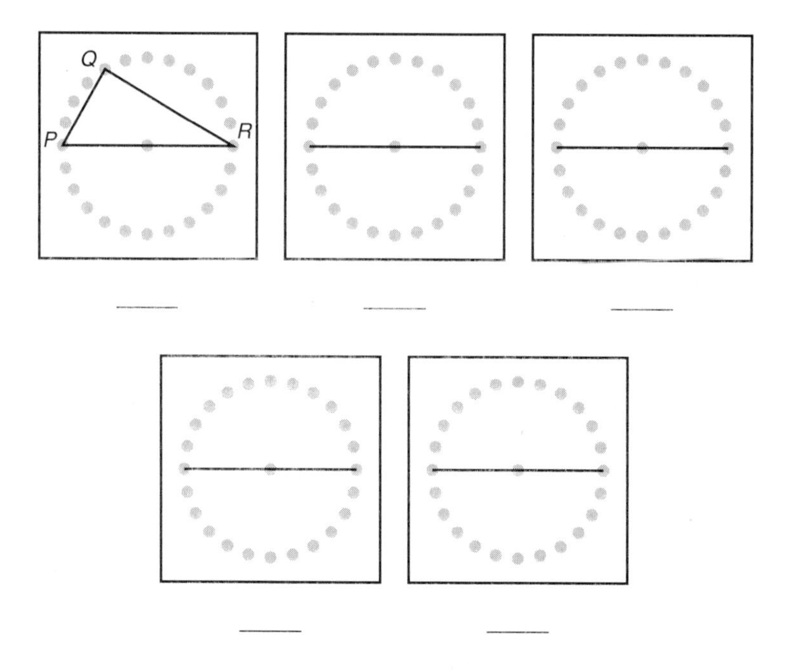

Conjecture:

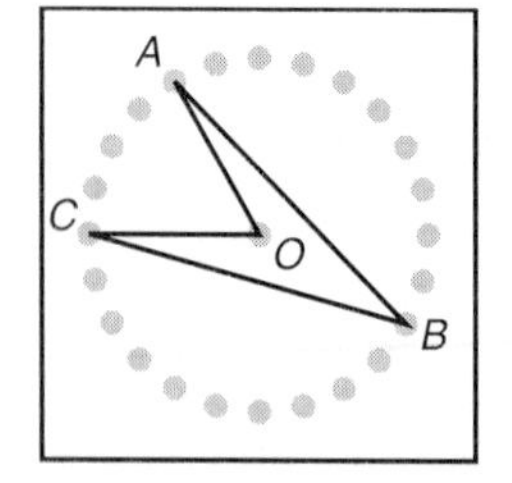

Angle AOC ______

Angle ABC ______

8. Construct central angle AOC and inscribed angle ABC on your circular geoboard (or draw them on your circular geoboard template). Determine the measure of central angle AOC. Measure angle ABC with your protractor. Record both measures beneath the figure.

★ **a.** Form each of the following inscribed angles on your geoboard, measure the angle with your protractor, and record the inscribed angle measure below the figure. Then, on each geoboard, determine the measure of the central angle that intercepts the same arc as the inscribed angle and record it.

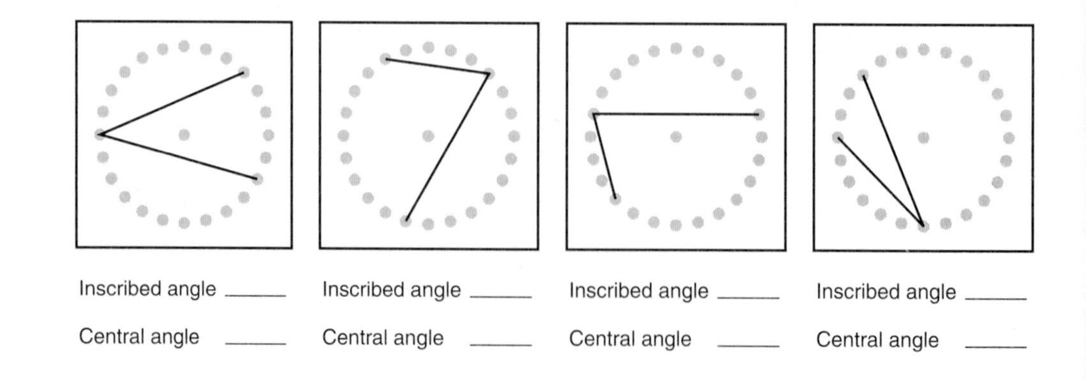

Inscribed angle ______ Inscribed angle ______ Inscribed angle ______ Inscribed angle ______

Central angle ______ Central angle ______ Central angle ______ Central angle ______

★ **b.** There is a relationship between the measure of an inscribed angle and that of its corresponding central angle. Write a conjecture about that relationship.

Conjecture:

JUST FOR FUN

TANGRAM PUZZLES

Tangrams are one of the oldest and most popular of ancient Chinese puzzles. The 7 tangram pieces are shown in a square. There are 5 triangles, 1 square, and 1 parallelogram. Fold a standard piece of paper to get a square and cut out the square. Then fold the paper to obtain the 7 tangram pieces and cut them out for the following activities.

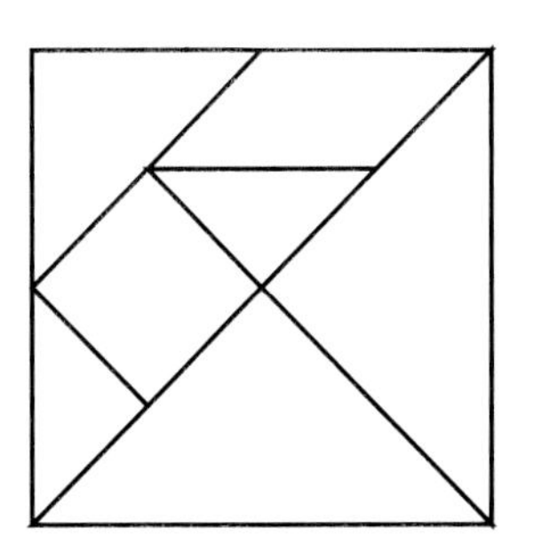

★**1.** The 7 pieces are used in Chinese puzzles called *tangram puzzles.* The object of tangram puzzles is to use all of the tangram pieces to form figures. The tangram pieces should be placed so that they do not overlap. Try to form the following figures.
There are many tangram figures.[2] You may wish to create and record a few of your own.

2. The tangram pieces below also can be used to create geometric shapes. For example:

a. Use the 2 small triangles to make a square; a larger triangle; a parallelogram.

b. Use the 2 small triangles and the medium-size triangle to make a rectangle; a parallelogram that is not a rectangle; a larger triangle.

3. The chart on the following page shows how a triangle can be created with 3 tangram pieces, a square with 5 tangram pieces, and a nonsquare rectangle with 4 pieces. Record your results from activity 2 and see if you can complete the chart. (There are no solutions for the entries that are crossed out.)

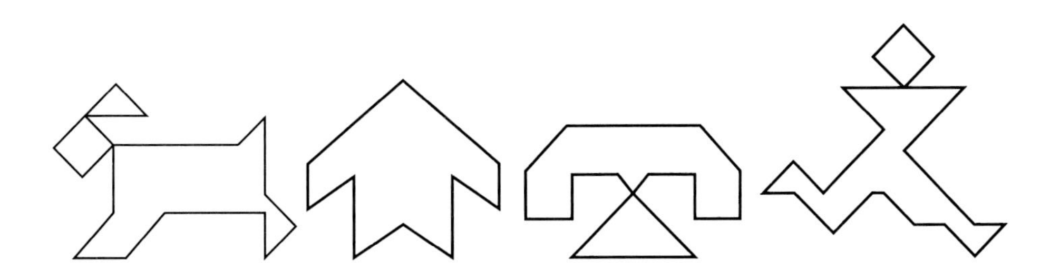

[2]R. Reed's *Tangrams* (New York: Dover, 1965) contains 330 tangram figures and their solutions.

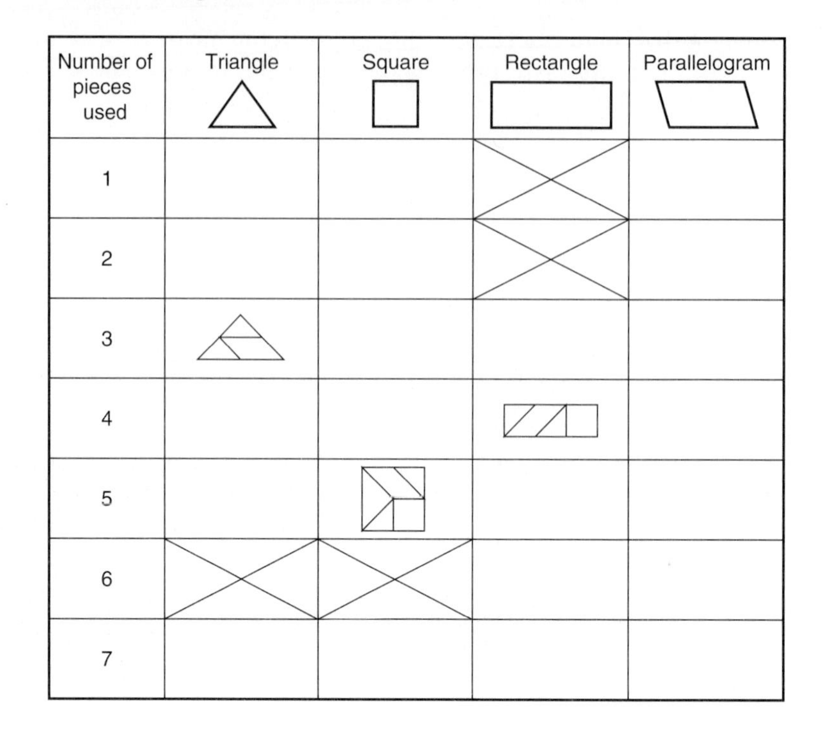

Number of pieces used	Triangle	Square	Rectangle	Parallelogram
1				
2				
3				
4				
5				
6				
7				

PUZZLER

Make one cut in this figure to create two congruent halves. (The cut does not have to be straight.)

ACTIVITY SET 9.2

REGULAR AND SEMIREGULAR TESSELLATIONS

Purpose To study regular and semiregular tessellations of the plane

Materials Pattern blocks and polygons for tessellations from the Manipulative Kit

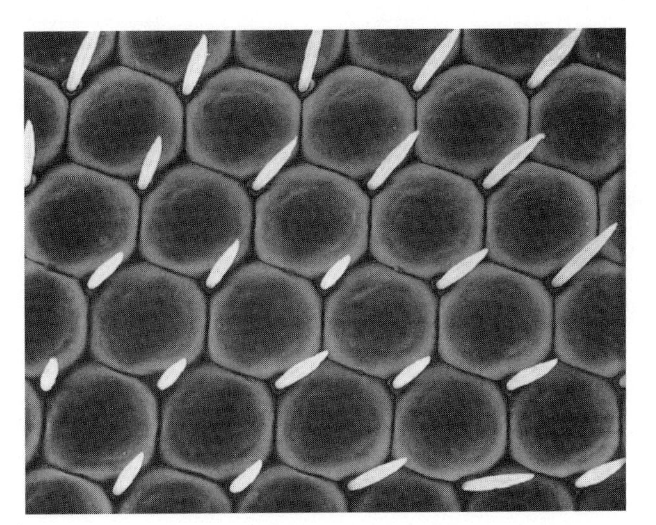

A portion of the eye of a fruit fly.

Activity The hexagonal shapes in the accompanying photograph show a small portion of the eye of a fruit fly. The partitioning of surfaces by hexagons or nearly hexagonal figures is common in nature.[3] A figure, such as a regular hexagon, that can be used repeatedly to cover a surface without gaps or overlaps is said to *tile* or *tessellate* the surface. The resulting pattern is called a *tessellation.* An infinite variety of shapes can be used as the basic figure for a tessellation. The famous Dutch artist Maurits C. Escher (1898–1972) is noted for his tessellations with drawings of birds, fish, and other living creatures. Techniques for creating Escher-type tessellations are developed in Activity Set 11.2.

Pattern Block Tessellations

1. Figure 1, on the next page, shows the beginnings of a tessellation with the yellow hexagon pattern block and figures 2 and 3 show the beginnings of two different tessellations with the red trapezoid.

[3]For some examples of hexagonal patterns and a discussion of why they occur in nature, see H. Weyl, *Symmetry* (Princeton: Princeton University Press, 1952), 83–89.

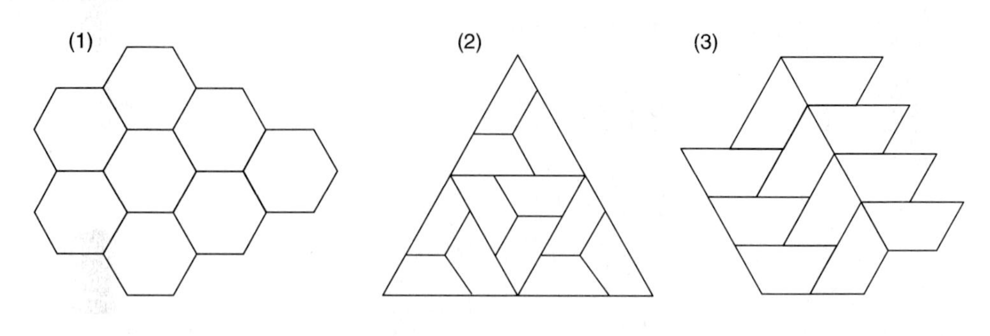

a. Find at least one other way to tessellate with the red trapezoid. Record your pattern.

b. All of the remaining pattern block pieces have more than one tessellation pattern like the trapezoid. For each of those, sketch enough of at least two patterns so that a reader will be able to extend them.

★ **2.** When 3 acute angles of white rhombuses are put together at a point, the sum of their measures is the same as one of the angles of an orange square. Since each angle of the square measures 90°, a small angle of the white rhombus measures 30°.

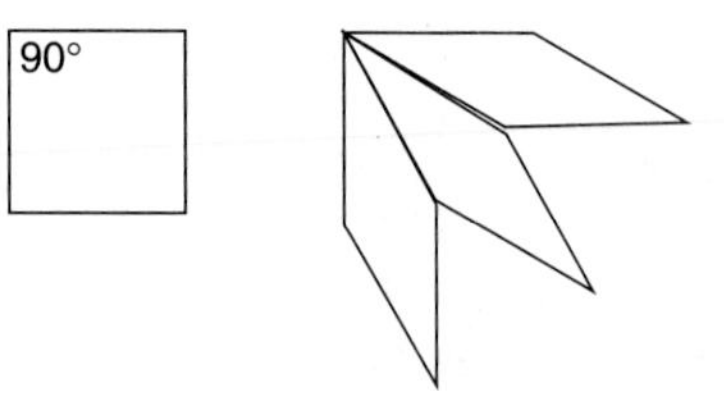

By comparing the angles of the various pieces (and using no angle-measuring instruments), determine the measures of the other seven angles of the pieces in the pattern block collection. Record the measures on the pattern block diagrams on the next page and describe or show briefly how you reached your conclusions.

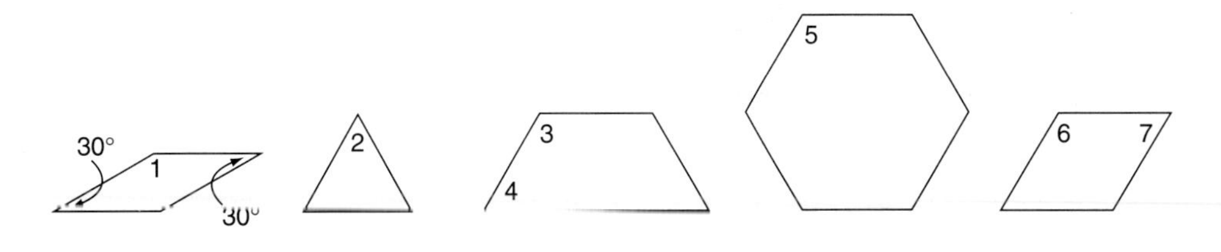

3. Both the figures below have been formed by taping two pattern blocks together to form new shapes. The first figure is a hexagon (6 sides and 6 interior angles) and the second is a pentagon (5 sides and 5 interior angles). On the diagrams record the measure of each interior angle in the new shapes.

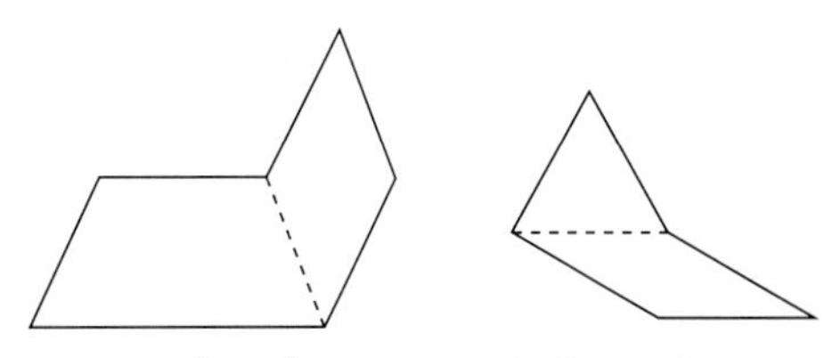

   a. Form these two shapes with your pattern blocks and test each of them to see if they will tessellate. If so, sketch enough of the beginnings of a pattern so the reader has sufficient information to extend it. If not, explain why it will not work.

   b. "Tape" two other pieces together to create a new shape that will not tessellate. Record your figure and explain why it will not tessellate.

Tessellations with Regular Polygons

4. Six regular polygons are shown at the top of the next page. For each polygon, trace additional copies around the enlarged vertex point, as if to begin a tessellation.

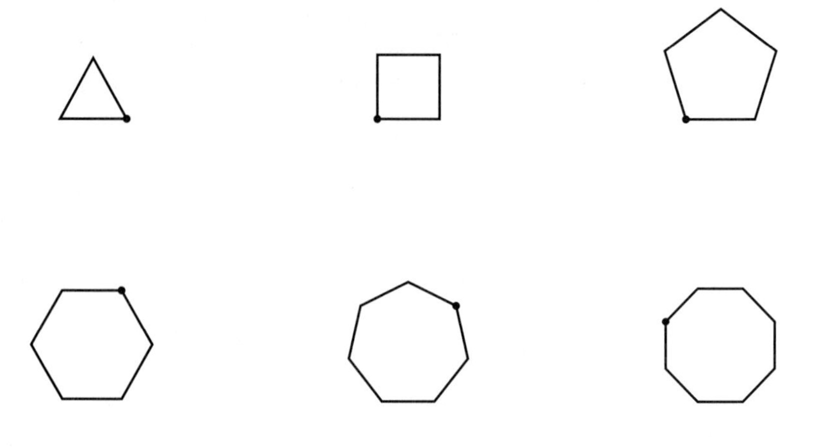

a. Which of these polygons will not tessellate?

b. What condition must the angles of a regular polygon satisfy in order for the polygon to tessellate?

★ c. Explain why regular polygons with more than 6 sides will not tessellate.

5. The equilateral triangle, square, and regular hexagon are the only regular polygons that tessellate. Although designs based on these patterns are common, more interesting tessellations can be formed by using two or three different regular polygons. A square and two octagons are around each vertex point of the tessellation shown here. Notice that the vertex angle of a square is 90° and of an octagon is 135°, and 90 + 135 + 135 = 360.

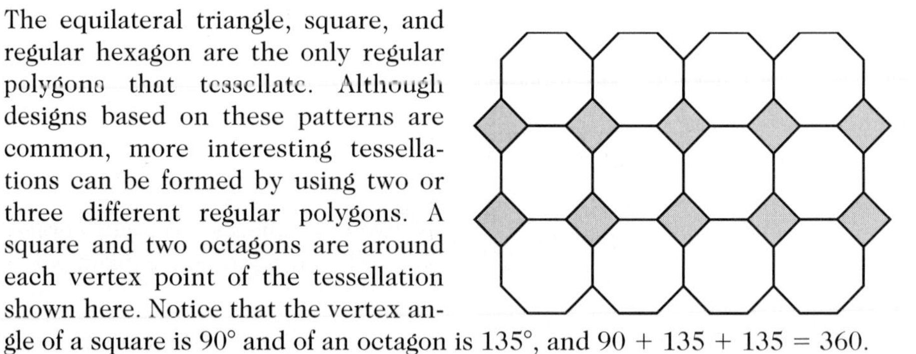

A tessellation that has two or more types of regular polygons arranged in the same order around every vertex point is called a *semiregular tessellation.* The five regular polygons that are used in semiregular tessellations are shown below. The number of degrees in a vertex angle of each polygon is recorded above the polygon. There are polygons for tessellating in the Manipulative Kit. Use them for the following activities.

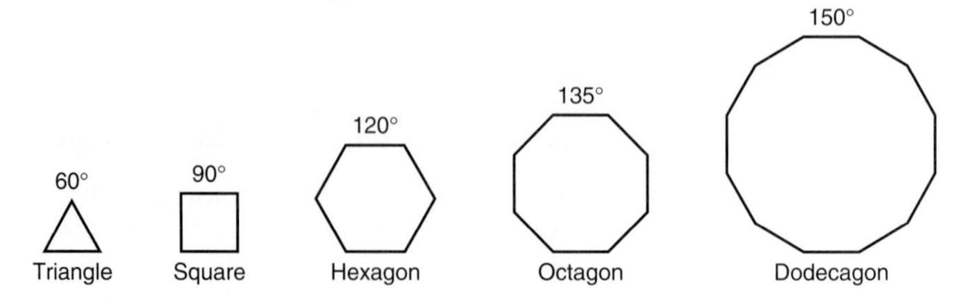

★ **a.** There are six semiregular tessellations that can be formed using combinations of two different regular polygons about each vertex. Find and sketch at least one of these tessellations. (*Hint:* Look for combinations of angles whose measures add up to 360°.)

★ **b.** There are two semiregular tessellations that use combinations of three different regular polygons about each vertex. Find and sketch one of these tessellations. (*Hint:* Look for combinations of angles whose measures add up to 360°.)

★ **c.** It is possible to tile a plane with combinations of polygons that do not form a semiregular tessellation. These tilings may have a regular pattern, but not every vertex point is surrounded by the same arrangement of polygons. Create such a tiling and sketch it here. (*Hint:* One such tiling uses dodecagons, triangles, and squares.)

JUST FOR FUN

THE GAME OF HEX (2 players)

Hex, a game of deductive reasoning, is played on a game board of hexagons similar to the 11 by 11 grid shown here (see Material Card 12). Player 1 and player 2 take turns placing their markers or symbols on any unoccupied hexagon. Player 1 attempts to form an unbroken chain or string of symbols from the left side of the grid to the right, and player 2 tries to form a chain of symbols from the top of the grid to the bottom. The first player to complete a chain is the winner. In this sample game, player 2 (using O's) won against player 1 (using X's).

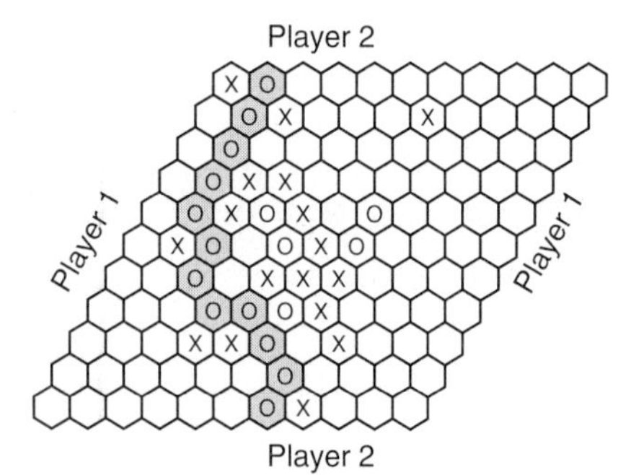

★***Winning Strategy:*** There will always be a winner in the game of Hex, since the only way a player can prevent the other from winning is to form an unbroken chain of markers from one side of the grid to the other. On a 3 by 3 game board, the person who plays first can easily win in 3 moves. What should be the first move? Find a first move on the 4 by 4 and 5 by 5 game boards that will ensure a win regardless of the opponent's moves.

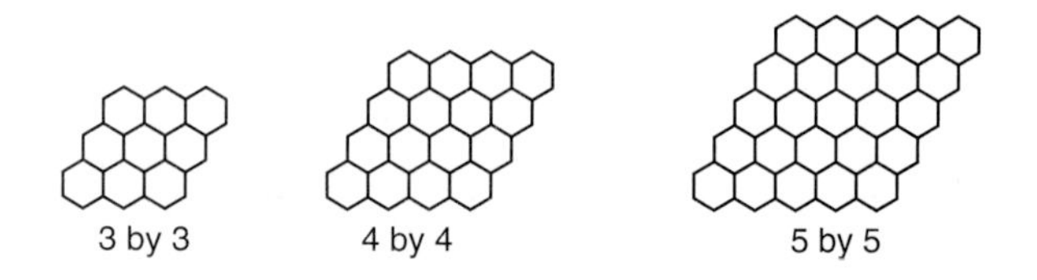

It is tempting to conclude that a winning strategy on an 11 by 11 game board begins with a first move on the center hexagon. However, there are so many possibilities for plays on a grid of this size that no winning strategy has yet been found. There is a type of proof in mathematics, called an *existence proof,* which shows that something exists even though no one is able to find it. It has been proved that a winning strategy does exist for the first player in a game of Hex on an 11 by 11 grid, but just what that strategy is, no one knows.[4]

PUZZLER

This 4 by 4 square has been divided into two congruent pieces. In how many different ways can this square be divided into two congruent pieces by drawing on the grid lines?

[4] For an elementary version of this proof, see M. Gardner, "Mathematical Games," *Scientific American* 197 (1) (1957): 145.

ACTIVITY SET 9.3

MODELS FOR REGULAR AND SEMIREGULAR POLYHEDRA

Purpose To construct and use models in order to observe and examine the properties of regular and semiregular polyhedra

Materials Patterns for constructing regular polyhedra are on Material Cards 34 and 35

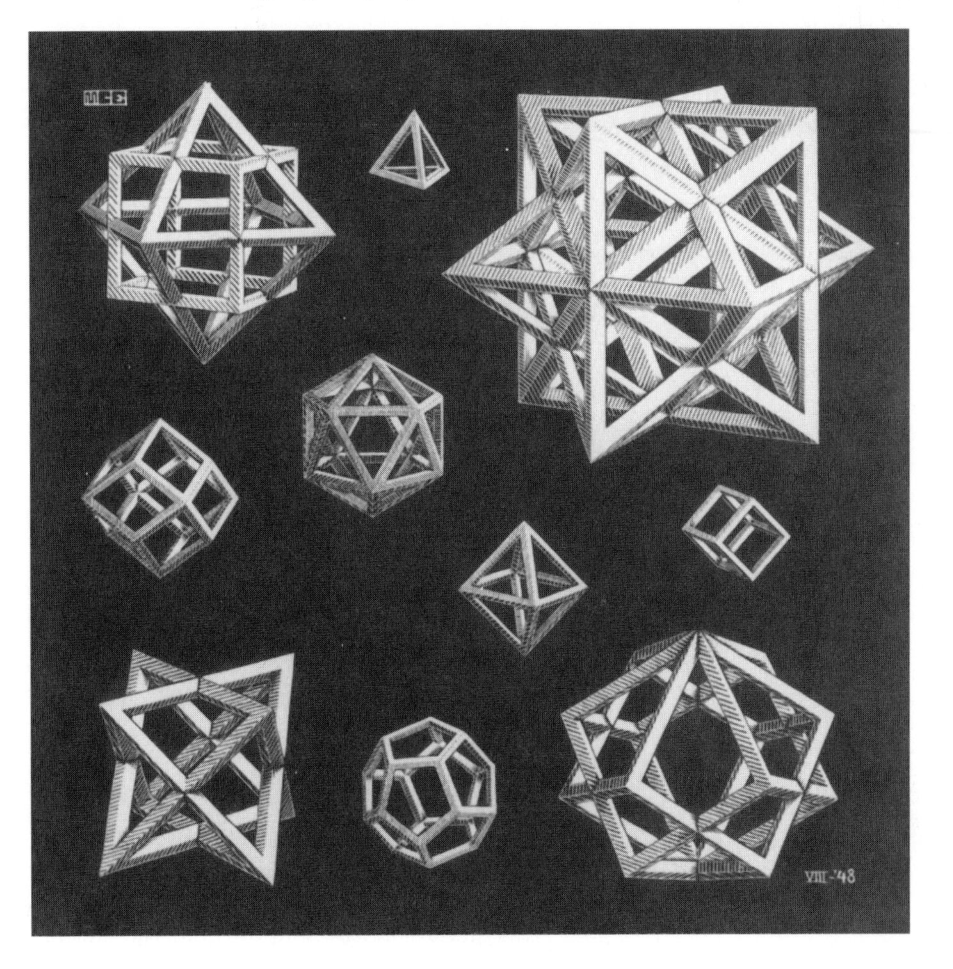

M. C. Escher's "Stars" © 1999 Cordon Art B. V.—Baarn-Holland. All rights reserved.

Activity The term polyhedron comes from Greek and means "many faces." A *regular polyhedron* is a convex polyhedron whose faces are regular polygons and whose vertices are each surrounded by the same number of these polygons. The regular polyhedra are also called *Platonic solids* because the Greek philosopher Plato (427–347 B.C.) immortalized them in his writings. Euclid (340 B.C.) proved that there were exactly five regular polyhedra.[5] Can you identify the Platonic solids in M. C. Escher's wood engraving *Stars*?

[5]For the proof that there are exactly five regular polyhedra, see P. G. O'Daffer and S. R. Clemens, *Geometry: An Investigative Approach* (Reading, MA: Addison-Wesley, 1976), 115–119.

Platonic Solids

1. There are several common methods for constructing the five Platonic solids. The polyhedra pictured here are see-through models similar to those in Escher's wood engraving. They can be made from drinking straws that are threaded or stuck together.

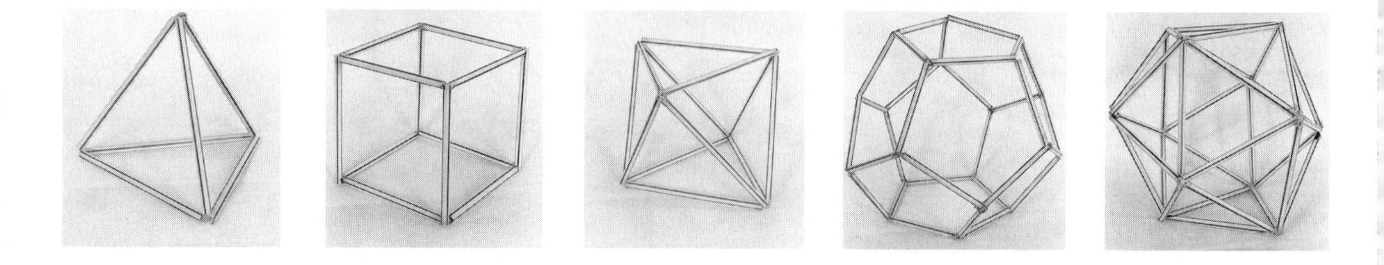

Another method is to cut out regular polygons and glue or tape them together. Material Cards 34 and 35 contain regular polygon patterns for making the Platonic solids. Cut out the patterns, use a ballpoint pen or sharp point to score the dashed lines, and assemble them by folding on the dashed lines and taping the edges. For each Platonic solid, record the number of faces (polygonal regions) and the name of the polygon used.

	NUMBER OF FACES	POLYGON
Tetrahedron	________	________
Cube	________	________
Octahedron	________	________
Dodecahedron	________	________
Icosahedron	________	________

2. The faces of the icosahedron and cube in the next photo were made from colorful greeting cards. (The triangular and square patterns for the faces are also shown. The corners of these patterns were cut with a paper punch and scissors, and the edges bent up along the dashed lines. The adjoining faces of these polyhedra are held together by rubber bands on their edges.)

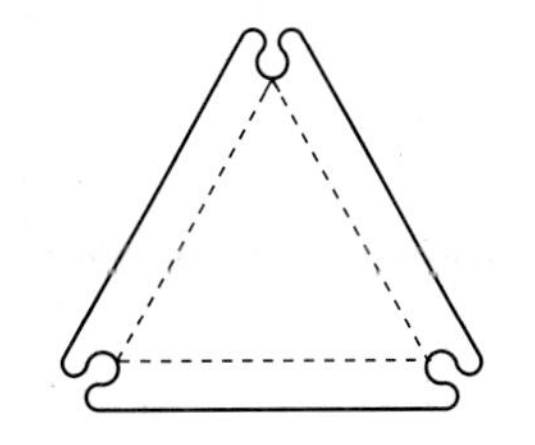

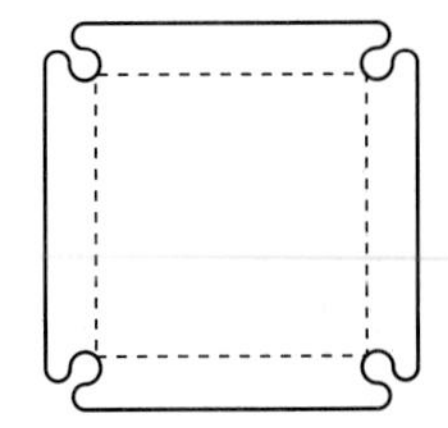

a. If you were to make your Platonic solids as described above, how many rubber bands would you need to make each one? (One band is used for each edge.)

Tetrahedron ________

Cube ________

Octahedron ________

Dodecahedron ________

Icosahedron ________

b. You may enjoy planning colors for your Platonic solids so that no two faces with a common edge have the same color. For this condition to be satisfied, what is the least number of colors needed for the octahedron?

3. Euler's Formula

★ **a.** In the following table, record the numbers of vertices, faces, and edges for each of the Platonic solids.

	VERTICES (V)	FACES (F)	EDGES (E)
Tetrahedron	________	________	________
Cube	________	________	________
Octahedron	________	________	________
Dodecahedron	________	________	________
Icosahedron	________	________	________

b. There is a relationship among the numbers of vertices, faces, and edges. Make a conjecture about this relationship. Try to express it in terms of the letters V, F, and E.

★ **c.** For each polyhedron shown here, count the vertices, faces, and edges. Test your conjecture from part a on these polyhedra, which are not Platonic solids. Record your results.

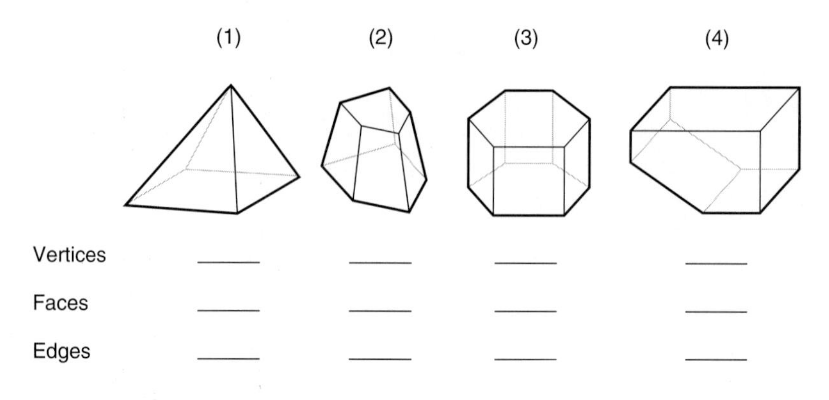

	(1)	(2)	(3)	(4)
Vertices	____	____	____	____
Faces	____	____	____	____
Edges	____	____	____	____

4. Patterns for Cubes: The patterns below were formed by joining 6 squares along their edges. Circle the patterns that can be folded into a cube. There are 11 such patterns. (You may wish to copy patterns, cut them out, and experiment by folding.)

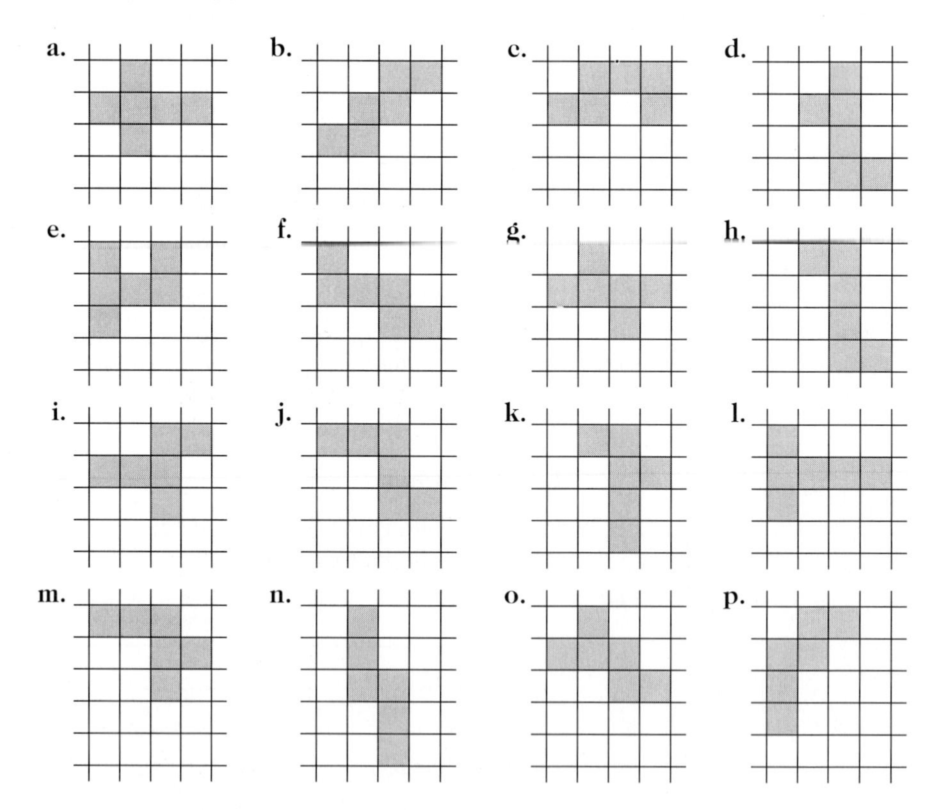

5. Archimedean Solids: A *semiregular polyhedron* is a polyhedron that has as faces two or more regular polygons and the same arrangement of polygons about each vertex. The 13 semiregular polyhedra shown on the next page are called *Archimedean solids.* These were known to Archimedes (287–212 B.C.) who wrote a book on these solids, but the book has been lost.

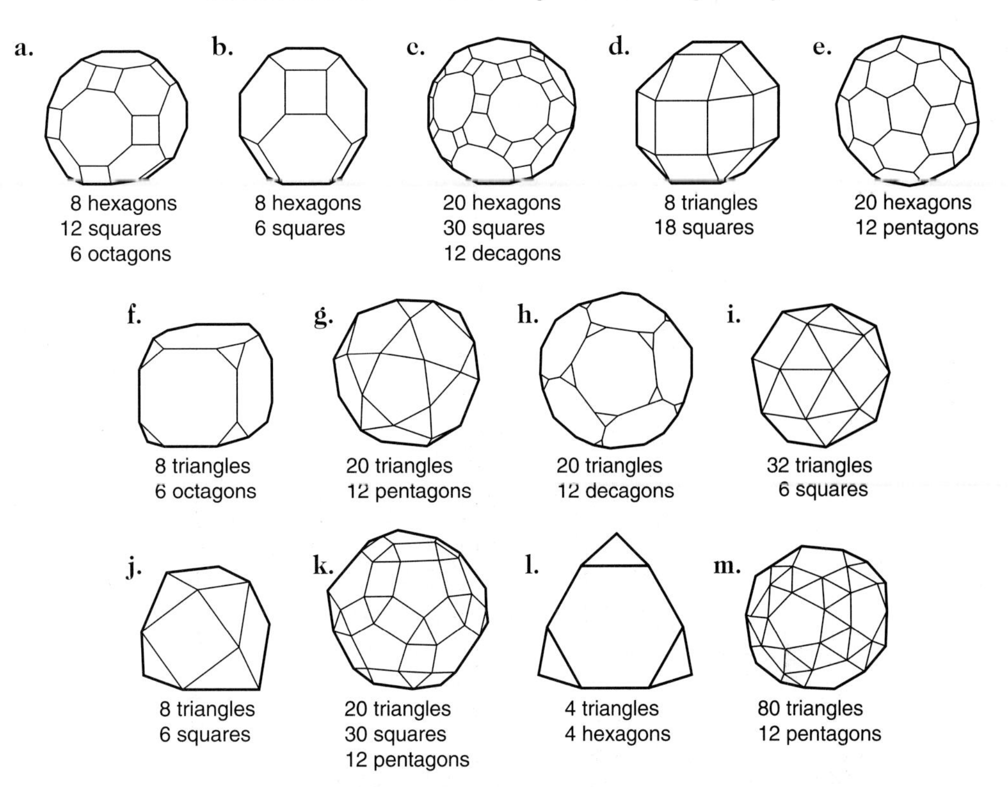

The seven Archimedean solids shown in b, e, f, g, h, j, and l can be obtained by truncating Platonic solids. For example, if each corner of a tetrahedron is cut off, as shown in the figure at the right, we obtain a truncated tetrahedron whose faces are regular hexagons and triangles, as shown in figure l above.

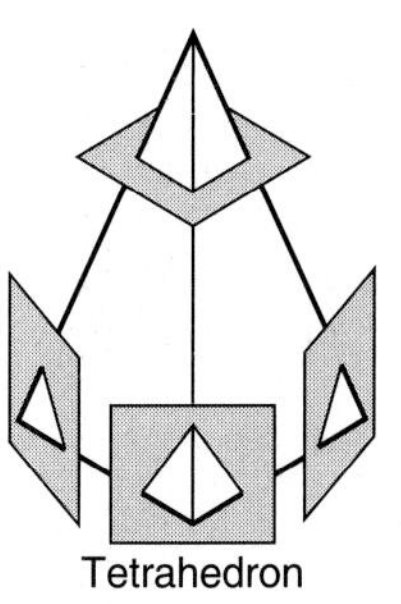

★ Identify which Platonic solid is truncated to obtain each of the semiregular solids in b, e, f, g, h, and j. (It may be helpful to look at the Platonic solids you constructed in activity 1.)

PUZZLER

One square is subdivided into 7 smaller squares. Subdivide the other square into 6 smaller squares.

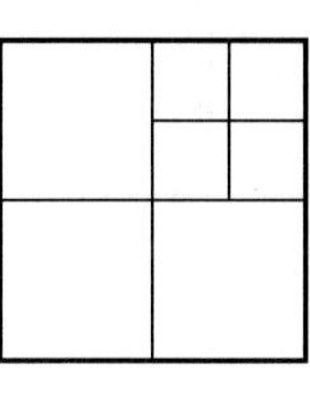

JUST FOR FUN

INSTANT INSANITY

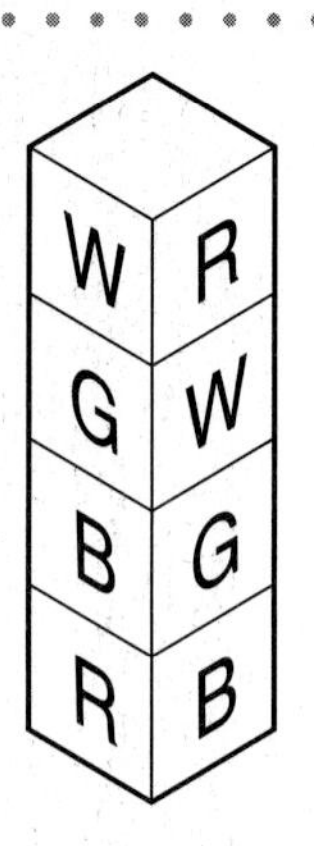

Instant Insanity is a popular puzzle that was produced by Parker Brothers. The puzzle consists of four cubes with their faces colored either red, white, blue, or green. The object of the puzzle is to stack the cubes so that each side of the stack (or column) has each of the four colors.

Material Card 36 contains patterns for a set of four cubes. Cut out and assemble the cubes and try to solve the puzzle.

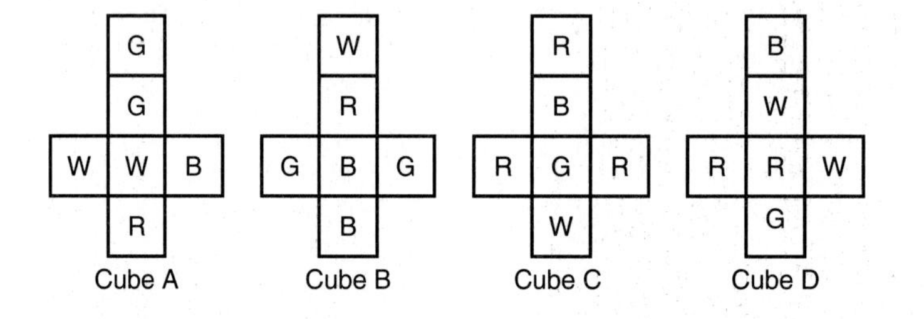

Robert E. Levin described the following method for solving this puzzle.[6] He numbered the faces of the cubes, using 1 for red, 2 for white, 3 for blue, and 4 for green, as shown below. To solve the puzzle you must get the numbers 1, 2, 3, and 4 along each side of the stack. The sum of these numbers is 10, and the sum of the numbers on opposite sides of the stack must be 20.

Levin made the following a table showing opposite pairs of numbers on each cube and their sums. For example, 4 and 2 are on opposite faces of cube A and their

G 4
G 4
W 2 W 2 B 3
R 1
Cube A

W 2
R 1
G 4 B 3 G 4
B 3
Cube B

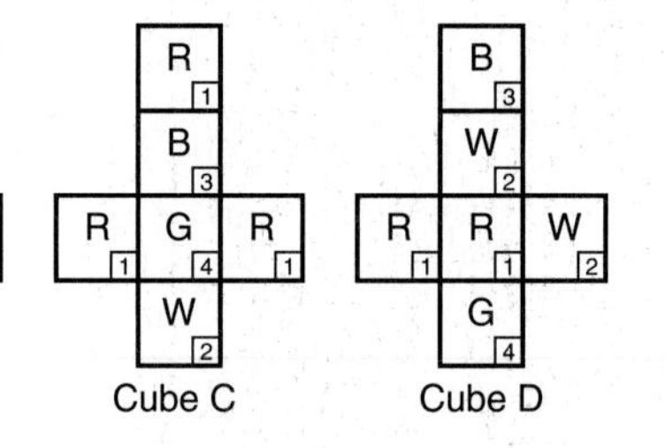

sum is 6. The bottom row of the table contains the sums of opposite faces for each cube. For one combination of numbers whose sum is 20, the sums have been circled. Notice that above these circled numbers, the numbers 1, 2, 3, and 4 each occur only twice. This tells you how to stack the cubes so that two opposite sides of the stack will have a total sum of 20. (Both of these sides will have all four colors.) The two remaining sides of the stack must also have a sum of 20. Find four more numbers from the bottom row (one for each cube) such that their sum is 20 and each of the numbers 1, 2, 3, and 4 occurs exactly twice among the faces. The remaining two faces of each cube will be its top and bottom faces when it is placed in the stack.

	CUBE A	CUBE B	CUBE C	CUBE D
Pairs of opposite faces	4 4 3 2 1 2	4 1 2 4 3 3	1 1 3 1 4 2	1 1 4 3 2 2
Sums of pairs	(6) 5 5	8 4 (5)	2 (5) 5	(4) 3 6

[6] R. Levin, "Solving Instant Insanity," *Journal of Recreational Mathematics* 2 (July 1969): 189–192.

ACTIVITY SET 9.4

CREATING SYMMETRIC FIGURES: PATTERN BLOCKS AND PAPER FOLDING

Purpose To introduce visual investigations of symmetry

Materials Pattern blocks from the Manipulative Kit, Material Card 13, scissors, and pieces of paper

Activity The *wind rose,* shown here, is a mariner's device for charting the direction of the wind. The earliest known wind rose appeared on the ancient sailing charts of the Mediterranean pilots, who charted eight principal winds. These are marked on our eight-pointed wind rose. Later, half-winds led to a wind rose with 16 points, and quarter-winds brought the total number of points to 32.[7]

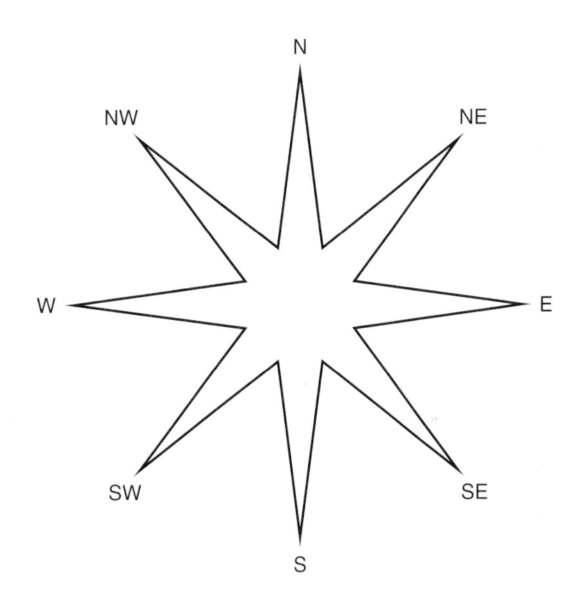

The eight-pointed wind rose is highly symmetric. It has eight lines of symmetry and eight rotational symmetries. For example, a line through any two opposite vertices of the wind rose is a *line of symmetry* because when the wind rose is folded on this line, the two sides will coincide. The wind rose has *rotational symmetries* of 45°, 90°, 135°, 180°, 225°, 270°, 315°, and 360° because, when rotated about its center through any one of these angles, the figure will coincide with the original position of the wind rose.

Activities 3 to 6 have several basic patterns for cutting out symmetric figures. The variety of figures that can be obtained from slight changes in the angles of the cuts is surprising. The eight-pointed wind rose can be cut from one of the patterns—try to predict which one as you do the folding and cutting.

[7]M. C. Krause, "Wind Rose, the Beautiful Circle," *Arithmetic Teacher* 20 (May 1973): 375.

Pattern Block Symmetries

1. Use the square and the white rhombus from your pattern block pieces to form figure 1 below. Notice that when another rhombus is attached as in figures 2 and 3, the new figures each have one line of symmetry. When the rhombus is attached as in figure 4, there are no lines of symmetry but there is rotational symmetry of 180°. Finally, when the rhombus is attached as in figure 5, there are no lines of symmetry or rotational symmetries.

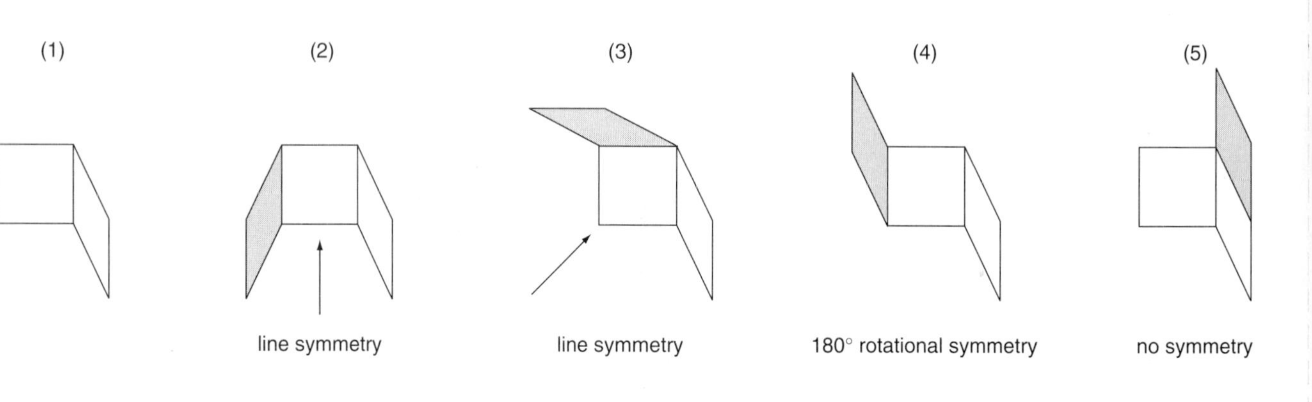

★ **a.** Form the following figure with your pattern block pieces. Determine the different ways you can attach another trapezoid to create figures with line symmetry. Use diagrams to record your results

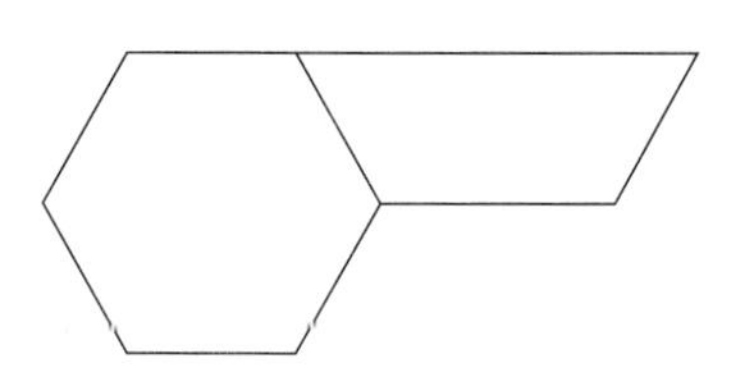

b. Do the same as in part a but find ways to create rotational symmetry (other than 360°).

★ **c.** By attaching 6 trapezoids to the hexagon, create figures that have the following properties. Make a sketch to record your results.

Figure 1: One line of symmetry and no rotational symmetries

Figure 2: Six rotational symmetries but no lines of symmetry

Figure 3: Three rotational symmetries and three lines of symmetry

Figure 4: No symmetries

2. a. Use your pattern blocks and Material Card 13 to build this design. Complete the design by adding additional pattern blocks so that the design is symmetric about both perpendicular lines. Sketch the completed design here. Does the completed design have any rotational symmetries? If so, how many?

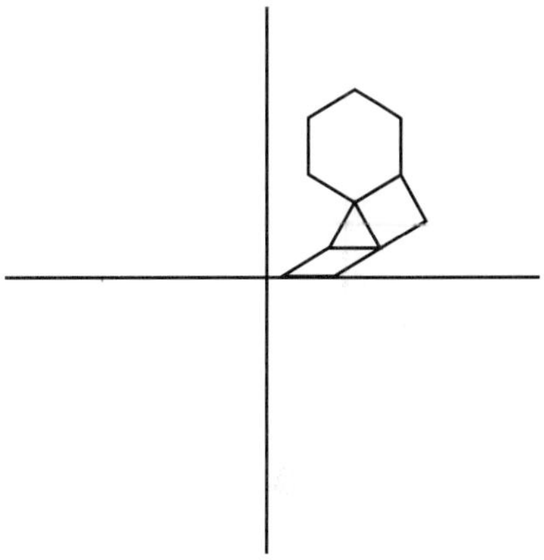

b. Build the design shown here and use additional pattern blocks to complete a design which has rotational symmetries of 90°, 180°, and 270° about the intersection of the perpendicular lines. Sketch the completed design on the figure. Does the completed design have any line symmetries?

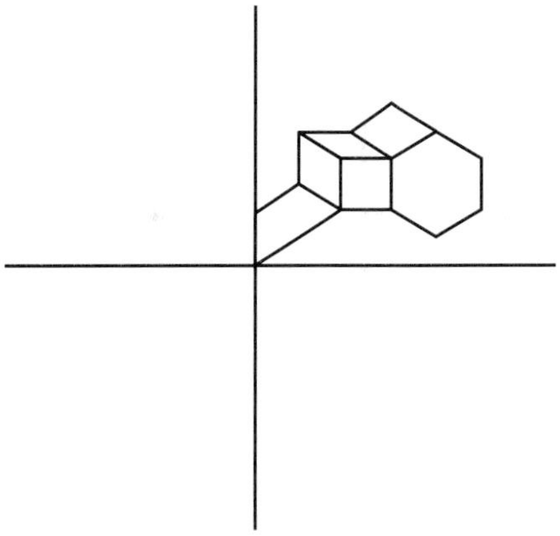

c. Arrange the pattern blocks as shown in this figure. Add additional pattern blocks so that the completed design has rotational symmetry of 90°, 180°, and 270°. Sketch the completed design here.

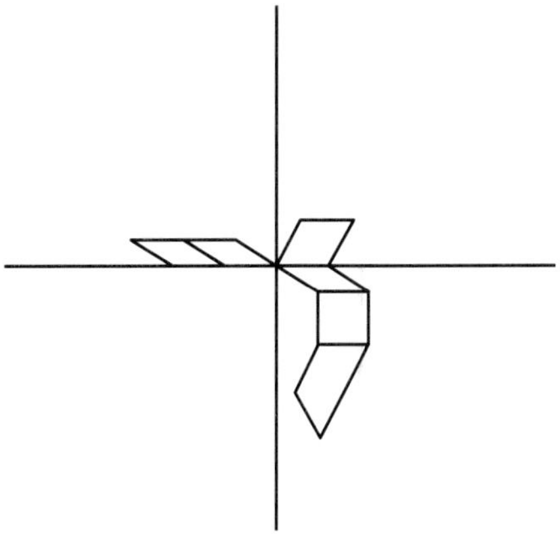

Paper-Folding Symmetries

3. Fold a rectangular piece of paper in half twice, making the second fold perpendicular to the first. Let C be the corner of all folded edges. Hold the folded paper at corner C and make one cut across as shown by segment AB on the diagram. Before opening the folded corner, predict what the shape will be and how many lines of symmetry it will have.

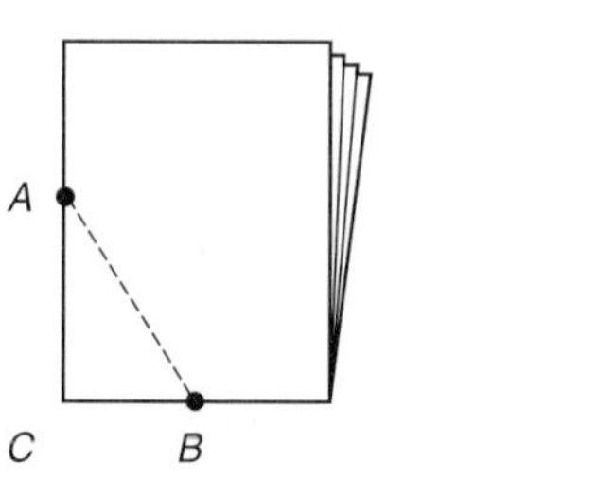

PREDICTED FIGURE	ACTUAL FIGURE

a. Is it possible to draw a segment AB and make one cut across the folds so that the piece will unfold to a square? If so, describe how you made your cut. If not, explain why it cannot be done.

b. Experiment with other single cuts that start at point A but which exit from any of the other three sides of the twice-folded paper. Continue to predict before you unfold. Make a list of the different types of figures you obtain.

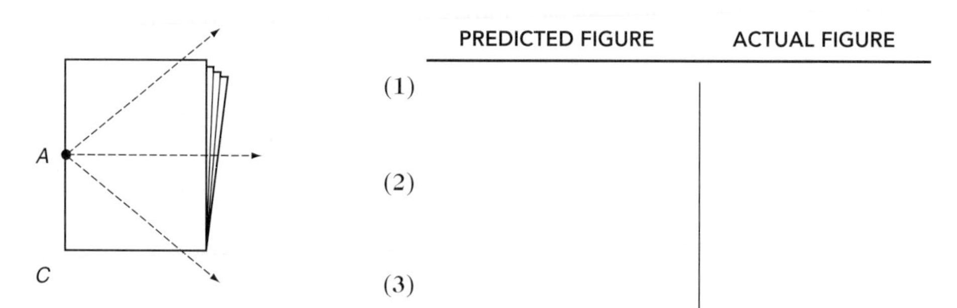

4. Fold a rectangular piece of paper in half twice, making the second fold perpendicular to the first. Let C be the corner of all the folded edges. Prepare to make two cuts from the edges to an inside point, as shown here. Before cutting the paper, predict what kind of polygon you will get, how many lines of symmetry the figure will have, and whether the figure will be convex or nonconvex. Cut and check your predictions.

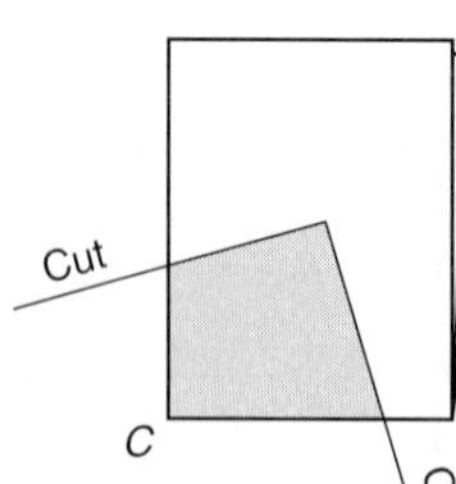

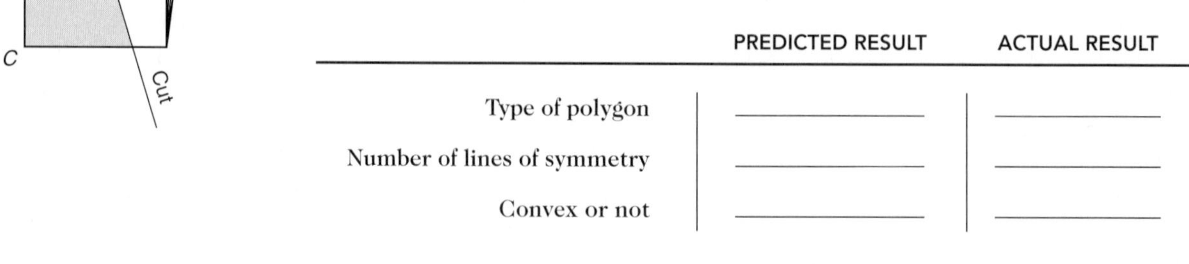

	PREDICTED RESULT	ACTUAL RESULT
Type of polygon	______	______
Number of lines of symmetry	______	______
Convex or not	______	______

★ **a.** Find a way to make two cuts into an inside point so that you get a regular octagon; a regular hexagon. Sketch the location of your cuts on these folded-paper diagrams.

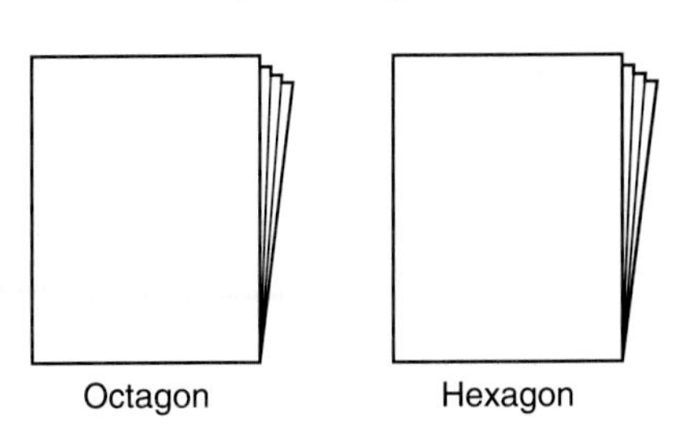

Octagon Hexagon

b. Make a sketch of the figure you think will result when you cut a piece of double-folded paper as shown below. Then cut the paper and sketch the result.

PREDICTED RESULT	ACTUAL RESULT

c. Vary the angle of the cuts in part b. Sketch lines to show what cuts will result in a rhombus inside a square; a square inside a rhombus. Sketch the resulting figures.

Cuts for rhombus inside square

Sketch of actual results

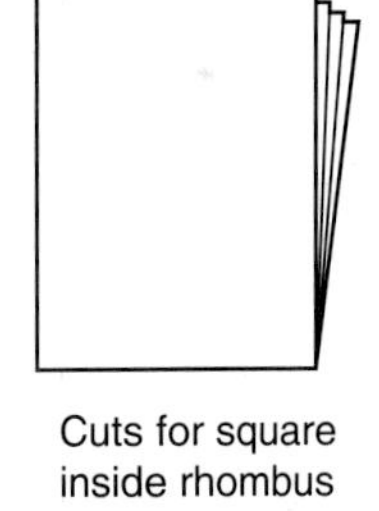

Cuts for square inside rhombus

Sketch of actual results

5. The following pattern leads to a variety of symmetrical shapes. Begin with a standard sheet of paper as in figure a, and fold the upper right corner down to produce figure b. Then fold the upper vertex R down to point S to obtain figure c. To get figure d, fold the two halves inward so that points S and T are on line l.

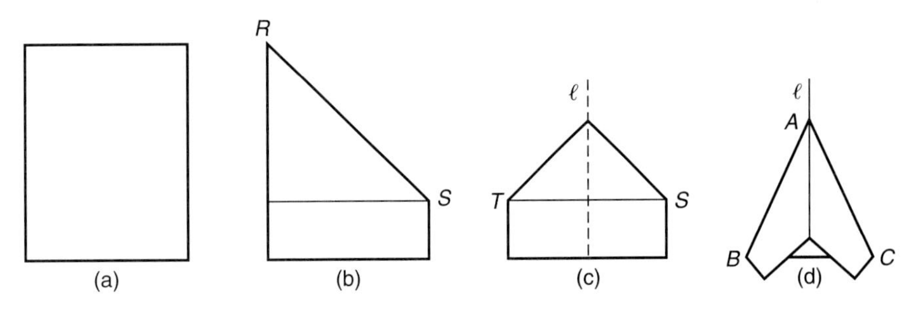

Make a horizontal cut, as indicated on this diagram, and draw a sketch of the resulting figure.

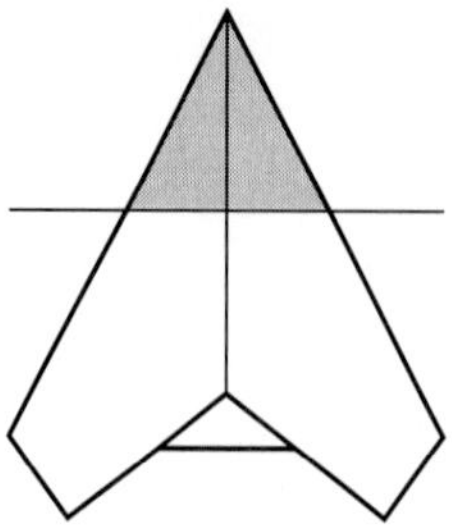

RESULTING FIGURE

For parts a–d, draw a sketch of the figure you predict will result from the indicated cut. Then fold the paper, make the cut, and draw the actual result. Determine the number of lines of symmetry for each.

a. Slanted cut

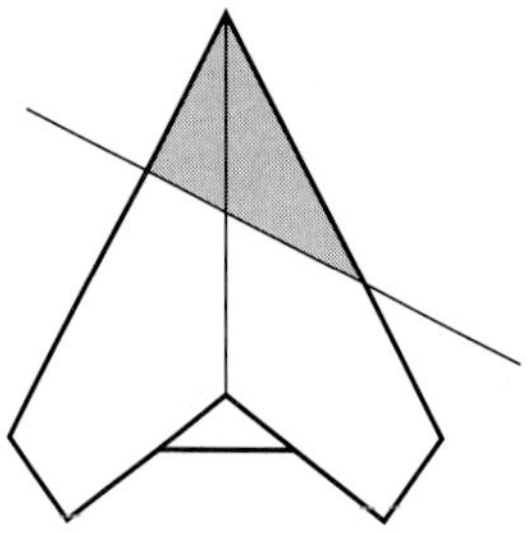

PREDICTED FIGURE	ACTUAL RESULT

b. Slanted cut (reversed)

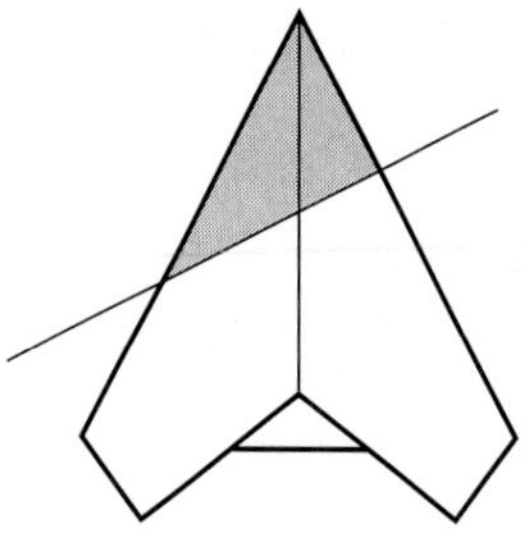

PREDICTED FIGURE	ACTUAL RESULT

c. Combination of two slanted cuts

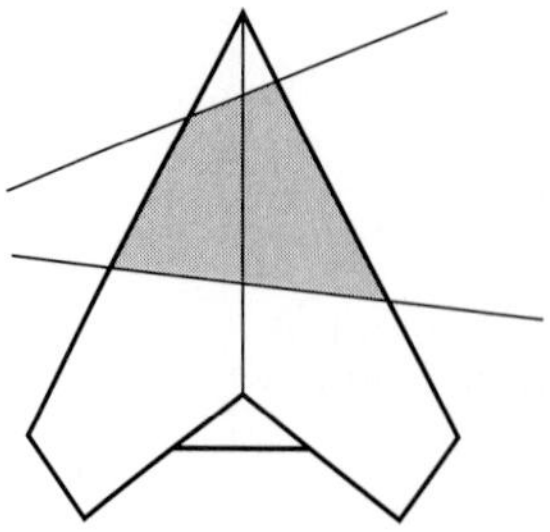

PREDICTED FIGURE	ACTUAL RESULT

★ **d.** Two cuts to an inside point on the center line

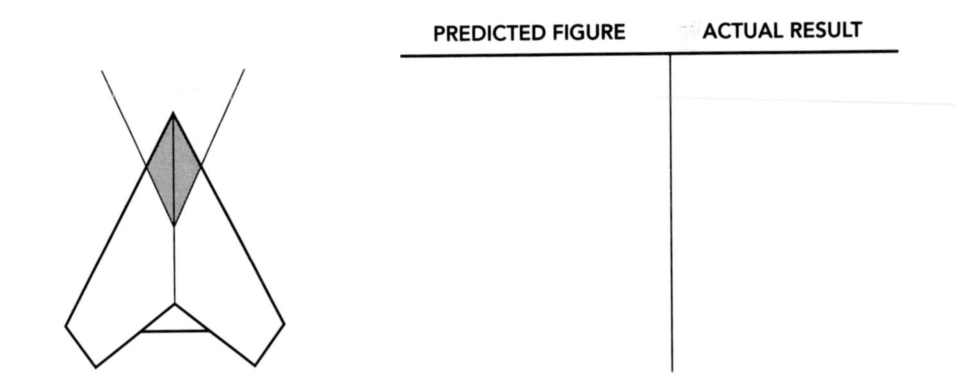

★ **e.** Find two cuts into a point on the center line that will produce a regular 16-sided polygon, and sketch them on this diagram.

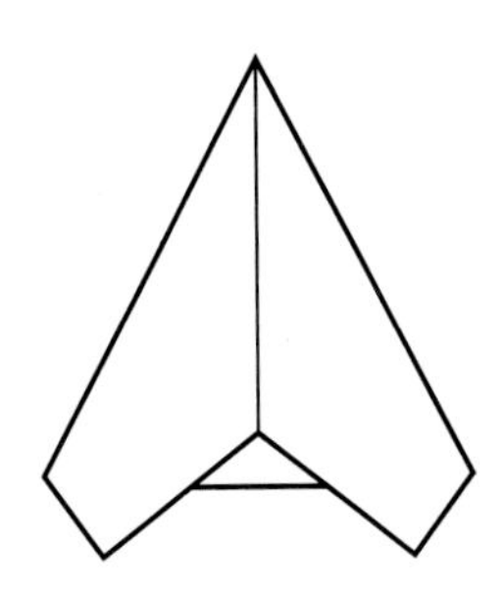

6. The five-pointed star has been used for badges and national symbols for centuries. It appears today on the flags of 41 countries and was once used on the back of a United States $4 piece.

   The four steps pictured in figures a through d illustrate a paper-folding approach to making a five-pointed star. To obtain figure a, fold a standard sheet of paper perpendicular to its longer side. Next, fold point C over to midpoint M of side $\overline{AD}$, as shown in figure b. To get figure c, fold up the corner containing point D and crease along $\overline{MZ}$. For the final step, fold the right side of figure c over to the left side so that edge $\overline{NZ}$ lies along edge $\overline{MZ}$.

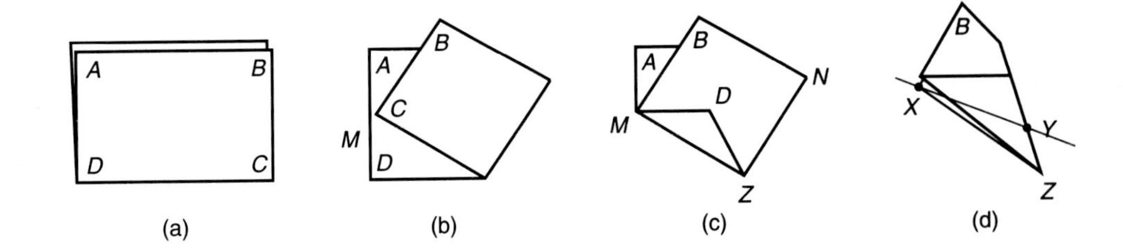

   a. Fold a piece of paper according to the instructions, make a cut from X to Y, and sketch the resulting figure.

b. Find a way to make a single cut so that the resulting figure is a regular pentagon; a regular decagon. Sketch the cut lines on the figures below.

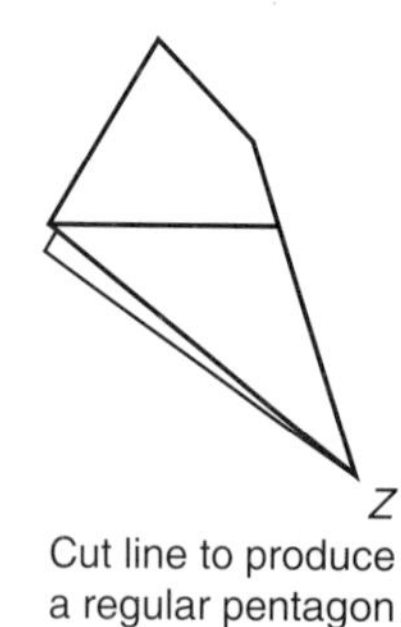

Cut line to produce a regular pentagon

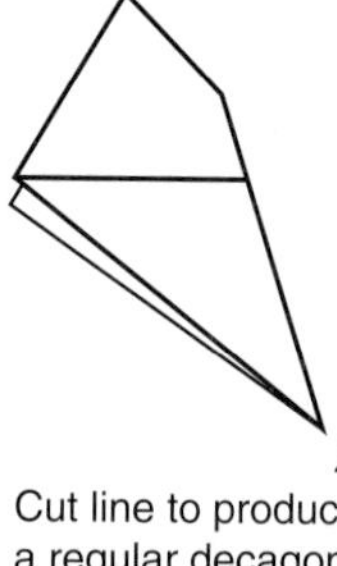

Cut line to produce a regular decagon

JUST FOR FUN

SNOWFLAKES

Snow is the only substance that crystallizes in many different figures. Yet in spite of the variations in the figures, all snow crystals have a common characteristic—their hexagonal shape.

A New England farmer, W. A. Bentley, began looking at snowflakes when he was given a microscope at the age of 15. A few years later he was given a camera, which he adapted to his microscope to take photographs of snowflakes. Shortly before his death in 1931, Bentley published a book containing 2500 pictures of snow crystals. His work has been used by artists, photographers, illustrators, jewelers, meteorologists, and crystallographers.[8]

To create your own snowflake, fold a standard-size sheet of paper according to the following directions. (Tracing paper or lightweight paper is easiest to fold and cut.)

1. Fold the paper in half twice, making the second fold perpendicular to the first. In figure a the corner C is also the center of the original piece of paper.
2. Fold to find the center line parallel to the longer edges, as indicated in figure b.
3. Bring corner X up to the center line and crease the paper along $\overline{CY}$, as shown in figure c.
4. Fold corner B back until side $\overline{CB}$ lies along side $\overline{CY}$, and crease along $\overline{CY}$, as in figure d.
5. Cut off the portion of paper above $\overline{XY}$, as shown in figure e.

Now cut out designs along sides $\overline{CY}$, $\overline{CX}$, and $\overline{XY}$. Unfold the finished product carefully to examine your snowflake design. What is its shape? What lines of symmetry and what rotational symmetries does it have? Experiment with other designs.

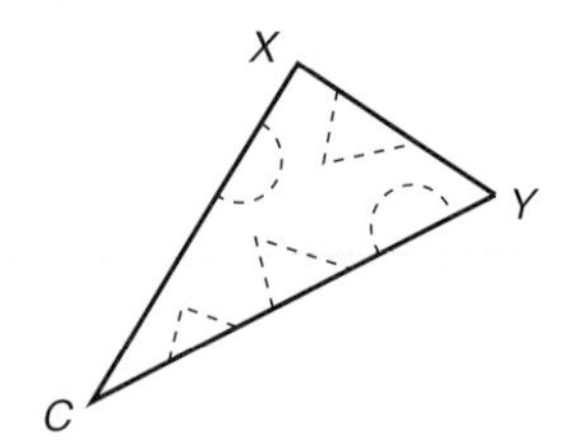

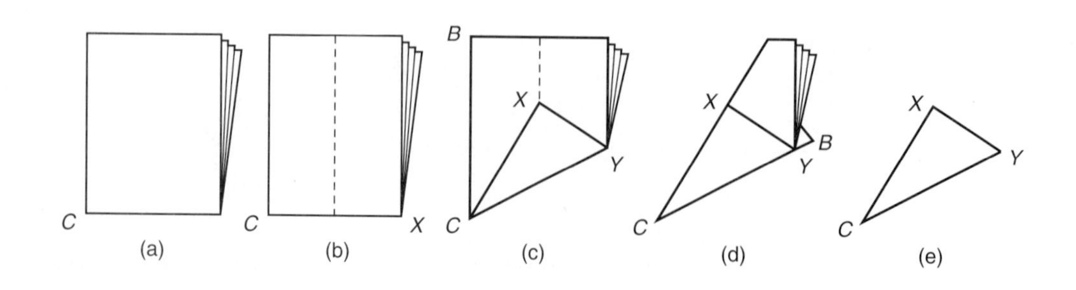

[8]F. Hapgood, "When Ice Crystals Fall from the Sky Art Meets Science," *Smithsonian* (January 1976): 67–72.

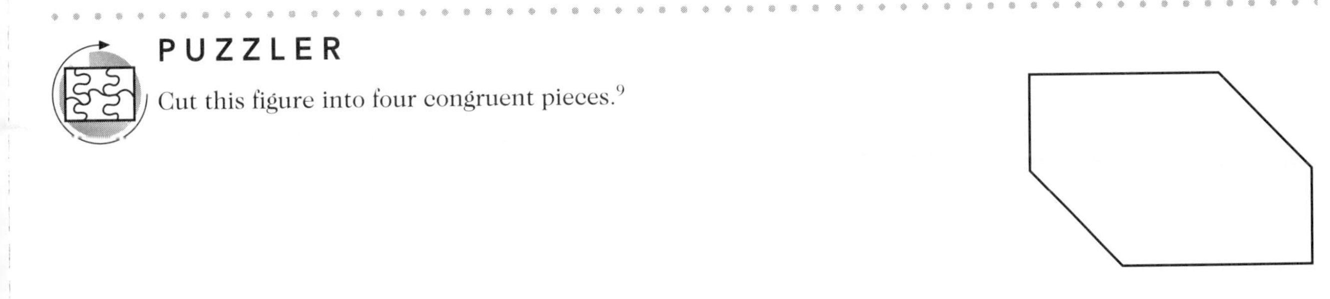

PUZZLER

Cut this figure into four congruent pieces.[9]

IDEAS FOR THE ELEMENTARY CLASSROOM

SUGGESTED CLASSROOM ACTIVITY: PAPER-FOLDING SYMMETRIES

Children can become actively involved in investigating symmetry through folding and cutting paper.

1. Single or double fold a sheet of paper, and have students cut a design and display the symmetries of the result.
2. Do as in activity 1, but ask the students to draw a sketch of what they think their design will look like before they unfold the paper.
3. Have the students double fold (folds perpendicular) a sheet of paper (a); draw a design on one side (b); draw a sketch to predict the outcome before cutting; and then cut, unfold, and compare the result to their prediction.

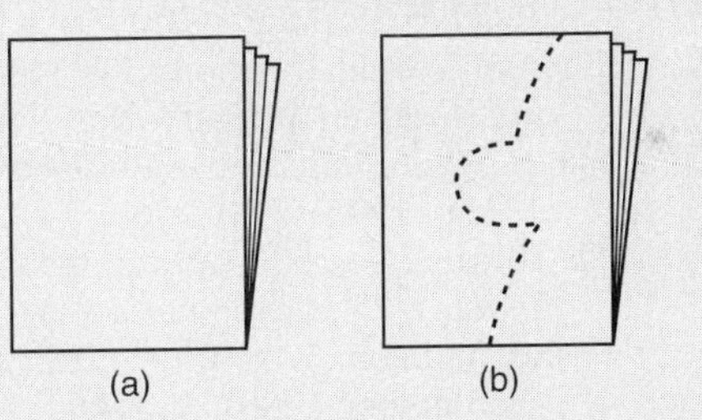

4. Show the students some symmetric patterns you have made and ask them to make similar patterns by folding and cutting.

Readings for More Classroom Ideas

Arcidiacono, Michael, D. Fielker, and E. Maier. "Seeing Symmetry," in *Math and the Mind's Eye* (Unit ME10). Salem, OR: The Math Learning Center, Box 3226, 1996.

Barson, Alan, and L. Barson. "Ideas: Geometry Activities." *Arithmetic Teacher* 35 (April 1988): 27–36.

Bennett, A., E. Maier, and L. Ted Nelson. "Looking at Geometry," in *Math and the Mind's Eye* (Unit ME5). Salem, OR: The Math Learning Center, Box 3226, 1988.

[9]From the "October Calendar" of *The Mathematics Teacher* (October 1988): 556–557.

Bidwell, James K. "Using Reflections to Find Symmetric and Asymmetric Patterns." *Arithmetic Teacher* 34 (March 1987): 10–15.

Carroll, William M. "Cross Sections of Clay Solids." *Arithmetic Teacher* 35 (March 1988): 6–11.

Carroll, William M. "Polygon Capture: A Geometry Game." *Mathematics Teaching in the Middle School* 4 (October 1998): 90–94.

Claus, Alison. "Exploring Geometry." *Arithmetic Teacher* 40 (September 1992): 14–17.

Clauss, Judith Enz. "Pentagonal Tessellations." *Arithmetic Teacher* 38 (January 1991): 52–56.

Cook, Marcy. "Ideas [Rectangle Activities]." *Arithmetic Teacher* 36 (March 1989): 27–32.

Del Grande, John, and Lorna Morrow. *Geometry and Spatial Sense,* in the *Curriculum and Evaluation Standards for School Mathematics Addenda Series: Grades K–6.* Reston, VA: National Council of Teachers of Mathematics, 1993.

Dutch, Steven I. "Folding *n*-pointed Stars and Snowflakes." *Mathematics Teacher* 87 (November 1994): 630–637.

Evered, Lisa J. "Folded Fashions: Symmetry in Clothing Design." *Arithmetic Teacher* 40 (December 1992): 204–206.

Flores, Alfinio. "Bilingual Lessons in Early-Grades Geometry." *Teaching Children Mathematics* 1 (March 1995): 420–424.

Fosnaugh, Linda S., and Marvin E. Harrell. "Covering the Plane with Reptiles." *Mathematics Teaching in the Middle School* 1 (January–February 1996): 666–670.

Geddes, Dorothy, et al. *Geometry in the Middle Grades,* in the *Curriculum and Evaluation Standards for School Mathematics Addenda Series: Grades 5–8.* Reston, VA: National Council of Teachers of Mathematics. 1992.

Happs, John, and Helen Mansfield. "Research into Practice: Estimation and Mental-Imagery Model in Geometry." *Arithmetic Teacher* 40 (September 1992): 44–46.

Kriegler, Shelley. "The Tangram—It's More Than an Ancient Puzzle." *Arithmetic Teacher* 38 (May 1991): 38–43.

Leeson, Neville J. "Improving Students' Sense of Three-Dimensional Shapes." *Teaching Children Mathematics* 1 (September 1994): 8–11.

Maupin, Sue. "Middle-Grades Geometry." *Mathematics Teaching in the Middle School* 1 (May 1996): 790–796.

McClintock, Ruth. "Animating Geometry Discussions with Flexigons." *Mathematics Teacher* 87 (November 1994): 602–606.

Miller, William A. "Puzzles That Section Regular Solids." *Mathematics Teacher* 81 (September 1988): 463–468.

Morrow, Lorna J. "Implementing the Standards: Geometry Through the Standards." *Arithmetic Teacher* 38 (April 1991): 21–25.

Naylor, Michael. "The Amazing Octacube." *The Mathematics Teacher* 92 (February 1999): 102–104.

Onslow, Barry, "Pentominoes Revisited." *Arithmetic Teacher* 37 (May 1990): 5–9.

Reys, Robert E. "Discovery with Cubes." *Mathematics Teacher* 81 (May 1988): 377–381.

Rubenstein, Rheta N., Glenda Lappan, Elizabeth Phillips, and William Fitzgerald. "Angle Sense: A Valuable Connector." *Arithmetic Teacher* 40 (February 1993): 352–358.

Serra, Michael. *Patty Paper Geometry.* Berkeley, CA: Key Curriculum Press, 1994.

Seymour, D., and J. Britton. "Creating Escher-Like Tessellations." *Introduction to Tessellations.* Palo Alto, CA: Dale Seymour Publications, 1989, 1–59.

Souza, Ronald. "Golfing with a Protractor." *Arithmetic Teacher* 35 (April 1988): 52–56.

Sundberg, Sue E. "A Plethora of Polyhedra." *Mathematics Teaching in the Middle School* 3 (March–April 1998): 388–391.

van Hiele, Pierre M. "Developing Mathematical Thinking through Activities That Begin with Play." *Teaching Children Mathematics* 5 (February 1999): 310–316.

Wheatley, Grayson H. "Research into Practice: Spatial Sense and the Construction of Abstract Units in Tiling." *Arithmetic Teacher* 39 (April 1992): 43–45.

Williams, Carol G. "Sorting Activities for Polygons." *Mathematics Teaching in the Middle School* 3 (March–April 1998): 441–445.

Wilson, Patricia, and Verna Adams. "A Dynamic Way to Teach Angle and Angle Measure." *Arithmetic Teacher* 39 (January 1992): 6–13.

Zaslavsky, Claudia. "Symmetry in American Folk Art." *Arithmetic Teacher* 38 (September 1990): 6–12.

CHAPTER 10

MEASUREMENT

Children can see the usefulness of measurement if classroom experiences focus on measuring real objects, making objects of given sizes, and estimating measurements. Textbook experiences cannot substitute for activities that use measurement to answer questions about real problems.[1]

ACTIVITY SET 10.1

MEASURING WITH METRIC UNITS

Purpose To use estimation and measurement activities to introduce the basic metric units

Materials A metric ruler is on Material Card 32, and a metric measuring tape is on Material Card 14

© 1974 United Feature Syndicate, Inc.

Activity The metric system of measurement is used by most nations of the world. The basic metric units for length, weight, and volume are the *meter* (a little longer than a yard), the *gram* (about the weight of a paper clip), and the *liter* (a little more

[1]*Curriculum and Evaluation Standards for School Mathematics* (Reston, VA: National Council of Teachers of Mathematics, 1989), 53.

than a quart). Larger and smaller measures are obtained by using the following prefixes. The four prefixes marked with an asterisk are most commonly used. There are additional prefixes which extend this system to even larger and smaller measures for scientific purposes.

kilo*	1000 times
hecto	100 times
deka	10 times
deci*	$\frac{1}{10}$
centi*	$\frac{1}{100}$
milli*	$\frac{1}{1000}$

The prefixes differ by powers of 10. For example,

10 millimeters (mm) = 1 centimeter (cm)
10 centimeters = 1 decimeter (dm)
10 decimeters = 1 meter (m)
1000 meters = 1 kilometer (km)

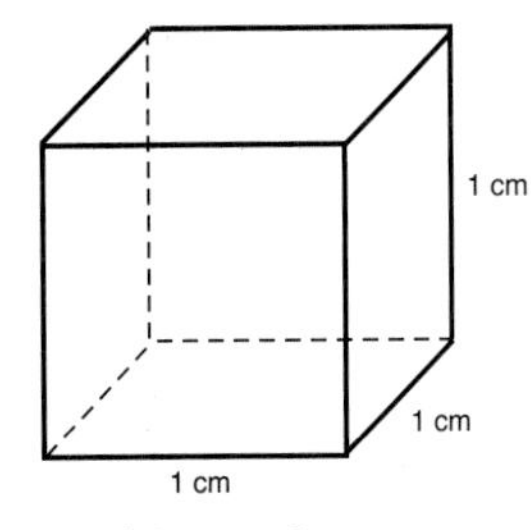

One cubic centimeter (1 cm^3) equals 1 milliliter (ml) and, if water, has a mass of 1 gram (g).

Here is one simple relationship among the metric units for length, weight, and volume: One cubic centimeter of water equals one milliliter of water and weighs about one gram.

The following activities involve the metric units for length and volume. If you are not familiar with the metric system and are unaccustomed to thinking of length in terms of centimeters and meters, the estimation and measuring game in activity 1 will give you expertise quickly. Another measuring game is in Just for Fun, at the end of this activity set.

1. **Centimeter and Meter Guessing Games (2 players):** One player marks two points on a sheet of paper. The opponent guesses the distance in centimeters between these two points. The distance is then measured with a ruler. If the guess is within 1 centimeter of being correct, the guessing player scores 1 point. If not, there is no score. The players alternate picking points and guessing distances until someone has scored 11 points.

   Challenging: If a player thinks an opponent's guess is not within 1 centimeter, that player may challenge the guess by giving his or her own estimate of the distance. If the challenger's estimate is within 1 centimeter and the opponent's is not, then the challenger receives 1 point and the opponent receives no score. However, if the opponent is within 1 centimeter, he or she receives 2 points for the insult, regardless of the challenger's estimate, and the challenger receives no score.

   Variation 1: Players estimate the lengths of objects around them. Since they will be estimating in centimeters, these objects should be reasonably small, such as pencils, books, and dollar bills.

   Variation 2: This game is called the Meter Guessing Game. On a player's turn, he or she selects an object in the room, and the opponent guesses its length in meters. If the guess is within $\frac{1}{2}$ meter, the guessing player scores 1 point. If not, there is no score. The players alternate picking objects and guessing lengths. As before, a player's guess may be challenged.

2. Your handspan is a convenient ruler for measuring lengths. Stretch out your fingers and use your metric ruler or metric measuring tape to measure the maximum distance from your little finger to your thumb in centimeters.

   a. Use your handspan to approximate the width and length of this page in centimeters. Record your results below and then check your approximations by measuring.

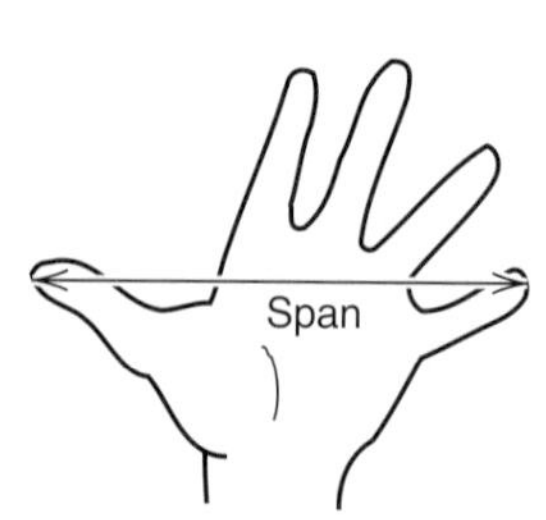

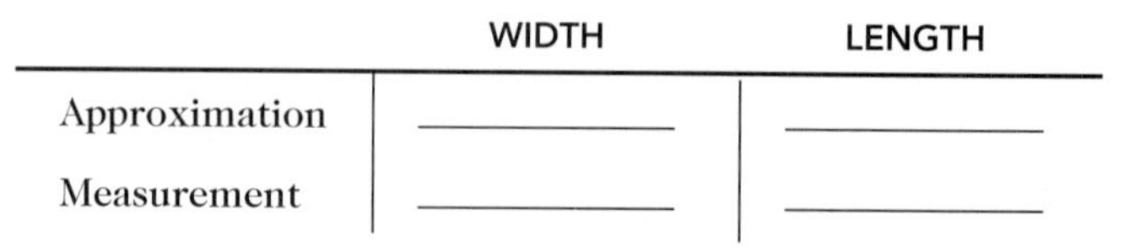

	WIDTH	LENGTH
Approximation		
Measurement		

   b. Select some objects around you and estimate their lengths in centimeters. Check your estimates by measuring. Repeat this activity 10 times, and see how many distances you can predict to within 1 centimeter. Record the objects, your estimates of lengths, and your measurements. In the last column record the difference between your estimation and measurement as a positive number, and then compute the average difference.

OBJECT	ESTIMATION	MEASUREMENT (nearest cm)	DIFFERENCE
		Average of differences	

3. A millimeter is the smallest unit on your metric ruler. Obtain the measurements of the objects in parts a through c to the nearest millimeter.

★ a. The length of a dollar bill

★ b. The diameter of a penny

   c. The width of one of your fingernails. Do you have one with a width of 1 centimeter?

d. What fraction of a millimeter is the thickness of one page of this book? (*Hint:* Measure several pages at once.)

4. This table contains some body measurements that are useful for buying and making clothes. Estimate or use your handspan to approximate these measurements. Then use your metric measuring tape and your metric ruler to obtain these measurements to the nearest centimeter.

	ESTIMATE	MEASUREMENT
Waist	________	________
Height	________	________
Arm length	________	________
Foot length	________	________
Width of palm	________	________
Circumference of head	________	________

5. Use your metric measuring tape (Material Card 14) for the following.

a. Measure the distance from the middle of your chest to your fingertips with your arm outstretched. Is this distance greater than or less than 1 meter? This body distance was once used by merchants to measure cloth. It is the origin of the English measure called the yard.

b. Measure the distance from the floor up to your waist. How does this distance compare with a meter?

c. Select some objects around you whose lengths are greater than 1 meter, and try to estimate their length to within $\frac{1}{2}$ meter (50 cm). For example, estimate the height, width, and length of a room and then measure these distances with your metric measuring tape. Repeat this activity for 10 different objects, and see how many distances you can predict to within $\frac{1}{2}$ meter in the following table. Record the objects, your estimates of lengths, and the measurements in the table on the next page. In the last column of the table, record the difference between your estimation and measurement, as a positive number, and then compute the average of the differences.

OBJECT	ESTIMATION	MEASUREMENT (nearest $\frac{1}{2}$ m)	DIFFERENCE
		Average of differences	

6. Your *pace* is the distance you cover in one normal step. To gather information about your pace, measure off a distance of 10 meters. Record the time in seconds and the number of paces it takes you to walk this distance.

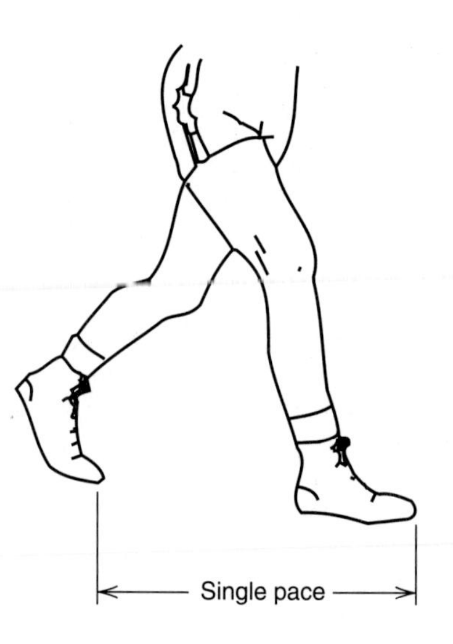

Number of paces needed to walk 10 meters: ________

Time, in seconds, needed to walk 10 meters: ________

Using this information, explain how to determine each of the following.

a. The length of a single pace, in centimeters

b. The number of paces it will take you to walk 1 kilometer

c. The time in minutes that it will take you to walk 1 kilometer

7. A liter is very close to the English measure the quart. A liter is equal to 1000 cubic centimeters, the volume of a 10 cm by 10 cm by 10 cm cube.

★ **a.** Use the dimensions on this drawing of a quart container to approximate the number of cubic centimeters in a quart. Which is greater, a liter or a quart?

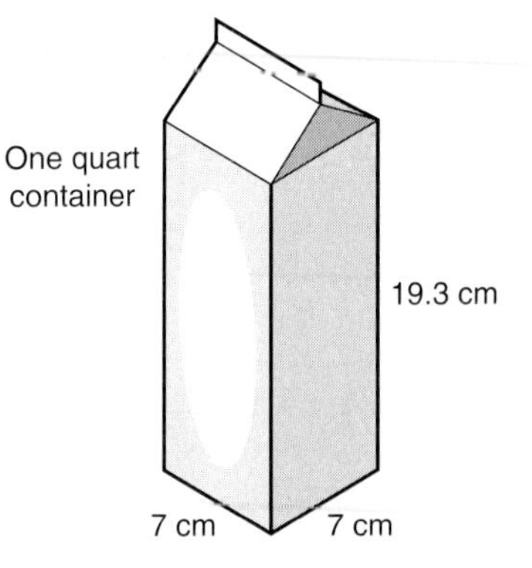

★ **b.** What height should be marked on a two-quart container to measure 1 liter?

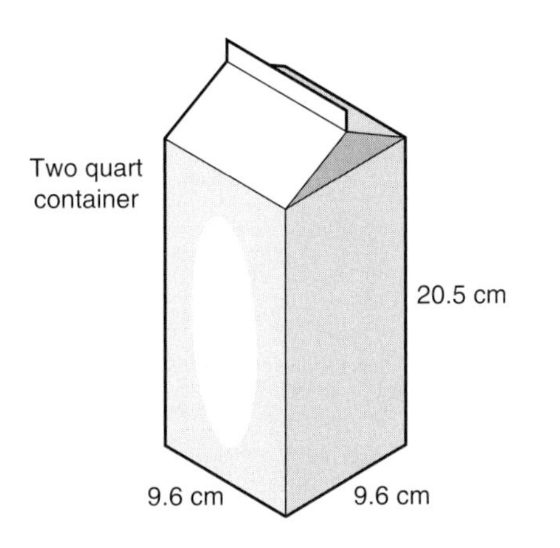

JUST FOR FUN

CENTIMETER RACING GAME (2 or more players)

This game is played on a racing mat (Material Card 15). The object of the game is to go from the center of the circle to a point inside the triangle, square, pentagon, and hexagon, in that order. Each player in turn rolls two dice. The two numbers may be added, subtracted, multiplied, or divided. The resulting computation is the number of centimeters the player moves. A metric ruler (Material Card 32) should be used to draw the line segment whose length is the player's number.

On each subsequent turn, players draw another line segment, beginning at the end of their last line segment. Before players can pass through a polygon or proceed from one polygon to the next, they must land inside the polygon with the end of a line segment. The first player to draw a line segment that ends inside the hexagon is the winner.

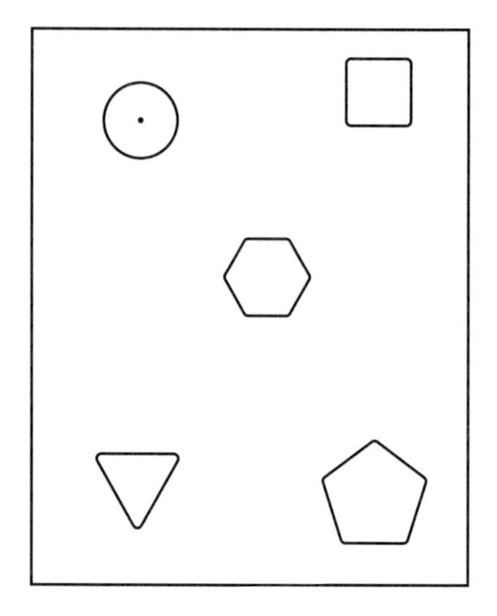

ACTIVITY SET 10.2

AREAS ON GEOBOARDS

Purpose To use rectangular geoboards to develop area concepts

Materials A rectangular geoboard and rubber bands. A template for making a rectangular geoboard is on Material Card 27.

Activity The rows and columns of nails on a rectangular geoboard, like the one shown at the left, outline 16 small squares. In this activity set, these small squares will serve as the unit squares for the areas of the geoboard figures. Finding the area of a geoboard polygon means determining the number of unit squares and partial squares needed to cover the polygon.

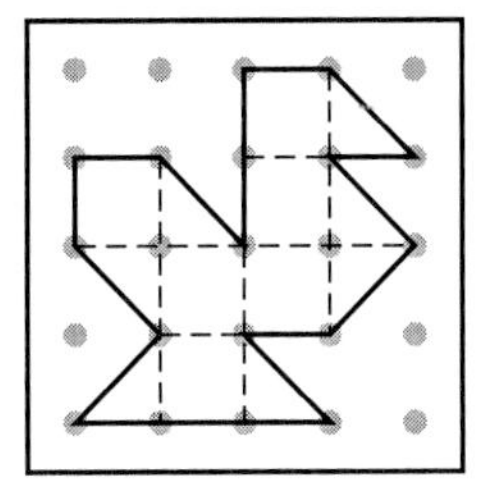

It takes 6 unit squares and 7 halves of unit squares to cover the figure on this geoboard, so its area is $9\frac{1}{2}$ unit squares.

In the following activities, you will examine different methods for determining areas of polygons and discover how the formulas for the areas of rectangles, parallelograms, and triangles are related.

1. Form each polygon on your geoboard, determine its area by counting squares and halves of squares, and record the area beneath the figure.

★ **a.** **b.** ★ **c.**

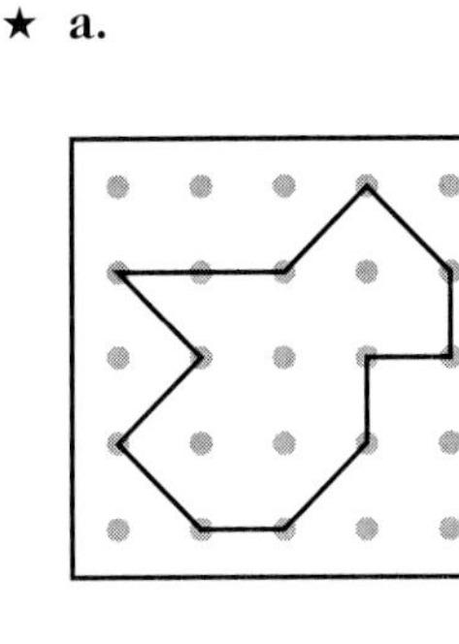

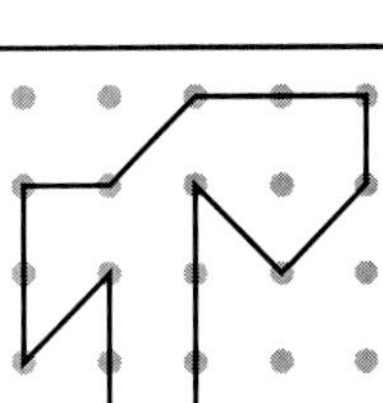

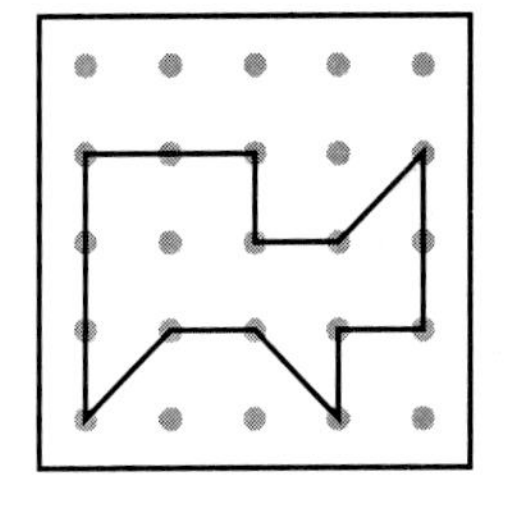

_____ _____ _____

d. ★ **e.** **f.**

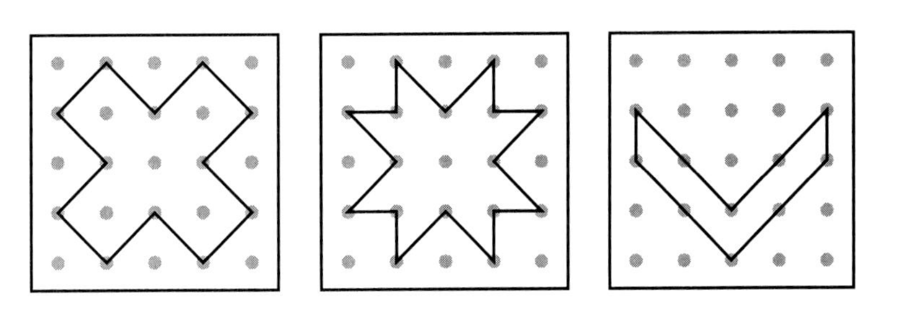

_____ _____ _____

2. In the figure for part a, the upper shaded triangle has an area of 1 square unit because it is half of a 2 by 1 rectangle. Similarly, the lower shaded triangle has an area of $1\frac{1}{2}$ square units because it is half of a 3 by 1 rectangle. Determine the areas of these figures by subdividing them into squares, halves of squares, and triangles that are halves of rectangles.

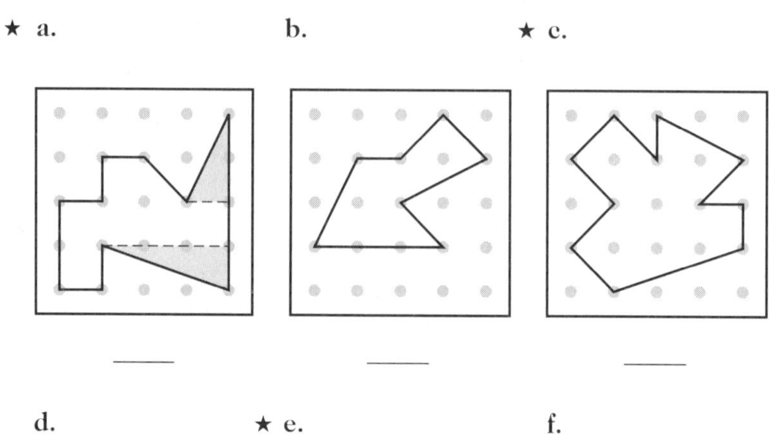

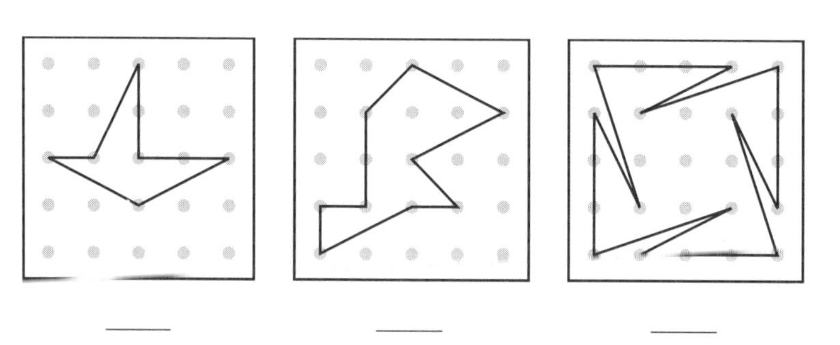

3. Sometimes it is easier to find the area outside a figure than the area inside. The hexagon in part a has been enclosed inside a square. What is the area of the shaded region? Subtract this area from the area of the 3 by 3 square to find the area of the hexagon. Use this technique to find the areas of the other figures.

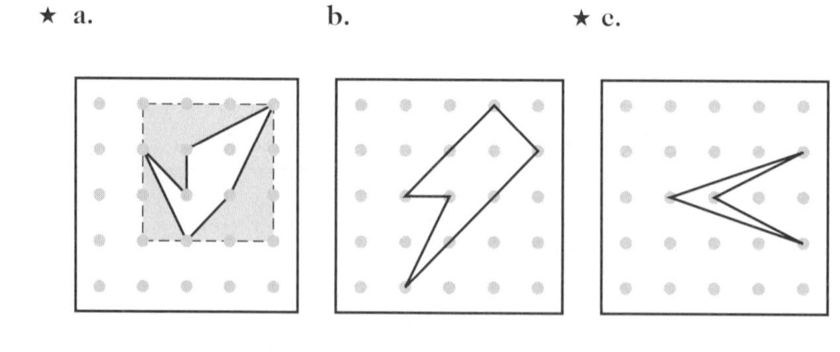

d. ★ e. f.

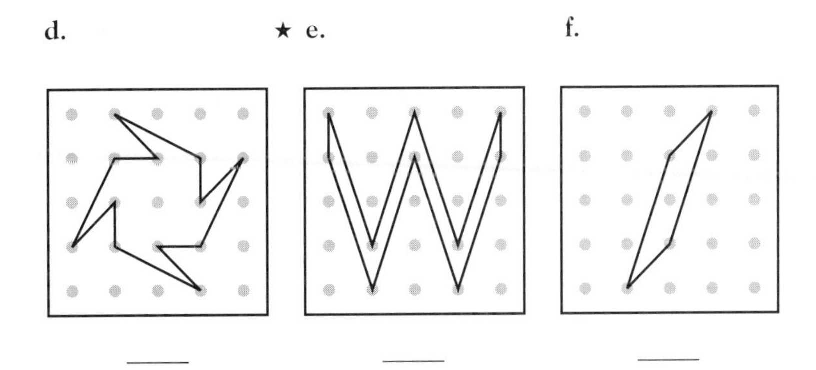

4. Determine the areas of the following figures using any technique or combination of techniques. Notice that the hexagon in part a has a vertical line of symmetry, so you can determine the area of half the region and double the result.

★ a. b. ★ c.

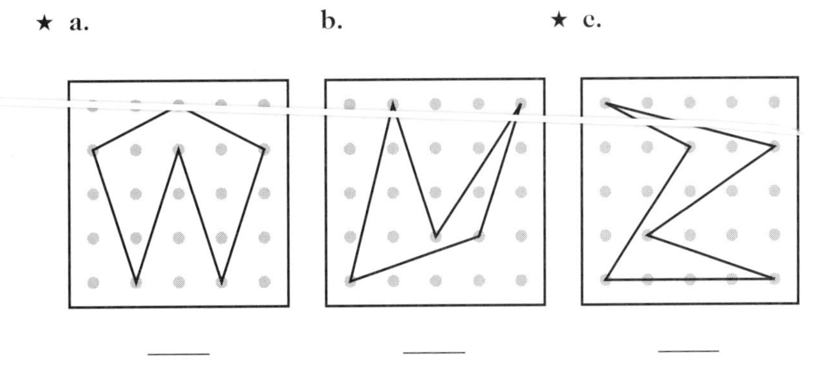

d. ★ e. f.

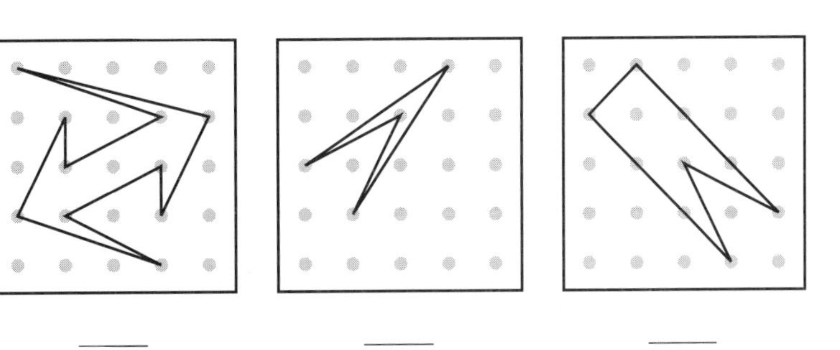

5. The *height,* or *altitude, of a triangle* is the perpendicular distance from a vertex to the line containing the opposite side. The *height of a parallelogram* is the perpendicular distance from a line containing one side to the line containing the opposite side. Determine the heights indicated below.

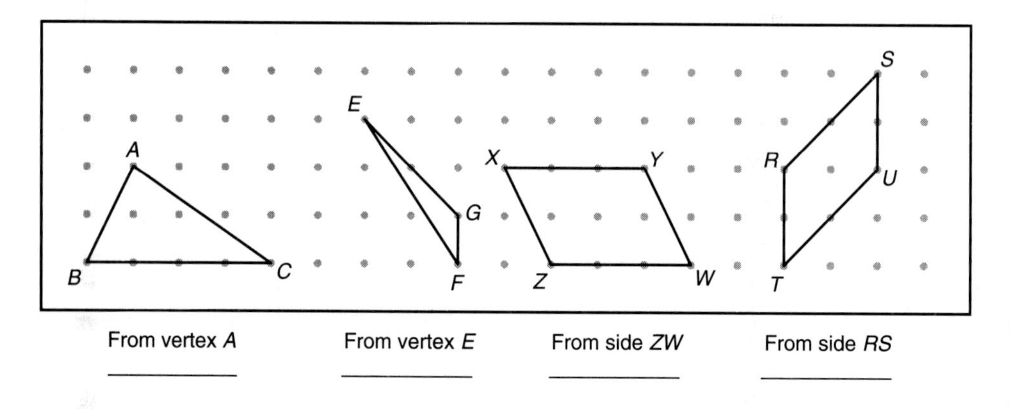

From vertex *A* ________ From vertex *E* ________ From side *ZW* ________ From side *RS* ________

6. Use your geoboard techniques to determine the area of each of these parallelograms.

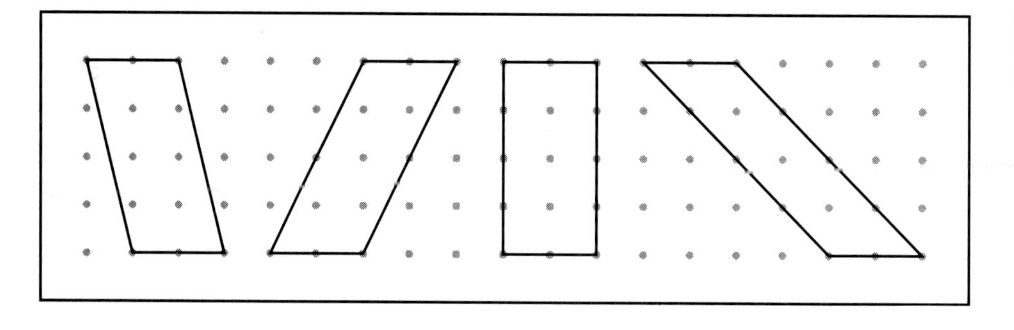

________ ________ ________ ________

Each of the preceding parallelograms has one pair of sides of length 2 and a common height. Make a conjecture about how to determine the area of a parallelogram if you know the length of a side and the height from that side. Sketch additional parallelograms below to test your conjecture. If your conjecture appears to be valid, record it at the top of the next page.

Conjecture:

7. Use geoboard techniques to determine the areas of these triangles.

________ ________ ________ ________

Make and test conjectures about triangles with bases of equal length and common heights. Record your conjecture below.

Conjecture:

8. The triangle in part a is enclosed in a parallelogram that shares two sides with it. Enclose each triangle in a parallelogram that shares two sides with it. Then, using geoboard techniques, record the areas of the triangle and its parallelogram.

★ **a.** **b.** ★ **c.** **d.**

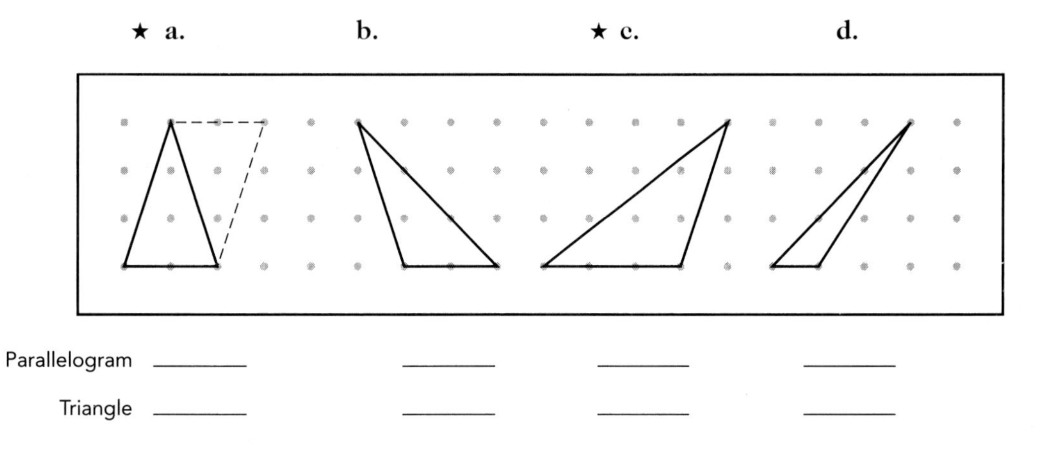

Parallelogram ________ ________ ________ ________

Triangle ________ ________ ________ ________

★ **9.** The formula for the area of a triangle is one-half of the base length times the height to that base [$A = (\frac{1}{2})bh$]. Based on the observations you have made in activities 5, 6, 7, and 8, write a statement that explains why this formula works.

JUST FOR FUN

PENTOMINOES

Pentominoes are polygons that can be formed by joining five squares along their edges. Surprisingly, there are 12 such polygons. Pentominoes can be used in puzzles, games, tessellations, and other problem-solving activities.

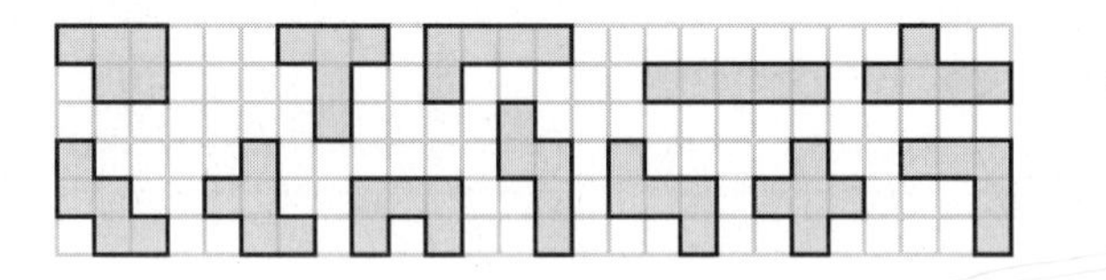

A complete set of pentomino pieces is on Material Card 37. Material Card 16 has a grid that can be used for activities 2, 3, and 4. *Note:* Pieces can be turned over.

1. Selecting the appropriate pieces from your pentomino set, form the 4 by 5 rectangular puzzle shown here. Outline this rectangle with pencil, and see how many other combinations of pieces will exactly fill the outline.

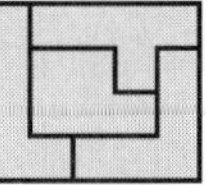

2. The best-known challenge is to use all 12 pieces to form a 6 by 10 rectangle. Supposedly there are over 2000 different solutions. Find one solution and record it on this grid.

3. It is possible to form several puzzles by constructing an 8 by 8 grid of squares and then removing four unit squares. In one of these, the unit squares are missing from the corners, as shown here. Use all 12 pieces to cover a grid with this shape. Record your solution here.

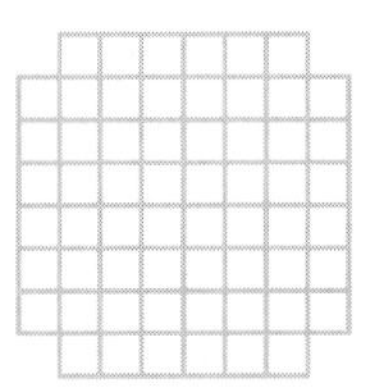

★ **4. Pentomino Game (2 players):** This game is played on an 8 by 8 grid. The players take turns placing a pentomino on uncovered squares of the grid. Play continues until someone is unable to play or all the pieces have been used. The player who makes the last move wins the game. The first seven plays of a game are shown on this board. Player A has made four moves, and player B has made three moves. Find a way player B can make the eighth move (with one of the remaining pentominoes) so that no more moves can be made. Play a few games and look for strategies.

★ **5.** It is possible to place 5 different pentomino pieces on an 8 by 8 grid so that no other piece from the same set can be placed on the grid. Can you find 5 such pieces?

PUZZLER

Make one straight cut so that the two resulting pieces can be put together to form a square.

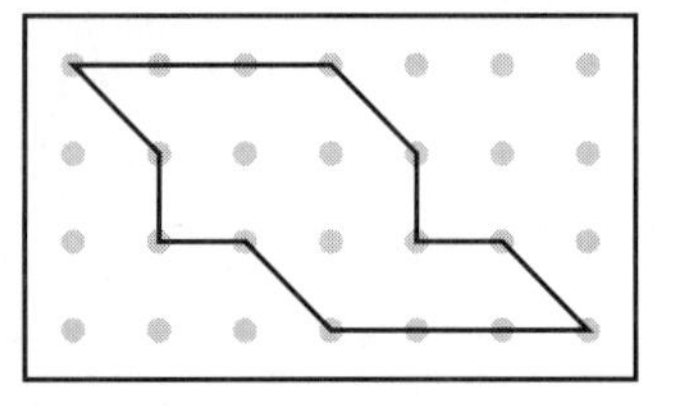

ACTIVITY SET 10.3

MODELS FOR VOLUME AND SURFACE AREA

Purpose To construct three-dimensional figures and measure their volume

Materials Patterns for a prism, pyramid, and cylinder are on Material Card 38, and a compass and protractor are on Material Card 32.

Activity One of the greatest mathematicians of all time was Archimedes (287–212 B.C.), a native of Syracuse on the island of Sicily. You may have heard the story of how Archimedes was able to solve a problem King Hieron had with his crown. The king had weighed out the exact amount of pure gold he wanted used for a crown, but upon receiving the crown he became suspicious that it was a mixture of gold and silver. The crown weighed as much as the original amount of gold, so there seemed to be no way of exposing the fraud. Archimedes was informed of the problem, and a solution occurred to him while he was taking a bath. He noticed that the amount of water that overflowed the tub depended on the extent to which his body was immersed. In his excitement to try his solution with gold and silver, Archimedes leaped from the tub and ran naked to his home, shouting, "Eureka! Eureka!"

As Archimedes discovered, the volume of a submerged object is equal to the volume of the displaced water. Finding the volumes of objects by submerging them in water is especially convenient with metric units, because the number of milliliters of water that is displaced is equal to the volume of the object in cubic centimeters. This relationship will be used in the following activities. You may also need the formulas for volumes of prisms and cylinders (area of base × height), pyramids and cones ($\frac{1}{3}$ × area of base × height), and spheres $[(\frac{4}{3})\pi r^3]$.

★ **1.** To understand Archimedes' solution, consider two cubes both weighing the same amount, one of pure gold and one of pure silver. Since gold is about twice as heavy as silver, the cube of gold will be smaller than the cube of silver. Therefore, when each is submerged in water, there will be more water displaced for the cube of silver. To apply his theory, Archimedes used the crown and an amount of pure gold weighing the same as the crown. He submerged them separately in water and measured the amount of overflow. There was more water displaced for the crown than for the pure gold. What does this show about the volume of the crown as compared to the volume of the gold? Was King Hieron cheated?

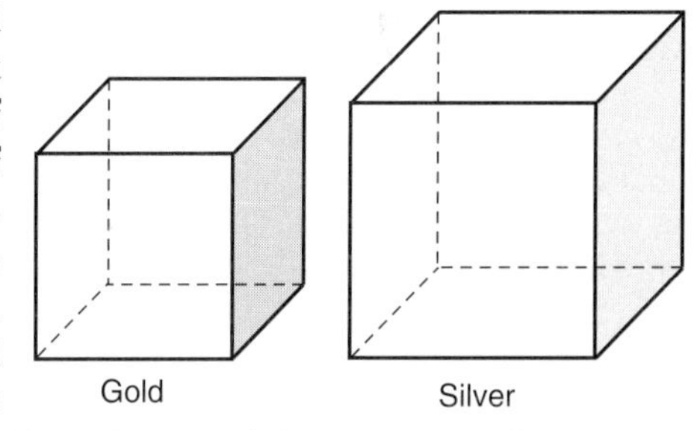

★ **2.** Patterns for prisms, pyramids, and cylinders are easy to make. Three such patterns are contained on Material Card 38. Cut them out and tape their edges. Complete the first two lines of the table on the next page and then determine the volumes and surface areas of these figures. You may wish to fill them with sand or salt to check your answers. (The area of the base of a cylinder is πr^2. Use 3.14 for π.)

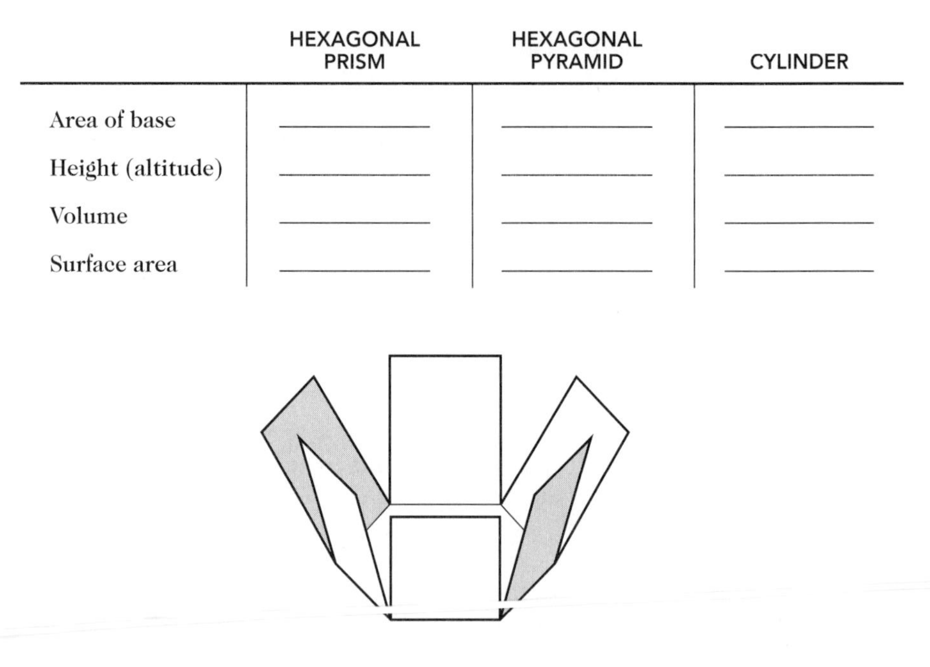

	HEXAGONAL PRISM	HEXAGONAL PYRAMID	CYLINDER
Area of base	______	______	______
Height (altitude)	______	______	______
Volume	______	______	______
Surface area	______	______	______

★ **3.** A sheet of paper can be made into a cylinder in the two ways shown in the following figures. Make a cylinder (without bases) from a standard sheet of paper so that its circumference is 21.5 centimeters and its height is 28 centimeters. Then use another sheet of paper to form a cylinder with a circumference of 28 centimeters and a height of 21.5 centimeters. Predict which cylinder has the greater volume. As you compute the volume of each cylinder, record your methods and answers below.

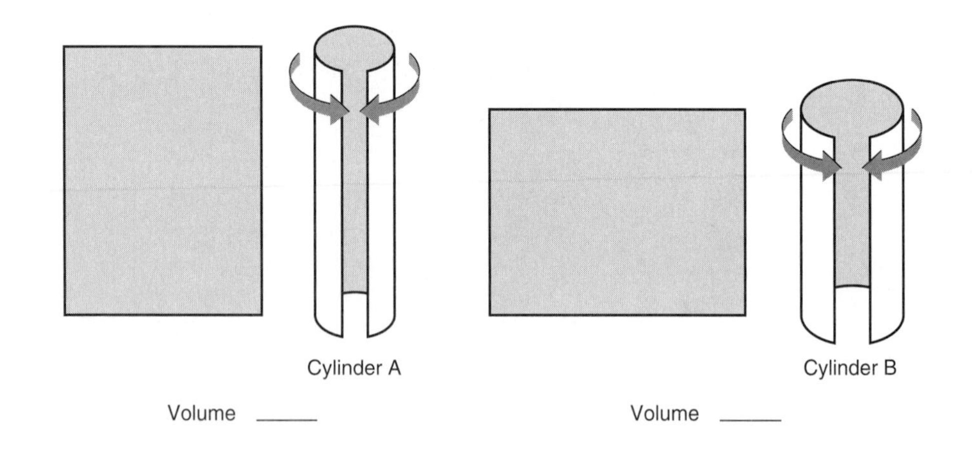

4. Cut a standard sheet of paper in half lengthwise and tape the two halves end to end, as shown on the next page. Make a cylinder without bases and compute its volume. Compare the volume of this cylinder with that of cylinder B in activity 3. What happens to the volume of a cylinder as the height is halved and the circumference (thus, radius and diameter) is doubled?

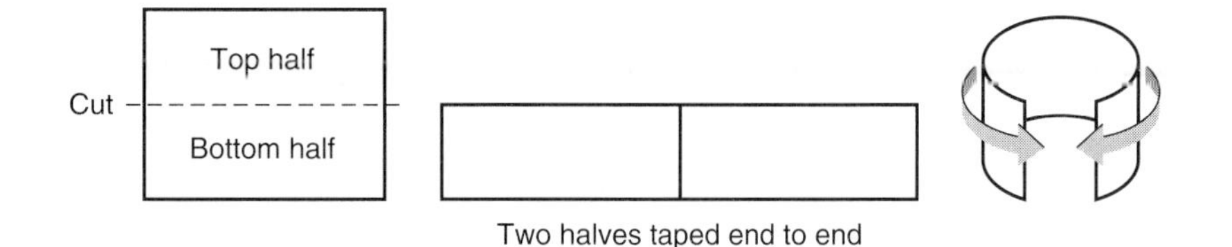

5. Make a set of six different cones by cutting out three disks, each of radius 10 centimeters. (There is a compass and protractor on Material Card 32.) Each disk can be used for making two cones, as shown here. Cut from these disks sectors with central angles of 45°, 90°, 135°, 225°, 270°, and 315°, and *number the sectors from 1 to 6 according to the size of the central angle.* After cutting each disk into two parts, bend it and tape the edges to form the cones.

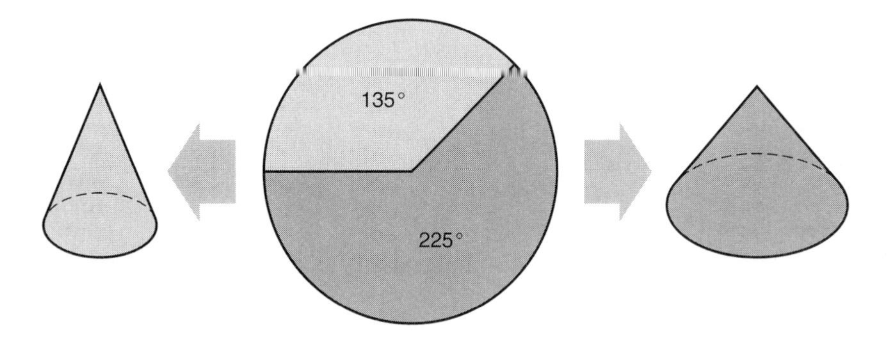

a. Without computing, line the cones up from left to right in what you estimate to be increasing volume. Record your estimated order by recording the numbers of the cones.

______ ______ ______ ______ ______ ______

Least volume Greatest volume

★ **b.** Compute the base areas and volumes (to the nearest two decimal places) of the cones, and enter them in the following table. Use the results to check the estimated order of the cones in part a. (The area of the base of a cone is πr^2. Use 3.14 for π.)

	1	2	3	4	5	6
Central angle of disk	45°	90°	135°	225°	270°	315°
Radius (cm) of base	1.25	2.50	3.75	6.25	7.50	8.75
Base area (cm^2)	______	______	______	______	______	______
Height (cm) of cone	9.92	9.68	9.27	7.81	6.61	4.84
Volume (cm^3) of cone	______	______	______	______	______	______

★ c. What will happen to the volume of the cone as sectors with central angles of less than 45° are used? as sectors with central angles of greater than 315° are used?

6. Choose one of your cones from activity 5, and place a mark on the surface halfway from the vertex to the base.

   a. If the cone were filled to this mark, estimate what fractional part of the total volume would be filled. Circle the fraction that best represents your estimate.

$$\frac{1}{8} \quad \frac{1}{7} \quad \frac{1}{6} \quad \frac{1}{5} \quad \frac{1}{4} \quad \frac{1}{3} \quad \frac{1}{2}$$

   b. To experimentally test your estimate, fill the cone up to the mark with water (or sand or rice) and then pour its contents into an identical cone, repeating the process until the second cone is full. How many "fillings" are required?

★ c. Compute the volume of your "half cone" and compare it to the volume of the whole cone. What do you discover?

7. A basketball displaces 7235 milliliters of water.

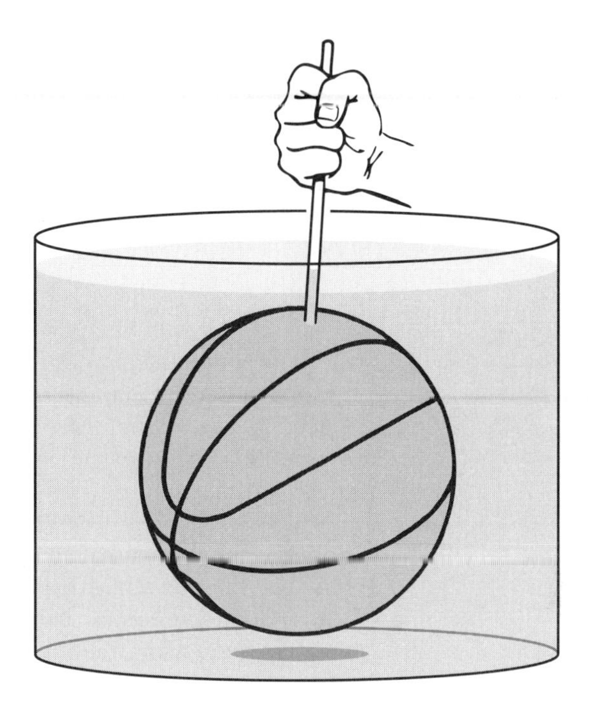

★ **a.** What is its volume in cubic centimeters?

b. What is the diameter of this ball? [Sphere volume = $(\frac{4}{3})\pi r^3$, where r is the radius. Use 3.14 for π.] Explain how you arrived at your conclusions.

c. Submerge a baseball in water and measure the displaced water in milliliters. According to this experiment, what is the volume of a baseball in cubic centimeters? A baseball has a diameter of approximately 7.3 centimeters. Compute its volume in cubic centimeters, and compare this answer with the results of your experiment. Record your results.

d. Submerge an unopened cylindrical can in water and measure the overflow. Then measure the diameter and height of the can and compute its volume. Compare the computed volume with the results of your experiment. Record your results.

8. Spherical objects such as Christmas tree ornaments and basketballs are packed in cube-shaped compartments and boxes.

a. If a sphere just fits into a box, estimate what fraction of the box is "wasted space." Circle the fraction that best represents your estimate.

$\frac{1}{6}$ $\frac{1}{5}$ $\frac{1}{4}$ $\frac{1}{3}$ $\frac{1}{2}$

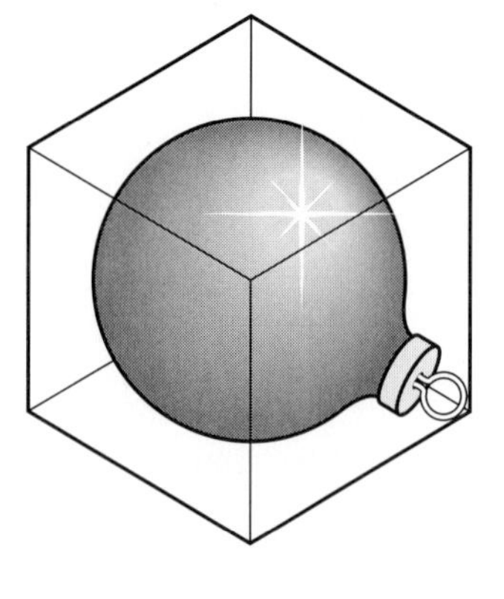

Compute the amount of wasted space when a sphere of diameter 25 centimeters is placed in a box. What fraction of the total volume of this box is wasted space?

b. Spheres are also packed in cylinders. A cylinder provides a better fit than a box, but there is still extra space. If 3 balls are packed in a cylindrical can whose diameter equals that of a ball and whose height is 3 times the diameter, estimate what fraction of the space is unused.

$\frac{1}{6}$ $\frac{1}{5}$ $\frac{1}{4}$ $\frac{1}{3}$ $\frac{1}{2}$

Compute the amount of unused space.

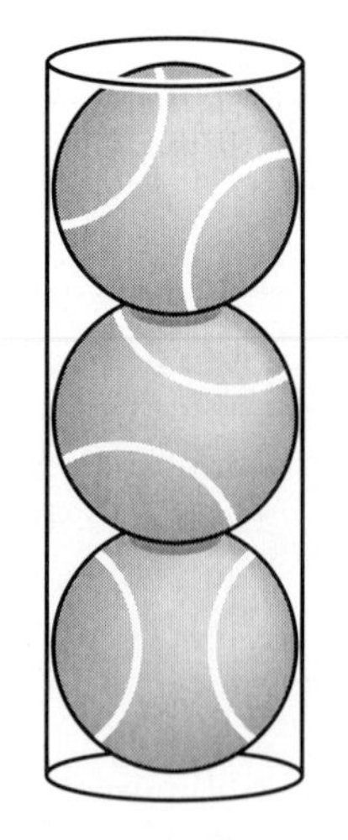

★ **c.** Tennis ball cans hold 3 balls. Determine the amount of wasted space in these cans if the balls have a diameter of 6.3 centimeters and the can has a diameter of 7 centimeters and a height of 20 centimeters. Check the reasonableness of your answer by experimenting with a can of balls and water.

JUST FOR FUN

SOMA CUBES

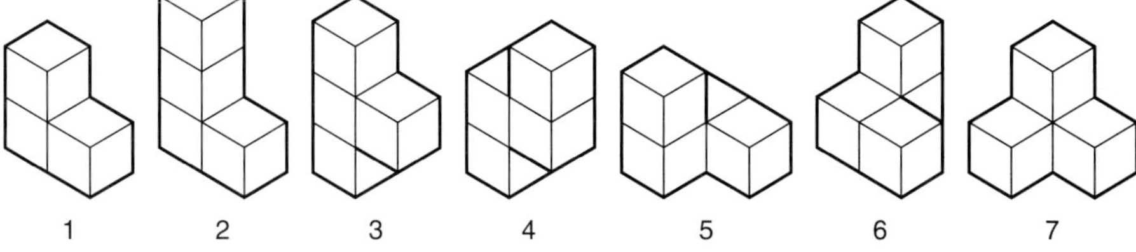

Soma cubes are a seven-piece puzzle invented by the Danish author Piet Hein. Piece 1 has 3 cubes, and the other pieces each have 4 cubes. Soma cubes can be purchased, or they can be constructed by gluing together cubes, such as sugar cubes, wooden cubes, or dice.

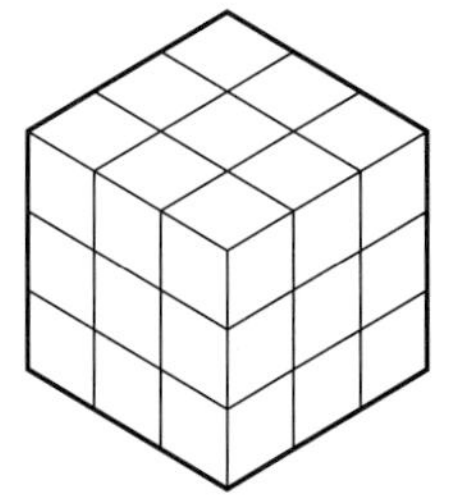

Surprisingly, these 7 pieces can be assembled to form a 3 by 3 by 3 cube. You may wish to try this challenging puzzle.

A common activity with soma cubes is constructing certain well-known figures. Each of the shapes in figure 1 below requires all 7 pieces. Try to build them.

There are several elementary techniques for determining when a figure cannot be constructed with soma cubes. The simplest of these is counting the number of cubes. For example, a figure with 18 cubes cannot be formed because there is no combination of 4s and one 3 that adds up to 18. Use this approach to find which of the shapes in Figure 2 cannot be formed with soma cubes.[2]

Figure 1

Stairs Pyramid Sofa

Figure 2

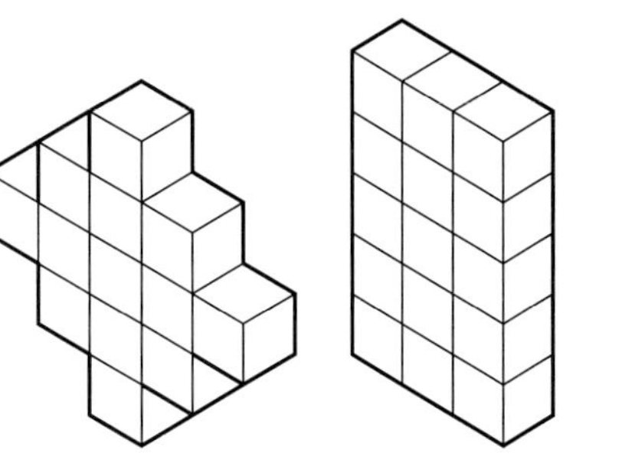

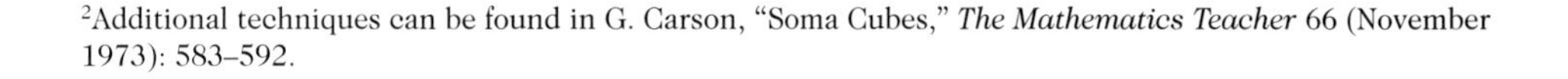

[2]Additional techniques can be found in G. Carson, "Soma Cubes," *The Mathematics Teacher* 66 (November 1973): 583–592.

PUZZLER

This solid is made from 1000 cubes. The solid is spray-painted on all sides, including the bottom. What percent of the cubes have no paint on them?

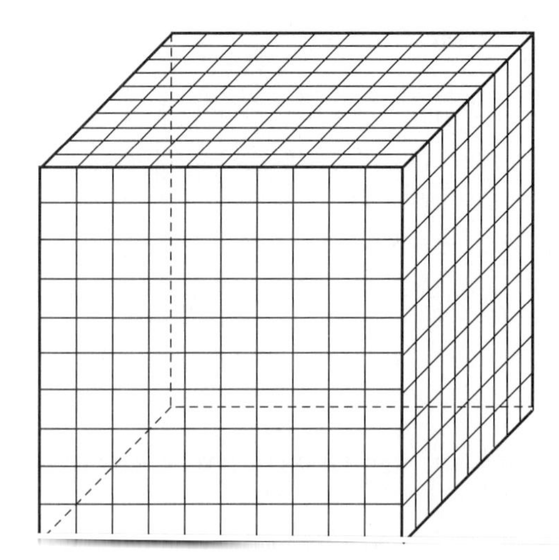

IDEAS FOR THE ELEMENTARY CLASSROOM

SUGGESTED CLASSROOM ACTIVITY: FOLDING BOXES

Building and filling boxes helps students understand the concepts of volume and surface area. By cutting squares from the corners of a grid sheet and folding and taping sides, students can construct boxes without tops. Seven boxes that can be formed by cutting whole numbers of squares from the corners of a 16 by 16 sheet of grid paper are shown here.

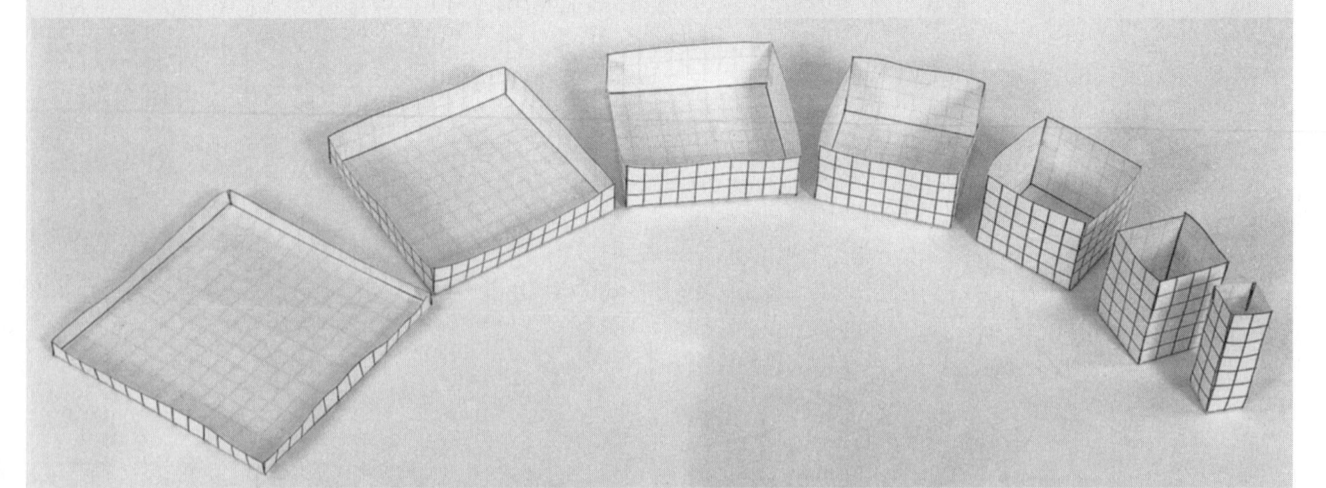

After the boxes have been constructed, have students do the following:

1. Arrange them in order from least capacity to greatest capacity by simply looking and comparing (estimating).

2. Order them by filling each with foam nuggets and comparing relative volumes.
3. Stack cubic centimeters or other cubical objects inside to compare volume.
4. (Older students) Compute the volume of each box once they understand the concept of volume.
5. Obtain the surface area of each box by counting squares or computing from the dimensions. Students may be surprised to discover that the surface area of the boxes from left to right in the photo decreases, whereas the volume increases and then decreases.
6. Use grid paper with other dimensions to construct boxes with different dimensions.
7. Cut patterns that will fold into boxes with tops.

Readings for More Classroom Ideas

Badger, Elizabeth. "More Than Testing." *Arithmetic Teacher* 39 (May 1992): 7–11.

Baker, Camille. "How Big was the Roman Empire?" *Mathematics Teaching in the Middle School* 1 (March–April 1996): 754–759.

Barson, Alan, and L. Barson. "Ideas: Measurement Activities." *Arithmetic Teacher* 35 (May 1988): 20–29.

Bay, Jennifer M., Ann M. Bledsoe, and Robert E. Reys. "State-ing the Facts: Exploring the United States." *Mathematics Teaching in the Middle School* 4 (September 1998): 8–14.

Chapin, Suzanne H. "Mathematical Investigations—Powerful Learning Situations." *Mathematics Teaching in the Middle School* 3 (February 1998): 332–338.

Cook, March. "Ideas: Rectangle Activities." *Arithmetic Teacher* 36 (March 1989): 27–32.

Duke, Charlotte. "Tangrams and Area." *Mathematics Teaching in the Middle School* 3 (May 1998): 485–487.

Fay, Nancy, and C. Tsairides. "Metric Mall." *Arithmetic Teacher* 37 (September 1989): 6–11.

Fitzgerald, William, and Janet Shroyer. *Mouse and Elephant: Measuring Growth.* Middle Grades Mathematics Project. Menlo Park, CA: Addison-Wesley Publishing Company, 1986.

Geddes, Dorothy, et al. *Measurement in the Middle Grades,* in the *Curriculum and Evaluation Standards for School Mathematics Addenda Series: Grades 5–8.* Reston, VA: National Council of Teachers of Mathematics, 1994.

Holcomb, Joan. "Using Geoboards in the Primary Grades." *Arithmetic Teacher* 27 (April 1980): 22–25.

Horak, Virginia M., and W. J. Horak. "Let's Do It: Making Measurement Meaningful." *Arithmetic Teacher* 30 (November 1982): 18–23.

Jensen, Rosalie, and D. R. O'Neil. "Let's Do It: Meaningful Linear Measurement." *Arithmetic Teacher* 29 (September 1981): 6–12.

Lamphere, Patricia. "Geoboard Patterns and Figures." *Teaching Children Mathematics* 1 (January 1995): 282–286.

Lindquist, Mary M. "Implementing the Standards: The Measurement Standards." *Arithmetic Teacher* 37 (October 1989): 22–26.

Reynolds, Anne, and Grayson H. Wheatley. "Third-Grade Students Engage in a Playground Measuring Activity." *Teaching Children Mathematics* 4 (November 1997): 166–170.

Rhone, Lynn. "Measurement in a Primary Integrated Curriculum," in *Connecting Mathematics Across the Curriculum,* 1999 Yearbook, edited by P. A. House and A. F. Coxford, VA: National Council of Teachers of Mathematics, 1995, 124–133.

Scavo, Thomas R., and Byron Petraroja. "Adventures in Statistics." [metric, area] *Teaching Children Mathematics* 4 (March 1998): 394–400.

Shaw, Jean M. "Let's Do It: Exploring Perimeter and Area Using Centimeter Squared Paper." *Arithmetic Teacher* 31 (December 1983): 4–11.

Shaw, Jean M. "Let's Do It: Student-Made Measuring Tools." *Arithmetic Teacher* 31 (November 1983): 12–15.

Shaw, Jean M. "Meaning Metrics: Measure, Mix, Manipulate, and Mold." *Arithmetic Teacher* 28 (March 1981): 49–50.

Stix, Andi. "Pic-Jour Math: Pictorial Journal Writing in Mathematics." *Arithmetic Teacher* 41 (January 1994): 264–269.

Trafton, Patricia A., and Christina Hartman. "Exploring Area with Geoboards." *Teaching Children Mathematics* 4 (October 1997): 72–75.

Whitin, David J., and Cassandra Gary. "Promoting Mathematical Explorations Through Children's Literature." *Arithmetic Teacher* 41 (March 1994): 394–399.

Zaslavsky, Claudia. "People Who Live in Round Houses." *Arithmetic Teacher* 37 (September 1989): 18–21.

CHAPTER 11

MOTIONS IN GEOMETRY

Students discover relationships and develop spatial sense by constructing, drawing, measuring, visualizing, comparing, transforming, and classifying geometric figures. Discussing ideas, conjecturing, and testing hypotheses precede the development of more formal summary statements.[1]

ACTIVITY SET 11.1

LOCATING SETS OF POINTS IN THE PLANE

Purpose To locate and sketch sets of points in a plane that satisfy given conditions

Materials A ruler, compass, and protractor from Material Card 32, an index card, and scissors

Activity Look at the two identical coins below. If the left coin is carefully rolled halfway around the other, what will be the position of the point A? Make a conjecture and test it by experimenting with two identical coins.

In this activity set, we will be locating the positions of points in the plane. Many geometric figures are described or defined as sets of points with certain

[1]*Curriculum and Evaluation Standards for School Mathematics* (Reston, VA: National Council of Teachers of Mathematics, 1989), 112.

conditions placed on them. Suppose, for example, that we are asked to locate and sketch *the set of points in the plane that are 2 cm from a given point P.* We can begin to identify points by using a ruler to locate points 2 cm from point P, as shown in figure a. As more and more points are added to the collection of points, we can see that a smooth curve drawn through the points, as in figure b, visually describes the set of all points 2 cm from P as a circle with center P and radius 2 cm.

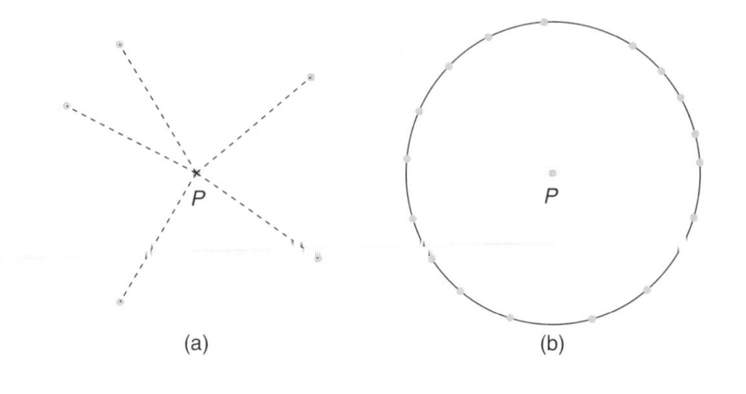

(a) (b)

1. For parts a and b, place on each figure 10 to 20 points that satisfy the stated condition. Then draw a smooth curve through all points satisfying the condition and describe the outcome in your own words.

   a. The set of all points in the plane that are the same distance from point K as from point L

   K •

   L •

★ b. The set of all points in the plane that are the same distance from line m as from line n

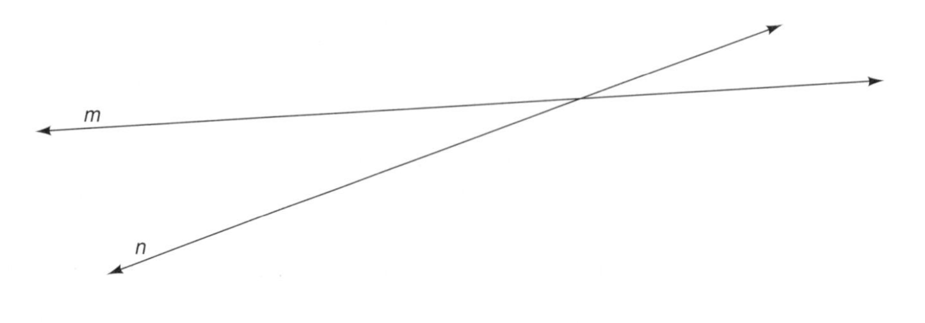

2. A chord is a line segment joining two points on a circle. Parts a through c involve chords of circles. Locate as many points as you need to feel confident about sketching the curve that satisfies the conditions. Describe each result in words, as specifically as possible.

★ **a.** The set of all points that are midpoints of chords of length 4 cm on this circle

P

b. The midpoints of all chords of this circle that have point R as one endpoint

P

R

★ **c.** The midpoints of all chords of this circle that pass through point S

S

P

3. Draw and cut out a circle with radius 1.5 cm.

   a. Suppose the circle rolls around the inside of the square in part b, always tangent to at least 1 edge. Sketch the path traced by the center of the circle and then describe it in words.

   b. Sketch and describe the path of the circle's center if the circle rolls around the outside of the square.

★ 4. Point B is the center of a circle with radius 1 cm, and point A is the center of a circle with radius 2 cm. Both X and Y, which are the intersections of these two circles, are twice as far from point A as from point B. Find more points that are twice as far from A as from B. (*Hint:* You may find it helpful to locate these points by drawing circles about A and B, with the radius about A twice the radius about B.) Sketch and describe the set of all points in the plane that satisfy the conditions.

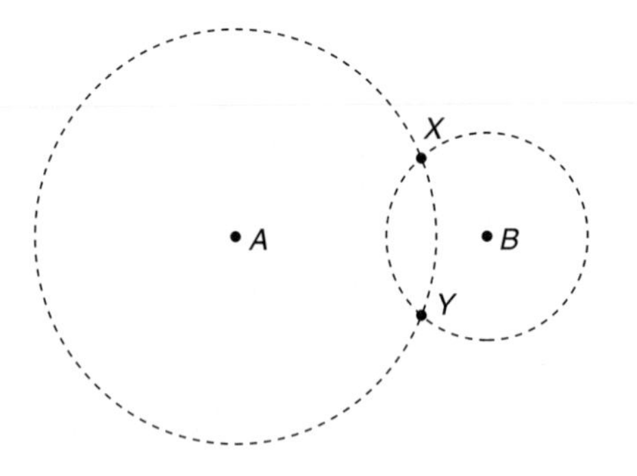

5. Cut out a narrow strip of paper having a length of 10 cm and punch a small hole at its midpoint. Place this strip on two perpendicular lines so that each end touches one of the lines. Place a pencil at the midpoint, M, and with the help of a classmate, move the strip so that its endpoints always touch the two perpendicular lines. What curve is traced by M?

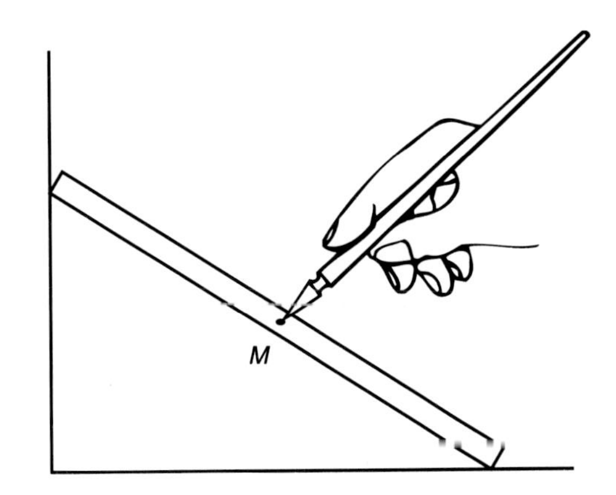

6. The line segment $\overline{PQ}$ is 7 cm long. Point X is the corner of an index card (any piece of paper with a square corner will work), placed on top of segment $\overline{PQ}$ so that its adjacent sides pass through points P and Q. It is possible to shift the index card to new positions so that P and Q are still on adjacent edges but the corner of the card is at a different point, such as point Y. Locate and mark several different corner points in this manner.

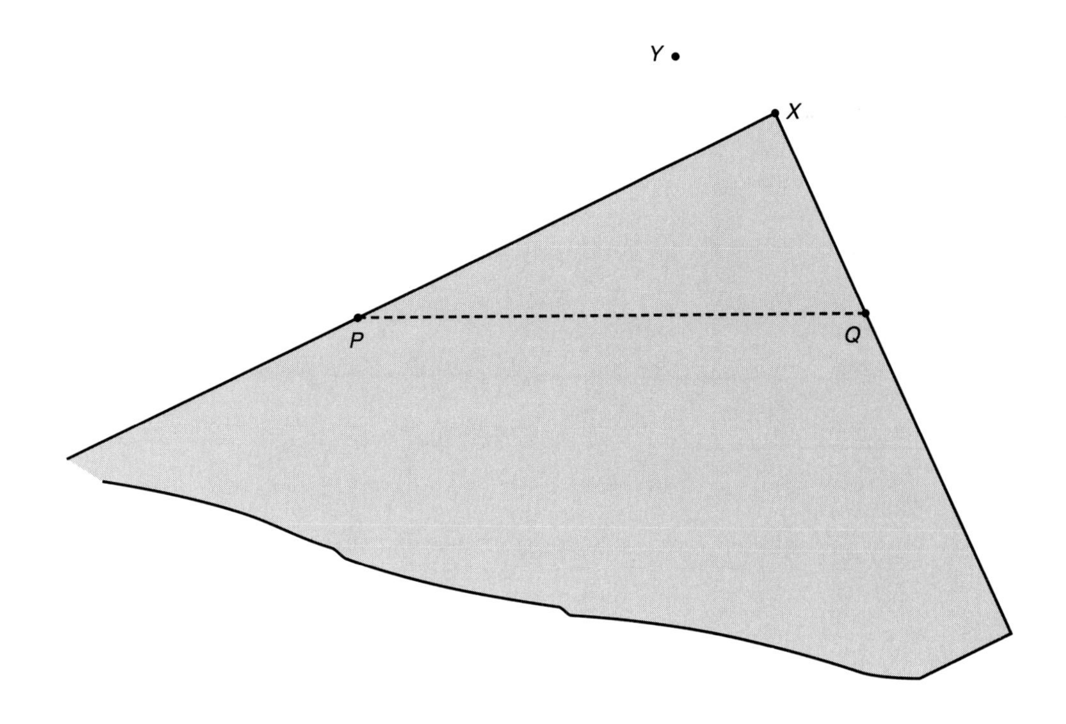

★ **a.** Describe what the curve made up of all the corner points will look like.

b. Bisect one corner of the index card, and cut along the angle bisector to obtain a 45° angle. The point W is the vertex point of the 45° angle. This card is placed on top of segment $\overline{RS}$ so that its sides pass through points R and S. Shift the card to locate other vertex points with R and S on the adjacent sides of the 45° angle. Describe the set of all such vertex points.

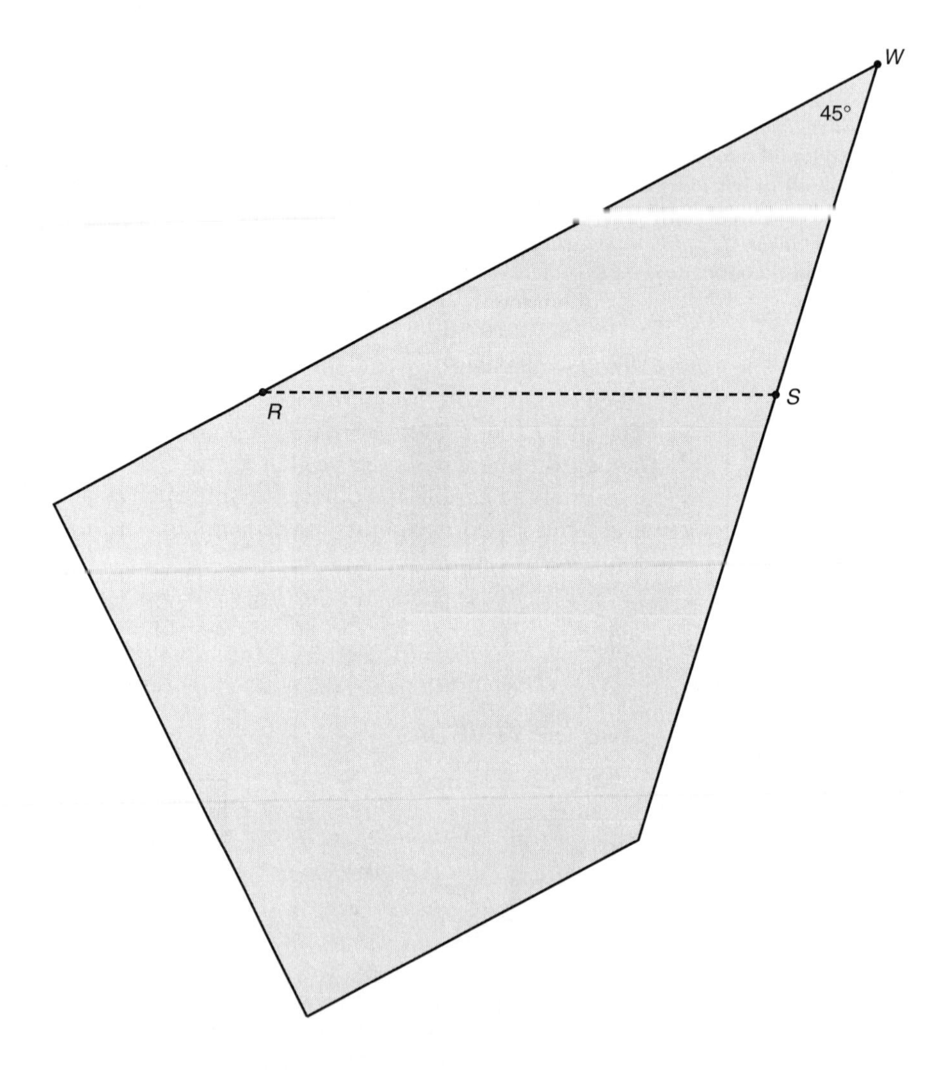

c. Suppose the vertex angle of the card described above were greater than 90°. Draw a sketch of what you think the set of vertex points would look like. Test your conjecture by cutting an index card angle greater than 90° and locating points.

JUST FOR FUN

LINE DESIGNS

Line designs are geometric designs formed by connecting sequences of points with line segments. One common method is to connect equally spaced points on the sides of an angle, as shown in figure a. These points are connected in the following order: A to a, B to b, C to c, and so on. The resulting design gives the illusion of a curve. This curve is a parabola. Experiment with different types of angles. For the obtuse angle in figure c, the curve will be closer to the angle than for the acute angle in figure b. Combining several angles, as in the figure above, produces beautiful designs.

Another way to create line designs is to draw chords in a circle according to certain rules. The design in Figure d is being formed in a circle whose circumference has been divided into 36 equal parts. The first chord connects point 1 to point 2; the second chord connects point 2 to point 4; the third chord connects point 3 to point 6; and so on. This figure shows the first 18 chords, with point 18 connected to point 36. Continue drawing chords by connecting point 19 to point 2, point 20 to point 4, etc., until you reach point 36. The resulting figure gives the illusion of a heart shape. A curve with this shape is called a cardioid. The front ends of the chords for this cardioid traveled twice as fast around the circle as the back ends of the chords.

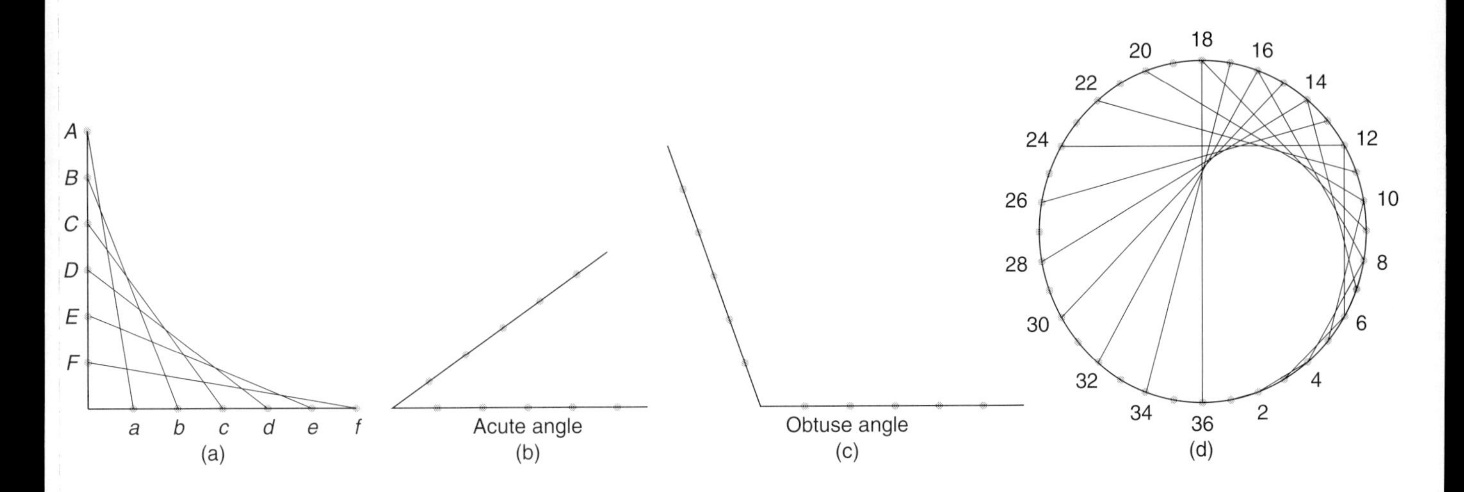

(a) (b) (c) (d)

In the figure below let the front ends of the chords travel 3 times as fast. That is, connect point 1 to point 3, 2 to 6, 3 to 9, and so forth. Continue drawing these chords until the back ends of the chords have made one revolution and the front ends have made three revolutions. The cardioid on the preceding page has 1 cusp (inward point). How many cusps will this curve have?

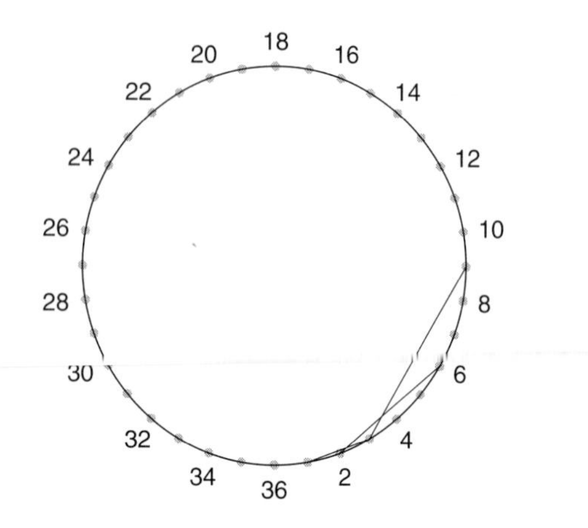

Experiment with some more patterns. For example, let the front ends of the chords move 4 times faster ($1 \to 4$, $2 \to 8$, $3 \to 12$, . . .) or 5 times faster ($1 \to 5$, $2 \to 10$, . . .). Can you predict the number of cusps in each case? (The 36 spaces around a circle will work for 2, 3, 4, or 6 cusps. For 5 cusps, use 40 spaces or some other multiple of 5 that conveniently divides into 360.) Many of these designs can be stitched with colored thread or yarn on cloth or posterboard or formed with nails and string or rubber bands. Below are some examples. The first one was formed by rubber bands and nails, the second by yarn and nails, and the third by white thread stitched into black silk.

PUZZLER

You ride your bike straight across a strip of fresh wet paint 10 inches wide. What pattern of marks will your bike tires leave?

ACTIVITY SET 11.2

DRAWING ESCHER-TYPE TESSELLATIONS

Purpose To create nonstandard tessellations and Escher-type drawings that illustrate motions in geometry

Materials A ruler, compass from Material Card 32, and tracing paper

M. C. Escher's "Alhambra Drawings" © *1999 Cordon Art B. V.—Baarn-Holland.* *All rights reserved.*

Activity The Dutch artist M. C. Escher was fascinated by the tile patterns he observed on the walls, floors, and ceilings of the Alhambra—a thirteenth-century Moorish palace in Granada, Spain. He considered the pure geometric tessellations his "richest source of inspiration." Escher thought it a "pity" that the talented Islamic artists were restricted, by religious beliefs, to geometric forms. His own works are filled with images of living things.

M. C. Escher's "Two Intersecting Planes" © *1999 Cordon Art B. V.—Baarn-Holland.* *All rights reserved.*

Escher-type drawings can be constructed by altering polygons that tessellate (triangles, quadrilaterals, and regular hexagons) by using congruence mappings (translations, rotations, reflections, and glide reflections). In this activity set, you will devise drawings and have the opportunity to create your own artistic work.

1. **Translation Tessellations:** The rectangle in figure a tessellates. In figure b, one side of the rectangle is modified by a jagged line design. The design is then *translated* to the opposite side of the rectangle as a modification in figure c.

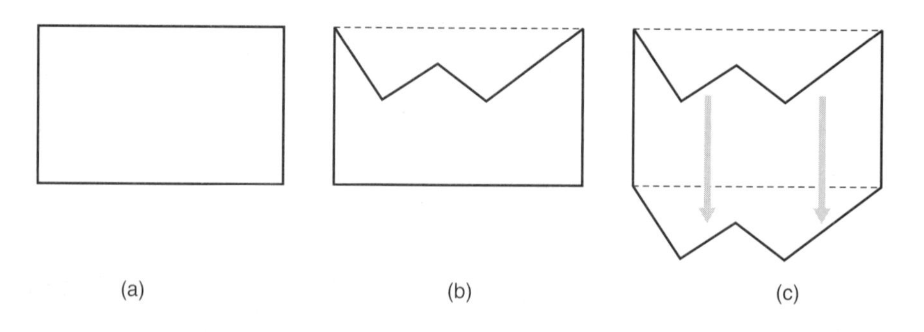

(a) (b) (c)

★ **a.** Explain why the new shape in figure c will also tessellate the plane

In figure d, one of the remaining parallel sides of the rectangle is altered by a different design. This design is then translated to the opposite side in figure e, producing the final shape in figure f.

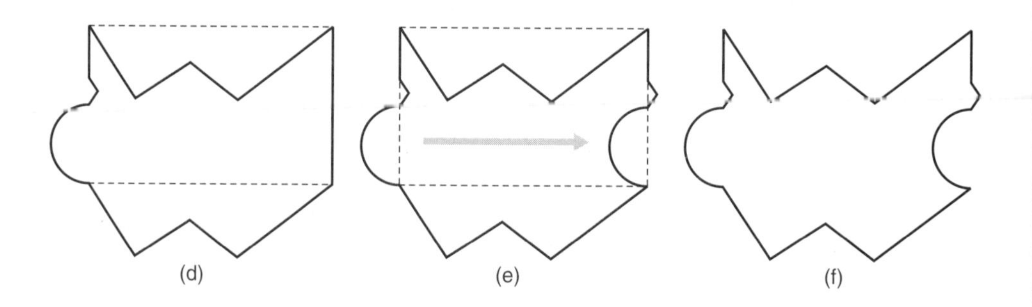

(d) (e) (f)

The tessellation formed by the modified rectangle is shown below. This tessellation becomes an Escher-type drawing when details and/or color are added to the basic design.

b. Create your own translation tessellation by modifying and translating the sides of this parallelogram.

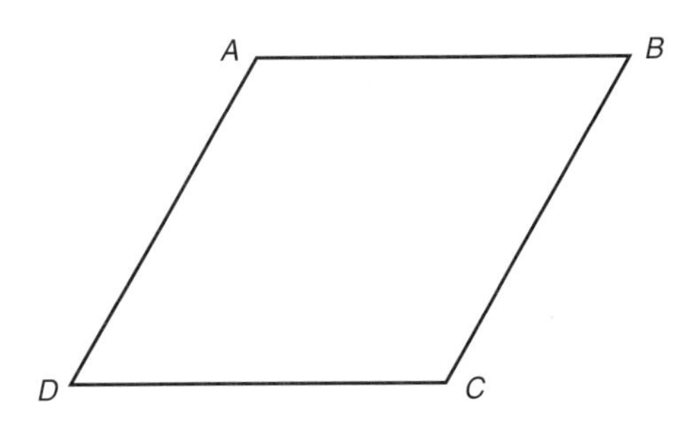

(1) Modify side $\overline{AB}$.

(2) Translate the design to $\overline{DC}$. Place a piece of tracing paper over the parallelogram. Mark points A and B and trace your modification for side $\overline{AB}$. Then slide the tracing paper so that point A coincides with D and point B coincides with C. Take a pointed object (ballpoint pen, compass point) and make enough imprints through the tracing paper so that when you remove the paper you can draw the translated design on $\overline{DC}$.

(3) Repeat steps 1 and 2 to translate a modification of side $\overline{AD}$ to side $\overline{BC}$.

c. Use the pattern you have just created to form a tessellation. An easy way to do this is to place a clean piece of tracing paper over the pattern you created and carefully trace the pattern. Then move the tracing paper up, down, right, or left, and trace additional patterns adjacent to your first copy on the tracing paper. Make an Escher-type drawing by sketching in details.

2. **Rotation Tessellations:** The following figures illustrate the modification and rotation of the sides of an equilateral triangle. First, side $\overline{AB}$ is modified as in figure a. Then the design is rotated about vertex A to adjacent side $\overline{AC}$, as shown in figure b, so that vertex B is mapped onto vertex C.

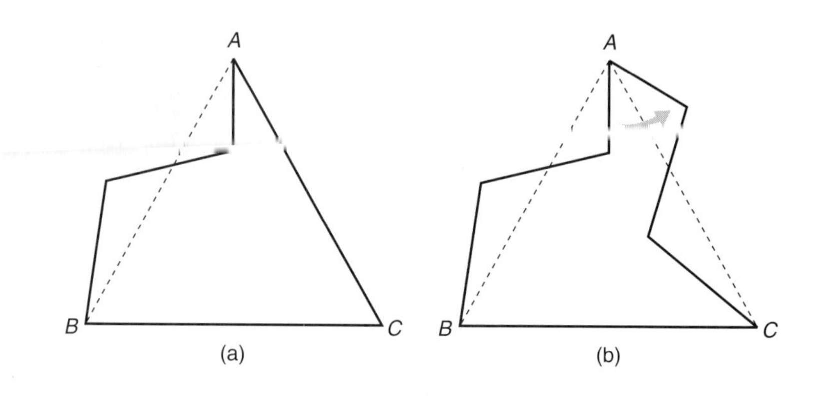

Side $\overline{BC}$ can be modified by rotating a design about its midpoint. In figure c, a design is created between the midpoint M and vertex C. The pattern is completed by rotating this design 180° about midpoint M so that vertex C is mapped onto vertex B, as shown in figure d.

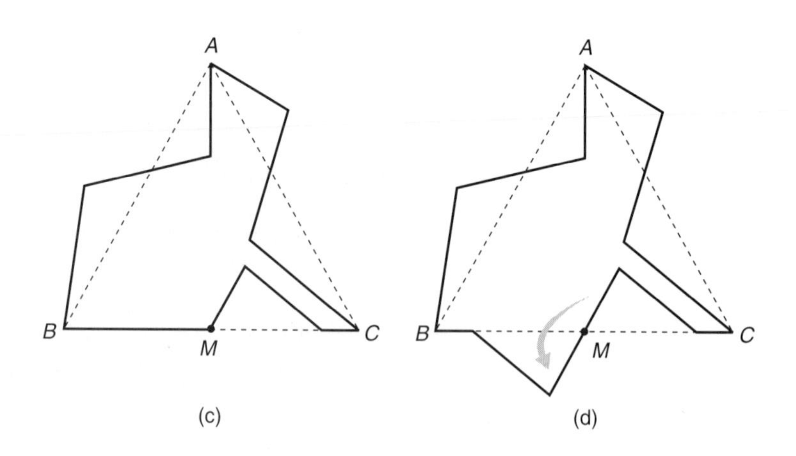

This pattern was used to create the following tessellation. Notice that this tessellation can be mapped onto itself by rotations about certain vertex points. List all the rotations about point P that will map this tessellation onto itself.

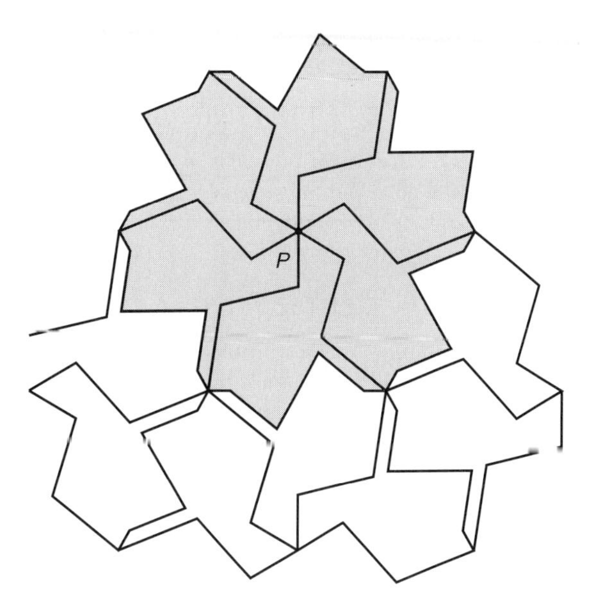

a. Create a rotation tessellation from square $ABCD$ by first modifying side $\overline{AB}$ and rotating the design about vertex B to adjacent side $\overline{BC}$ and then modifying side $\overline{AD}$ and rotating the design about vertex D to adjacent side $\overline{DC}$. Use your pattern to form a tessellation.

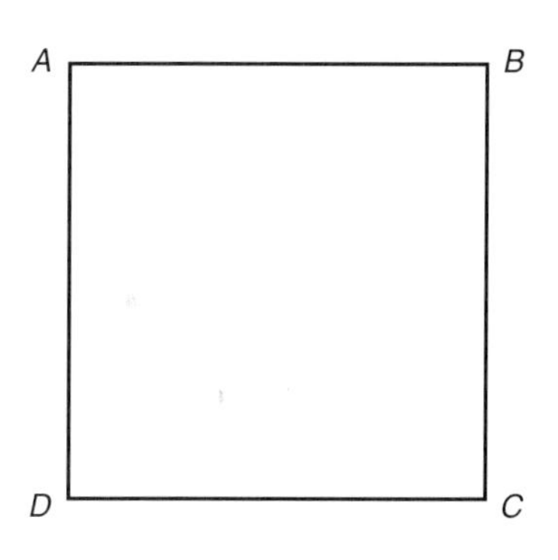

b. Use the following scalene triangle to design a rotation tessellation figure. Because no two sides have equal length, you cannot rotate one side to another about a vertex, as in figure b at the beginning of activity 2. However, half a side can be rotated about the midpoint of that side, as was done in figures c and d of activity 2. The midpoint of each side has been marked. For each side, create a design between a vertex and the midpoint, and then rotate the design 180° about the midpoint. Show that this pattern tessellates.

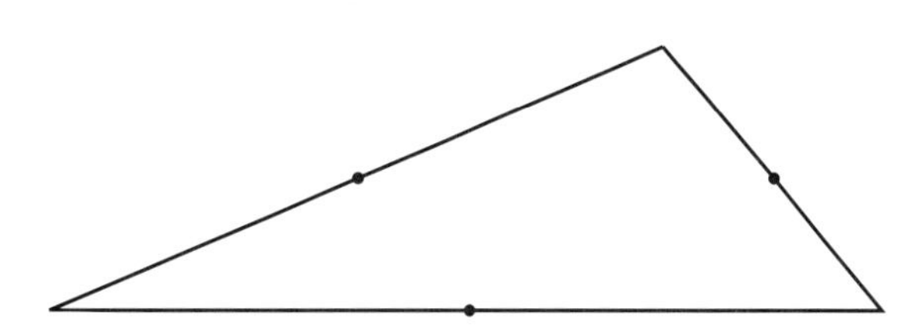

c. Convert one or both of your tessellations to an Escher-type drawing.

3. Glide-Reflection Tessellations: A *glide reflection* can be used to create a tessellating figure from parallelogram $ABCD$. Side $\overline{AD}$ is modified in figure a, and the design is translated to side $\overline{BC}$ in figure b. Next, side $\overline{DC}$ is modified as shown in figure c.

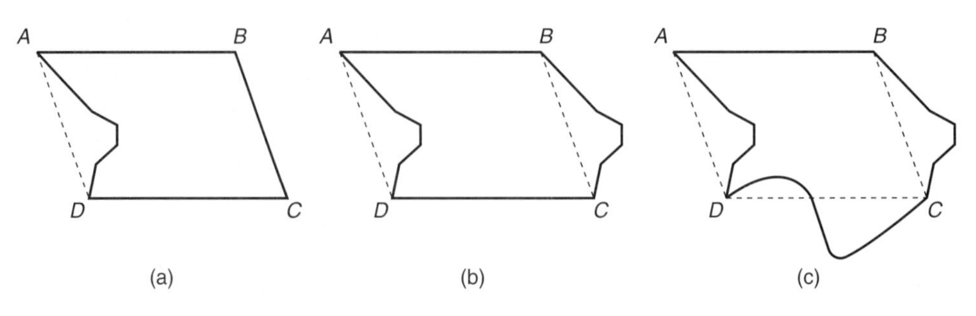

The design on $\overline{DC}$ is then translated to side $\overline{AB}$, as shown in figure d (this is the *glide part*). Finally, the design along side $\overline{AB}$ is *reflected* about a line l perpendicular to the midpoint of $\overline{AB}$, as illustrated in figure e. The two motions shown in figures d and e comprise a *glide reflection*.

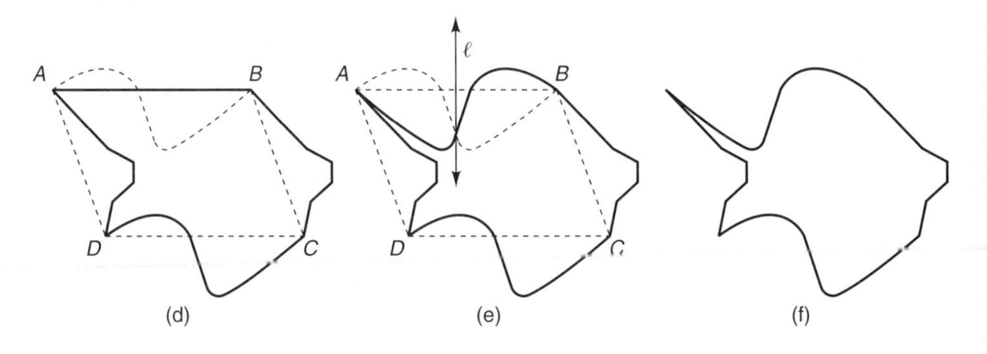

a. The pattern created in figures a–f produces the following tessellation. What mapping will map the top row of this tessellation onto the second row? The top row onto the third row?

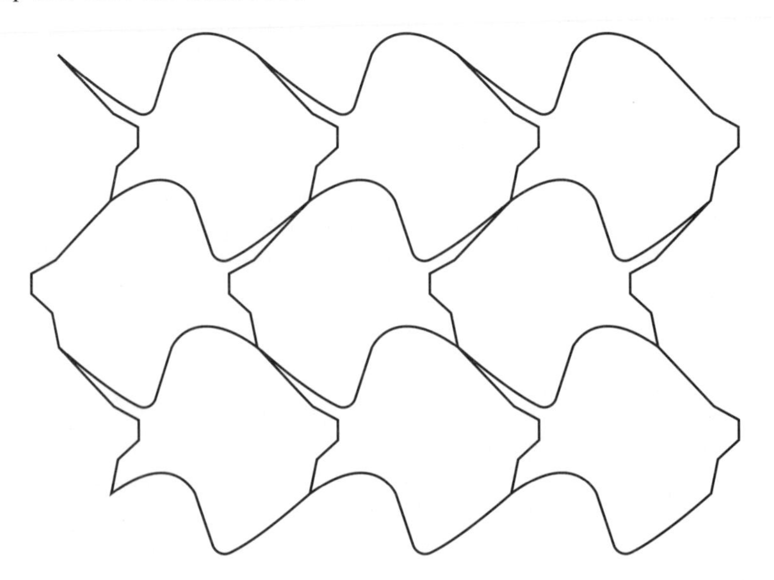

b. Modify the square *ABCD*, using two glide reflections to produce a figure that tessellates. Create a design along side $\overline{AD}$, translate the design to side $\overline{BC}$, and then reflect it about a line perpendicular to the midpoint of $\overline{BC}$. Similarly, modify side $\overline{AB}$ and translate that design to side $\overline{DC}$ through a glide reflection. (*Suggestion:* Use tracing paper to perform the glide and reflection.)

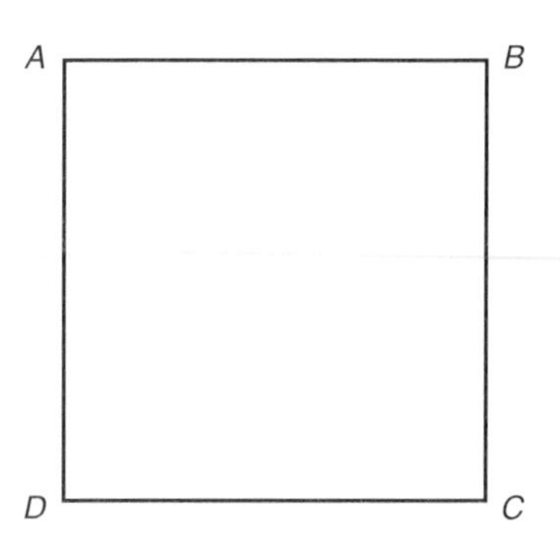

JUST FOR FUN

PAPER PUZZLE 1: PAPER-STRIP CUBE

Draw a strip of 8 squares that each measure 3 cm on a side. Cut out the strip of squares, and fold it to make a cube that measures 3 cm on each edge.

PAPER PUZZLE 2: FOLDING NUMBERS IN ORDER

The following rectangle with the eight numbered squares can be folded so that the numbers are in order from 1 through 8 by carrying out the three folds shown in figures a, b, and c. (*Note:* The numbers will be in order, but not all will be facing up.)

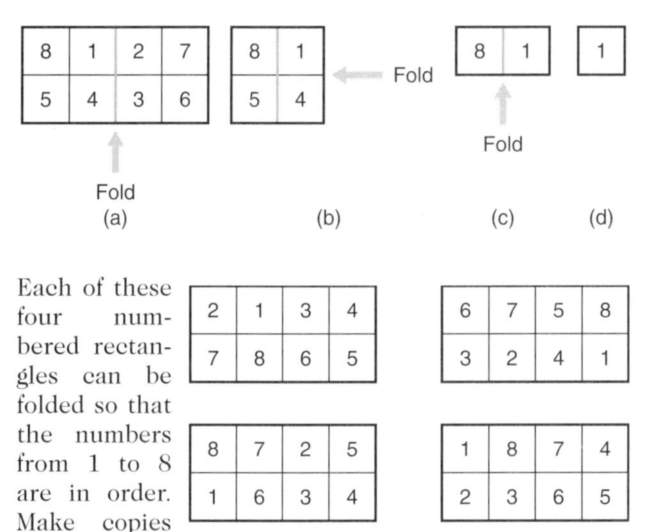

Each of these four numbered rectangles can be folded so that the numbers from 1 to 8 are in order. Make copies of the rectangles, and fold them to get the numbers in order. Devise a way to record your solutions so that another person can understand your procedures.

Design a similar puzzle. One method is to *work backwards* by creatively folding a rectangle with blank squares, numbering the squares in order, and then unfolding it to write the numbers right side up on the rectangle.

PAPER PUZZLE 3: STEPPING THROUGH A PIECE OF PAPER

It is possible to cut a piece of $8\frac{1}{2}$-by-11-inch paper in such a way that you create a hole large enough to step through. A paper and scissors are all you need. No taping, gluing, or refastening the paper is needed or permitted. Try it. (*Hint:* Fold a sheet of paper and mark three points, *A, B,* and *C,* on the fold line. Make three cuts, as indicated by the dotted lines. Then cut along the fold from *A* to *C.* To create a larger hole, extend this procedure by making more cuts.)

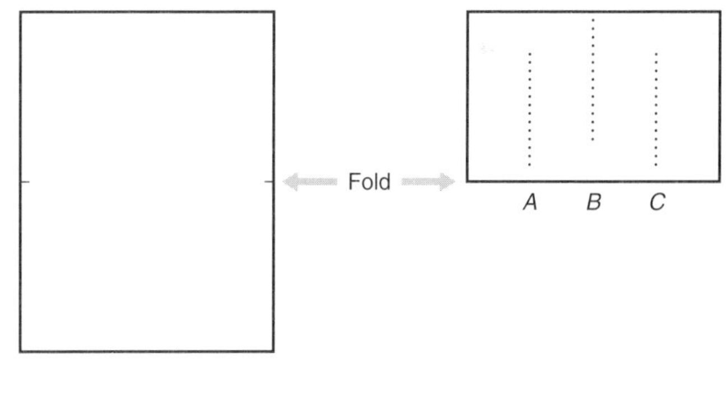

ACTIVITY SET 11.3

DEVICES FOR INDIRECT MEASUREMENT

Purpose To devise methods and simple instruments for measuring distances based on knowledge of similar triangles

Materials Tape, drinking straw, paper-towel tube, hypsometer-clinometer from Material Card 39 and ruler and protractor from Material Card 32 (depending on which instruments you make)

Activity It was in the second century B.C. that the Greek astronomer Hipparchus applied a simple theorem from geometry to measure distances indirectly. He computed the radius of the earth, the distance to the moon, and many other astronomical distances. The theorem that Hipparchus applied states that if two triangles are *similar* (have the same shape), their sides are proportional. This important theorem is used in surveying, map-making, and navigation.

All that is needed to establish that two triangles are similar is that two angles of one are congruent to two angles of the other. The two triangles shown below are similar because $\angle A$ is congruent to $\angle D$ and $\angle C$ is congruent to $\angle F$. (When two angles of one triangle are congruent to two angles of another, the remaining angles also must be congruent. Why?)

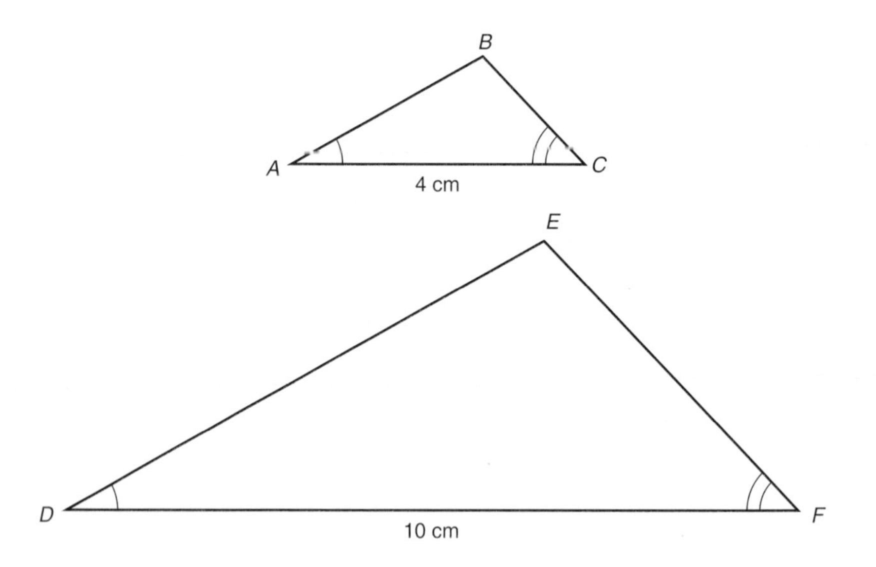

What Hipparchus knew about two such similar triangles is that their side lengths are proportional. For example, side $\overline{DF}$ corresponds to side $\overline{AC}$ and is 2.5 times as long. You can check by measuring that this same relationship holds for the other corresponding sides. That is,

$$DF = 2.5 \times AC \qquad FE = 2.5 \times CB \qquad ED = 2.5 \times BA$$

Notice also that the ratios of the lengths of sides in one triangle are equal to the corresponding ratios in a similar triangle. For example,

$$\frac{AB}{AC} = \frac{DE}{DF} \quad \text{since} \quad \frac{AB}{AC} = \frac{3}{4} \quad \text{and} \quad \frac{DE}{DF} = \frac{2.5 \times 3}{2.5 \times 4}$$

Similarly, the following ratios are also equal.

$$\frac{AB}{BC} = \frac{DE}{EF} \quad \text{and} \quad \frac{BC}{AC} = \frac{EF}{DF}$$

All of the estimating methods and instruments in this activity set use the fact that the length of corresponding sides of similar triangles are proportional.

1. **Sighting Method:** Artists hold out their thumbs to estimate the sizes of objects they are painting. Let's examine the theory behind this method. Stretch your arm out in front of you, holding a ruler in a vertical position. Select some distant object such as a person and estimate his or her height. Hold the ruler so that the top is in line with your eye and the top of the person's head. Position your thumb on the ruler so that it is in line with the person's feet.

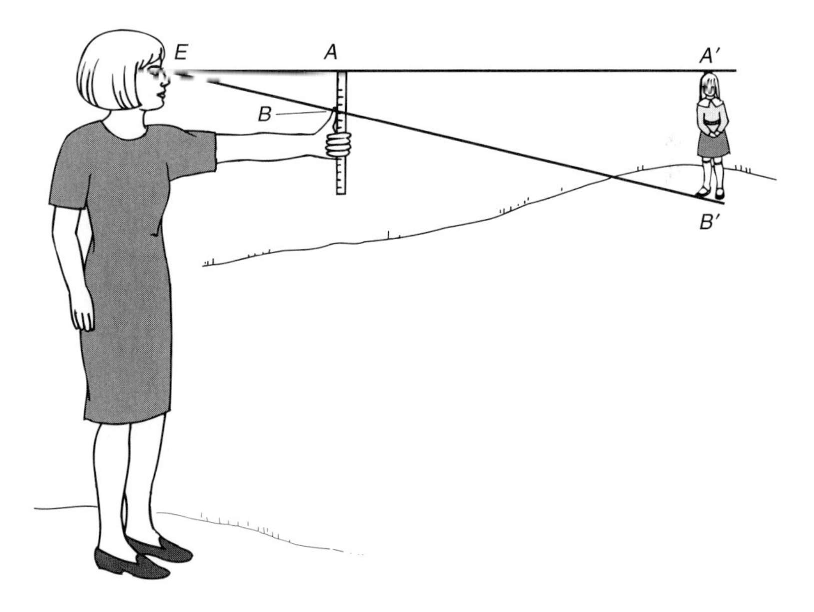

★ a. The lines of sight from your eye (point E) to the top of your ruler and your thumb form triangle EAB. The lines of sight from E to the head and feet of the observed person form triangle $EA'B'$. *Why are these triangles similar?*

b. Obtain the length of $\overline{EA}$ by measuring the distance from your eye to a ruler, as shown in the figure.

c. The two similar triangles are represented separately in the diagram below.

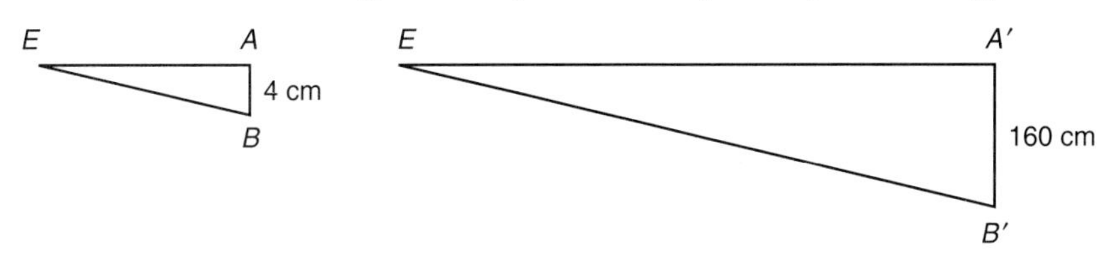

Since triangle EAB is similar to triangle $EA'B'$, their sides are proportional. Assume that AB is 4 cm and the height of the person is 160 cm. Use your measurement for $\overline{EA}$ from part b to determine the distance EA' between the two people.

2. **Scout Sighting Method:** Here is an old scout trick for getting a rough estimate of the distance to an object. This variation of the sighting method does not require a ruler. The scout stretches out his or her arm and, with the left eye closed, sights with the right eye along the tip of one finger to a distant object. Next the scout closes the right eye and sights with the left eye along this finger to another object. The scout then estimates the distance between the two objects and multiplies by 10 to get the distance from these objects.[2]

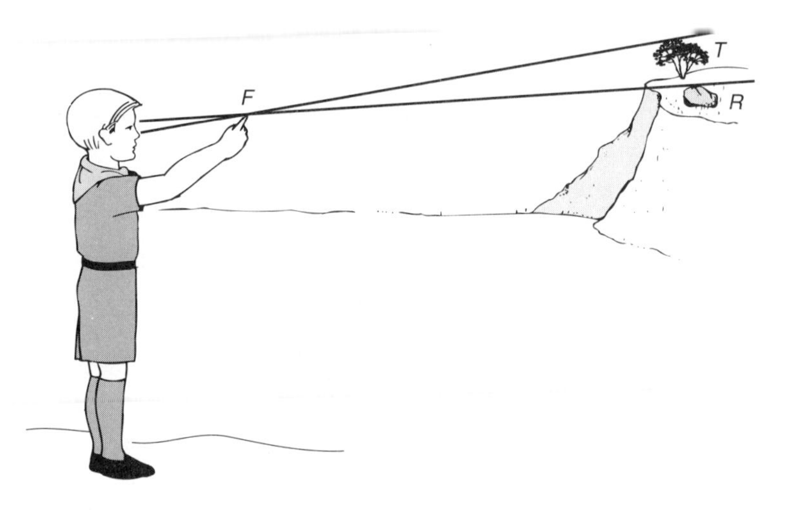

   a. If the scout estimates the distance between the tree (T) and the rock (R) to be 90 m (about the length of a football field), how far is the scout from the tree?

   b. In the following diagram, E_1 and E_2 represent the eyes, F the finger, T the tree, and R the rock. The two isosceles triangles that are formed are similar. What is the ratio, E_1F/E_1E_2, for the measurements on this diagram?

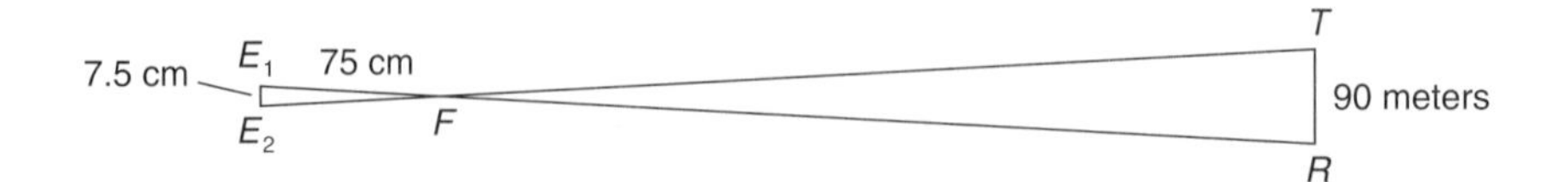

[2]For several more elementary sighting methods, see C. N. Shuster and F. L. Bedford, *Field Work in Mathematics* (East Palistine, OH: Yoder Instruments, 1935), 56–57.

★ **c.** Explain why 10×90 gives the distance from F to T in meters.

d. The scout sighting method is based on the assumption that E_1F (the distance from eye to fingertip) is 10 times E_1E_2 (the distance between the eyes). Stretch your arm out in front of you and measure the distance from your eye to your raised index finger. Then measure the distance between your eyes. Should you be using the number 10 for the scout sighting method? If not, what number should you be using?

★ **e.** When you measure your distance from an object you are fairly close to, such as something in a room, you should add the distance from eye to finger to the result obtained from the scout sighting method. In approximating greater distances, you can ignore the distance from eye to finger, E_1F. Explain why.

3. Stadiascope: The stadiascope works on the same principle as the sighting method with the ruler (see activity 1). This instrument has a peep sight at one end and cross threads at the other. The stadiascope shown here was made from a paper-towel tube having a length of 27 cm. At one end of the tube, six threads have been taped in horizontal rows. These threads are 1 cm apart. A piece of paper has been taped over the other end, and a peep sight has been punched on the same level as the lowest thread.

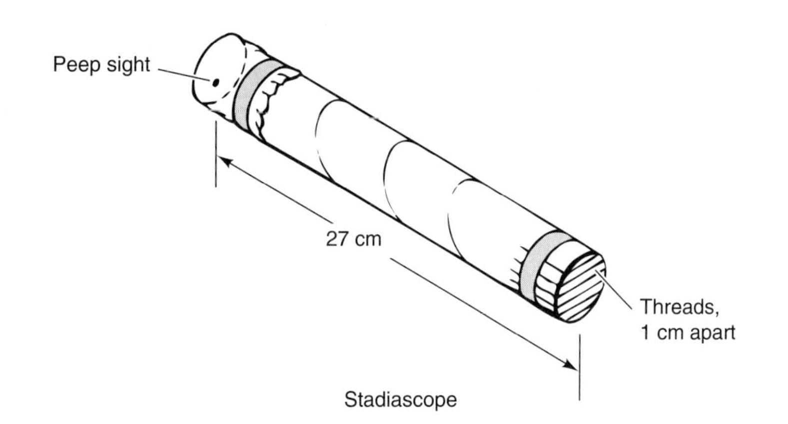

Stadiascope

The following diagram shows a stadiascope being used to measure the distance to a truck. The truck fills 4 spaces at the end of the stadiascope. Let's assume that the truck has a height of 3 m.

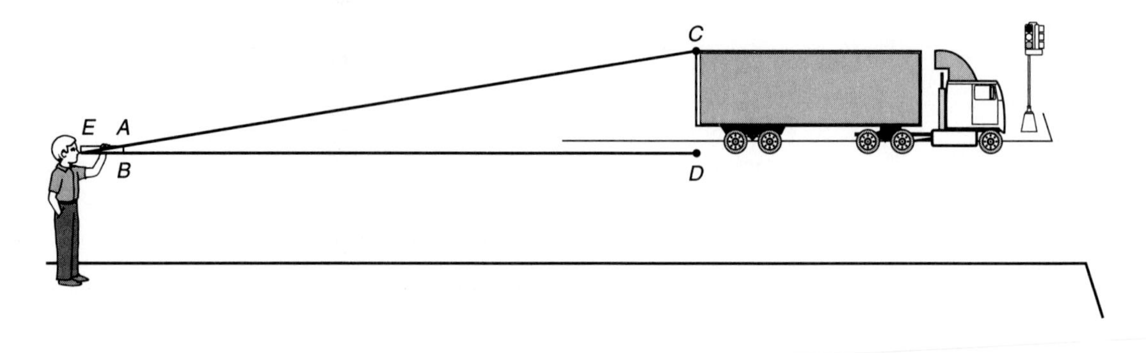

a. Point E is the peep sight, and points A and B mark the upper edge and lower edge of the 4 spaces of the stadiascope in which the truck is seen. Triangle EBA and triangle EDC are right triangles. Explain why they are similar.

★ **b.** Since triangle EBA is similar to triangle EDC, the ratios of their corresponding sides are equal. Use the ratios given here to compute the length of $\overline{ED}$ in meters. (*Note:* EB and AB can be left in centimeters and CD in meters.)

$$\frac{ED}{EB} = \frac{CD}{AB}$$

★ **c.** Suppose you sight a person who fills 2 spaces of the stadiascope and you estimate this person's height to be 180 cm. How many meters is this person from you?

d. Make a stadiascope and use it to measure some heights or distances indirectly. Check your results with direct measurements.

4. Clinometer: The clinometer, used to measure heights of objects, is an easy device to construct (see Material Card 39). It is a simplified version of the quadrant, an important instrument in the Middle Ages, and the sextant, an instrument used for locating the positions of ships. Each of these devices has arcs that are graduated in degrees for measuring angles of elevation. The arc of the clinometer is marked from 0 to 90°. When an object is sighted through the straw, the number of degrees in angle BVW can be read from the arc. Angle BAC is the angle of elevation of the clinometer. What will happen to angle BVW as angle BAC increases?

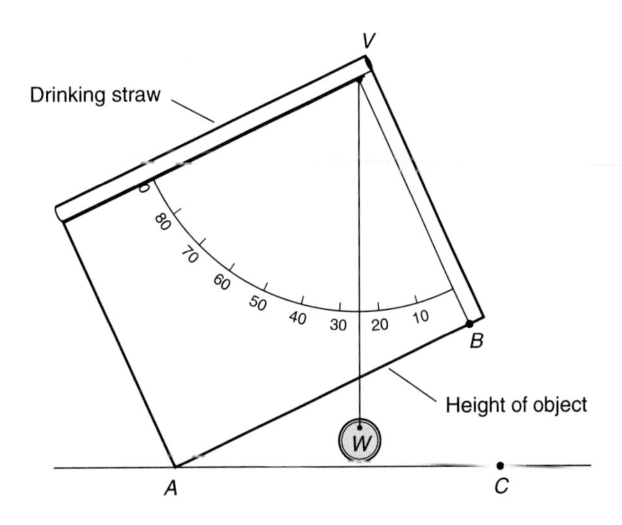

Angle BVW on the clinometer is equal to the angle of elevation of the clinometer, angle BAC. In the next figure, the clinometer was used to find the angle of elevation from eye level to the top of the tree. This angle is 18°. The distance from the person's feet to the base of the tree is 60 m.

a. Right triangle ETR in the drawing is similar to the real triangle that the observer sees outdoors. Explain why.

★ **b.** Use a ruler to set up a scale on $\overline{ET}$. For example, $\overline{ET}$ can be marked off in millimeters. If $\overline{ET}$ represents a distance of 60 m, what distance does each millimeter unit of your scale represent? Use your scale to find the length of $\overline{TR}$.

c. The observer's eye is 150 cm above the ground. What is the height of the tree?

d. Make a clinometer by pasting Material Card 39 onto a solid backing. Use this device to measure the heights of some objects. (*Note:* Draw a similar triangle and set up a scale.)

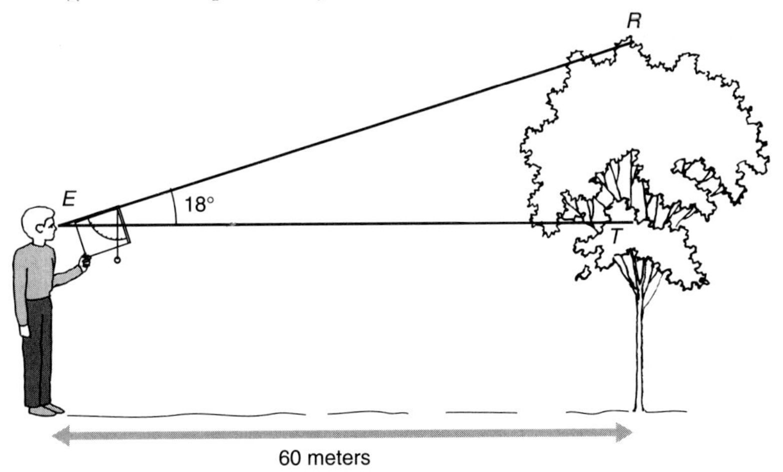

5. **Hypsometer:** The hypsometer is used for the same purpose as the clinometer, but it eliminates the need to measure angles (see Material Card 39). The grid is used to set up a scale on the sides of the hypsometer. For example, suppose that you sight to the top of a building and that the string of the hypsometer falls in the position shown here. If you are standing 55 m from the base of the building, this distance is represented on the right side of the hypsometer by letting each space be 10 m. By beginning at distance 55 on the right side of the hypsometer and following the arrow in to the plumb line (weighted string) and then down to the lower edge of the grid, you can see that the height of the building is 25 m plus the distance from the ground to the viewer's eye.

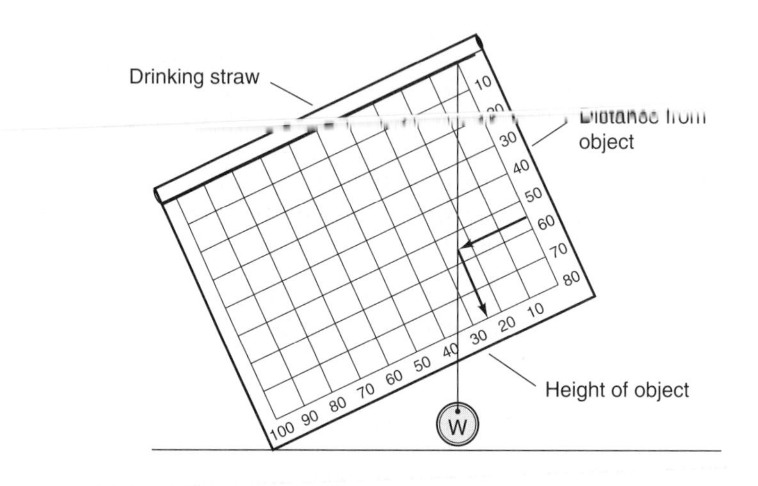

★ **a.** Given the position of the string shown on this hypsometer, determine the height of an object that is at a distance of 70 m.

b. If you know the height of an object, you can use the hypsometer to find the distance to its base. Given the position of the string on this hypsometer, determine the distance to the base of a tree that is 30 m tall.

c. Make a hypsometer from Material Card 39 and use it to measure the heights of some objects.

6. **Transit:** The transit (or theodolite) is the most important instrument for measuring horizontal and vertical angles in civil engineering. In its earliest form, this instrument was capable of measuring only horizontal angles, much like the simplified version shown on the next page. To make a transit, tape a protractor (Material Card 32) to one end of a meterstick (or long thin board) and pin a drinking straw at the center point of the protractor.

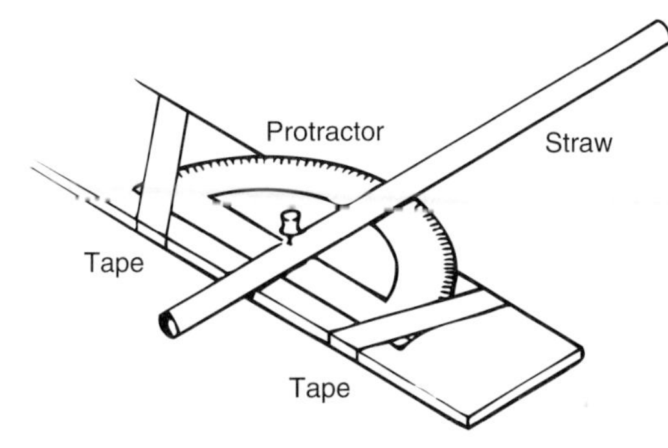

The next diagram shows how the transit can be used to find the distance between two points. Line segment $\overline{AB}$ has been marked by posts on the left bank of the river. By holding the stick of the transit parallel to $\overline{AB}$, you can measure angle *FAB* and angle *FBA*. The distance between the posts can be found by using a tape measure or by pacing off. This distance is 500 m. On paper, a triangle can be drawn (like the one in the figure) to represent the real triangle formed by the fort and the posts. This small triangle is similar to the real triangle because angle *A* and angle *B* have been drawn to be congruent to the angles measured with the transit.

a. Use a ruler to set up a scale on $\overline{AB}$ in the diagram. The actual distance from A to the fort can be found by using your scale. What is this distance?

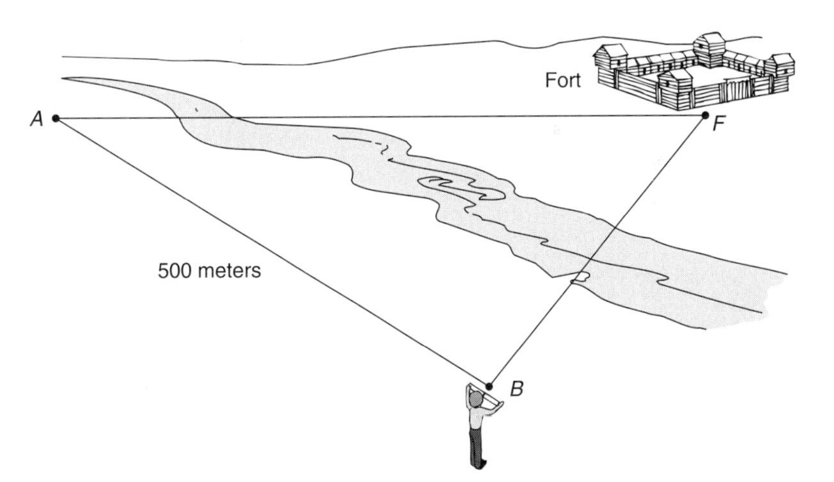

★ **b.** Explain how you would use the transit to make a map of the objects in this field. What is the least number of angles and distances you would have to measure?

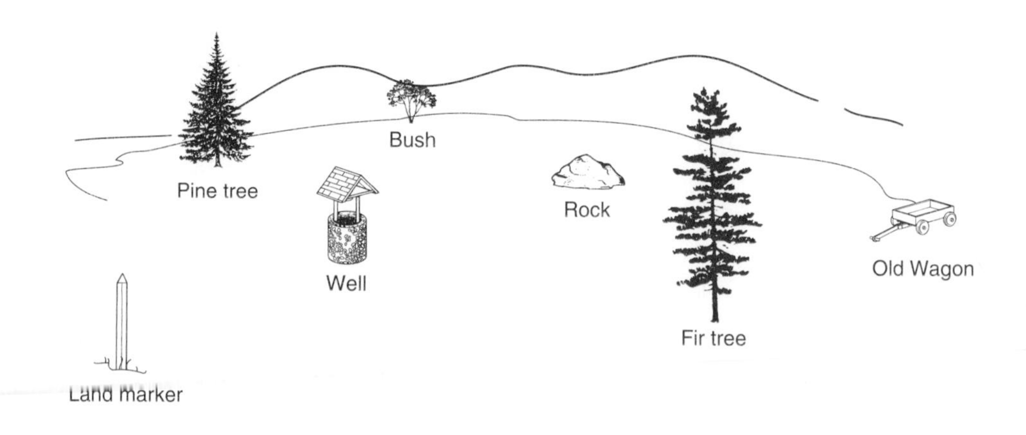

JUST FOR FUN

ENLARGING DRAWINGS

Figure a shows a drawing of a bunny. Suppose you wish to copy and double the size of the drawing. Here is one method that works when you do not have immediate access to a copy machine that automatically enlarges. Draw or place a transparent grid over the bunny as shown in figure b.

(a)

(b)

A grid similar to the grid in figure b, but whose edges are 2 times as long as those in figure b, is drawn below. The sketch of the bunny is then copied square by square from the small grid to the larger grid. On the grid below, five squares have been copied from figure b above. Complete the enlargement by copying the remaining squares.

Choose a picture, cartoon, or drawing of your own and enlarge it using the grid method. You can make it as large as you wish by choosing an appropriate grid to copy upon.

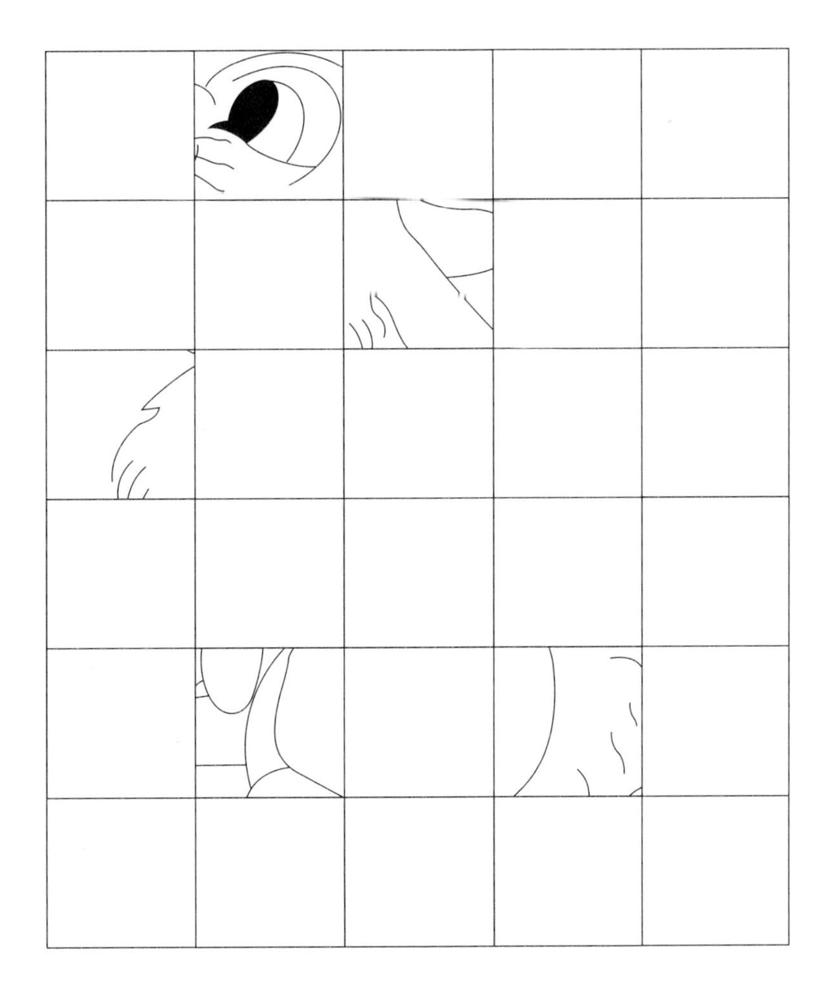

Ideas for the Elementary Classroom

SUGGESTED CLASSROOM ACTIVITY: ANALYZING SHAPES

Once students are able to recognize geometric shapes in different contexts, they can analyze properties of particular geometric shapes. Provide pairs of scissors, rulers, and copies of large parallelograms.

1. Ask the students to investigate the parallelograms and make a list of everything they observe. Encourage them to measure, cut (along diagonals and through the center), fold, and roll up their parallelograms. Make a master list of all observations on the chalkboard or butcher paper.
2. Discuss whether all the properties listed are true for all parallelograms.
3. Do a similar investigation of squares, rectangles, rhombuses, trapezoids, triangles, and other shapes.

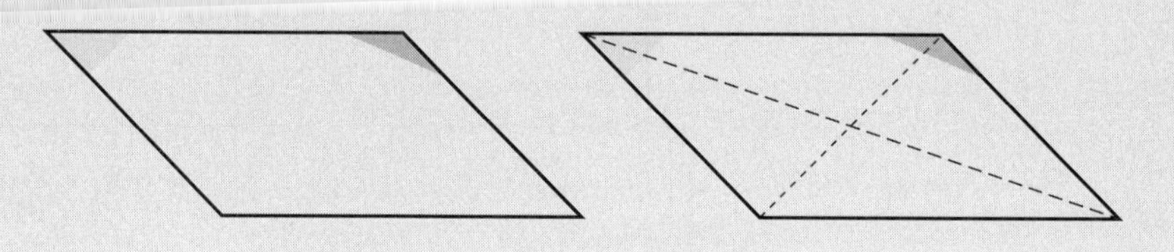

Readings for More Classroom Ideas

Bidwell, James K. "Using Reflections to Find Symmetric and Asymmetric Patterns." *Arithmetic Teacher* 34 (March 1987): 10–15.

Britton, Jill, and Walter Britton. *Teaching Tessellating Art.* Palo Alto, CA: Dale Seymour Publications, 1992.

Geddes, Dorothy, et al. *Geometry in the Middle Grades,* in the *Curriculum and Evaluation Standards for School Mathematics Addenda Series: Grades 5–8.* Reston, VA: National Council of Teachers of Mathematics, 1992.

Hirsch, Christian R. "Graphs, Games, and Generalizations." *Mathematics Teacher* 81 (December 1988): 741–745.

Horak, Virginia M., and W. J. Horak. "Let's Do It: Using Geometry Tiles as a Manipulative for Developing Basic Concepts." *Arithmetic Teacher* 30 (April 1983): 8–15.

Kaiser, Barbara. "Explorations with Tessellating Polygons." *Arithmetic Teacher* 36 (December 1988): 19–24.

Phillips, Elizabeth, Glenda Lappan, Mary J. Winter, and William Fitzgerald. *Similarity and Equivalent Fractions,* Middle Grades Mathematics Project. Menlo Park, CA: Addison-Wesley Publishing Company, 1986.

Sawada, Daiyo. "Symmetry and Tessellations from Rotational Transformations on Transparencies." *Arithmetic Teacher* 33 (December 1985): 12–13.

Schattschneider, Doris. *Visions of Symmetry: Notebooks, Periodic Drawings, and Related Work of M. C. Escher.* New York, NY: W. H. Freeman and Company, 1990.

Seymour, D. and J. Britton. "Creating Escher-like Tessellations," in *Introduction to Tessellations.* Palo Alto, CA: Dale Seymour Publications, 1989, 181–236.

Sovchik, Robert, and L. J. Meconi. "Ideas: Measurement." *Arithmetic Teacher* 41 (January 1994): 253–262.

Woodward, E., V. Gibbs, and M. Shoulders. "A Fifth-Grade Similarity Unit." *Arithmetic Teacher* 39 (April 1992): 22–25.

Woodward, Ernest, and P. G. Buckner. "Reflections and Symmetry—A Second-Grade Miniunit." *Arithmetic Teacher* 35 (October 1987): 8–11.

Zaslavsky, Claudia. "Networks—New York Subways, a Piece of String, and African Traditions." *Arithmetic Teacher* 29 (October 1981): 42–47.

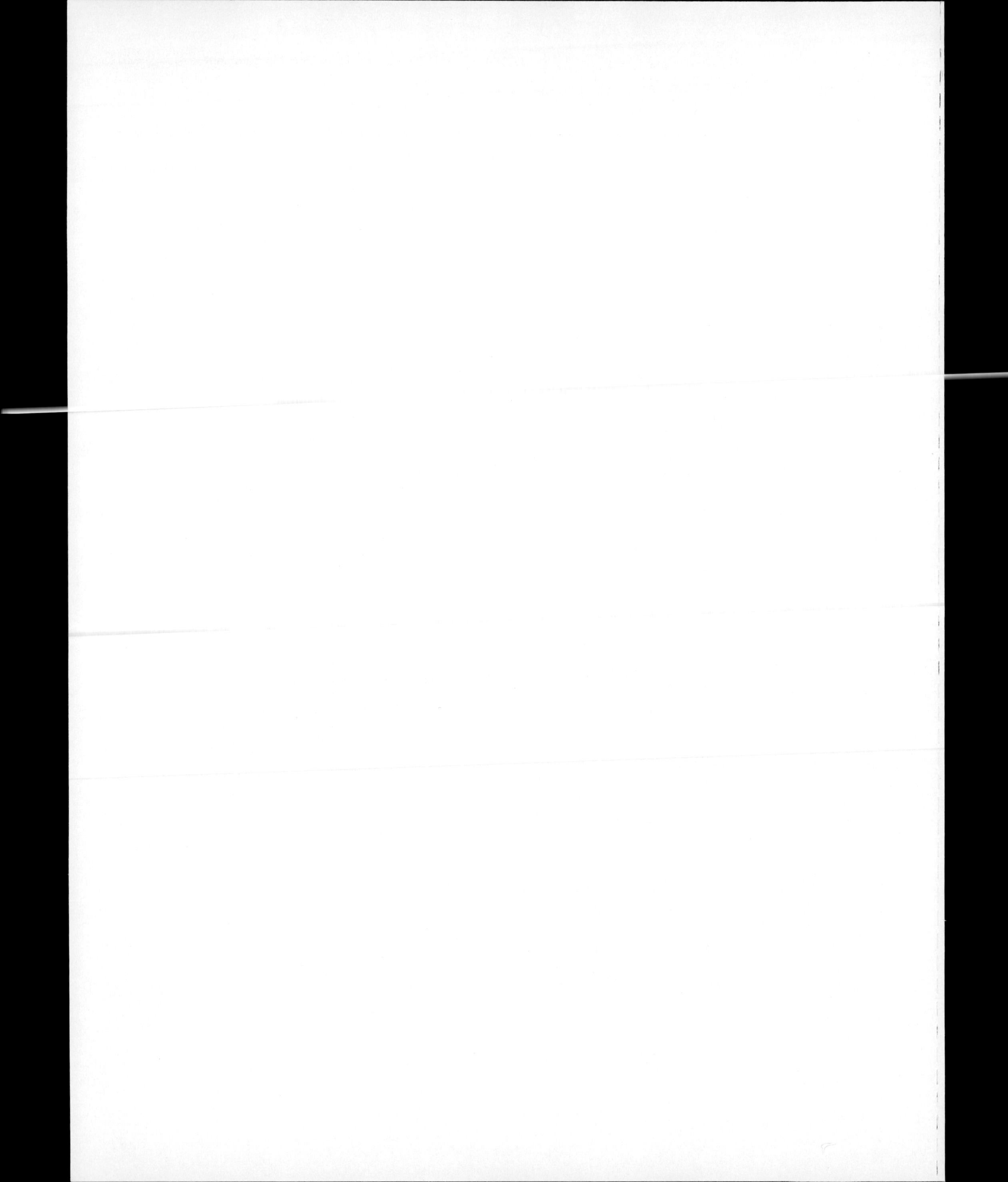

Answers to Puzzlers

P. 7 Move toothpicks A and B as indicated.

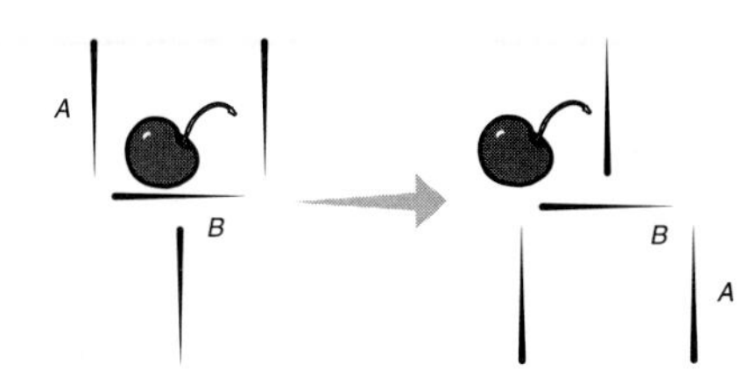

P. 12 Move toothpicks A, B, and C as shown here.

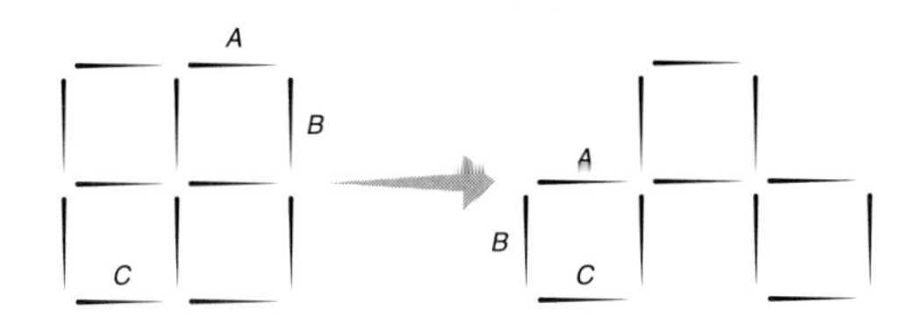

P. 12 Move toothpicks A and B as shown here.

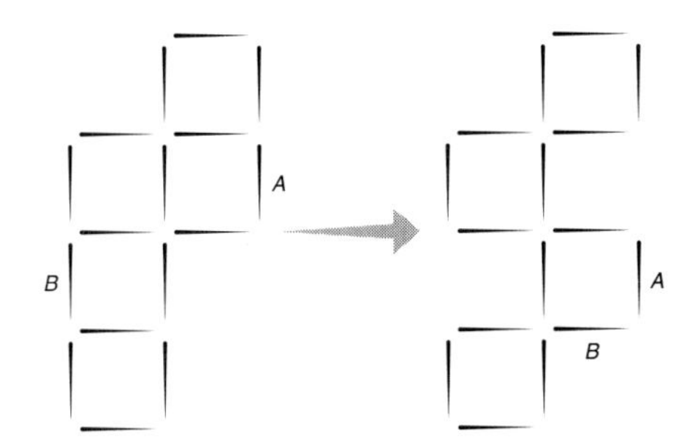

P. 18 You have 7 pennies and your friend has 5.

P. 27 Draw one coin from the container marked "Nickel and Penny." If it is a nickel, then that container must contain two nickels and the container marked "Two Nickels" must contain two pennies—otherwise the third container would be labeled correctly. If a penny is drawn from the "Nickel and Penny" container, then that container holds two pennies and the container marked "Two Pennies" contains two nickels.

P. 34 30 inches. *Hint:* Set one table with its block on top of the other table to determine the heights of the two tables.

P. 40

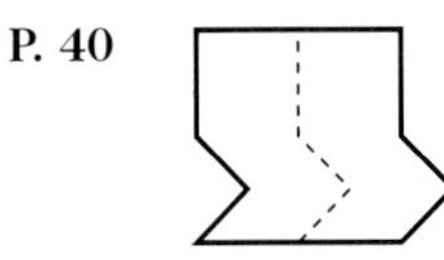

P. 54 On the back of the disk marked 1 is a 5; on the back of 2 is a 4; on the back of 3 is a 4.

P. 67 9 coins

P. 77 Here is one way:

Start		H	T	H	T	H	T				
Move	1	H	T	H	_	_	T	T	H		
Move	2	_	_	H	H	T	T	T	H		
Move	3	_	_	_	_	T	T	T	H	H	H

P. 87 The same number of committees of 2 as committees of 8.

P. 103 120 first-year students

P. 113 All the listed lengths can be obtained.

P. 141 The 5 by 10 rectangle

P. 149

P. 158 They are equally likely.

P. 177 Here is one way. There are many others.

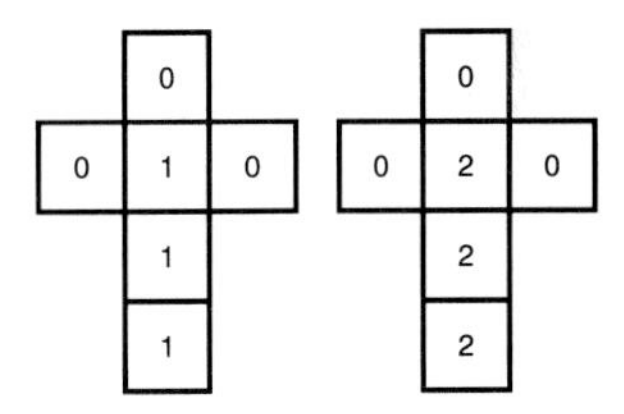

P. 186 Greater than 50%. The probability of obtaining one big loop is $\frac{2}{3}$, or 67%.

P. 198 $\frac{2}{3}$

P. 208

P. 214 Six different (noncongruent) ways

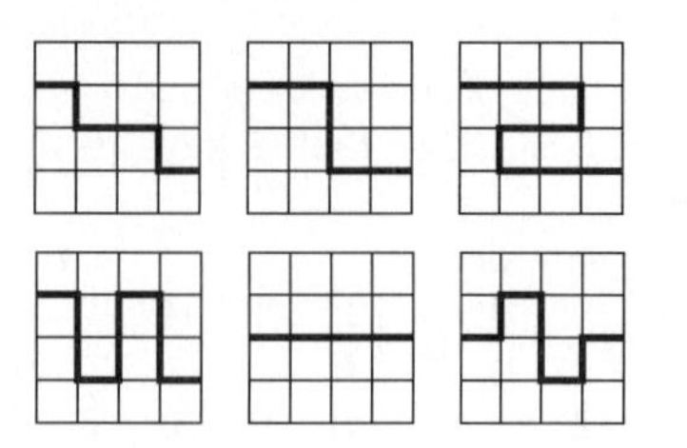

P. 219

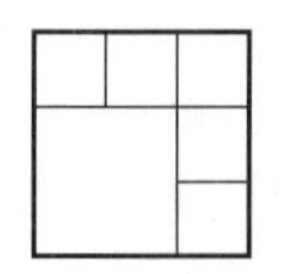

P. 229 2-dimensional solution

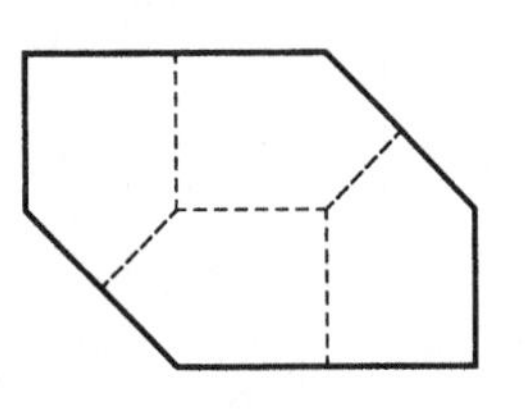

3-dimensional solution

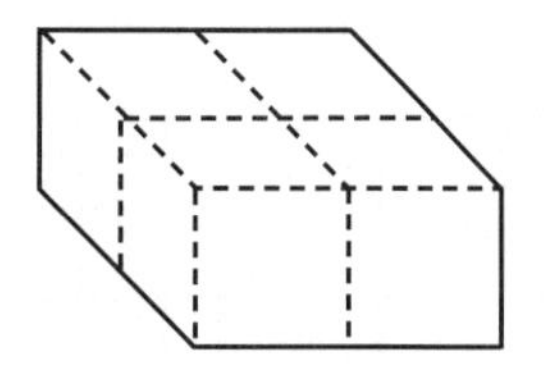

P. 244

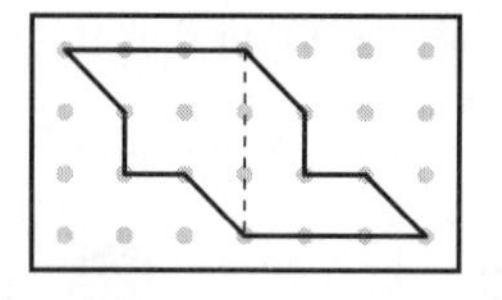

P. 252 51.2%

P. 262 Strips of paint 10 inches long, where the distance from the beginning of one strip to the beginning of the next is the circumference of the circle.

Answers to Selected Activities

ACTIVITY SET 1.1

1. a. Here are some of the many ways they differ:
   (1) The 7th pattern has one more triangle than the 6th.
   (2) The 6th has the same number of triangles pointing up as pointing down.
   (3) The 6th figure forms a parallelogram, the 7th a trapezoid.

   b. The 15th figure would contain 15 green triangles in a row. It would have the shape of a trapezoid with 8 triangles pointing up and 7 triangles pointing down.

3. a. New figures are created in this sequence by adding a shape onto the right side of the previous figure. The 1st figure and every 3rd figure after that adds a yellow hexagon. The 2nd figure and every 3rd figure after that adds a white rhombus. The 3rd figure and every 3rd figure after that adds an orange square.

   b. A rhombus. The pattern repeats after 3 steps so every 3rd figure has a square on the right. The 15th figure has a square on the right, the 16th a hexagon on the right, and the 17th a rhombus.

   c. There are 7 hexagons, 7 rhombuses, and 6 squares in the 20th figure. Multiples of 3 have the same number of each shape. The 18th figure has 6 of each shape. A hexagon is added on for the 19th and a rhombus for the 20th.

   d. 19 of each shape.

6. Here is one way. There are other possibilities. Sequence I: Add one square to the top of each column in the previous figure.

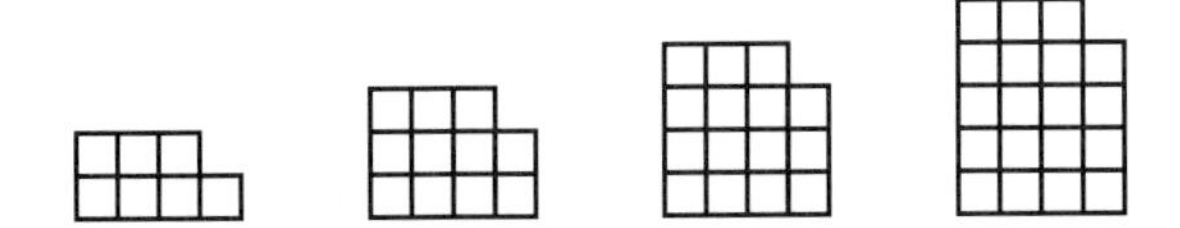

7. Sequence II: 10 squares, 19 triangles, and 9 rhombuses, with a triangle on the right end. The 36th figure looks like 9 copies of the 4th figure so it contains 9 squares, 18 triangles, and 9 rhombuses. To get the 38th figure we need 1 more square and 1 more triangle.

JUST FOR FUN 1.1

1. With 1 red peg and 1 black peg and 3 holes, it takes 3 moves. With 2 red pegs and 2 black pegs and 5 holes, it takes 8 moves.

ACTIVITY SET 1.2

1. a. The 10th odd number (that is the number of tiles in the 10th figure) is 19.

   b. There are many ways to describe it. Here is one way: The 20th figure looks like the letter L made with tiles. Twenty tiles are used to make the vertical part and nineteen additional tiles are needed to complete the horizontal part. Thus, there is a total of 20 + 19 = 39 tiles, and the 20th odd number is 39.

   c. Here is one way to describe it: The 50th figure will have a horizontal row of 50 tiles with a vertical column of 49 tiles on its left end. It will contain a total of 49 + 50 = 99 tiles.

   d. Here are a few methods.
   (1) Add the figure number plus the figure number minus 1.
   (2) Multiply the figure number by 2 and subtract 1.
   (3) Square the figure number and subtract the figure number minus 1 squared.

3. b. The fourth term in the sequence is 20 and the tenth term is 44. Given any figure number, multiply it by 4. Then add 4 to that result to get the total number of tiles.

5. a. The 5th stairstep has 15 tiles and the 10th stairstep has 55 tiles.

   b. The 50th stairstep has 1275 tiles.

   c. The 100th stairstep

   d. $(100 \times 101) \div 2 = 5050$

   e. The sum of consecutive whole numbers from 1 to any specified number (let's call it the last number) can be represented by a stairstep of tiles. This number of tiles in the stairstep is one half of a rectangle with dimensions last number by last number + 1. So,

$$1 + 2 + 3 + 4 + \cdots + \text{last number} = (\text{last number}) \times (\text{last number} + 1) \div 2$$

JUST FOR FUN 1.2

1. If the sequence begins with "loves me," the last response will be "loves me" for daisies with an odd number of petals.

ACTIVITY SET 1.3

1. a. The length of the rectangle is twice its width.

   b. Six variable pieces represent the perimeter of the rectangle, so each variable piece represents 9 units ($54 \div 6$). Therefore, the length is 18 units and the width 9 units.

   c.

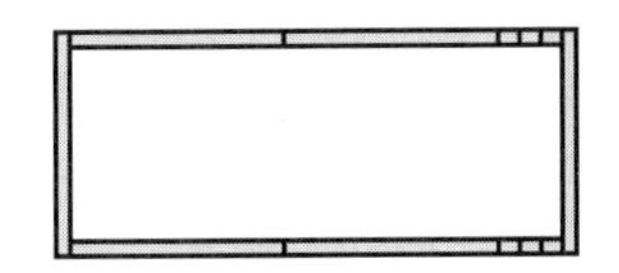

d. Since 6 variable pieces and 6 unit pieces represent the perimeter of 90 units, the 6 variable pieces must represent 84 units (90 − 6). So each variable piece represents 14 units (84 ÷ 6) and the rectangle has a width of 14 units and a length of 31 units.

3. a.

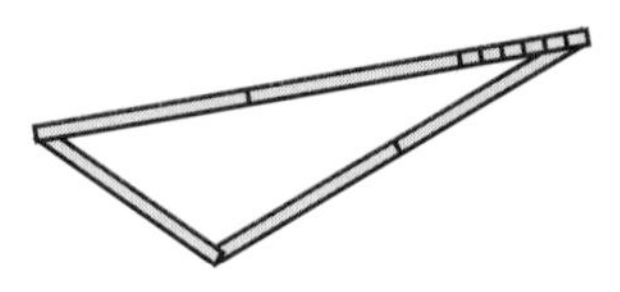

b. The perimeter of the triangle can be represented with 5 variable pieces and 6 units. The perimeter is 66, so the 5 variable pieces represent 60 units. Therefore, each single variable piece represents 12 units. The sides of the triangle are 12 units, 24 units, and 30 units.

5.

Short piece of rope

Long piece of rope

End to end

End to end, the length is 75 m. Then 2 variable pieces represent 68 m (75 − 7) and one variable piece is 34 m. So the short piece is 34 m and the long piece is 41 m.

7.

Number of dimes Greg has

Number of nickels Andrea has

Suppose Greg exchanges his dimes for nickels.

Number of nickels Greg will have

Number of nickels Andrea will have

Andrea has 80 cents more than Greg, so 2 variable pieces represent 80 cents, which is the amount of money that Greg has.

9.

Number of men

Number of women

Double number of men

Increase women by 9

Suppose each variable piece represents one-fifth of the students. Then 2 variable pieces represent the total number of men and 3 pieces the number of women. The equality given by doubling the number of men and increasing the women by 9, shows that each variable piece represents 9 people. The original class has 45 people, 18 men and 27 women.

ACTIVITY SET 2.1

2. a. By asking yes or no questions that refer to a single attribute, a player can identify any piece in at most 6 guesses. For example: Is it large? Is it red? Is it blue? Is it triangular? Is it square? Is it circular?

4. b. LRC, SRT, SBT, LRS, LRH, SRC, SRS, SRH, LRT, SYT, LYT, LBT

5. a. LRC, LYC, SRC, SYC

6. b. LARGE RED pieces, or LARGE AND RED pieces

c. LYT, SYT, LBT, LBS, LBC, LBH, SBT, SBS, SBC, SBH, LRT, LRH, LRS, LRC, SRT, SRH, SRS, SRC

7. b. (2) $(H \cap NY) \cap L$: LBH, LRH

(5) $H \cap (NY \cup L)$: SBH, SRH, LBH, LRH, LYH

ACTIVITY SET 2.2

1. All but d and h can be completed.

3. i and k are complete after 4 passes, j is complete after 2 passes, and l is not complete.

5. Sequences c, d, and f complete after 2 passes; a completes after 4 passes; b and e do not complete.

a.

E	S	W	N
1	2	6	1
2	6	1	2
6	1	2	6
9	9	9	9

b.

E	S	W	N
1	2	1	4
1	2	1	4
1	2	1	4
3	6	3	12

c.

E	S	W	N
5	2	1	7
3	4	5	2
1	7	3	4
8	13	8	13

d.

E	S	W	N
1	1	2	2
3	3	1	1
2	2	3	3
6	6	6	6

e.

E	S	W	N
8	7	6	5
4	3	2	1
8	7	6	5
4	3	2	1
24	20	16	12

f.

E	S	W	N
1	2	3	4
5	5	4	3
2	1	1	2
3	4	5	5
4	3	2	1
15	15	15	15

g. All spirolaterals of odd order and spirolaterals of even order which are not multiples of 4.

h. Odd order spirolaterals require 4 passes and even order spirolaterals which are not multiples of 4 require 2 passes.

ACTIVITY SET 2.3

3. Math—*Robinson,* English—*Smith,* French—*Brown,* Logic—*Jones*

4. Pilot—*Jones,* Copilot—*Smith,* Engineer—*Robinson*

ACTIVITY SET 3.1

1. c. 0 long-flats, 0 flats, 4 longs, 4 units

2. c. 1 long-flat, 0 flats, 0 longs, 1 unit

f. 453 unit squares

3. c. 1001_{five}

f. 3303_{five}

4. b. 142 unit squares

5. a. base three

Flat Long Unit

6. b. 425_{seven}. Total number of unit squares: 215

7. b.

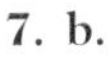

123_{nine}

JUST FOR FUN 3.1

Mind-Reading Cards

2. 27 = 1 + 2 + 8 + 16, cards 1, 2, 8, and 16

4. A number k is on a card if and only if the number of the card is a term in the sum of the binary numbers that equals k. So the sum of the numbers of the cards on which a person's age appears is the person's age.

5. Since 44 = 32 + 8 + 4, the number 44 must be written on cards 4, 8, and 32.

Game of Nim

3. Crossing out 1 stick in the second row leaves an even number of binary groups.

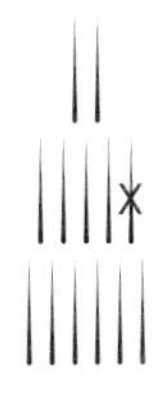

4. It is not possible in 1 turn to change an even situation into another even situation. Therefore, if you leave your opponent with an even situation, there will be an odd situation on your turn.

ACTIVITY SET 3.2

2. **b.** Since the maximum amount you can win in a roll is 2 longs and 2 units, it will take you at least 11 turns to win the game. (If you win the maximum amount on each turn, then in 2 turns you will be 1 unit short of a flat. So in 10 turns you will be 5 units short of a long-flat, and it will take at least 1 more turn to win.)
   d. Your winnings of 4 flats, 2 longs, and 1 unit are 2 longs and 4 units short of a long-flat. You cannot win on the next turn because the maximum amount you can win on 1 turn is 2 longs and 2 units.
3. **a.** 122_{five} **c.** 1021_{five}
4. **a.** 1010_{three} **c.** 433_{six}
6. **b.** Comparing the base pieces for 313_{five} and 242_{five} shows that the difference is 2 longs and 1 unit. So the player must have rolled 5 on one die and 6 on the other.
8. **a.** 424_{five} **c.** 3414_{five}
9. **a.** 1022_{three} **c.** 101_{two}

JUST FOR FUN 3.2

If you cross off 16, then no matter what your opponent does, you will be able to cross off numbers through 20 on your next turn and win the game. To ensure that you are able to cross off 16, you must be the player to cross off 12. Similarly, you can be sure of crossing off 12 if you cross off 8, and you can be sure of crossing off 8 if you cross off 4. The only way you can be sure of crossing off 4 is if your opponent begins the game.

ACTIVITY SET 3.3

2. Any two-digit number can be represented by, at most, 9 longs and 9 units. Multiplication of a one-digit number by a two-digit number can be done concretely by laying out the specified number (1 to 9) of copies of longs and flats for the two-digit number. The product of the numbers can then be determined by regrouping the base pieces. Since 10 units can be regrouped as 1 long, 10 longs as 1 flat, and so on, it is not necessary to count beyond 10.
3. **b.** 483

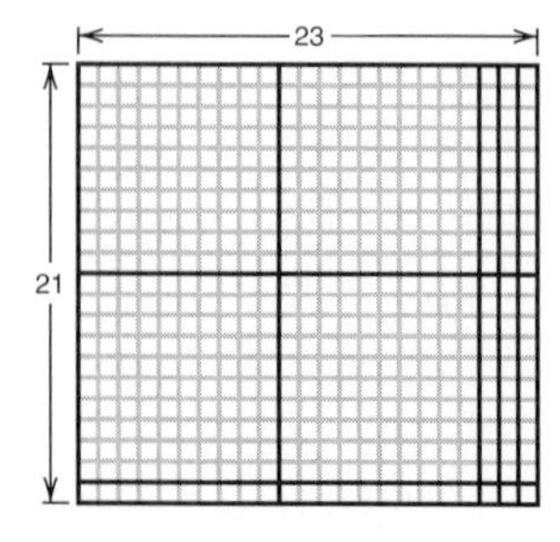

4. Products can be determined by counting the numbers of flats, longs, and units in rectangular arrays. Use the numbers being multiplied to outline a rectangle. Then fill in the rectangle with your base pieces. Regroup the pieces to determine the product. The following example shows this procedure for 21 × 33.

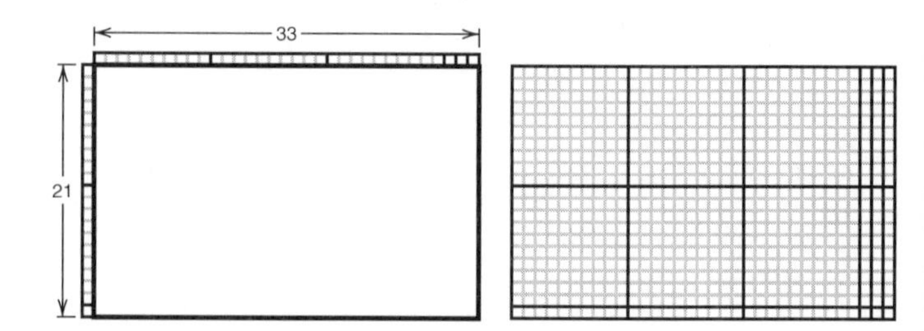

So 21 × 33 is represented by 6 flats, 9 longs, and 3 units, and 21 × 33 = 693.

5. **a.**

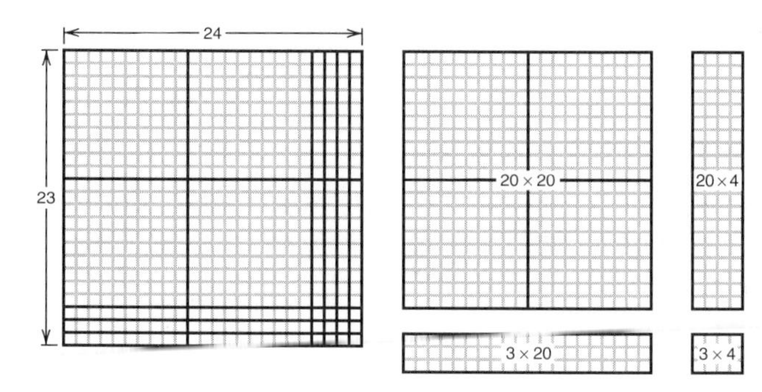

6. **b.**

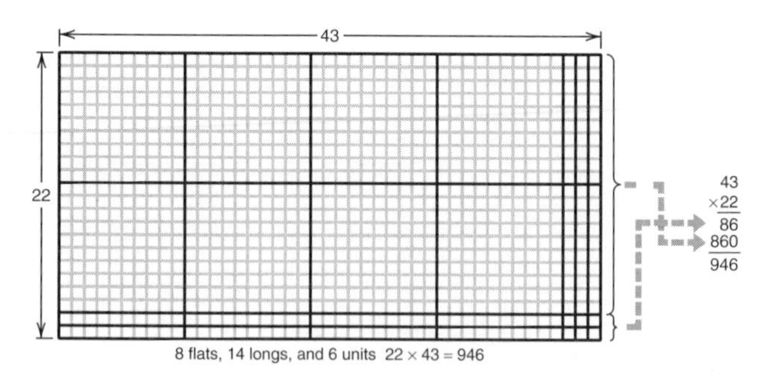

8 flats, 14 longs, and 6 units 22 × 43 = 946

ACTIVITY SET 3.4

1. **b.**

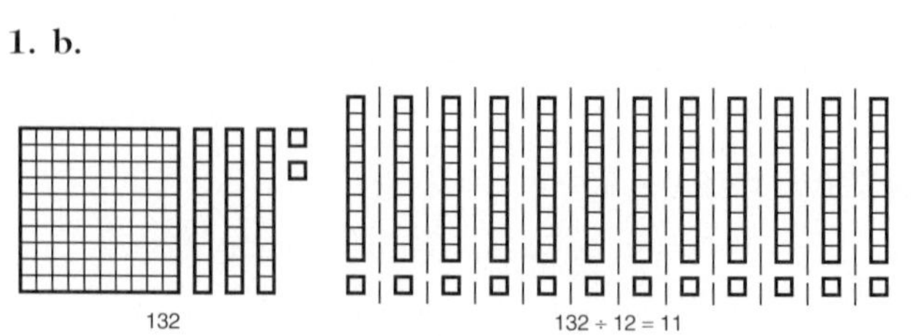

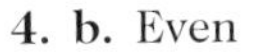

2. b.

114 ÷ 12 = 9 with a remainder of 6

3. b.

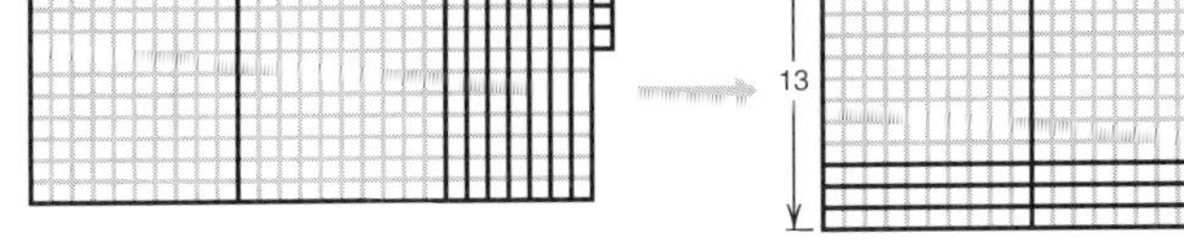

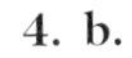
4. b.

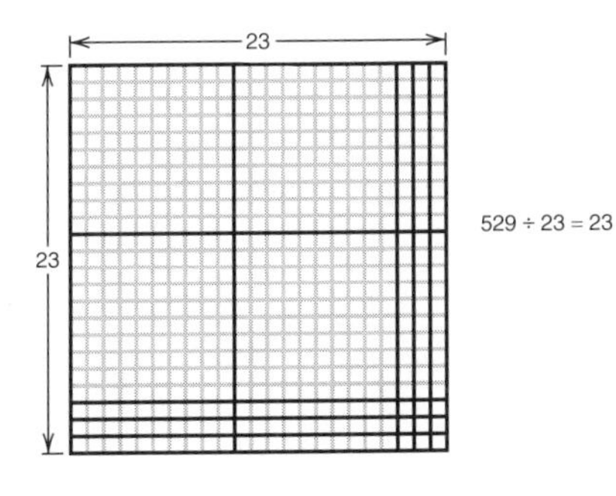

6. c.

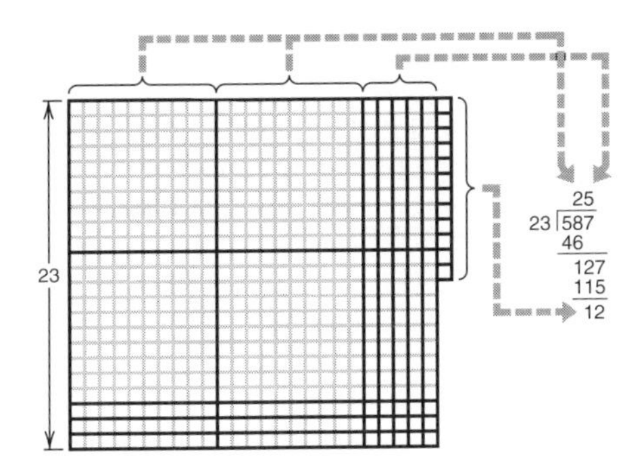

ACTIVITY SET 4.1

2. b. The 125th even number has two rows of 125 tiles. The 125th even number is 250.
3. b. The 15th odd number has a horizontal row of 15 tiles on the bottom and a row of 14 tiles on the top. The 15th odd number is 29.
   d. 79 is the 40th position; 117 is in the 59th position.
4. b. Even e. Even
5. b. The L-shapes representing the first 10 consecutive odd numbers form a 10 by 10 square, so the sum of the first 10 consecutive odd numbers is 100.
8. b. 1 is the only nonzero whole number with exactly 1 factor.
   e. Numbers with an odd number of factors have a square among their rectangles. That is, they are square numbers.

ACTIVITY SET 4.2

1. b. 1, 2, 3, 6, 9, 18
2. b. GCF(18, 25) = 1
4. a. 8
5. Suppose you have two rods m and n of different length where $m > n$. When m and n are cut into rods of common length m is always divided at a point coinciding with the length of n. So the difference rod $(m - n)$ and the shorter rod (n) have the same rods of common length as have m and n. In other words, GCF(m, n) = GCF$(m - n, n)$.

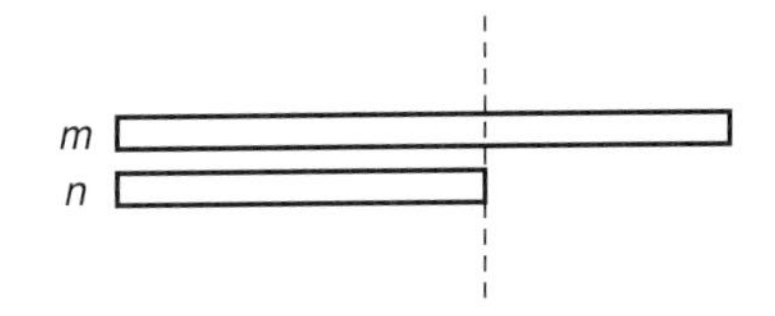

6. b. GCF(280, 168) = GCF(112, 168) = GCF(112, 56) = GCF(56, 56) = 56
   d. GCF(306, 187) = GCF(119, 187) = GCF(119, 68) = GCF(51, 68) = GCF(51, 17) = GCF(34, 17) = GCF(17, 17) = 17
7. b. LCM(14, 21) = 42
   d. LCM(8, 10) = 40

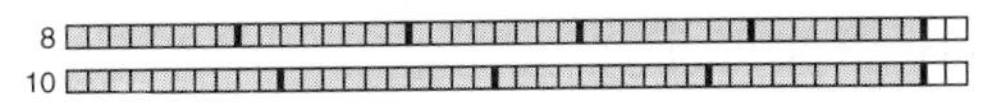

8. a.

A	B	GCF(A, B)	LCM(A, B)
(3) 14	21	7	42
(5) 8	10	2	40

9. a. GCF(9, 15) = 3 and 9 × 15 = 135, 135 ÷ 3 = 45. So LCM(9, 15) = 45.
   c. GCF(14, 35) = 7 and 14 × 35 = 490, 490 ÷ 7 = 70. So LCM(14, 35) = 70.

JUST FOR FUN 4.2

1. Yes

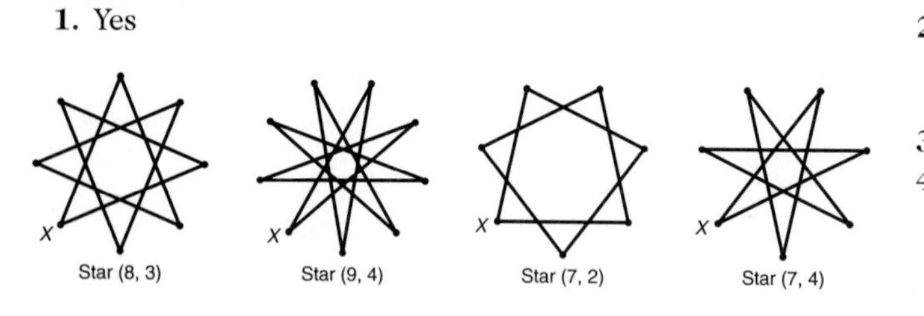

2. Star (n, s) is congruent to (has the same size and shape as) star (n, r).

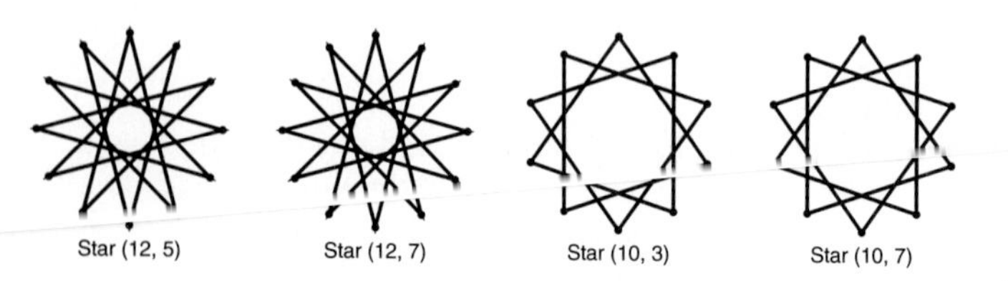

3. Star (5, 1) has one path; star (10, 4) has 2 paths; star (12, 4) has 4 paths; and star (6, 3) has 3 paths.

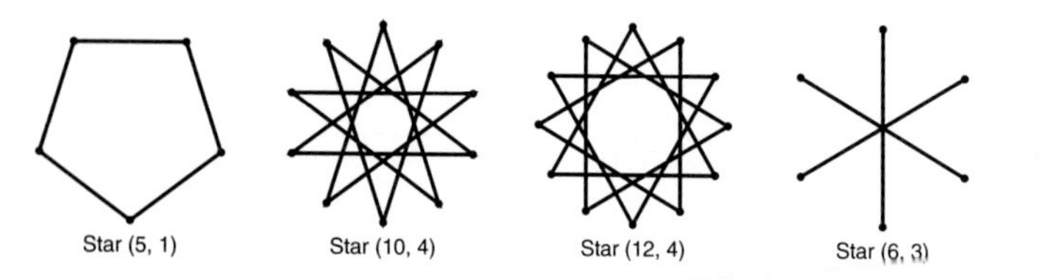

4. n and s must be relatively prime; that is, GCF$(n, s) = 1$. These stars have 1 continuous path: (1) star (15, 2) = star (15, 13); (2) star (15, 4) = star (15, 11); and (3) star (15, 7) = star (15, 8).

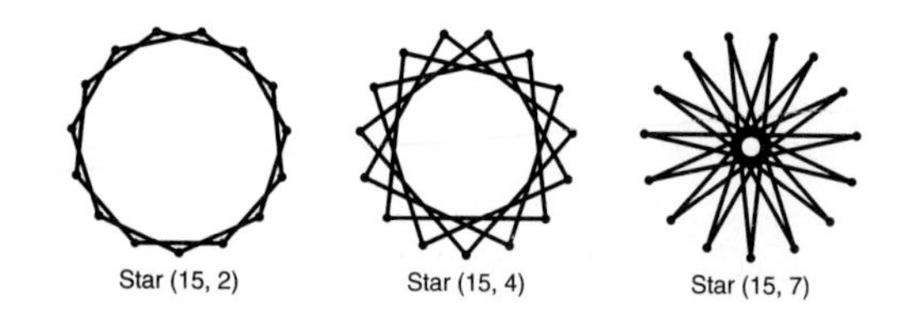

5. Star (9, 2) has 1 path and requires 2 orbits. Star (7, 3) has 1 path and requires 3 orbits. Star (14, 4) has 2 paths and each path requires 2 orbits.

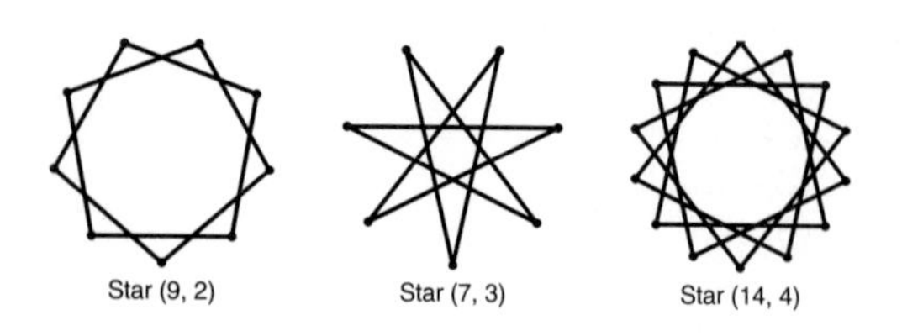

6. b. The number of orbits for star (n, s) is LCM(n, s) divided by n.

7. b. 15

ACTIVITY SET 5.1

2. a. 37   8   29   21
c. 20   6   14   8
f. 24   13   11 $^-2$

3. c. 13; $^-59$; $^-62$

4. b. If there are more red tiles than black tiles, subtract the number of black tiles from the number of red tiles and make the answer negative. If the number of black tiles is greater than or equal to the number of red tiles, subtract the number of red tiles from the number of black tiles.

5. b.

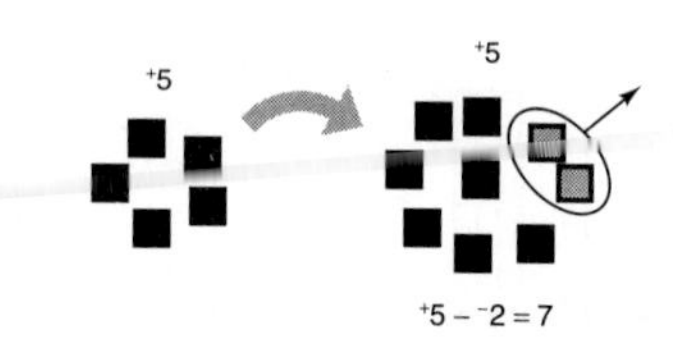

6. Notice that in each of the preceding examples, taking out a given number of tiles of one color is equivalent to putting in the same number of tiles of the opposite color. For example, in 5b, 2 red tiles and 2 black tiles were put in the set and then 2 red tiles were withdrawn from the set. The same result could be accomplished by simply adding 2 black tiles to the set.

7. b.

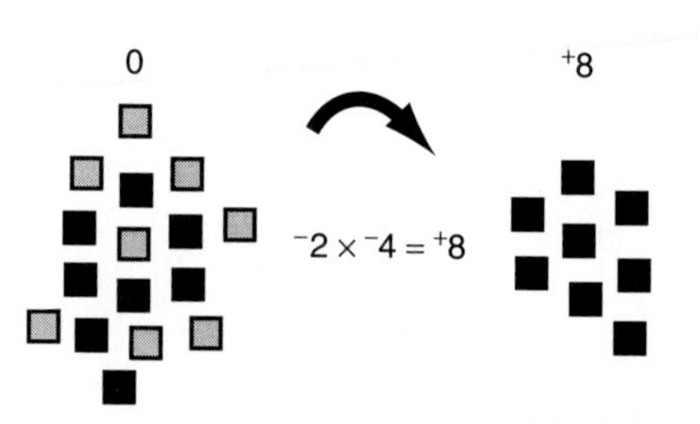

9. c. In the sharing approach, $^-3$ represents the number of groups, which is not possible. In the measurement approach, groups of 3 red tiles cannot be formed from 12 black tiles.

10. c. $91 \div {}^-7 = {}^-13$ ${}^-4 \div {}^-4 = {}^+1$
${}^-1001 \div 11 = {}^-91$ ${}^-221 \div {}^-17 = {}^+13$
$0 \div {}^-9 = 0$

ACTIVITY SET 5.2

1. a.

FRACTION	NO. OF BARS WITH EQUAL FRACTIONS
$\frac{0}{4}$	5
$\frac{1}{2}$	4
$\frac{2}{3}$	3
$\frac{6}{6}$	5
$\frac{1}{4}$	2

2. a. $\frac{1}{12} < \frac{1}{6} < \frac{1}{4} < \frac{1}{3} < \frac{5}{12} < \frac{1}{2} < \frac{7}{12} < \frac{2}{3} < \frac{3}{4} < \frac{5}{6} < \frac{11}{12}$

3. b. Here are 10 inequalities; there are others.

$\frac{1}{2} < \frac{2}{3}, \frac{1}{2} < \frac{3}{4}, \frac{1}{2} < \frac{4}{6}, \frac{1}{2} < \frac{5}{6}, \frac{1}{3} < \frac{2}{4}, \frac{1}{3} < \frac{3}{4},$

$\frac{1}{4} < \frac{1}{2}, \frac{1}{4} < \frac{1}{3}, \frac{1}{4} < \frac{2}{6}, \frac{1}{4} < \frac{3}{6}$

4. b. $\frac{5}{11}$ is less than $\frac{1}{2}$ and $\frac{4}{6}$ is greater than $\frac{1}{2}$, so $\frac{5}{11}$ is less than $\frac{4}{6}$.

d. Cutting a bar into 50 pieces produces smaller pieces than does cutting the same bar into 30 pieces, so $\frac{1}{50}$ is less than $\frac{1}{30}$.

5. a. Nearest whole number, 4

Diagram:

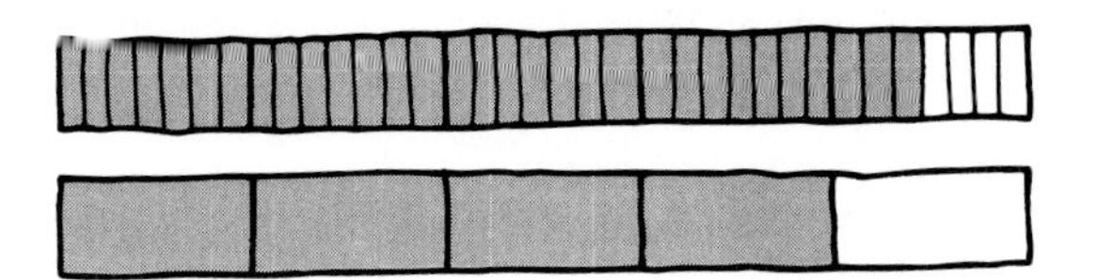

c. Nearest whole number, 1

Diagram:

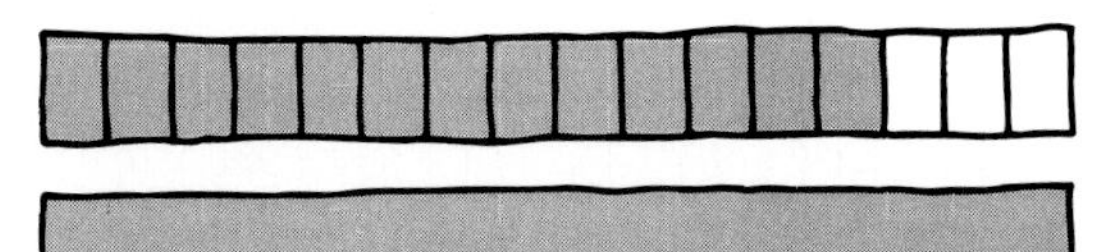

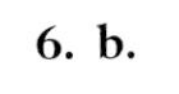

6. b.

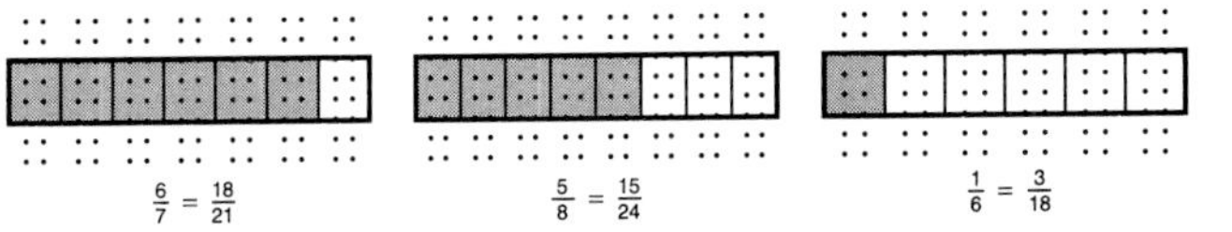

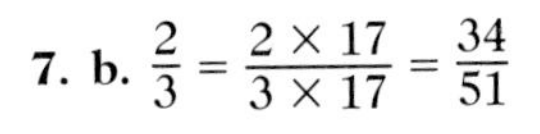

7. b. $\frac{2}{3} = \frac{2 \times 17}{3 \times 17} = \frac{34}{51}$

d.

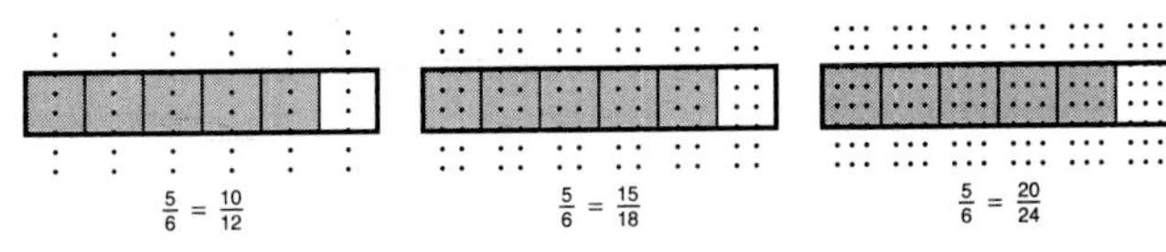

8. a.

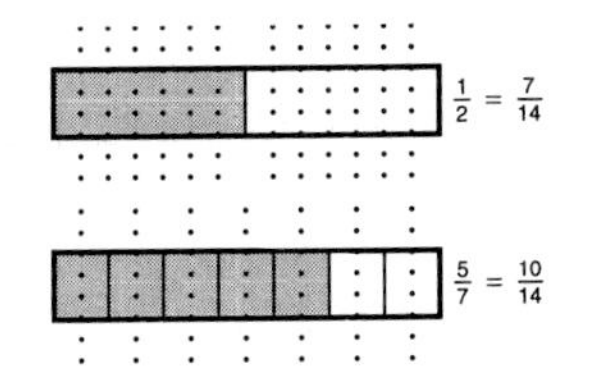

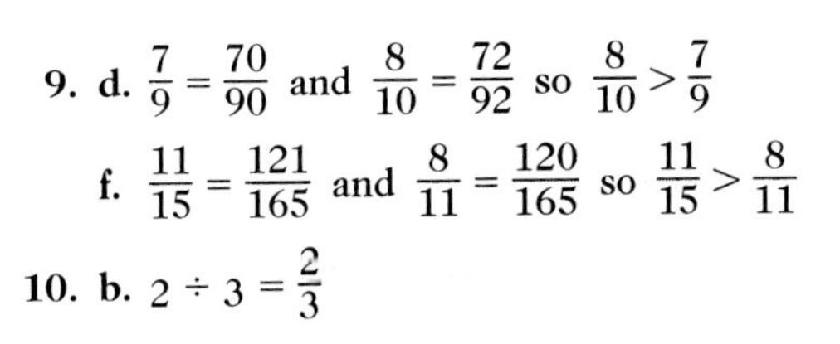

9. d. $\frac{7}{9} = \frac{70}{90}$ and $\frac{8}{10} = \frac{72}{92}$ so $\frac{8}{10} > \frac{7}{9}$

f. $\frac{11}{15} = \frac{121}{165}$ and $\frac{8}{11} = \frac{120}{165}$ so $\frac{11}{15} > \frac{8}{11}$

10. b. $2 \div 3 = \frac{2}{3}$

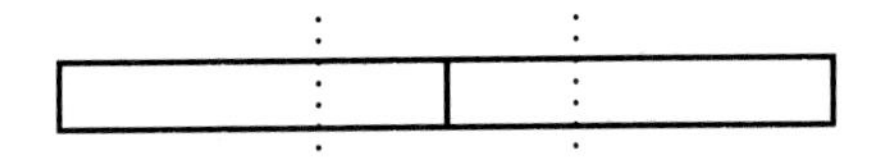

ACTIVITY SET 5.3

1. a. c. e.

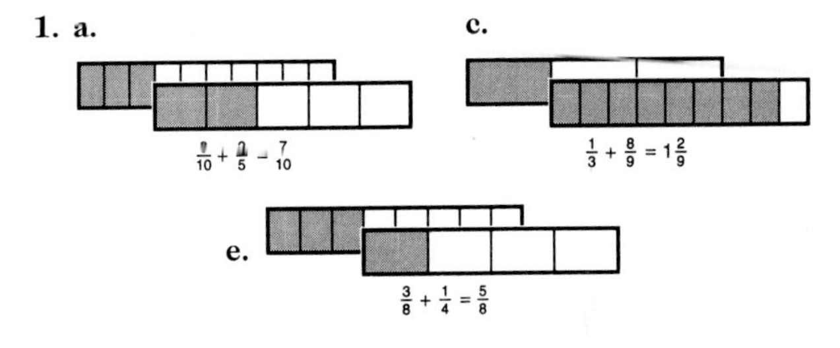

3. a.

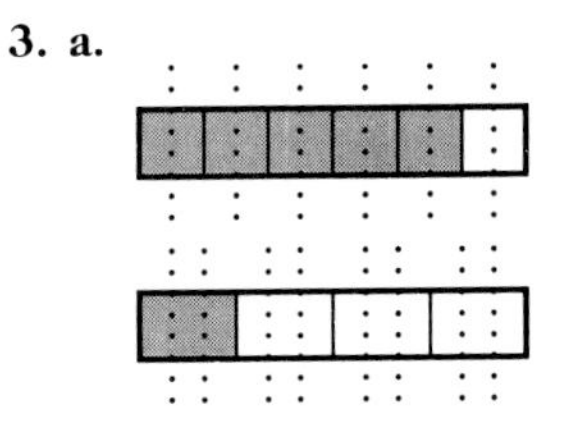

d. $\frac{13}{12} = 1\frac{1}{12}$

g. $\frac{7}{12}$

4. a. $\frac{5}{9} + \frac{1}{3} = \frac{8}{9}$

c. $\frac{2}{5} - \frac{1}{3} = \frac{1}{15}$

5. b.

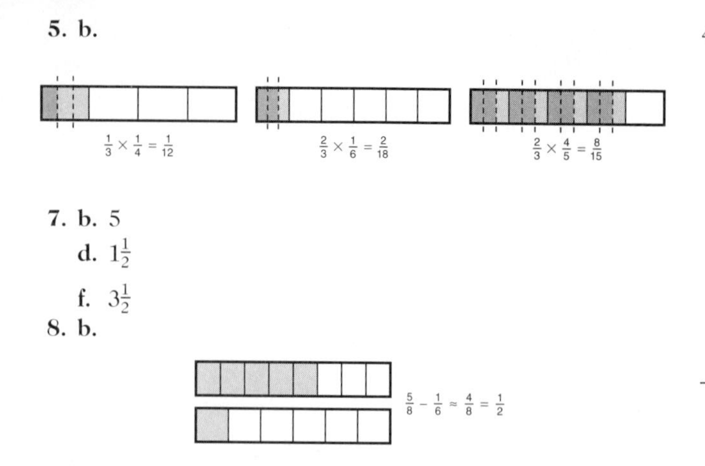

$\frac{1}{3} \times \frac{1}{4} = \frac{1}{12}$
$\frac{2}{3} \times \frac{1}{6} = \frac{2}{18}$
$\frac{2}{3} \times \frac{4}{5} = \frac{8}{15}$

7. b. 5

d. $1\frac{1}{2}$

f. $3\frac{1}{2}$

8. b.

d.

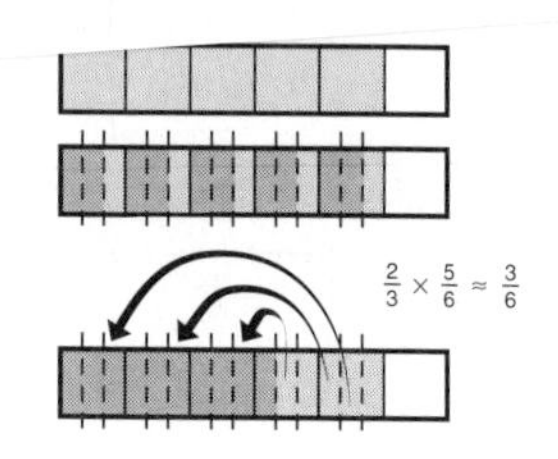

9. a. $\frac{1}{6} + \frac{1}{4} = \frac{5}{12} \quad 3 \times \frac{3}{12} = \frac{3}{4} \quad \frac{2}{4} - \frac{1}{2} = \frac{0}{3} \quad \frac{2}{3} \div \frac{2}{6} = 2$

or

$\frac{2}{4} + \frac{0}{3} = \frac{1}{2} \quad 3 \times \frac{1}{4} = \frac{3}{4} \quad \frac{5}{12} - \frac{3}{12} = \frac{1}{6} \quad \frac{2}{3} \div \frac{2}{6} = 2$

ACTIVITY SET 6.1

1. d. **f.**

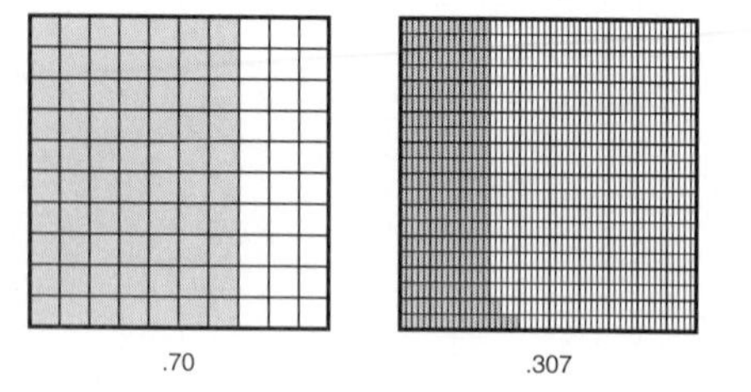

2. b. .33 **f.** .09 **h.** .090

3. a. 3 parts out of 10 is equal to 30 parts out of 100; .3 = .30

c. 470 parts out of 1000 is equal to 47 parts out of 100; .470 = .47

e. 1 part out of 100 is equal to 10 parts out of 1000; .01 = .010

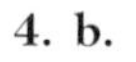

4. b.

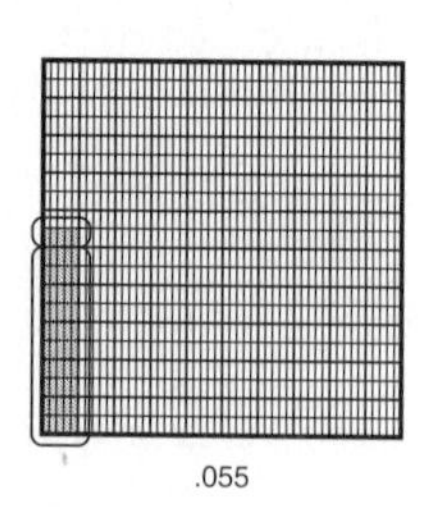

TENTHS	HUNDREDTHS	THOUSANDTHS
0	5	5

d. Every collection of 10 like parts (thousandths, hundredths, or tenths) can be regrouped to form 1 larger part and is recorded in the adjacent column, to the left.

5. b. Each part would be $\frac{1}{100,000}$ of the unit square. The decimal is written as .00001 and named 1 hundred-thousandth.

6. b. .042 > .04

e. (.0420) .0047 .0400

8. c.

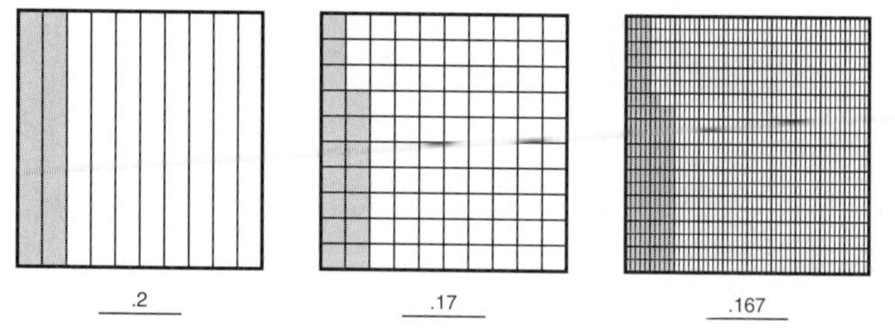

ACTIVITY SET 6.2

2. a. .4 + .29 = .69

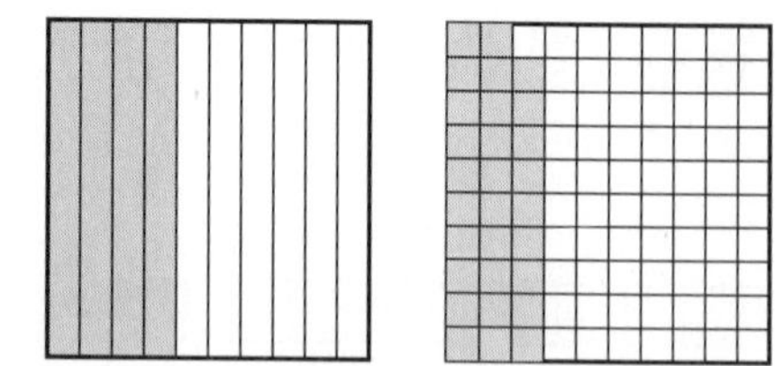

c. .07 + .605 + .2 = .875

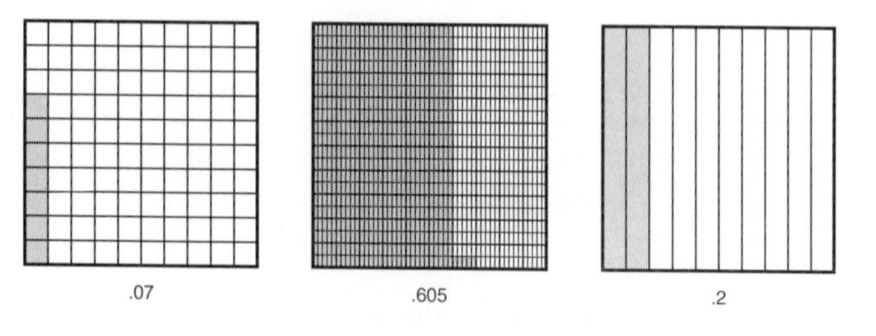

4. a. $.5 - .07 = .43$. If 3 hundredths and 4 tenths are added to the square for .07, you get the square for .5.

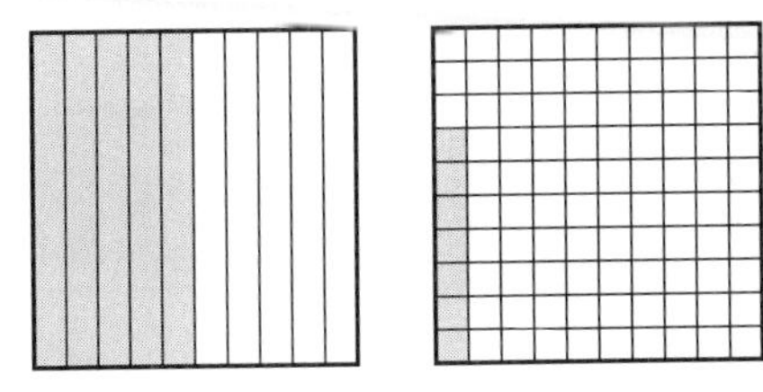

c. $.435 - .3 = .135$. Add 1 tenth, 3 hundredths, and 5 thousandths to .3.

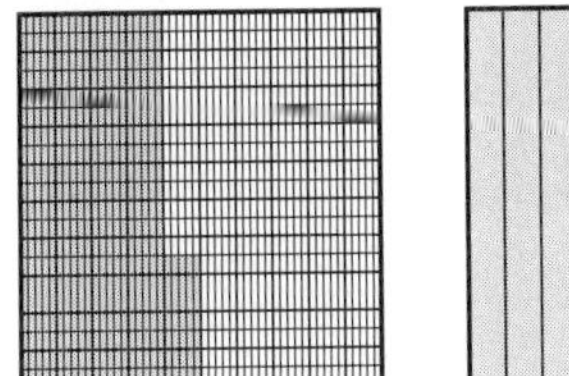

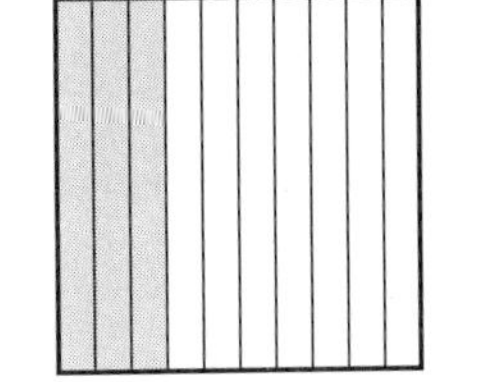

5. b. $4 \times .725 = 2.9$ because 4 thousandth squares, each with 725 parts shaded, yield a total of 2900 shaded parts. Since there are 1000 parts in 1 whole square, the total shaded amount would be equal to 2 whole squares and 900 parts shaded out of 1000.

d. Multiplying by 10 makes each shaded tenth column of a decimal square equal to a unit square. Similarly, multiplying by 10 makes each hundredth square equal to a tenth column and each thousandth piece equal to a hundredth square.

6. c.

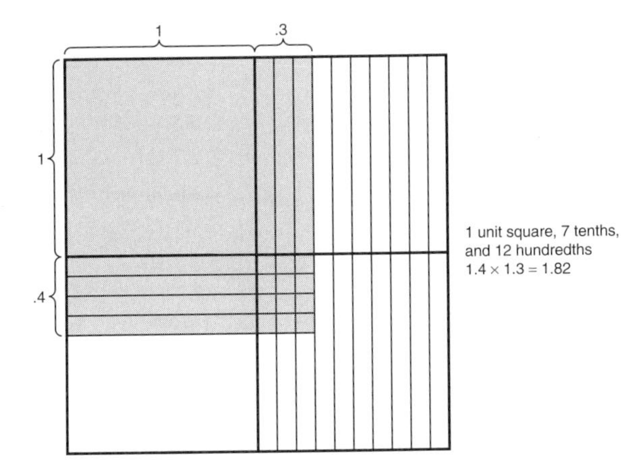

9. The decimal 2.87 can be represented by 2 whole squares and a square with 87 hundredths shaded (8 tenth columns and 7 hundredth squares). Dividing a unit square by 10 results in a region equivalent to 1 tenth column. Dividing a tenth column by 10 results in a region equal to 1 hundredth square, and dividing a hundredth square by 10 results in a region equal to 1 thousandth piece. When the squares representing 2.87 are divided by 10, the result is 2 tenth columns, 8 hundredth squares, and 7 thousandth pieces.

10. b. $.70 \div .05 = 14$

d. Five tenths can be measured off or subtracted from each unit twice. Since there are 250 units, .5 can be subtracted 500 times.

ACTIVITY SET 6.3

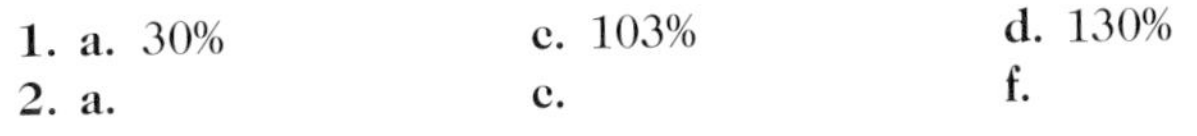

1. a. 30% c. 103% d. 130%
2. a. c. f.

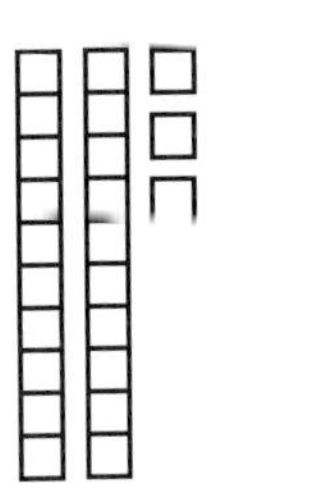

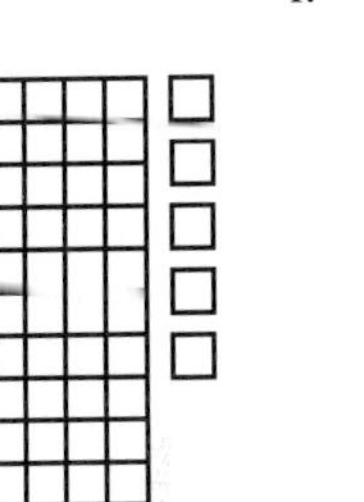

3. a. The value is 45. One strip (10 squares) has value 30 and one-half strip has value 15.

c. The value is 288. Each small square has value 3 and $300 - 12 = 288$.

4. a. Value 37.5: each strip (10%) has value 15 and one-half strip (5%) has value 7.5.

c. Value 10.5: every pair of small squares has value 3, and $3 + 3 + 3 + 1.5 = 10.5$.

5. a. 270 is 60% of 450: each strip has value 45, so 6 strips (60%) have value 270.

c. 441 is 98% of 450: each small square has value 4.5, so 98 (98%) small squares have value 441.

6. a. Value of grid is 400: since 6% has value 24, 1% has value 4, and the value of the 10 by 10 grid is 400.

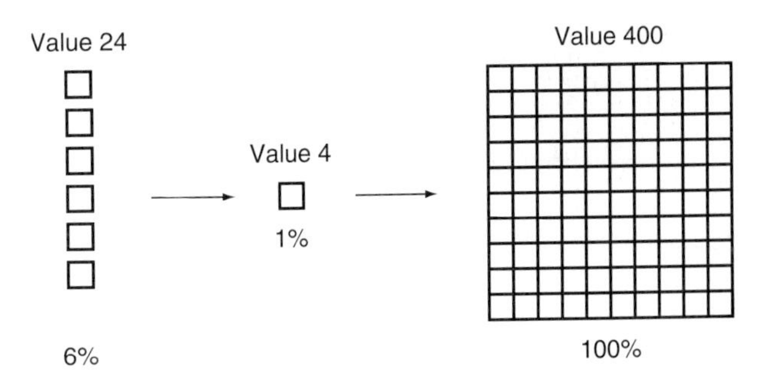

c. Value of the grid is 240: since 55% has value 132, 1% has value 2.4 ($132 \div 55$), and 100% has value 240.

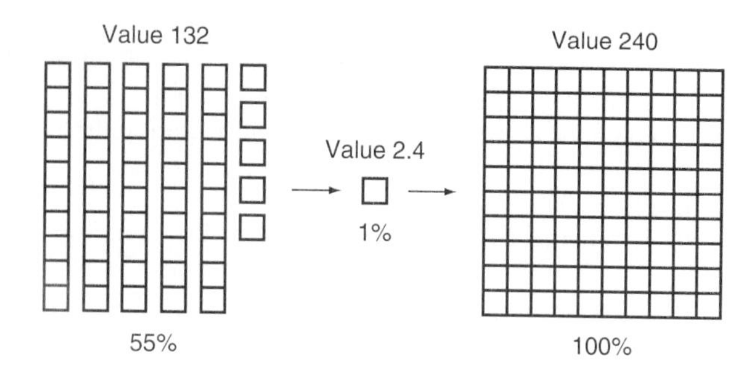

7. a. Some possible observations: there are 208 women employees; 35% of the employees are men; there are 112 men employees.

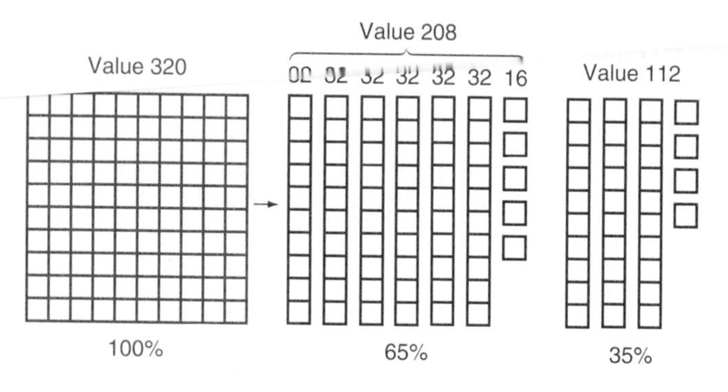

b. Some observations: 35% of the students were absent; 65% of the students were present.

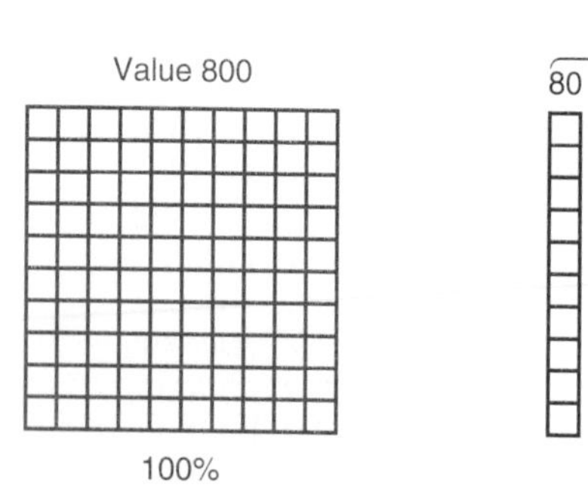

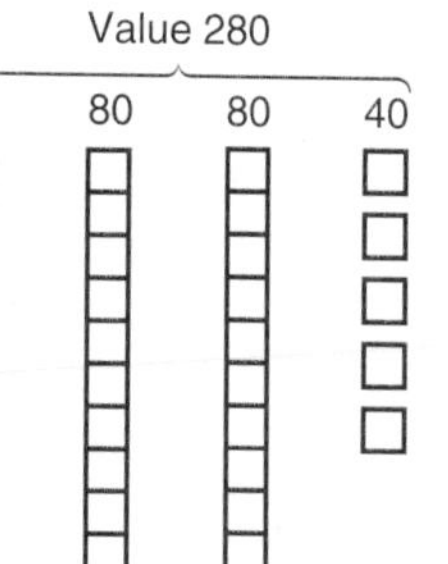

8. b. $\frac{2}{5} = 40\% = .4$

20 parts 20 parts 20 parts 20 parts 20 parts

d. $\frac{13}{20} = 65\% = .65$

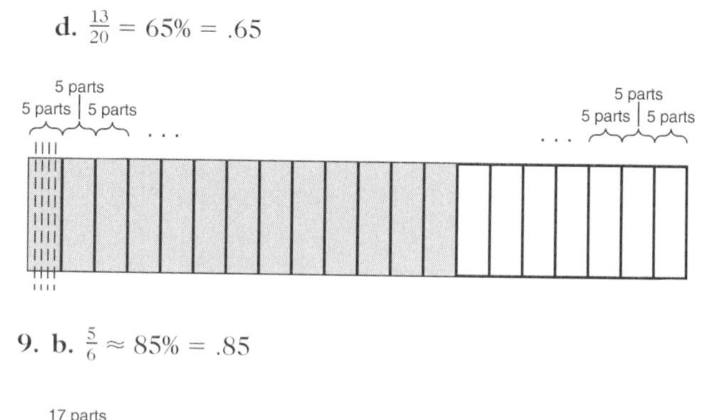

9. b. $\frac{5}{6} \approx 85\% = .85$

17 parts

ACTIVITY SET 6.4

1. b. 7 square units **e.** 4 square units **h.** 2 square units

2. b. 4 square units **e.** 2 square units **g.** 10 square units

4. a. $\sqrt{13}$ **b.** 5 **c.** $\sqrt{17}$

6.

	SQUARE A	SQUARE B	SQUARE C
b. Fig. 2	9	4	13
d. Fig. 4	45	20	65

7. b.

	AREA OF SQUARE ON HYPOTENUSE	LENGTH OF HYPOTENUSE
(2) Fig. 2	26	$\sqrt{26}$
(4) Fig. 4	58	$\sqrt{58}$

ACTIVITY SET 7.1

4. b. \$117,333.33 is the average for the 6 greatest incomes; \$18,666.67 is the average for the 6 smallest.

d. \$73,166.67 is the greatest average and \$41,833.33 is the least. Both are closer to the mean of \$56,444.44 than the greatest and least from part b.

ACTIVITY SET 7.2

1. a. Shortest, 60 inches; tallest, 68.5 inches

d. For one couple the woman was taller. One couple had the same height.

2. This trend line has a positive slope. It seems to indicate that for this data set, the taller the father the taller the mother.*

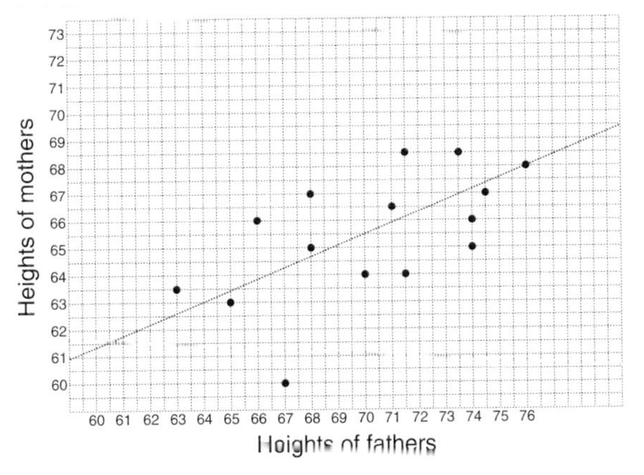

3. a. The trend line for this data has a negative slope. It indicates that the taller the father the shorter the mother.

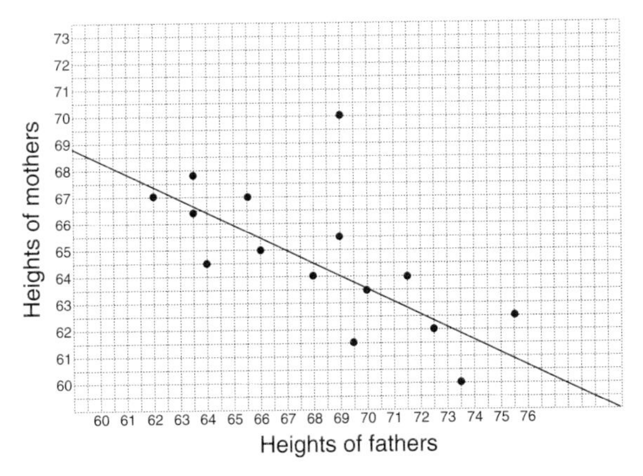

b. The horizontal trend line for this data has zero slope. Short fathers are paired with short and tall mothers and the same is true for taller fathers. There seems to be no apparent trend.

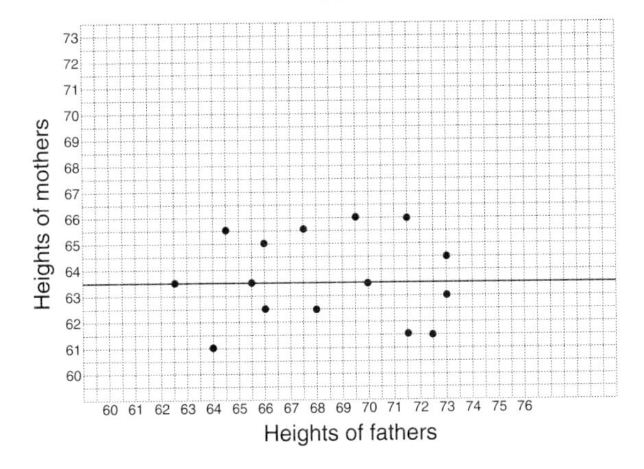

ACTIVITY SET 7.3

1. c. *At least 3* means 3, 4, 5, or 6 girls. Count the number of outcomes in the chart that have 3, 4, 5, or 6 heads and divide that number by 64.

2. a. The pulse rate distribution comes close to satisfying this property because 68.3% of the population is within ± 1 standard deviation of the mean, and 95% is within ± 2 standard deviations of the mean.

3. a.

Letter	*e*	*s*	*c*	*w*	*k*
Percent	16.3%	7.2%	2.6%	1.7%	.6%

c. This assignment of code symbols requires 43 time units, as compared to 47 time units for Morse code.

e	t	a	o	n	i	s
•	—	• •	• —	— •	• • •	— —

s	r	h	h
• • —	• — •	• • • •	— • •

4. c. There are 2,776,984 numbers from 7,092 to 2,784,075. Eighty percent begin with the digits 1, 2, 3, or 4. If the populations are randomly distributed you would expect a greater number of those populations to start with digits 1, 2, 3, or 4.

JUST FOR FUN 7.3

Here is a frequency distribution for the letters that occur in the message.

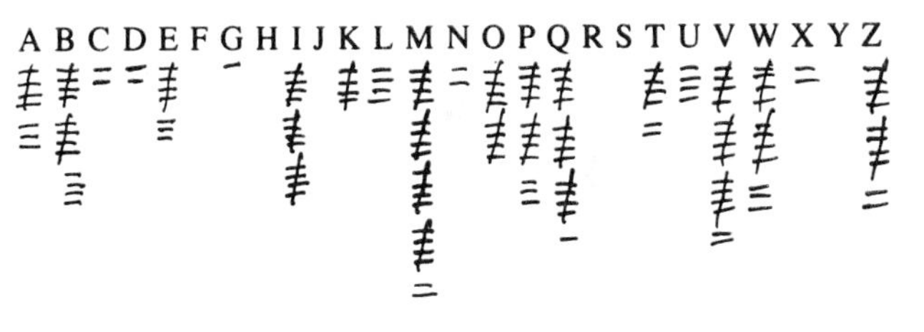

The letter M occurs most frequently, so it is reasonable to guess that this is the enciphered letter for E. If M represents E, then I represents A and Q represents I. This agrees with one of the patterns for the distribution of letters in large samples. Also notice that the fifth and sixth letters beyond M did not occur in the message. This also agrees with the fact that the fifth and sixth letters beyond E are the low-frequency letters J and K. It appears that the message has been coded by replacing each letter by the letter that is 8 letters beyond. The message can be decoded by reversing that process. The result is:

"Galileo showed men of science that weighing and measuring are worthwhile. Newton convinced a large portion of them that weighing and measuring are the only investigations that are worthwhile."

Charles Singer

*A more accurate indication of the strength of these relationships is indicated by the correlation coefficient, which can be obtained by entering the data into a graphing calculator.

ACTIVITY SET 8.1

2. d. $\frac{1}{4}$ **e.** No. There is only a 25% chance of winning.

3. b. $\frac{4}{9}$

4. a. 2 **c.** A_1 A_2 A **e.** No
A_2 A_1 B

ACTIVITY SET 8.2

1. b. Probability of 2 of different colors: $\frac{3}{5}$

Probability of 2 of the same color: $\frac{2}{5}$

d. From the probability tree, we see that the probability of drawing first a red and then a green is $\frac{1}{2} \times \frac{3}{5} = \frac{3}{10}$, and the probability of drawing first a green and then a red is $\frac{1}{2} \times \frac{3}{5} = \frac{3}{10}$. So the probability of drawing 2 squares of different colors is $\frac{3}{10} + \frac{3}{10} = \frac{3}{5}$.

2. c. $\frac{6}{16}$ or $\frac{3}{8}$

4. b. $\frac{49}{100}$

5. a. $\frac{1}{36}$ **c.** $\frac{35}{36} \times \frac{35}{36}$ or $\left(\frac{35}{36}\right)^2$

e. Approximately $1 - .51 = .49$

6. d. The probability of choosing a junk door on your first pick is $\frac{2}{3}$. But, using the switch strategy, you automatically win if your first choice is a junk door (the host must open the remaining junk door, so when you switch you win). So the probability of winning with the switch strategy is $\frac{2}{3}$.

The probability of winning with the stick strategy is $\frac{1}{3}$.

JUST FOR FUN 8.2

2. $\frac{2}{3}$ **3.** $\frac{1}{3}$ **4.** $\frac{1}{3}$ **5.** $\frac{2}{3}$

ACTIVITY SET 9.1

1. b. All triangles that have their third vertex on the same row parallel to the base have the same area.

c. There are 12 noncongruent triangles that can be formed on the given base.

4.

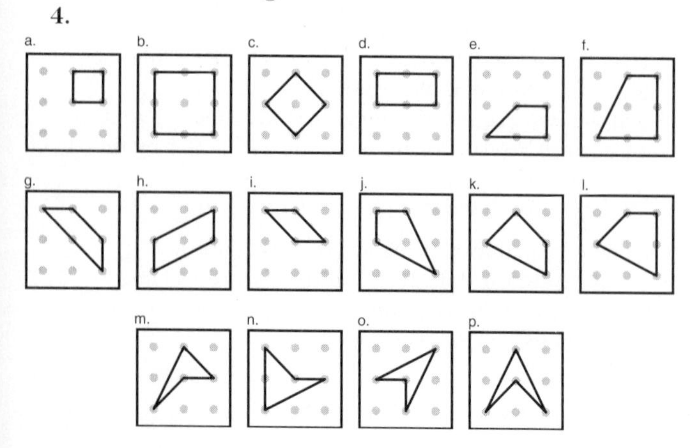

Figures a, b, and c are squares; figure d is a rectangle but not a square; figures h and i are parallelograms but not rectangles; figures e, f, and g are trapezoids; figures a, b, c, d, e, f, g, h, i, j, k, and l are convex; figures m, n, o, and p are nonconvex.

6. a. 75°; 90°; 210°; 120°

b.

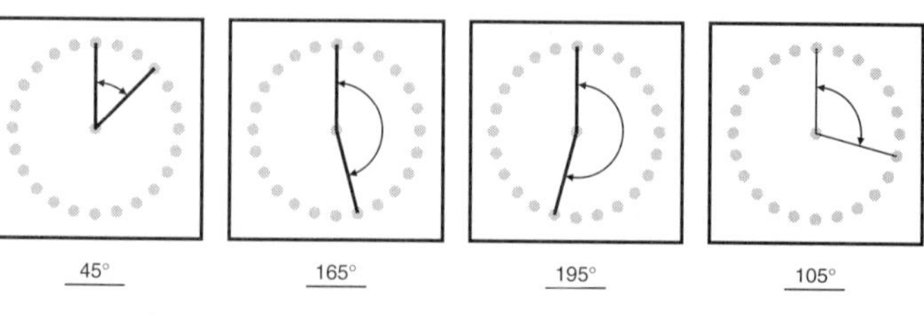

7. a. 45°; 30°; 45°; 30°; 45°

b. Inscribed angles that intercept arcs of equal length in the same or congruent circles have the same measure.

c. Inscribed angles whose sides intersect the ends of a diameter are right angles.

8. a.

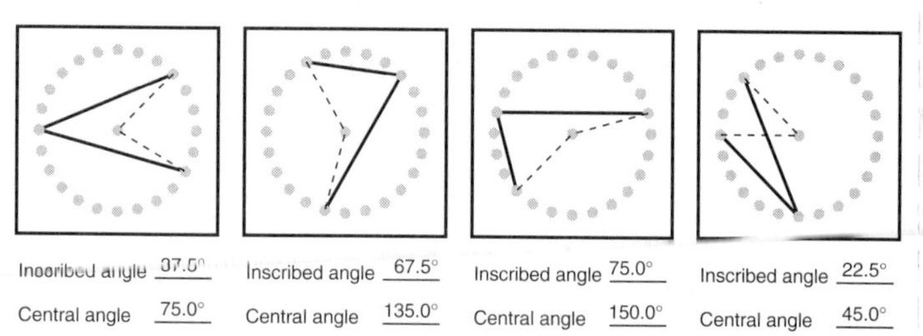

b. The measure of an inscribed angle is half the measure of the central angle that intercepts the same arc.

JUST FOR FUN 9.1

1. The 7 tangram pieces have been numbered (see the square below) in the solutions to the puzzles.

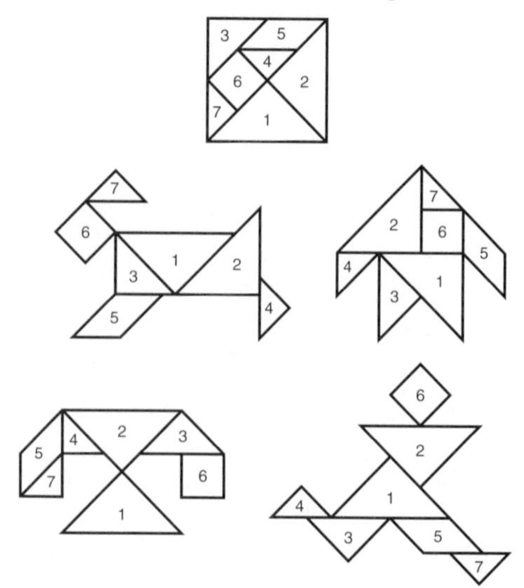

ACTIVITY SET 9.2

2.

ANGLE	MEASURE
1	150°
2	60°
3	120°
4	60°
5	120°
6	120°
7	60°

4. c. Each interior angle of a regular polygon with more than 6 sides is greater than 120° and less than 180°. The measures of such angles are not factors of 360.

5. a.

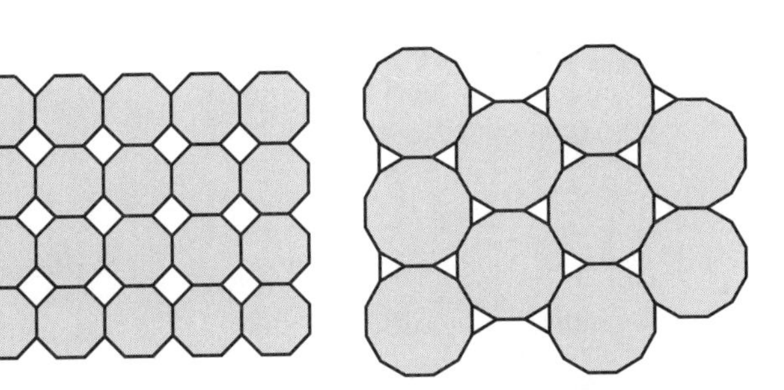

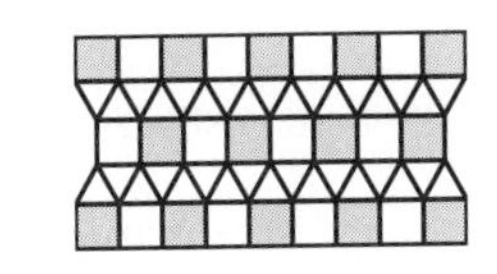

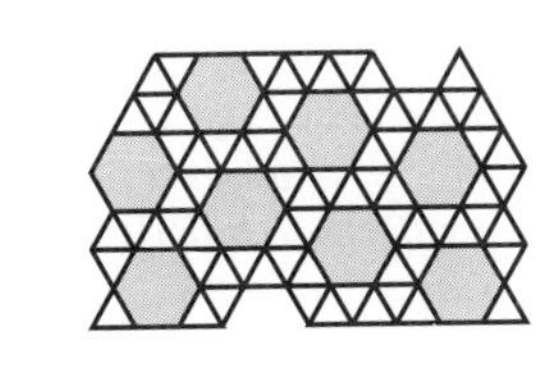

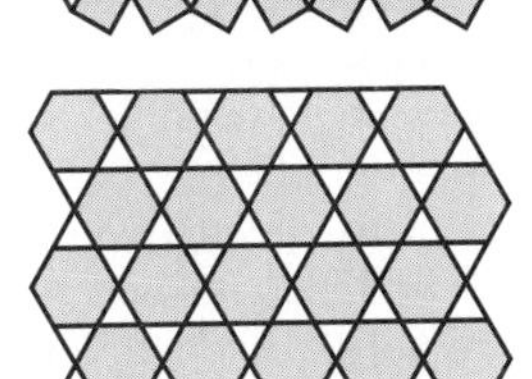

b.

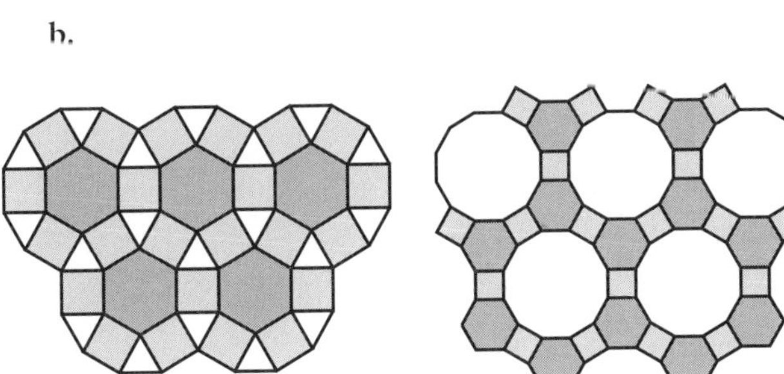

c. Here is one example. There are others.

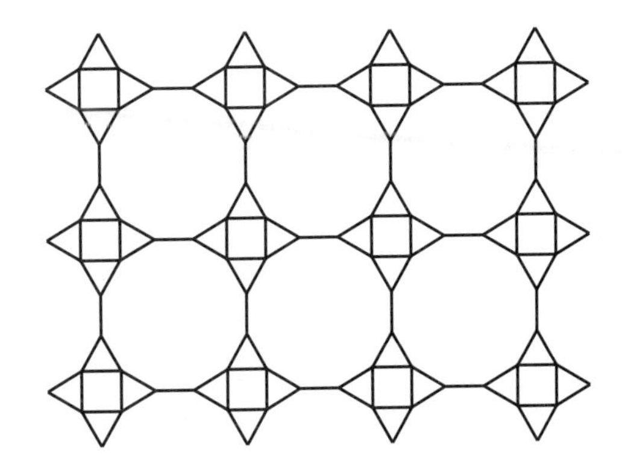

JUST FOR FUN 9.2

The winning strategy on a 3 by 3 or 5 by 5 grid is to occupy the center hexagon on the first move. On a 4 by 4 grid, the winning first move is any of those marked here.

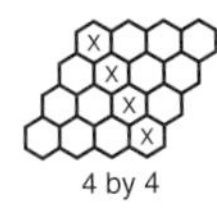

ACTIVITY SET 9.3

3. a.

	VERTICES (V)	FACES (F)	EDGES (E)
Tetrahedron	4	4	6
Cube	8	6	12
Octahedron	6	8	12
Dodecahedron	20	12	30
Icosahedron	12	20	30

c.

	(1)	(2)	(3)	(4)
Vertices	5	10	12	10
Faces	5	7	8	7
Edges	8	15	18	15

5. b. Octahedron e. Icosahedron f. Cube
g. Icosahedron or dodecahedron
h. Dodecahedron j. Cube or octahedron

ACTIVITY SET 9.4

1. a.

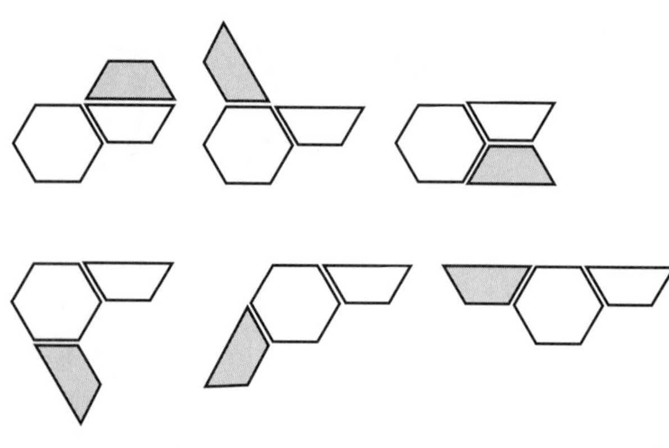

c. There are other ways besides the ones shown.

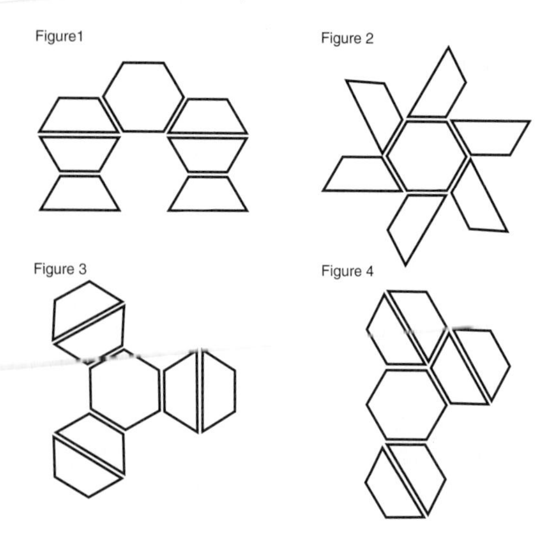

4. a. Angle *ABC* has a measure of 135°; $AD = DC$ and $AB = BC$.

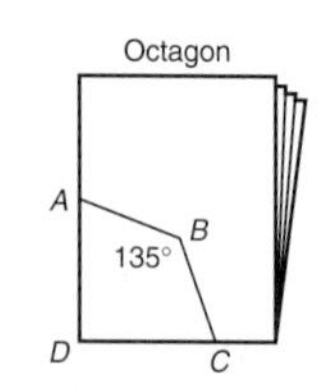

Angle *EFG* has a measure of 120°; $FG = HG = HF$ and $EF = \frac{1}{2}$ of *FG*.

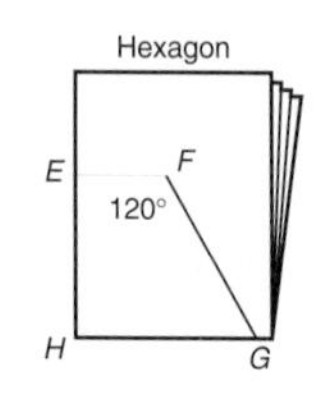

5. d. The eight-pointed wind rose.
   e. A regular 16-sided polygon can be obtained by cutting a 157.5° angle as shown here, with $AB = AC = AD$. *Note:* Some of the lines of symmetry are not crease lines.

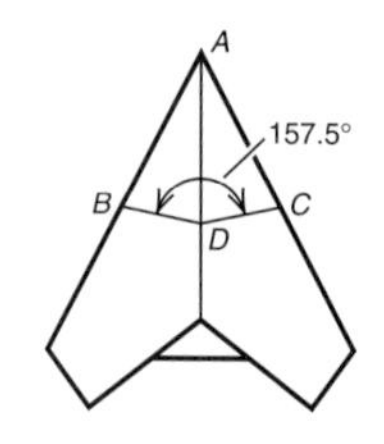

ACTIVITY SET 10.1

3. a. 156 mm b. 19 mm
7. a. A quart contains approximately 946 cm^3. A liter contains slightly more than a quart.
   b. Approximately 10.9 cm

ACTIVITY SET 10.2

1. a. 9 sq. units c. $7\frac{1}{2}$ sq. units e. 8 sq. units
2. a. 9 sq. units c. $10\frac{1}{2}$ sq. units e. 6 sq. units
3. a. 3 sq. units c. 1 sq. unit e. 4 sq. units
4. a. 8 sq. units c. $5\frac{1}{2}$ sq. units e. 1 sq. unit
8. a. Parallelogram, 6 square units; triangle, 3 square units.
   c. Parallelogram, 9 square units; triangle, $4\frac{1}{2}$ square units.
9. The area of a parallelogram is equal to the length of a side multiplied by the height from that side. A triangle with base of length b units and height of length h units has half the area of a parallelogram with base length b and height of length h from that side.

JUST FOR FUN 10.2

4. The squares with an X in the following figure show one possible final move in the pentomino game.

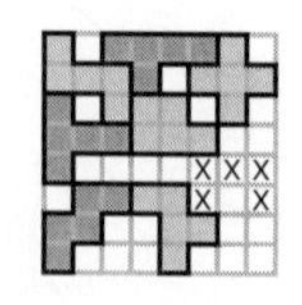

5. Here are 5 pieces that work.

ACTIVITY SET 10.3

1. The volume of the crown was greater than the volume of the gold. King Hieron was cheated.
2.

	HEXAGONAL PRISM	HEXAGONAL PYRAMID	CYLINDER
Area of base	9.1 cm^2	9.1 cm^2	12.6 cm^2
Height (altitude)	5.0 cm	5.0 cm	3.0 cm
Volume	45.5 cm^3	15.2 cm^3	37.7 cm^3
Surface area	75.2 cm^2	38.7 cm^2	62.8 cm^2

3. Cylinder A: 1030 cm^3; cylinder B: 1342 cm^3
5. b.

	1	2	3	4	5	6
Central angle of disk	45°	90°	135°	225°	270°	315°
Radius (cm) of base	1.25	2.50	3.75	6.25	7.50	8.75
Base area (cm^2)	4.91	19.63	44.16	122.66	176.63	240.41
Height (cm) of cone	9.92	9.68	9.27	7.81	6.61	4.84
Volume (cm^3) of cone	16.24	63.34	136.45	319.32	389.17	387.86

c. For central angles of less than 45°, the smaller the angle is, the smaller the volume is. For central angles of more than 315°, the larger the angle is, the smaller the volume is. The maximum volume of approximately 403 cm^3 occurs with a central angle of 294°.

6. c. A cone with a base of radius r and a height of h has a volume of $\frac{1}{3} \times \pi r^2 h$. The "half cone" has a base of radius $\frac{r}{2}$ and a height of $\frac{h}{2}$. Its volume is $\frac{1}{3} \times \pi\left(\frac{r}{2}\right)^2\left(\frac{h}{2}\right)$, which is equal to $\frac{1}{3} \times \pi r^2 h \times \frac{1}{8}$. Therefore, the volume of the "half cone" is $\frac{1}{8}$ of the volume of the original cone.
7. a. 7235 cm^3
8. c. The volume of the 3 balls is $3 \times \frac{4}{3} \times \pi(3.15)^3 \approx 394$ cm^3. The volume of the can is $20 \times \pi(3.5)^2 \approx 769$ cm^3. Therefore, approximately $\frac{1}{2}$ of the space is wasted.

ACTIVITY SET 11.1

1. b. The points form 2 perpendicular lines that pass through the intersection of lines m and n and bisect the angles formed by the intersection of m and n.
2. a. The points form a circle with P as center and radius equal to the distance from P to the midpoint of the chords.

c. The points form a circle whose center is the midpoint of segment $\overline{PS}$ and whose diameter is $\overline{PS}$.

4. A circle
6. a. A semicircle on diameter $\overline{PQ}$ (a circle if you reverse the card and repeat the process on the other side of segment $\overline{PQ}$)

ACTIVITY SET 11.2

1. a. The new figure will tessellate because the exact shape that has been adjoined to one edge of the rectangle was deleted from the opposite side of the rectangle.

ACTIVITY SET 11.3

1. a. Both triangles contain the angle at E, and the angles at A and A' are right angles. When two angles of one triangle are congruent to two angles of another, the triangles are similar.
2. c. Triangle E_1FE_2 is similar to triangle RFT. Therefore,

$$\frac{75}{7.5} = \frac{FT}{90} \quad \text{or} \quad FT = 10 \times 90$$

e. When you are estimating large distances, the distance from eye to finger is relatively small and ignoring it will not significantly affect the estimate. However, when you are estimating small distances, which may be only a few times greater than the distance from eye to finger, ignoring the distance from the eye to the finger will give a larger error. In the latter case, the distance should be added to the estimated distance.

3. b. 20.25 m c. 24.3 m
4. b. $\overline{ET}$ in the figure measures 80 mm. So 80 mm represents 60 m, or 1 mm represents .75 m. The height of the tree from T to R in the diagram is 24 mm, so the height of the actual tree between those two points is $24 \times .75 = 18$ m.
5. a. Approximately 33 m plus the height of the hypsometer above the ground.
6. b. Use a line from the land marker to the pine tree as a base line and measure the distance between these two points. From the pine tree, measure the 5 angles between the base line and the remaining 5 objects. From the landmark, measure the 5 angles between the base line and the remaining 5 objects. With the 10 angles, a map can be made of the locations of each object. Using the distance from pine tree to the landmark, a scale can be set up to determine all distances.

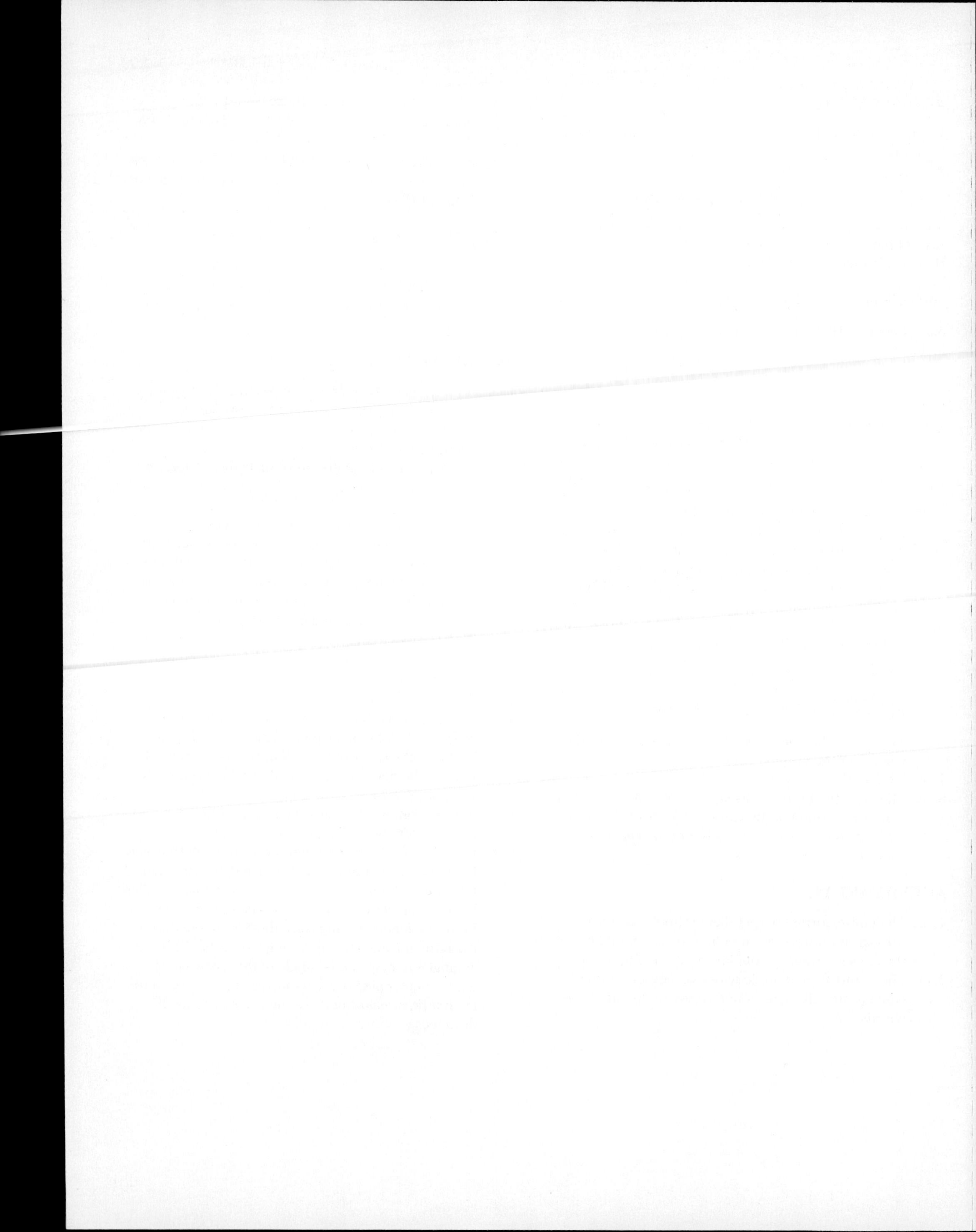

CREDITS

PHOTOS

Chapter 1

Page 8: Courtesy of British Information Service; **p. 13:** Ron Bergeron; **p. 13:** *Patterns in Nature* by Peter S. Stevens, Atlantic Monthly/Little Brown, 1974. Reprinted by courtesy of Peter S. Stevens.

Chapter 4

Page 84: Photo by Ron Bergeron; **p. 87:** Talbot Lovering

Chapter 7

Page 151: *Peanuts* reprinted by permission of United Feature Syndicate, Inc.; **p. 166:** Library of Congress

Chapter 8

Page 180: *Peanuts* reprinted by permission of United Feature Syndicate, Inc.; **p. 187:** Ron Bergeron

Chapter 9

Page 209: Courtesy of Museum of Comparative Zoology, Harvard Univ.; **p. 216, 217:** Talbot Lovering

Chapter 10

Page 232: *Peanuts* reprinted by permission of United Feature Syndicate, Inc.; **pp. 248, 252:** Talbot Lovering

Chapter 11

Page 262: Talbot Lovering

ILLUSTRATIONS AND TEXT

Chapter 5

Page 97: Fraction Bar model from the Fraction Bars® materials by Albert B. Bennett, Jr., and Patricia Davidson © 1981. Permission of Scott Resources, Inc. Fort Collins, Colorado.

Chapter 6

Page 115: Decimal Squares model from the Decimal Squares® materials by Albert B. Bennett, Jr.,© 1981. Permission of Scott Resources, Inc. Fort Collins, Colorado.

Chapter 7

Page 167: Graphs from *A Guide to the Unknown*, by Frederick Mosteller et al., copyright 1972. Reprinted by courtesy of Holden-Day, Inc.

Chapter 11

Page 261: Line designs from *Line Designs* by Dale Seymour, Linda Libey, and Joyce Snider. Reprinted by permission of Creative Publications, Palo Alto.

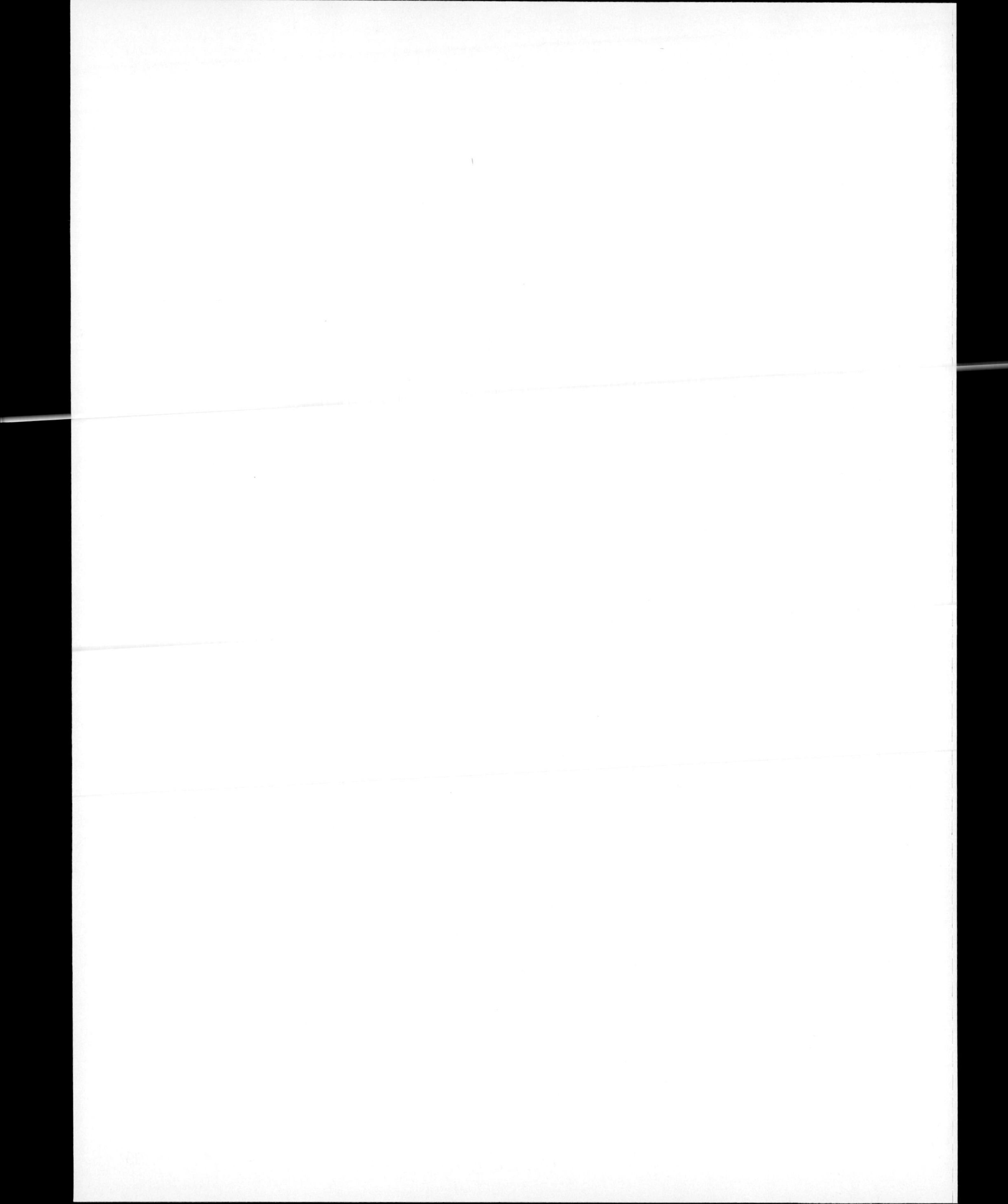

INDEX

A

Acute, 261
Addition
  algorithm, 49
  decimals, 125–126, 132–133
  fractions, 104–106
  integers, 91–92
  models, 49–51, 91–92, 104–106, 125–126
  whole numbers, 49–51
Algebraic Expressions Game, 18–19
Algebra pieces, 14–19
Algebra story problems, 14–19
Algorithm. *See* Addition; Division; Multiplication; Subtraction
Altitude
  cone, 247
  cylinder, 245–247
  parallelogram, 242
  prism, 245–246
  pyramid, 245–246
  triangle, 242
Angle
  acute, 261
  central, 204–205
  horizontal, 276
  inscribed, 205–206
  measurement, 204–207, 210–211, 274–278
  obtuse, 261
  right, 202
  vertex, 204
  vertex point, 211–212
  vertical, 276
Approximation, 122
Arabian Nights Mystery, 67
Arc, circle, 204
Archimedean solids, 218–219
Archimedes, 245
Area
  plane, 143–148, 201, 239–244
  surface, 245–251
Art and Techniques of Simulation, The, 157
Attribute Grid Game, 23–24
Attribute Guessing Game, 23
Attribute Identity Game, 27
Attribute pieces, 21–27
Average
  mean, 152
  median, 152
  mode, 152

B

Bar graph, 167
Base (numeration systems)
  five, 42–45, 46
  nine, 46
  seven, 45, 46
  ten, 45, 46, 55–59, 60–68
  three, 45, 46
Base-ten decimal model, 150
Bedford, F.L., 272
Bell-shaped curve, 168
Bennett, A.B., Jr., 68, 84, 88, 150
Bentley, W.A., 228
Binary numbers, 47–48
Black and red tile model, 89–96
Body measurements, 235
Bradford, C.L., 78

C

Caesar, Julius, 175
Caesar cipher, 175–176
Calculator(s)
  Cross-Number Puzzle, 59
  decimals, 124
  Keyboard Game, 67
  number tricks, 67
  random digits, 152–154
  square root, 146
  whole numbers, 59, 67
Cardioid, 261–262
Carpenter, T.P., 104
Carson, G., 251
Centimeter, 233
  cubic, 233
Centimeter Guessing Game, 233
Centimeter Racing Game, 235
Central angle, 204–205
Chinese
  Game of Nim, 48
  model for integers, 89
  tangram puzzle, 207–208
Chip trading, 43
Chi-square test, 154
Circle
  arc, 204
  chord, 257
  radius, 257
Circular geoboard, 204–207
Clemens, S.R., 215
Clinometer, 274–275
Codes, 173, 175–176
Color tiles, 8–13
Commensurable, 142
Common denominator, 105–106
Communication, and reasoning, 40
Comparison model for subtraction, 54
Complementary probability, 193
Composite number, 75, 76, 77
Compound probability. *See* Multi-stage probability
Cones
  base area, 247
  height (altitude), 217
  radius, 247
  volume, 245, 247–248
Conjecture, 84–87
Continuous path, 84–87
Convex, 204
Cooperative learning groups, 36–39
Cooperative logic problems, 36–39
Coordinate Guessing (game), 34
Corbitt, M.K., 104, 116
Corresponding sides, 270
Cross-Number Puzzle, 59
Cryptanalysis, 175–176
Cubes, 216–217, 218, 245, 251, 252
Cubic centimeter, 233, 249
Curriculum and Evaluation Standards for School Mathematics, 1, 21, 22, 42, 70, 89, 115, 151, 179, 180, 200, 232, 255
Curves
  cardioid, 261–262
  parabola, 261
Cylinders
  area, 245
  circumference, 246–247
  volume, 245, 246–247

D

Davis, P. J., 180
Decimal Bingo, 123
Decimal Place Value Game, 123
Decimals, 115–150
  addition, 125–126
  approximation, 122
  calculator, 124
  division, 131–132
  equality, 118–119
  games, 123, 133
  inequality, 121–122

Decimals—*Cont.*
- infinite nonrepeating, 142
- models, 115–123, 150
- multiplication, 128–130
- part-to-whole concept, 117–118
- place value, 119–120, 123
- subtraction, 127–128
- unit, 116

Decimal squares, 124–133
Decimal Squares Blackjack, 133
Decimal Squares Model, 115–123
Decimeter, 233
Decipher, 175–176
Deductive reasoning, 36–39
Degree, angle
- central, 204–205
- inscribed, 205–206
- polygon, 210–212
- rotation, 221
- tessellation, 212
- triangle, 212

Denominator, 97
Dependent event, 187–188
Descartes Dream, 180
Diameter, 249
Dienes blocks, 43
Distribution
- normal, 168
- uneven, 172
- uniform, 167

Division
- algorithm, 64–66
- decimals, 131–132, 133–134
- fractions, 108
- integers, 94
- long division, 60
- measurement (subtractive), 60–62, 68, 108, 132
- models, 60–68, 94, 108, 131–132
- quotient, 61, 63, 64, 108
- regrouping, 61
- sharing (paritive), 60–62, 68, 131
- whole numbers, 60–68

Division concept of fractions, 97, 102
Dodecagon, 212
Dodecahedron, 216–217
Driscoll, M.J., 104
Drizigacker, R., 103

E

Elementary Cryptanalysis, 176
Encipher, 175–176
Enlarging drawings, 278–279
Equality
- decimals, 118–119
- fractions, 97–103

Equilateral triangle, 212
Equivalent sets, 43
Eratosthenes, 78
Escher, M.C., 209, 263
Escher-type tessellations, 263–269
Estimation, 109
Euclid, 14
Euler's formula, 217–218
Even number, 71–74, 87–88
Event
- dependent, 187–188
- independent, 187
- multi-stage, 187–197

Everybody Counts: A Report to the Nation on the Future of Mathematics Education, 124
Existence proof, 214
Experimental probability, 179–186, 189–196

F

Factors, 74–77
Fair game, 181–182
Fechner, Gustav, 181–182
Fibonacci numbers, 13
Field Work in Mathematics, 272
Five House Puzzle (logic problem), 38
Five-pointed star, 227
Force Out (game), 54
Formulas
- circular area, 245
- cone volume, 245
- cylinder volume, 245
- prism volume, 245
- pyramid volume, 245
- sphere volume, 245
- triangular area, 244

Fraction Bar Blackjack, 110
Fraction bars, 97–103
Fraction Bingo, 103
Fractions
- addition, 104–106
- as approximate percent, 140–141
- common denominator, 105–106
- denominator, 97
- division, 108
- division concept, 97, 102
- equality, 97–103
- estimation, 109
- inequality, 97–103
- models, 97–103
- multiplication, 107
- numerator, 97
- part-to-whole concept, 97
- as percent, 138–139
- subtraction, 104–106

Frequency distribution, 166
FRIO (Fractions in Order), 103

G

Gadsby, 173
Game(s)
- for addition, 49, 95–96

Game(s)—*Cont.*
- Algebraic Expressions Game, 18–19
- Attribute Grid Game, 23–24
- Attribute Guessing Game, 23
- Attribute Identity Game, 27
- for binary numbers, 47–48
- for calculators, 67
- Centimeter Guessing Game, 233
- Centimeter Racing Game, 235
- Coordinate Guessing, 34
- Decimal Bingo, 123
- Decimal Place Value Game, 123
- for decimals, 123, 133
- Decimal Squares Blackjack, 133
- for division, 96, 111
- for equality, 103, 123
- Force Out, 54
- Fraction Bar Blackjack, 110
- Fraction Bingo, 103
- for fractions, 103, 110
- FRIO, 103
- Game of Interest, 141
- for geometry, 244
- Greatest Difference, 133
- Greatest Quotient, 111, 133
- Hex, 214
- Hide-a-region, 35
- for inequality, 96, 103
- for integers, 95–96
- Keyboard Game, 67
- Meter Guessing Game, 233
- for metric units, 233, 235
- for multiplication, 96
- Nim, 48
- Page Guessing, 165
- Pentominoes, 244
- Pica-Centro, 39
- for place value, 123
- for probability, 181–183, 185–186
- Racetrack Game, 185–186
- for reasoning, 22, 23, 27, 39, 214
- for sets, 23, 27
- Simulated Racing Game, 158
- for subtraction, 52, 54, 67, 95–96, 111
- Three-Penny Grid, 183
- Trading-Down Game, 52
- Trading-Up Game, 49
- Two-Penny Grid, 181–182

Gardner, M., 214
Gauss, Karl Friedrich, 8
GCF. *See* Greatest common factor
Geoboard
- areas on, 239–244
- circular, 204–207
- irrational numbers on, 142–149
- rectangular, 201–204

Geometric figures, 200–229
Geometric model, solving algebra story problems with, 14–19
Geometric patterns, 1–7, 8–13

Geometry: An Investigative Approach, 215
Geometry, motions in, 255–280
Giambrone, T., 18
Glide-reflection, 268
Gnanadesikan, M., 157
Golden Mean, The, 149
Golden ratio, 149
Golden rectangles, 148–149
Gram, 232
Graph
  bar, 167
  frequency distribution, 167
  spirolaterals, 28–33
Greatest common divisor (GCD), 80
Greatest common factor (GCF), 79–82
Greatest Difference (game), 133
Greatest Quotient (game), 111, 133
Greeks
  Archimedes, 245
  Eratosthenes, 78
  goldent rectangle, 148
  Hipparchus, 270
  Hippasus, 142
  Plato, 216
  Pythagoreans, 14
Grossman, A., 122

H

Handshake problem, 19
Handspan, 234, 235
Hapgood, F., 228
Hein, Piet, 14
Hersh, R., 180
Hex, game of, 214
Hexagon, 212
Hide-a-region, 35
Hipparchus, 270
Hippasus, 142
Hypotenuse, 147
Hypsometer, 276

I

Icosahedron, 216–217
Ideas for the Elementary Classroom, 19, 40, 68, 87–88, 111–112, 150, 177, 198, 229, 252–253, 280
Incommensurable, 142
Independent event, 187
Indirect measurement
  clinometer, 274–275
  hypsometer, 276
  ratios, 270
  scout sighting method, 272–273
  sighting method, 271–272
  similar triangles, 271
  stadiascope, 273–274
  transit, 276–278
Inductive reasoning, 36
Inequality
  decimals, 121–122
  fractions, 97–103
  integers, 96
Infinite nonrepeating decimals, 142
Inscribed angle, 205–206
Instant Insanity (puzzle), 219–220
Integer balloon, 111–112
Integers
  addition, 91–92
  division, 94
  inequality, 96
  models, 89–96
  multiplication, 93
  number line, 95
  subtraction, 92–93
Intercepted arc, 204
Intersection of sets, 24–25
Inverse operations, 49
Irrational numbers, 142–149
Isometric grid, 32–33
Isosceles triangle, 202

J

Johnson, David, 36
Johnson, Roger, 36

K

Kepner, H.S., 104, 116
Keyboard Game, 67
Kilometer, 233
Krause, M.C., 221

L

Lady Luck, 173
Lappan, Glenda, 40
LCM. *See* Least common multiple
Least common multiple (LCM), 79, 82–84
Length, 146–148, 202, 233–236
Levin, R., 220
Lindquist, M.M., 104, 116
Line designs, 261–262
Line of symmetry, 221
Line segments, 202
Linn, C.F., 149
Liter, 232, 237
Locus of points, 255–260
Logic problems, 36–39
Long division algorithm, 60, 64–65

M

Magic Formulas, 67
Maier, E., 68, 88, 150
Mapping
  glide-reflection, 268
Mapping—*Cont.*
  rotation, 266–267
  similarity, 278–279
  translation, 264–266
Math and the Mind's Eye, 68, 88, 150
Mean, 152
Measurement, 232–253
  angle, 204–207, 210–212, 247, 274–278
  body, 235
  geoboards, 239–244
  indirect, 270–280
  length, 146–148, 202, 233–236
  with metric units, 232–239
  plane area, 143–148, 201
  surface area, 245–251
  volume, 232, 245–251
  weight, 232–233
  width, 234
Measurement approach to division, 60–62, 68, 108, 132
Median, 152
Meter, 232, 235
Meter Guessing Game, 233
Metric prefixes, 233
Metric system, 232–239
Metric units
  gram, 232
  liter, 232
  meter, 232
Millimeter, 233, 234–235
Mind-reading cards, 47
Mode, 152
Models for equality, 90, 97–98, 100–102, 118–119
Models for inequality, 98–99, 102, 121–122
Models for numbers
  decimals, 115–123, 150
  even numbers, 71–74, 87–88
  fractions, 97–103
  integers, 89–96
  irrational numbers, 142–149
  odd numbers, 71–74, 87–88
  whole numbers, 42–48
Models for number theory
  factors and primes, 74–77
  greatest common factor, 80–82
  least common multiple, 82–84
Models for operations
  addition, 49–51, 91–92, 104–106, 125–126
  division, 60–68, 94, 108, 131–132
  multiplication, 55–59, 93, 107, 128–130
  subtraction, 52–54, 92–93, 104–106, 127–128
Models for percents, 134–141
Morse, Samuel, 173
Morse code, 173
Mosteller, F., 166

Multibase pieces
  addition with, 49–51
  base five, 42–45, 46
  base nine, 46
  base seven, 45, 46
  base ten, 45, 46, 55–59, 60–68
  base three, 45, 46
  division with, 60–68
  models for even numbers, 71–74
  models for factors, 74–77
  models for numeration with, 42–48
  models for odd numbers, 71–74
  models for primes, 74–76
  multiplication with, 55–59
  subtraction with, 52–54
Multiples, 79, 82
Multiplication
  algorithm, 57–58
  decimals, 128–130, 132–133
  fractions, 107
  integers, 93
  models, 55–59, 93, 107, 128–130
  whole numbers, 55–59
Multi-stage probability, 187–197

N

National Assessment of Educational Progress (NAEP), 115
National Research Council, 124
Negative numbers, 89
Nelson, L. Ted, 68, 88, 150
Net value, 89
New Directions for Elementary School Mathematics, 36, 40
Nim, game of, 48
Nine in a row (logic problem), 37
Nonconvex, 204
Normal curve (bell-shaped),168
Normal distribution, 168
Number(s)
  binary, 47–48
  composite, 75, 76, 77
  decimals, 115–133
  even, 71–73
  Fibonacci, 13
  fractions, 97–111
  integers, 89–96
  irrational, 142–149
  negative, 89
  odd, 71–73
  positive, 89
  prime, 74–76
  signed, 89
  square, 76
  whole, 42–46
Number chart primes and multiples, 78
Number line, 95
Number patterns, 8–13
Number theory, 70–88
Numeration, 42–48, 89–90, 97, 116–118
Numerator, 97

O

Obtuse, 261
Octagon, 212
Octahedron, 216–217
O'Daffer, P.G., 215
Odd-even class models, 87–88
Odd numbers, 71–74, 87–88
On the Shoulders of Giants: New Approaches to Numeracy, 28
Opposites, 90
Orbit, 86–87
Order of spirolateral, 28

P

Pace, 236
Page guessing, 165
Pagni, David, 177
Paper-folding
  lengths, 113
  symmetries, 224–228, 229
  tangram pieces, 207–208
Paper puzzles, 269
Parabola, 261
Parallelogram
  geoboard, 242–243
  height (altitude), 242
Partitive approach to division. *See* Sharing approach to division
Part-to-whole concept
  decimals, 117–118
  fractions, 97
Pattern blocks, 1–7
Pattern block symmetries, 222–223
Pattern block tessellations, 209–211
Patterns
  algebra pieces, 14–19
  artistic, 84–87, 261–262
  geometric, 1–7, 8–13
  graphical, 28–35
  number, 8–13
  symmetric, 224–229
Peg-Jumping Puzzle, 7, 19
Pentominoes, 244
Percent
  fractions as, 138–139
  fractions as approximate, 140–141
  models for, 134–141
  part and whole, 134, 135, 136
  percent and part, 134, 137
  percent and whole, 134, 135, 136
Perfect squares, 76
Perpendicular, 223, 259
Piaget, Jean, 104
Pica Centro (game), 39
Pi (π), 245
Place value
  decimals, 119–120, 123
  whole numbers, 43
Plato, 216
Platonic solid, 216–219
Polygons
  angles, 210–212
  area, 143–145, 239–244
  regular, 211–213
  vertices, 212
Polyhedra, 215–220
Positional numeration
  decimals, 119–120, 123
  whole numbers, 44–45
Positive numbers. *See* Integers
Prime numbers, models for, 74–76
Prisms
  altitude, 245–246
  surface area, 245–246
  volume, 245–246
Prizes (logic problem), 38
Probability, 179–198
  complementary, 193
  dependent event, 187–188
  experimental, 179–186, 189–196
  independent event, 187
  multi-stage (compound), 187–197
  theoretical, 180, 182–184, 189–192
Probability tree, 188, 190, 191
Problem solving, 1–19
Proof, existence, 214
Proportional, 271
Protractor, 205
Puzzle(s)
  Cross-Number, 59
  Five House Puzzle, 38
  Folding Numbers in Order, 269
  Instant Insanity, 219–220
  Paper-Strip Cube, 269
  Peg-Jumping Puzzle, 7, 19
  Pentominoes, 244
  Soma Cubes, 251
  Stepping Through a Piece of Paper, 269
  Tangram, 207–208
Puzzlers, 7, 12, 18, 27, 34, 40, 54, 67, 77, 87, 103, 113, 141, 149, 158, 177, 186, 198, 208, 214, 220, 229, 244, 252, 262
Pyramids
  height (altitude), 245–246
  surface area, 245–246
  volume, 245–246
Pythagoreans, 14

Q

Quadrilateral
  parallelogram, 204
  rectangle, 204

Quadrilateral—*Cont.*
square, 204
trapezoid, 204
Quart, 237
Quotient, 61, 63, 64, 108

R

Racetrack Game, 185–186
Racetrack probability, 198
Radius, 247, 257
Random digits, 152–154
Ratio, 270
Reasoning, 21–40
and communication, 40
deductive, 36–39
inductive, 36
Rectangle, 204, 208
Rectangular array, 55
Rectangular geoboard, 201–204
Reed, R., 107
Reflection, 268
Regrouping whole numbers
addition, 49–51
division, 63
multiplication, 55–56
numeration, 43–44
subtraction, 52
Regular polyhedron, 216
Regular tessellation, 211–212
Research within Reach: Elementary School Mathematics, 104
Reys, R.E., 104, 116
Rhombus, 225
Right angle, 202
Right triangle, 147, 202
Rotation, 266–267
Rotational symmetry, 221

S

Sampling, 154–155
Scale, 277
Scalene triangle, 202
Scatter plots, 159–165
Scheaffer, R., 157
Schram, Pamela, 40
Scout sighting method, 272–273
Sector of a circle, 247
Semicircle, 206
Semiregular polyhedron, 218–219
Semiregular tessellation, 212–213
Sequences, 2–7, 8–11
Sets
attribute pieces, 21–27
intersection, 24–25
union, 25–26
Venn diagram, 24–26
Sextant, 274
Sharing approach to division, 60–62, 68, 131
Sholes, Christopher, 172
Shuster, C.N., 272
Sighting method, 271–272
Signed numbers, 89
Similarity mapping, 278–279
Similar triangle, 271
Simulated Racing Game, 158
Simulation, 156–157
Sinkov, A., 176
Snowflakes, 228
Solitaire, 111
Soma Cubes, 251
Spheres, volume, 245, 249–250
Spirolaterals, 28–35
Square, 212
Square numbers, 76
Square root, 145–146
Stadiascope, 273–274
Standard deviation, 171
Star polygons, 84–87
Statistics, 151–177
mean, 152
median, 152
mode, 152
random digits, 152–154
sampling, 154–155
scatter plots, 159–165
simulation, 156–157
stratified sampling, 155–156
Statistics: A Guide to the Unknown, 166
Steen, Lynn A., 28
Stochastic, 179
Story problems, 14–19
Stratified sampling, 155–156
String art, 87, 262
Student-centered data collection, 177
Subtraction
algorithm, 49
comparison concept, 54
decimals, 127–128, 132–133
fractions, 104–106
integers, 92–93
models, 52–54, 92–93, 104–106, 127–128
take away concept, 54
whole numbers, 52–54
Surface area, 245–251
Swift, J., 157
Symmetry, 221–229
Symmetry, 209

T

Take-away concept subtraction, 54
Tangram puzzle, 207–208
Tangrams, 207
Tennis Team (logic problem), 37
Tessellation(s)
Escher-type, 263–269
glide-reflection, 268
Tessellation(s)—*Cont.*
pattern block, 209–211
regular, 211–212
with regular polygons, 211–213
rotation, 266–267
semiregular, 212–213
translation, 264–266
Tetrahedron, 216–217, 219
Theodolite, 276–278
Theoretical probability, 180, 182–184, 189–192
Three-Penny Grid (game), 183
Tile. *See* Tessellation(s)
Trading-Down Game, 52
Trading-Up Game, 49
Transit, 276–278
Translation, 264–266
Trapezoid, 204, 210
Tree Diagram, 188, 190, 191
Trend line, 160
Triangle
area, 244
equilateral, 212
geoboard, 242–243
height (altitude), 242
isosceles, 202
right, 202
scalene, 202
similar, 271
Trick Dice, 197
Tricks
Arabian Nights Mystery, 67
Magic Formula, 67
Mind-reading cards, 47
Trick Dice, 197
Truncated tetrahedron, 219
Two-Penny Grid (game), 181–182

U

Uniform distribution, 167
Union of sets, 25–26
Unit, 116, 143
Unit of measure
area, 143
length, 232
volume, 232
weight, 232

V

Venn diagram, 24–25, 26
Vertex
angle, 204
polygon, 242
polyhedron, 217–218
tessellation, 213–214
Visualization of basic operations, 68
Volume
cones, 245, 247
cylinders, 245, 246–247

Volume—*Cont.*
- models for, 245–251
- prisms, 245–246
- pyramids, 245–246
- spheres, 245, 249–250

W

Weaver, W., 173
Weight, 232
Weyl, H., 209
Whole numbers, 42–68
- addition of, 49–51
- calculator, 59, 67
- composite, 75, 76, 77
- division of, 60–68
- even, 71–74
- models for, 42–48
- multiple, 79, 82
- multiplication of, 55–59
- numeration, 42–48
- odd, 71–74
- perfect squares, 76
- place value, 43
- prime, 74–76
- subtraction of, 52–54

Who's Who (logic problem), 38
Wiebe, James H., 134
Wind rose, 221

Material Cards

1. Rectangular Grid (2.2)
2. Isometric Grid (2.2)
3. Attribute-Game Grid (2.1)
4. Two-Circle Venn Diagram (2.1)
5. Three-Circle Venn Diagram (2.1, 9.1)
6. Pica-Centro Recording Sheet (2.3)
7. Coordinate Guessing and Hide-a-Region Grids (2.2)
8. Table of Random Digits (7.1, 7.3, 8.1)
9. Two-Penny Grid (8.1)
10. Three-Penny Grid (8.1)
11. Geoboard Recording Paper (9.1)
12. Grids for Game of Hex (9.2)
13. Perpendicular Lines for Symmetry (9.4)
14. Metric Measuring Tape (10.1)
15. Centimeter Racing Mat (10.1)
16. Pentomino Game Grid (10.2)
17. Attribute Label Cards (2.1)
18. Logic Problem Clue Cards and People Pieces (Problem 1) (2.3)
19. Logic Problem Clue Cards (Problems 2 and 3) (2.3)
20. Logic Problem Clue Cards (Problems 4 and 5) (2.3)
21. Object Pieces for Logic Problem 5 (2.3)
22. Mind-Reading Cards (3.1)
23. Decimal Squares* (6.1, 6.2)
24. Decimal Squares* (6.1, 6.2)
25. Decimal Squares* (6.1, 6.2)
26. Decimal Squares* (6.1, 6.2)
27. Rectangular Geoboard Template (6.4, 9.1, 10.2)
28. Algebra Pieces (1.3)
29. Algebraic Expression Cards (1.3)
30. Simulation Spinners (8.2)
31. Trick Dice (8.2)
32. Metric Ruler, Protractor, and Compass (9.1, 10.1, 10.3, 11.1, 11.2, 11.3)
33. Circular Geoboard Template (9.1)
34. Regular Polyhedra (9.3)
35. Regular Polyhedra (9.3)
36. Cube Patterns for Instant Insanity (9.3)
37. Pentominoes (10.2)
38. Prism, Pyramid, and Cylinder (10.3)
39. Hypsometer-Clinometer (11.3)
40. Interest Gameboard (6.3)

*Decimal Squares is a registered trademark of Scott Resources.

Rectangular Grid (Activity Set 2.2)

MATERIAL CARD 1

Copyright © 2001 by The McGraw-Hill Companies, Inc. All Rights Reserved.

MATERIAL CARD 2

Copyright © 2001 by The McGraw-Hill Companies, Inc. All Rights Reserved.

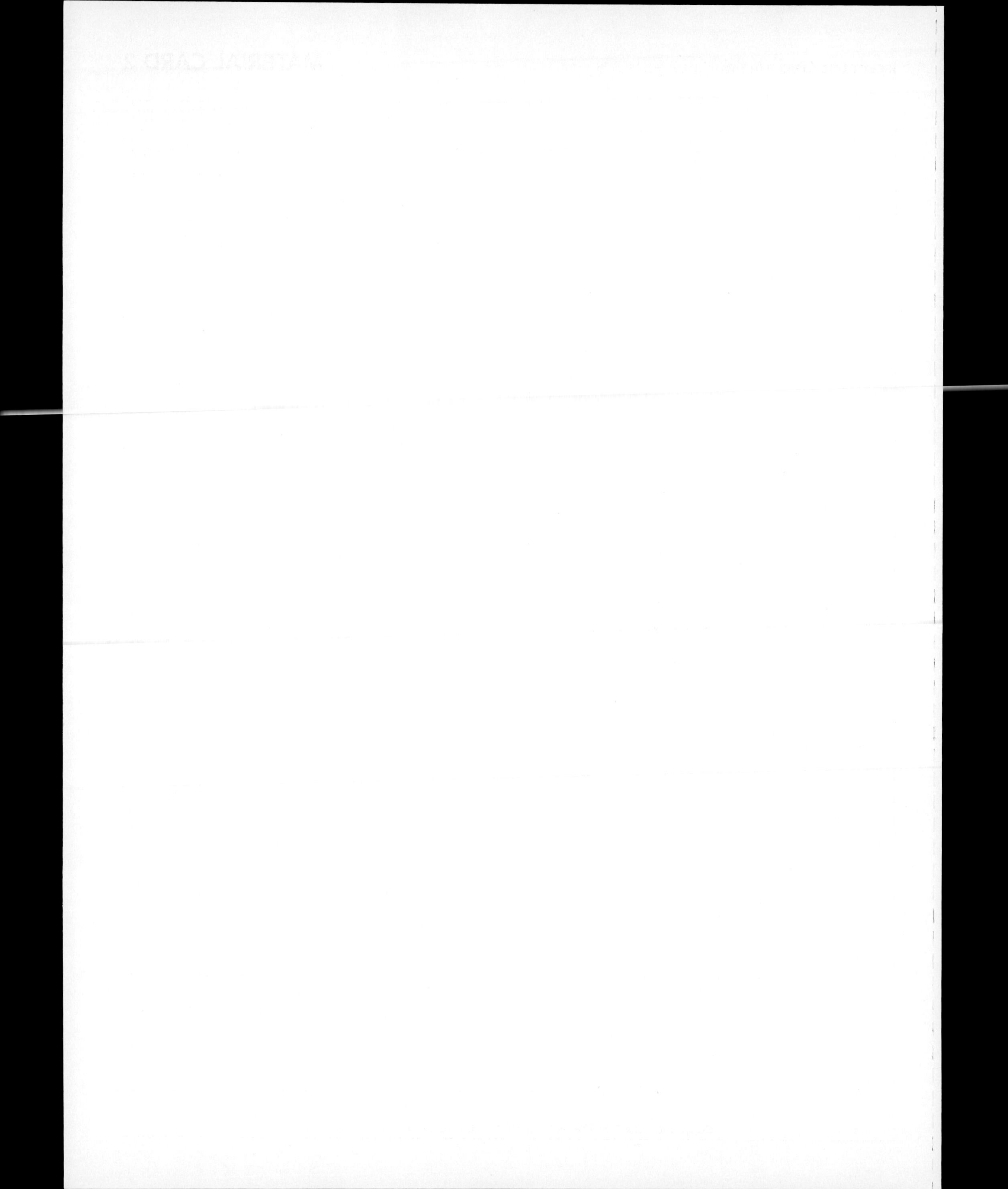

Attribute-Game Grid (Activity Set 2.1)

MATERIAL CARD 3

Rows, 1 difference (1 point) Columns, 2 differences (2 points) Diagonals, 3 differences (3 points)

		Place attribute piece here to start game		

Copyright © 2001 by The McGraw-Hill Companies, Inc. All Rights Reserved.

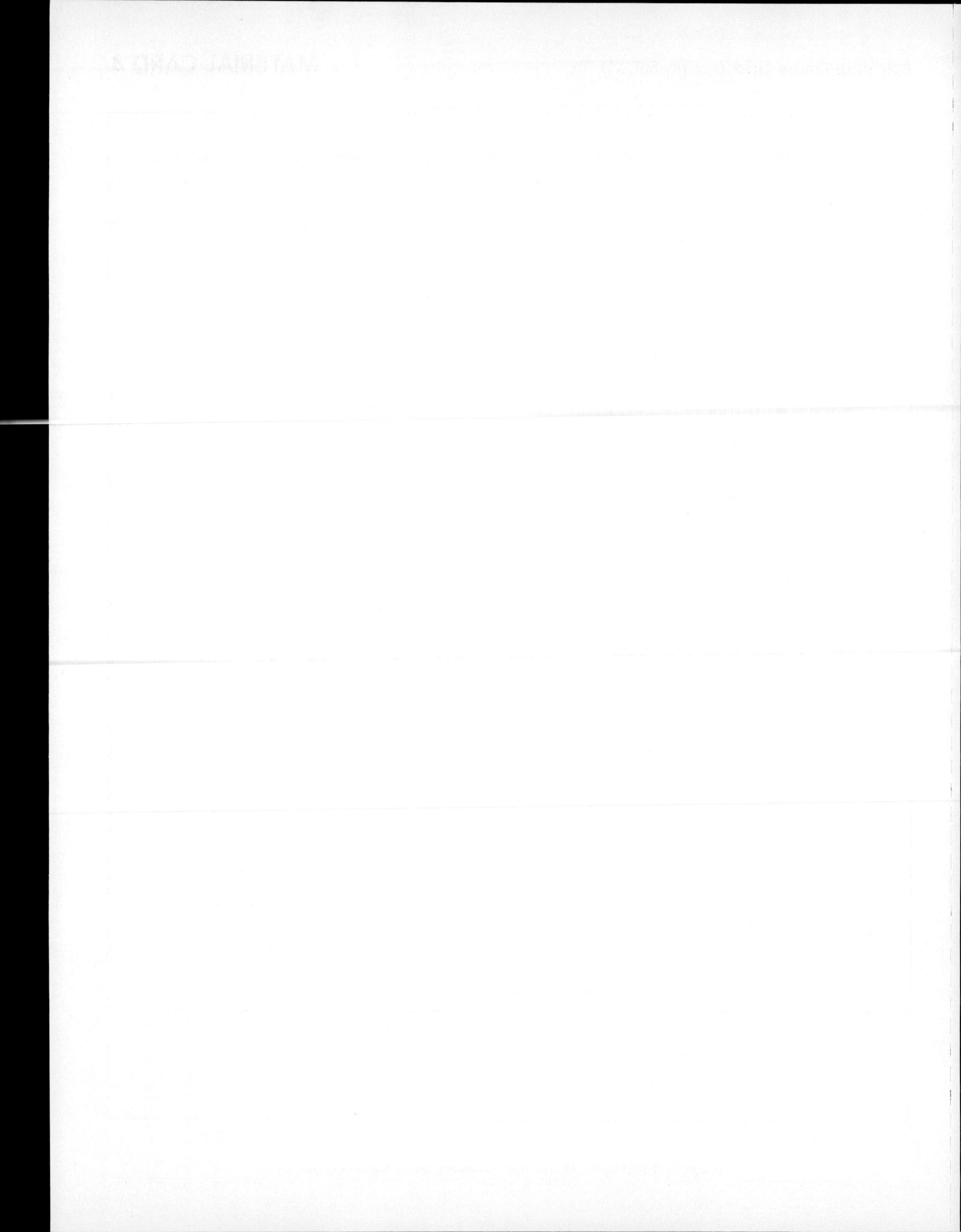

Two-Circle Venn Diagram (Activity Set 2.1)

MATERIAL CARD 4

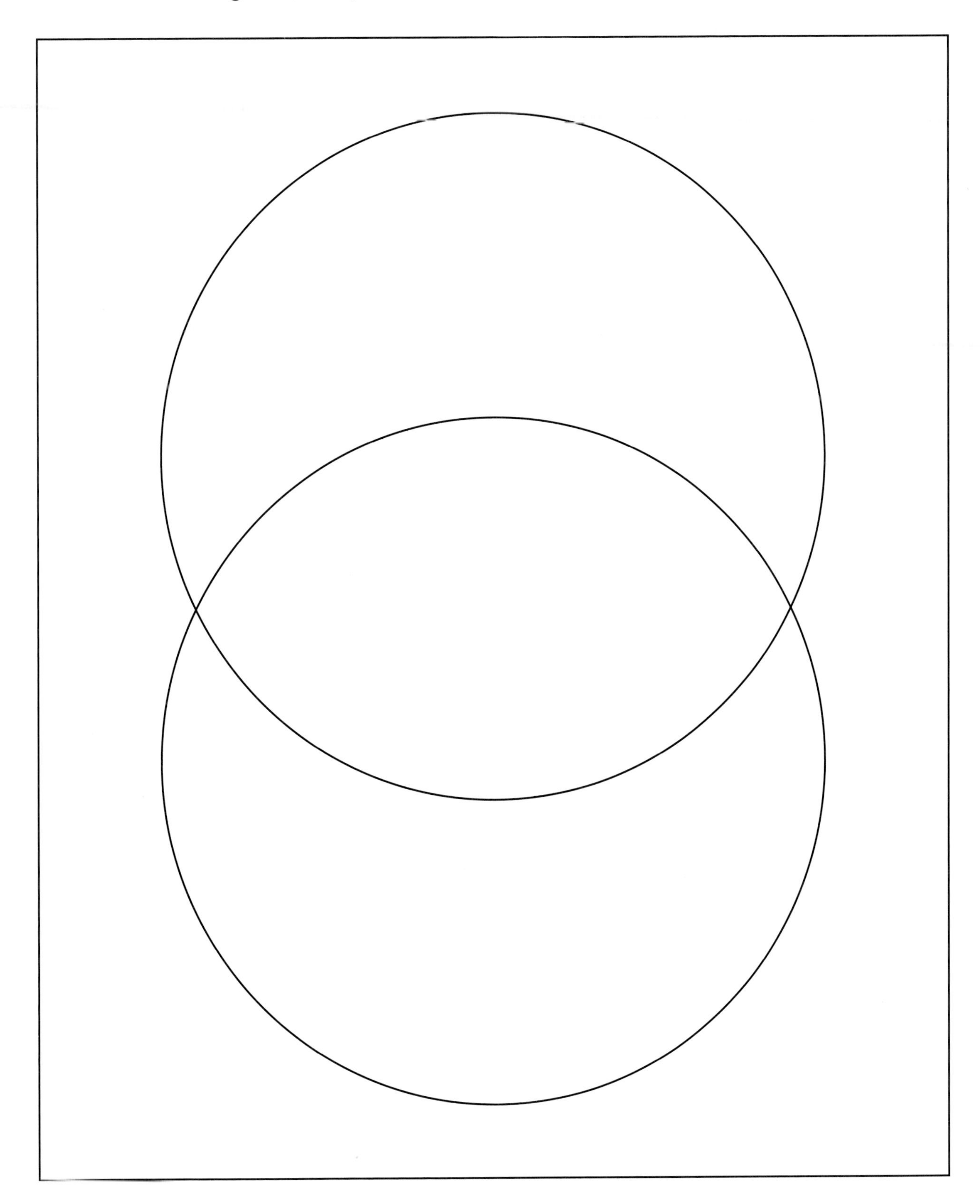

Copyright © 2001 by The McGraw-Hill Companies, Inc. All Rights Reserved.

Three-Circle Venn Diagram (Activity Sets 2.1 and 9.1)

MATERIAL CARD 5

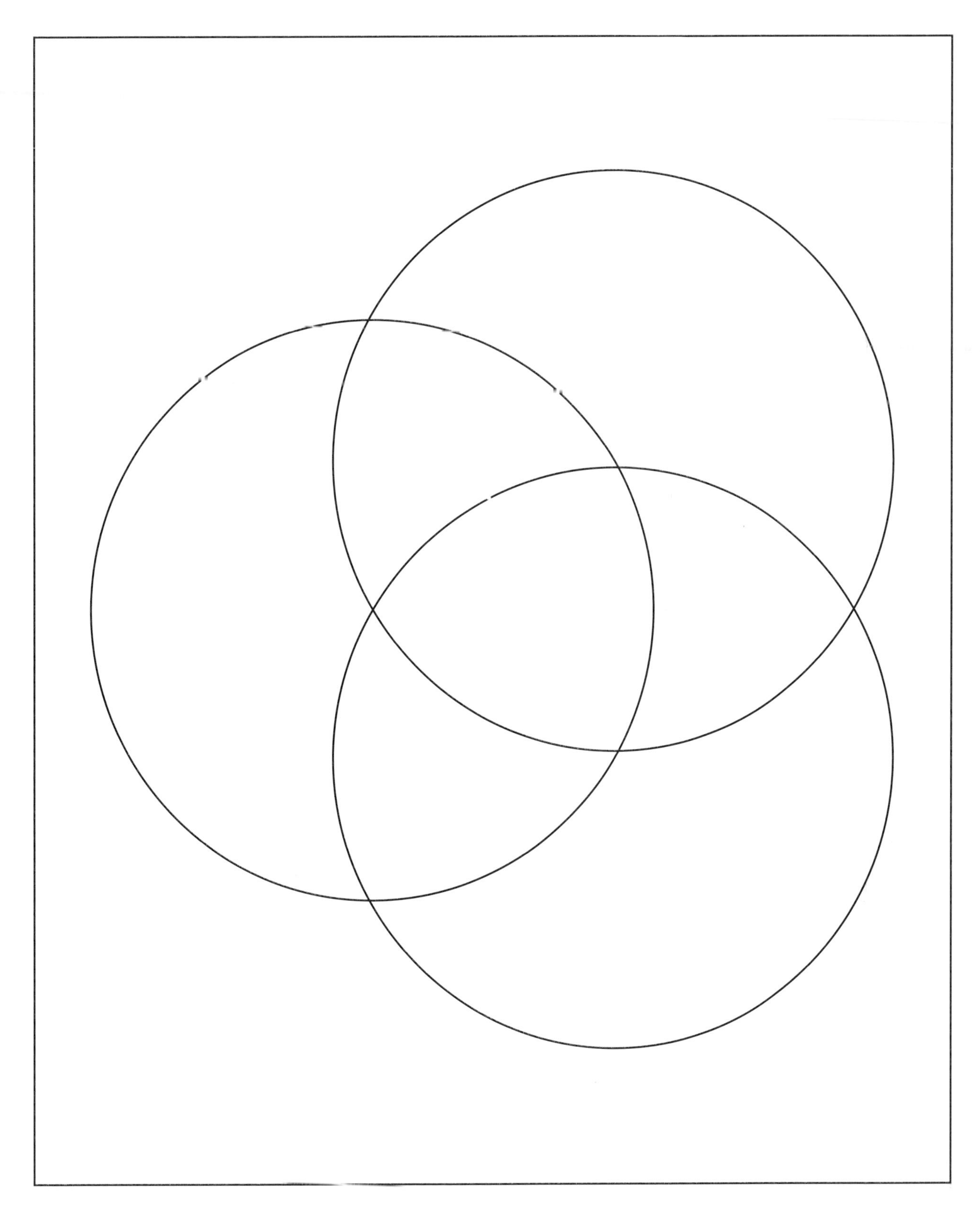

Copyright © 2001 by The McGraw-Hill Companies, Inc. All Rights Reserved.

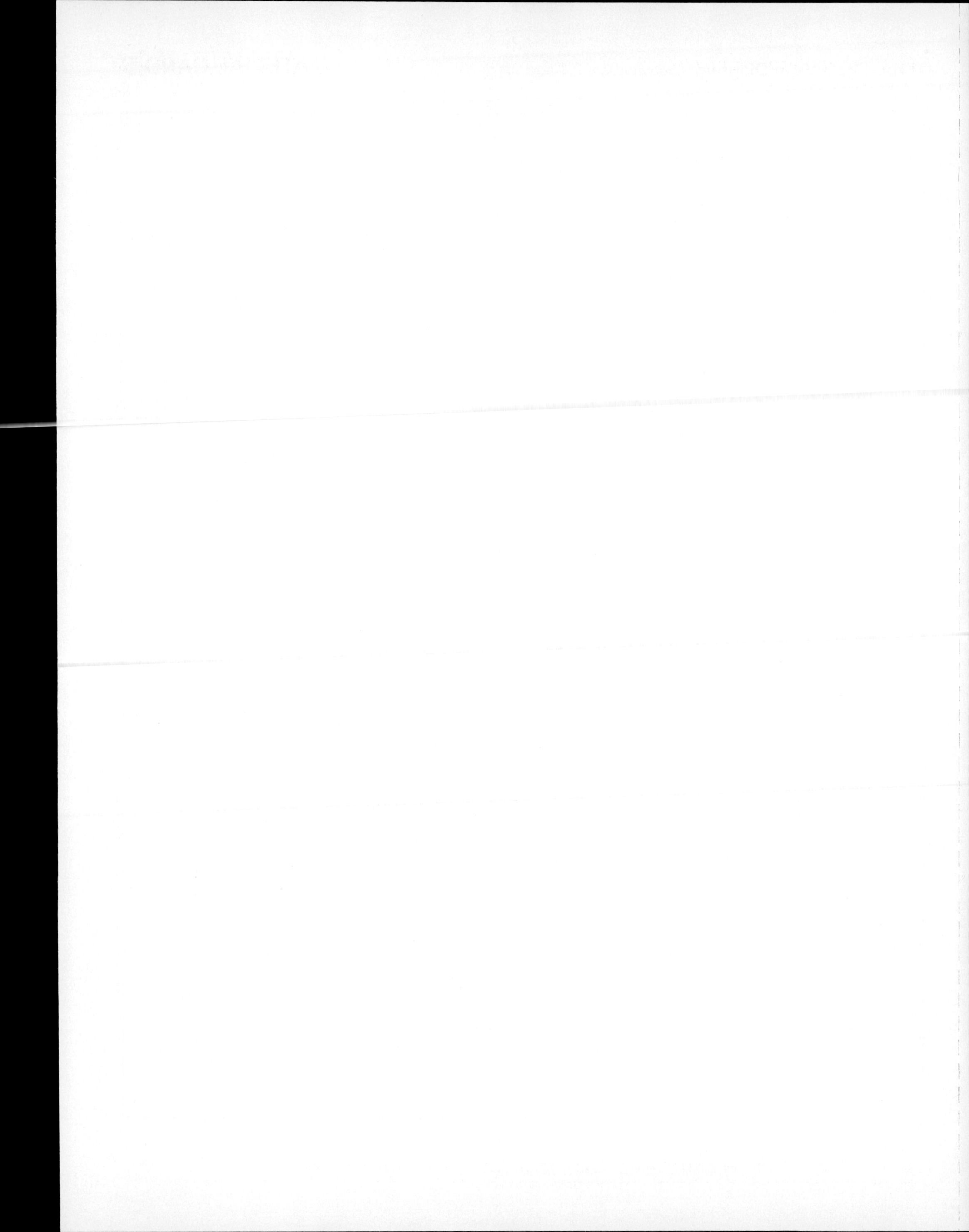

Pica-Centro Recording Sheet
(Activity Set 2.3 JUST FOR FUN)

MATERIAL CARD 6

THREE-DIGIT RECORDING SHEETS

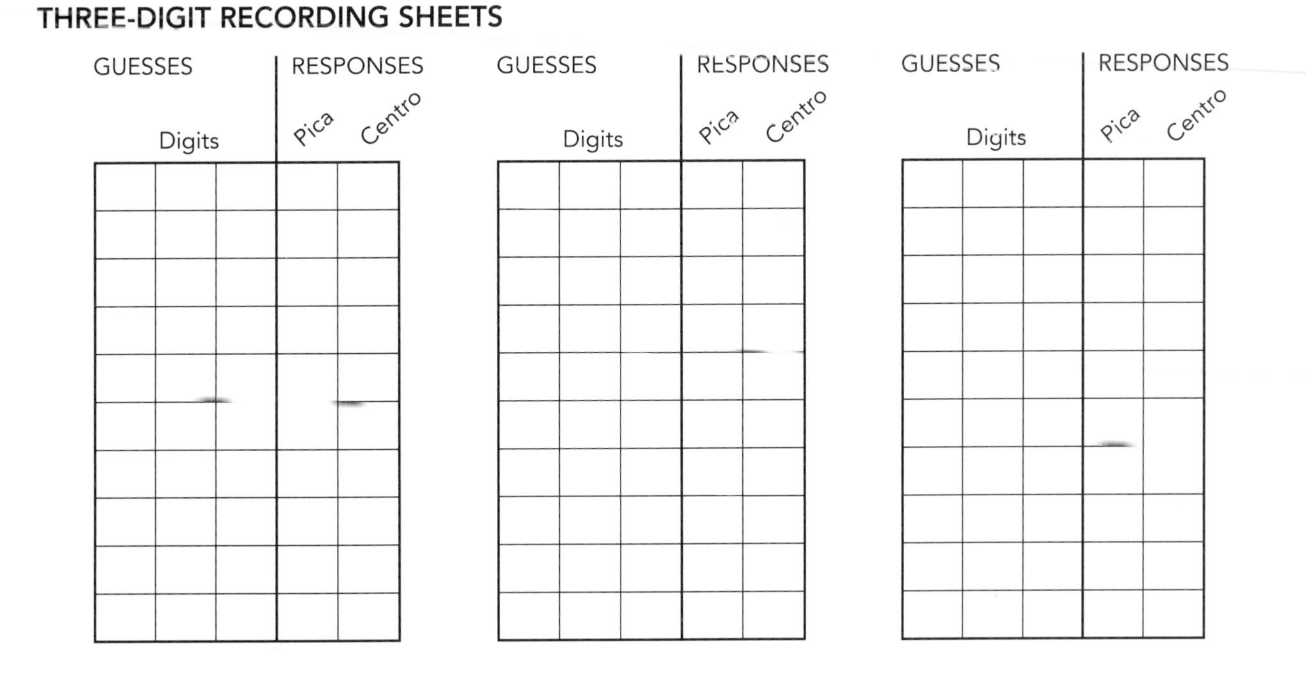

FOUR-DIGIT RECORDING SHEETS

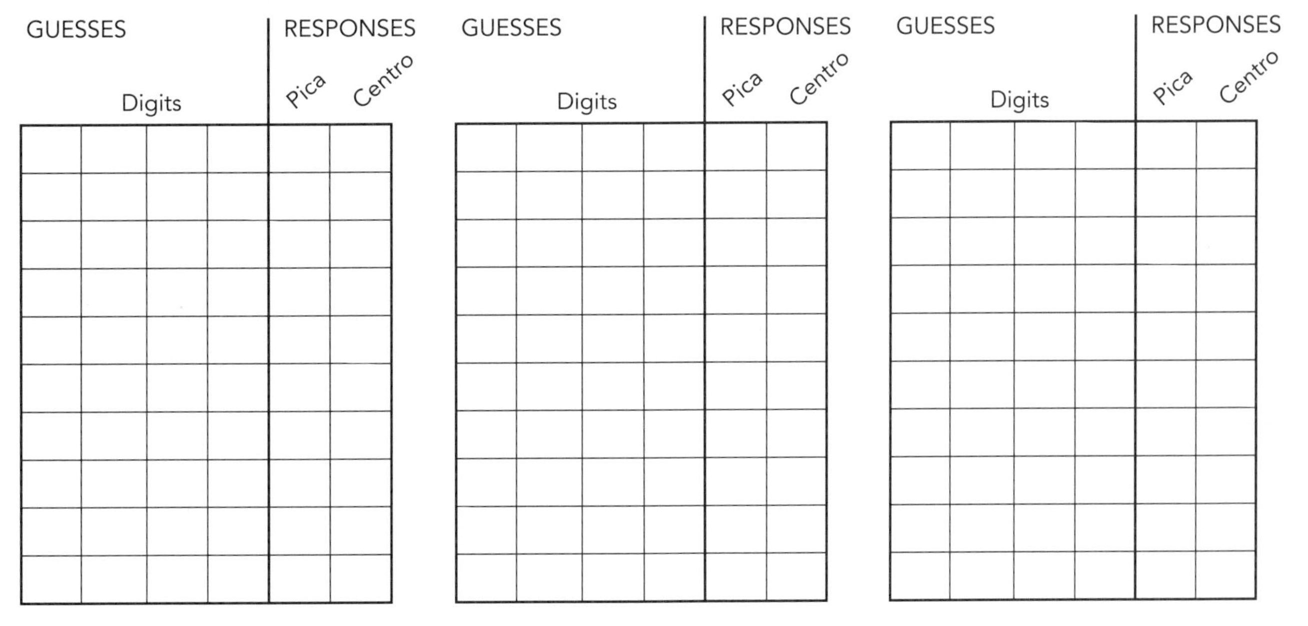

Copyright © 2001 by The McGraw-Hill Companies, Inc. All Rights Reserved.

Coordinate Guessing and Hide-a-Region Grids

(Activity Set 2.2)

Copyright © 2001 by The McGraw-Hill Companies, Inc. All Rights Reserved.

Table of Random Digits (Activity Sets 7.1, 7.3, and 8.1)

MATERIAL CARD 8

69588	18885	24831	28185	84019	91168	62187	69079	86492	24195
19797	27708	70288	72541	10576	79693	28704	97678	95777	46759
50212	65574	17333	21694	47617	71881	88513	62695	76222	42544
16380	57486	86835	82724	23964	77540	93652	75555	77283	29473
29826	51673	28082	85640	43062	21997	77247	14653	13471	03865
51427	10863	96787	43214	68955	13735	72107	63399	15951	53968
36037	55838	30244	28819	19794	72893	43642	89344	72686	17378
52421	74281	44363	37701	25103	06701	33100	00483	38208	57711
64936	77559	27432	27412	89263	50899	68524	54274	35905	40343
05013	22060	17792	96536	74156	88924	19652	84311	92236	18008
69970	37423	22837	28337	06347	84836	68812	06121	23216	64020
96601	64066	87785	41608	32193	07050	46509	70020	18435	75427
30984	17658	07126	96234	66401	05691	55657	29217	61037	83086
93775	97887	39692	84605	98008	43174	07226	17637	63106	11806
10975	47371	37928	71999	19339	87802	88208	23697	57359	64825
34487	51270	55303	80666	63115	97793	01883	46650	10668	04238
77876	02108	02420	23983	49776	29281	01665	04908	78948	09727
38781	58184	25008	00962	80002	22962	92127	49898	24259	61862
83788	62480	44711	21815	84629	06939	67646	06948	04529	45214
89704	30361	15988	83311	78147	00323	49940	68820	24291	44077
24113	92774	25852	01814	44175	64734	25392	58525	88314	97199
23041	81334	43410	28285	59247	22853	24395	17770	61573	38097
43123	25825	91840	79914	41137	99165	33796	95374	25960	35135
20585	11696	84826	43878	74239	98734	35854	82082	97416	20039
45699	87633	29489	57614	52384	88793	24228	60107	83502	28129
42060	77709	09509	58712	21415	38415	10841	93372	53568	70864
23038	92956	27304	20825	70350	33151	93654	81710	34692	58456
72936	52909	92458	61567	72380	00931	49994	38711	12237	49788
24631	18803	35690	70000	10040	38204	67146	85913	48332	86757
62397	54028	77523	24891	27989	02289	74513	26279	85382	82180
75724	78027	60675	01294	37607	84053	41144	94743	85633	50356
44997	49702	68073	99166	05851	22239	76340	79458	25968	75875
20733	42897	05445	46195	97046	16120	41546	04463	34074	05515
90448	49154	27693	90607	98476	50924	25067	01233	64588	31287
50151	76497	10878	87213	84640	35943	46320	03267	15324	95455
33944	22308	60879	07308	10759	75179	05130	66184	69554	65891
42837	55847	75893	54245	38752	68095	33158	81860	56469	84190
39357	40544	25447	96260	87283	22483	78031	91393	38040	82382
45673	87347	44240	61458	87338	62039	71825	52571	27146	86404
97714	52538	91976	52406	19711	99248	23073	43926	81889	47540

Copyright © 2001 by The McGraw-Hill Companies, Inc. All Rights Reserved.

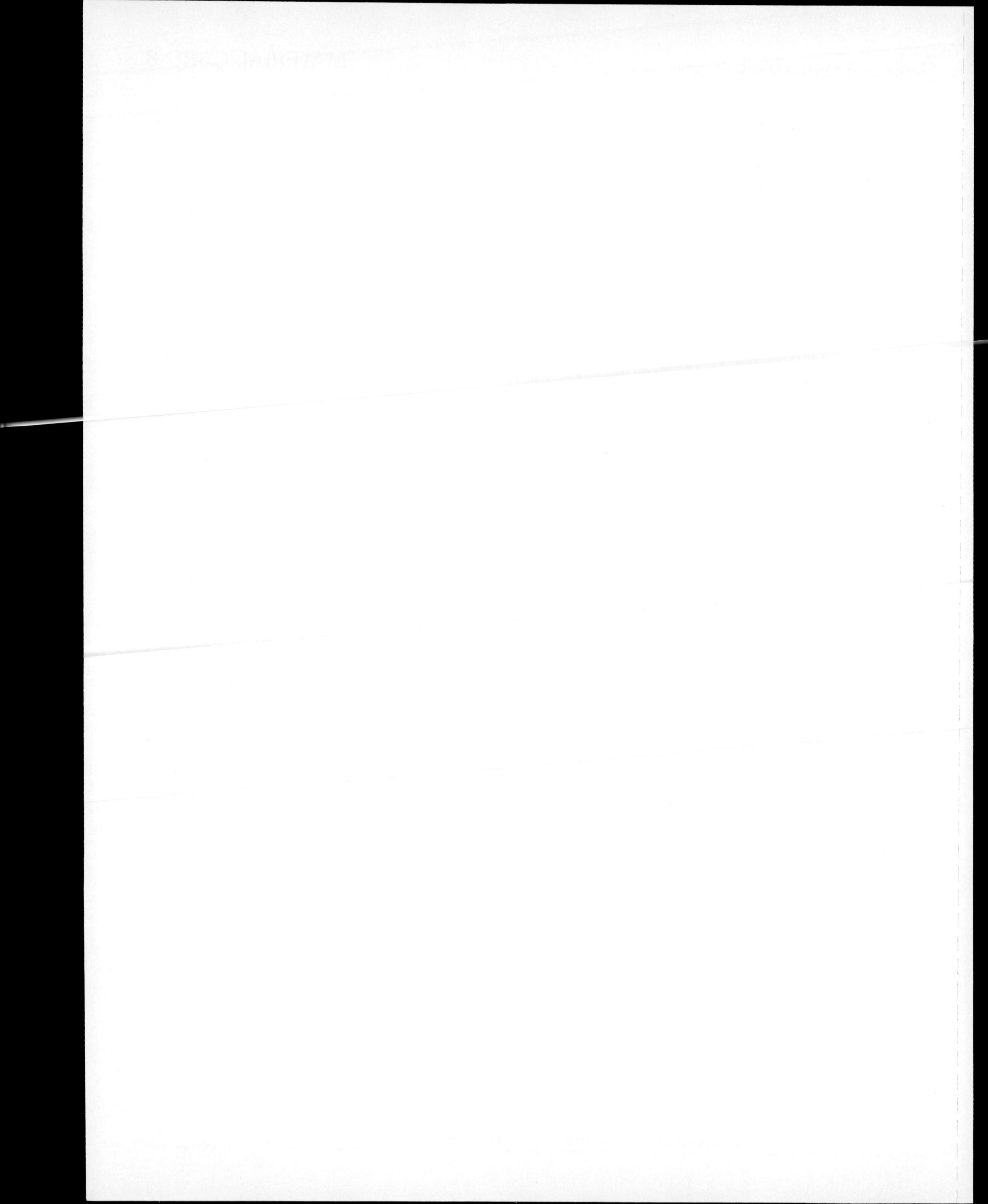

Two-Penny Grid (Activity Set 8.1)

MATERIAL CARD 9

Copyright © 2001 by The McGraw-Hill Companies, Inc. All Rights Reserved.

Three-Penny Grid (Activity Set 8.1)

MATERIAL CARD 10

Copyright © 2001 by The McGraw-Hill Companies, Inc. All Rights Reserved.

Copyright © 2001 by The McGraw-Hill Companies, Inc. All Rights Reserved.

Grids for Game of Hex
(Activity Set 9.2 JUST FOR FUN)

MATERIAL CARD 12

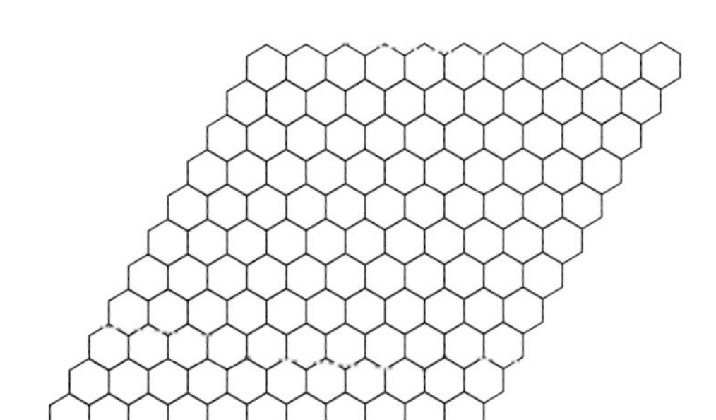

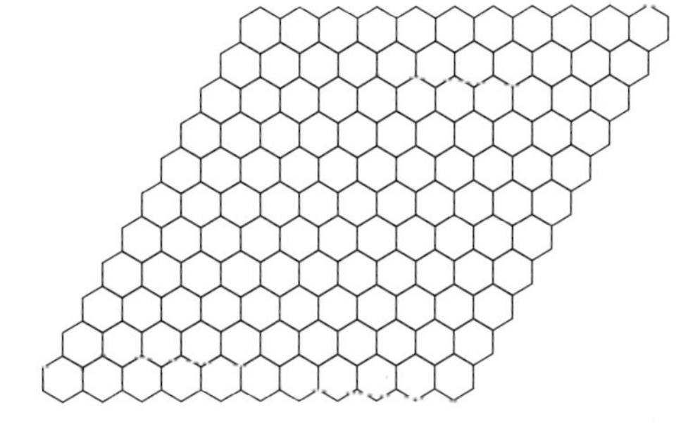

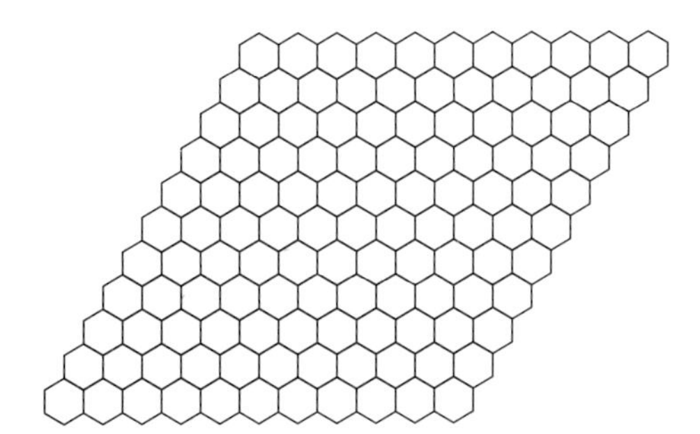

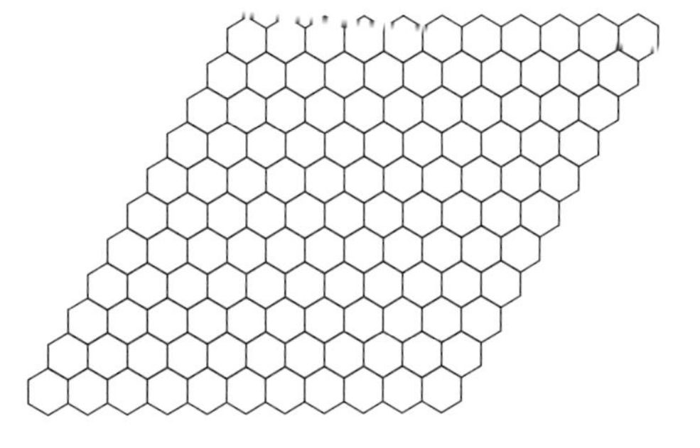

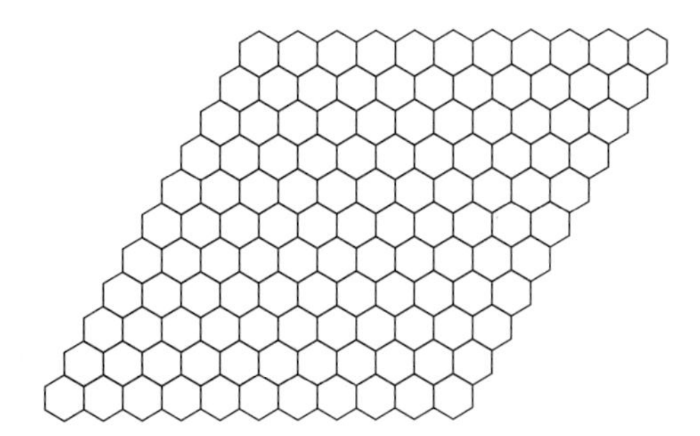

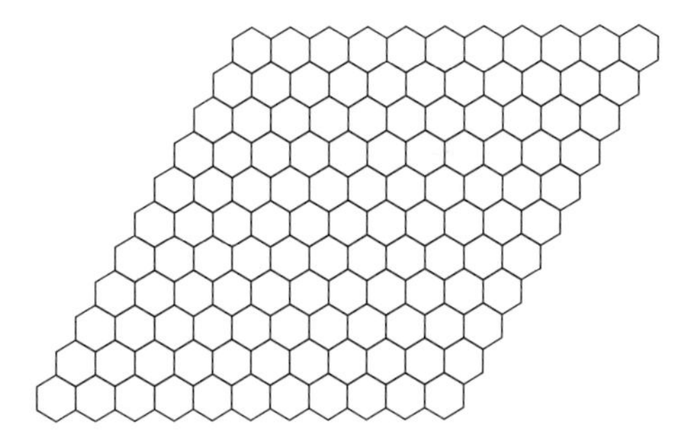

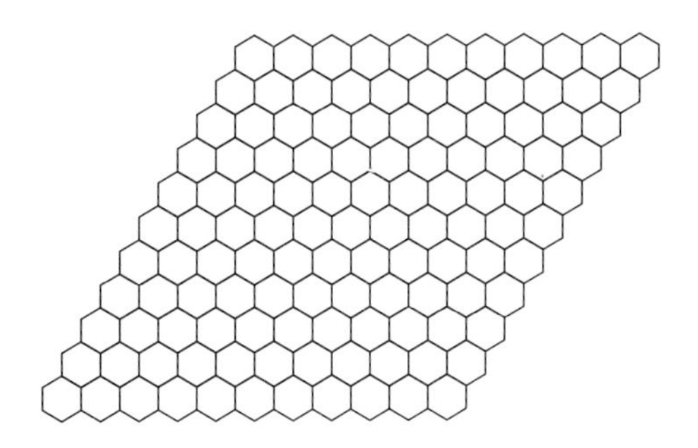

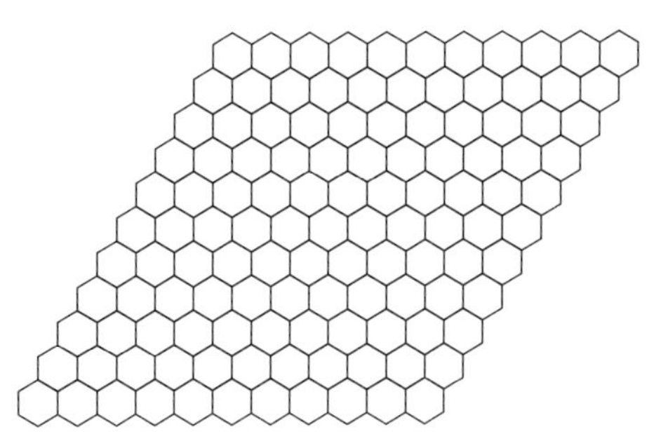

Copyright © 2001 by The McGraw-Hill Companies, Inc. All Rights Reserved.

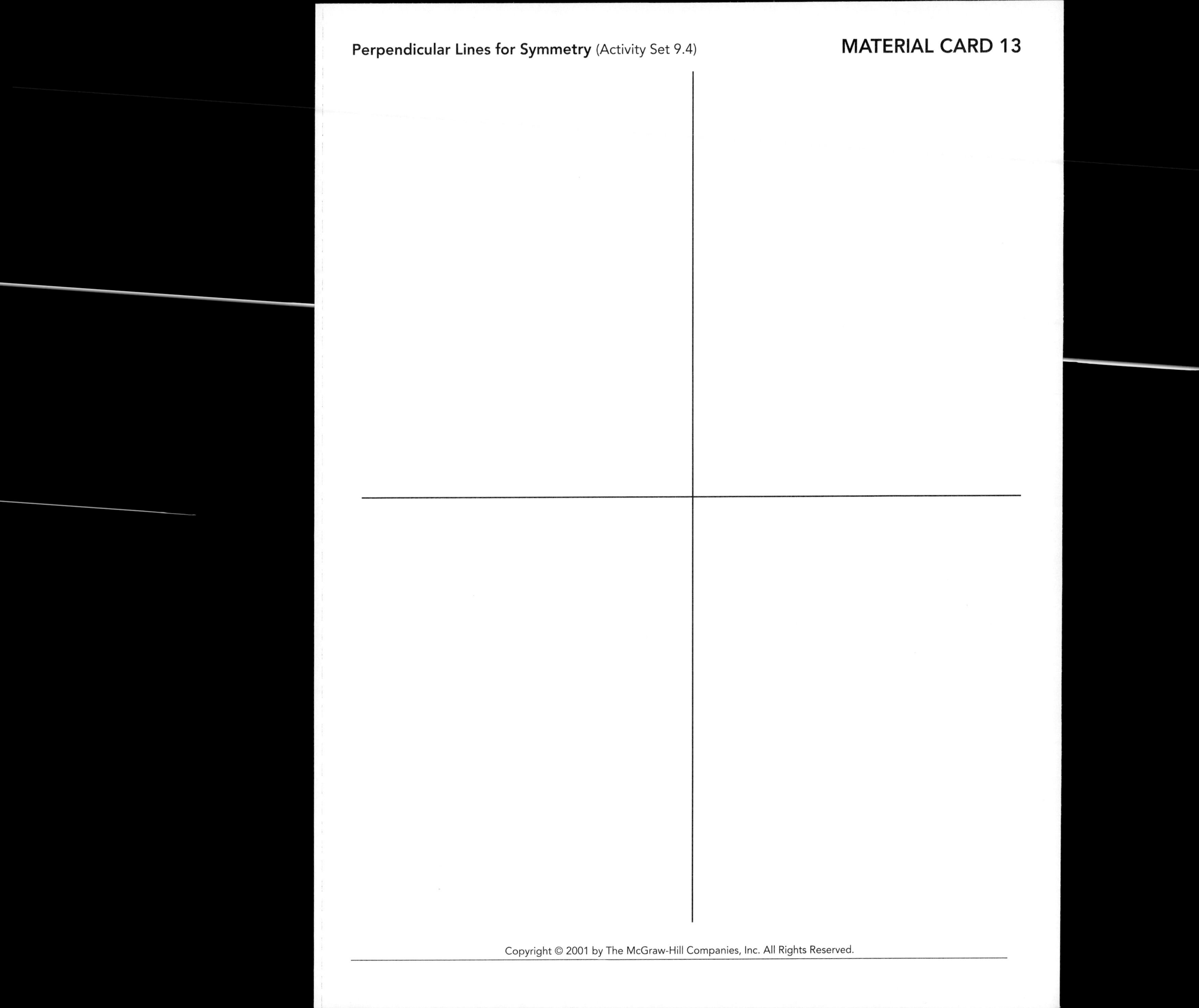

Copyright © 2001 by The McGraw-Hill Companies, Inc. All Rights Reserved.

Metric Measuring Tape (Activity Set 10.1)

MATERIAL CARD 14

Cut out the five vertical strips and then tape or glue tab 2 under the edge labeled 2, tab 4 under the edge labeled 4, tab 6 under the edge labeled 6, and tab 8 under the edge labeled 8, to form a metric tape that is 100 centimeters long.

2
4
6
8
100 cm

15 cm
35 cm
55 cm
75 cm
95 cm

10 cm
30 cm
50 cm
70 cm
90 cm

5 cm
25 cm
45 cm
65 cm
85 cm

20
40
60
80

Tabs

20
40
60
80

Copyright © 2001 by The McGraw-Hill Companies, Inc.
All Rights Reserved.

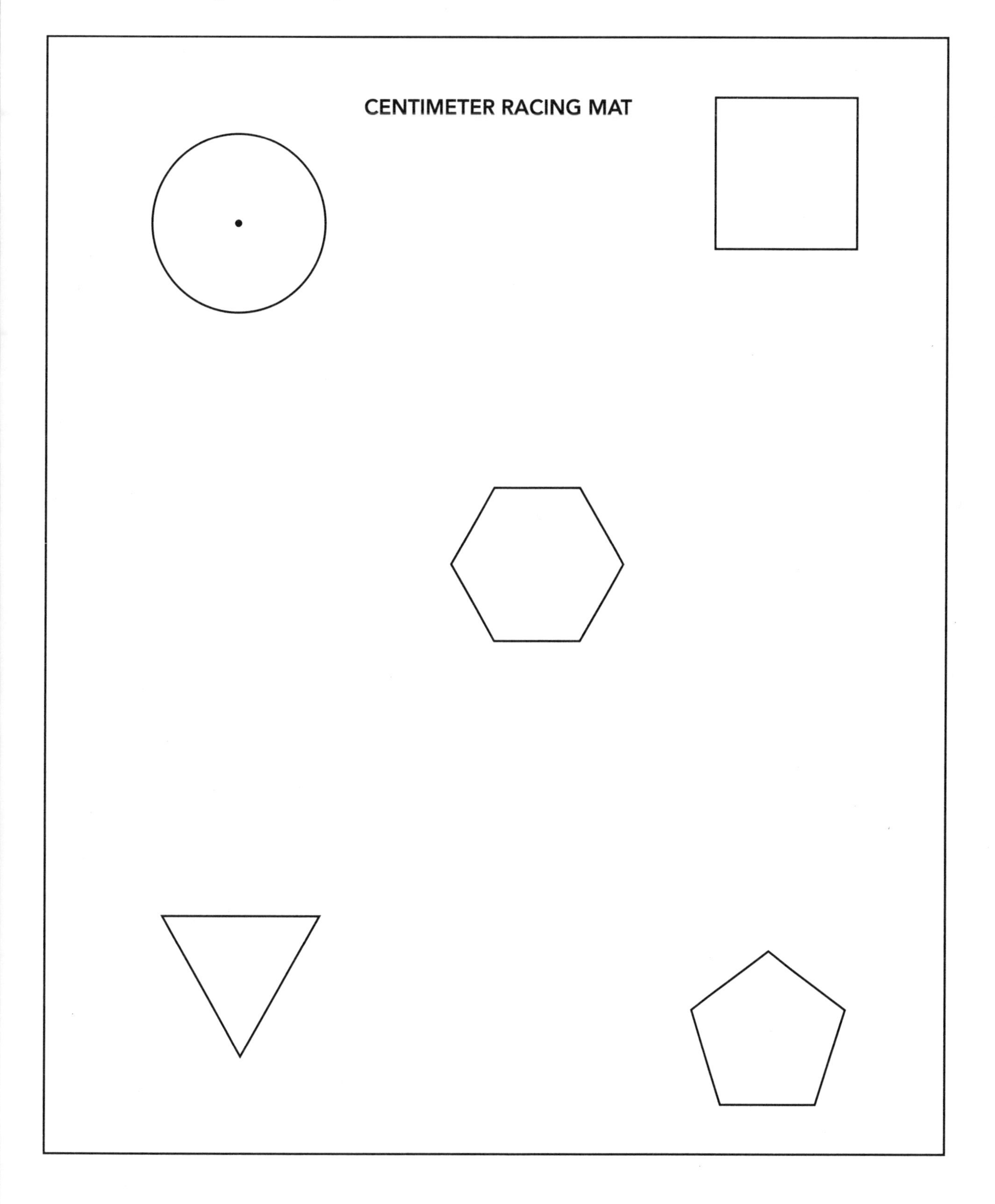

Copyright © 2001 by The McGraw-Hill Companies, Inc. All Rights Reserved.

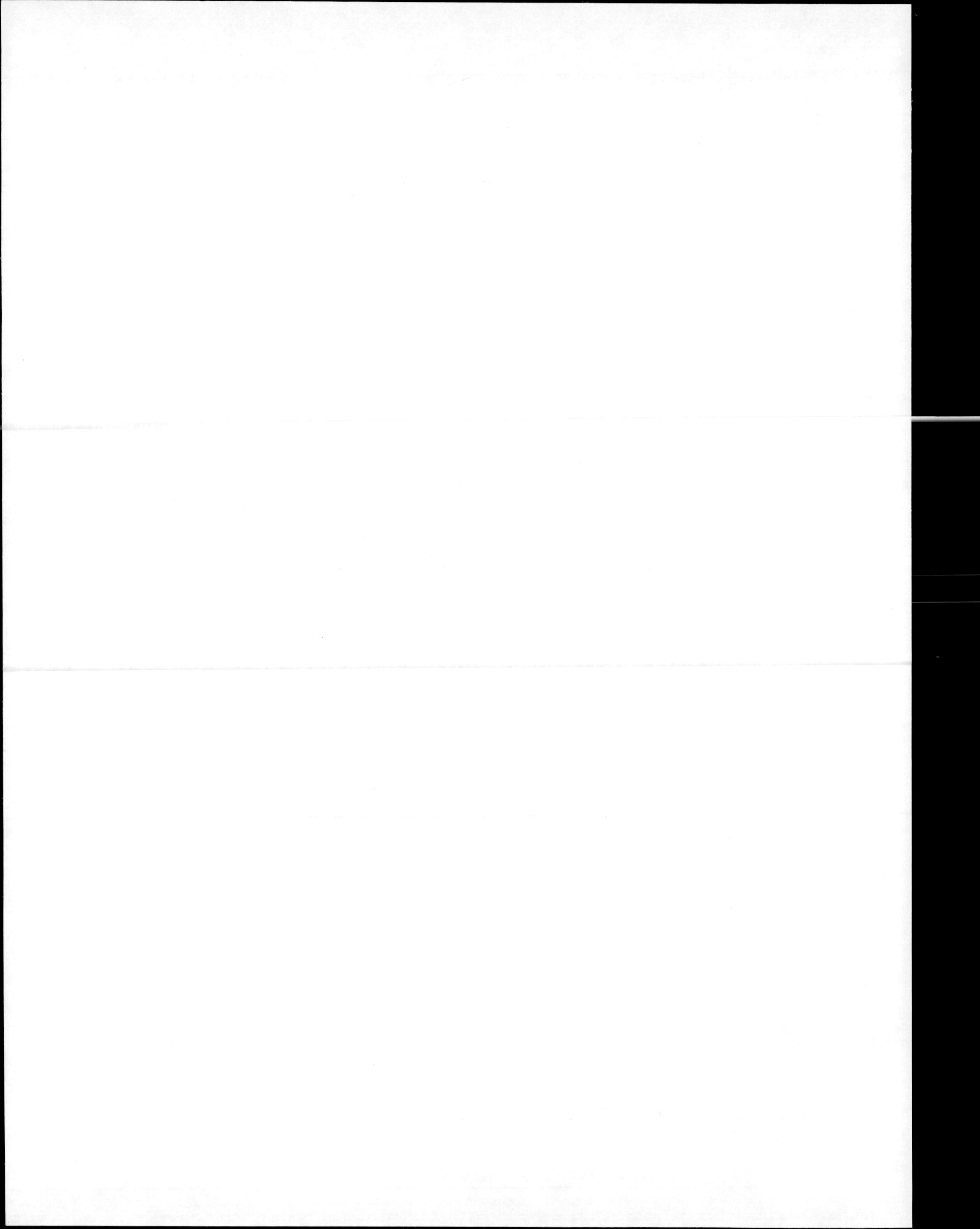

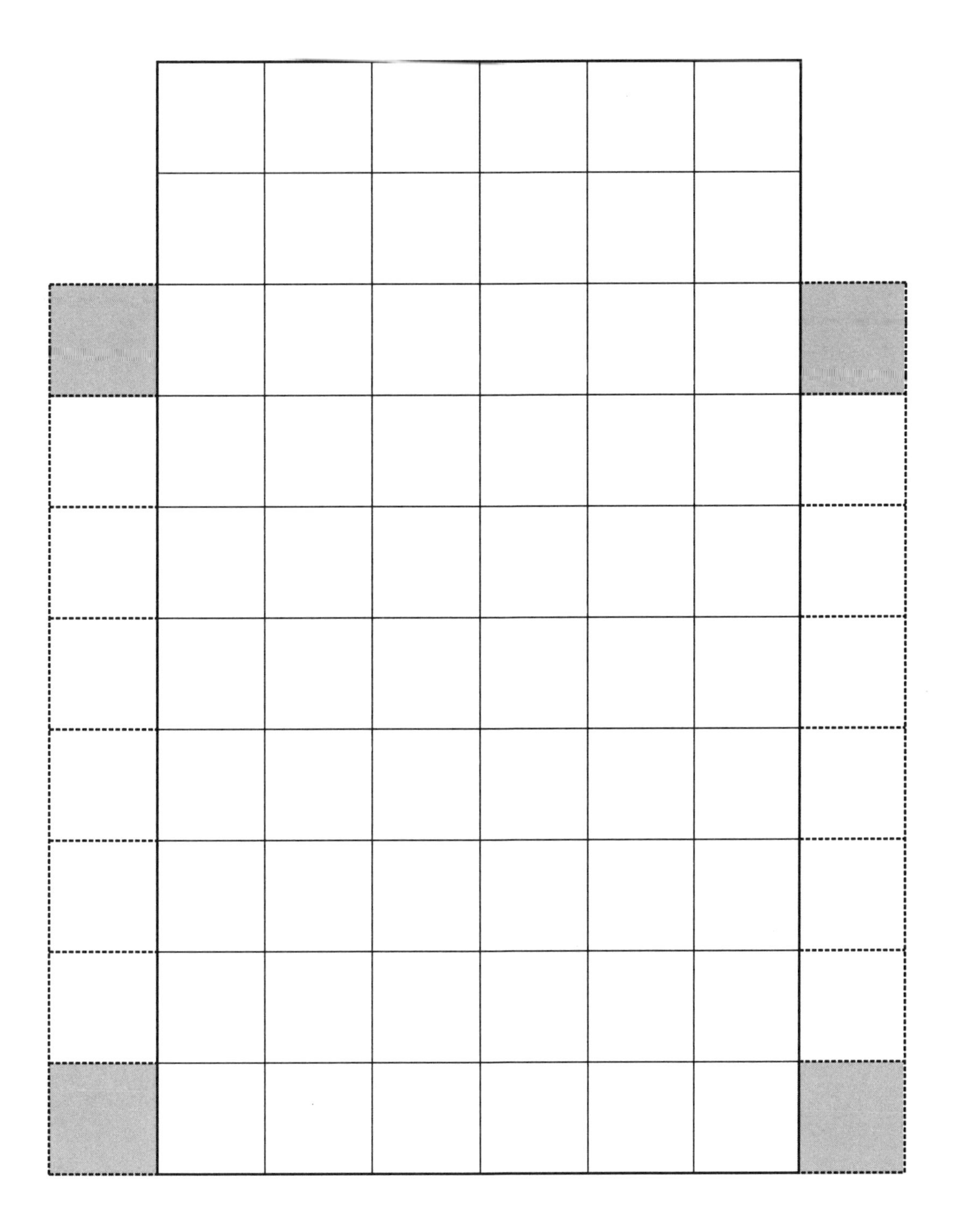

Copyright © 2001 by The McGraw-Hill Companies, Inc. All Rights Reserved.

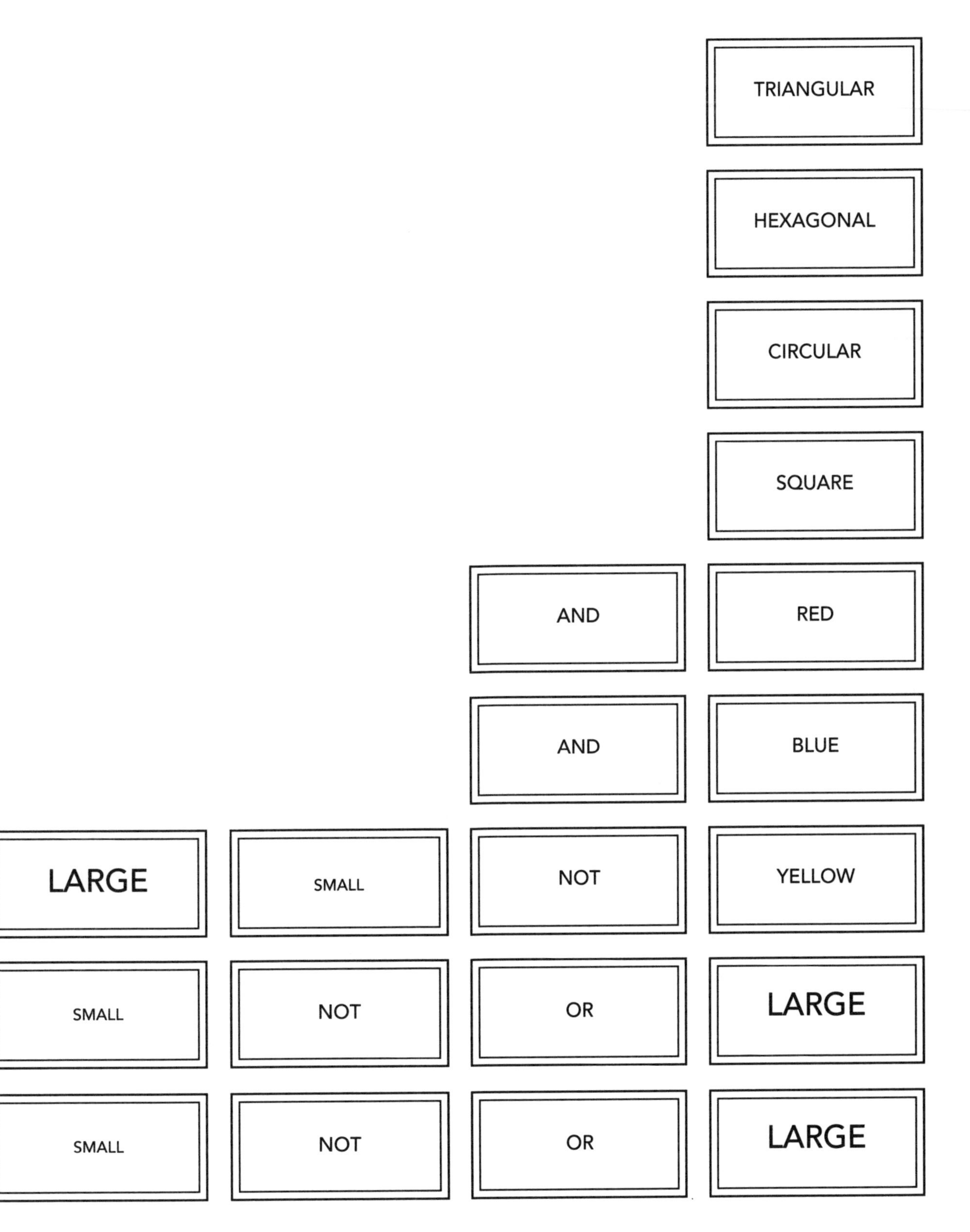

Copyright © 2001 by The McGraw-Hill Companies, Inc. All Rights Reserved.

Logic Problem Clue Cards and People Pieces

MATERIAL CARD 18

PROBLEM 1 (Activity Set 2.3)

Paul, Nicky, Sally, Uri, Jane, Veneta, Monty, Marc, and Edith are standing in a row according to height, but not necessarily in the order given.

Edith is closer to Sally than she is to Nicky.

LOGIC PROBLEM 1 CARD A

Veneta, Edith, Marc, Sally, Monty, and Jane are shorter than Paul, Nicky, and Uri.

There is just one person between Uri and Paul.

Name the 9 people in order from shortest to tallest.

LOGIC PROBLEM 1 CARD B

Edith, Marc, Monty, Jane, and Veneta are shorter than the other four.

Uri is between Sally and Nicky.

Monty and Paul have 7 people between them.

LOGIC PROBLEM 1 CARD C

Uri, Nicky, Jane, Paul, and Sally are taller than the others.

Veneta and Monty are the only two people shorter than Edith.

LOGIC PROBLEM 1 CARD D

Nicky

Paul

Sally

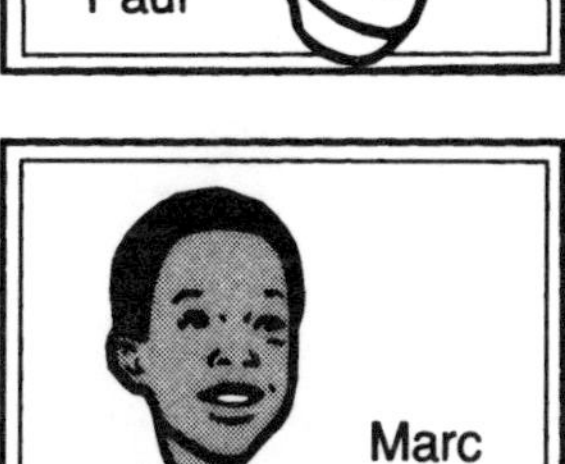

Uri

Copyright © 2001 by The McGraw-Hill Companies, Inc. All Rights Reserved.

Logic Problem Clue Cards

MATERIAL CARD 19

PROBLEMS 2 AND 3 (Activity Set 2.3)

Mr. Racquet is coach of the tennis team.

Susan said, "I won't play if Mike is on the team."

LOGIC PROBLEM 2 CARD A

John, Susan, Mike, Derek, Linda, and Anne play mixed doubles.

Mike said, "I won't play if either Derek or Linda is chosen."

LOGIC PROBLEM 2 CARD B

How can Mr. Racquet select four compatible players (two men and two women) for a road trip?

Derek said, "I'll only play if Anne plays."

LOGIC PROBLEM 2 CARD C

John said, "I'll only play if Susan plays."

Anne had no likes or dislikes.

LOGIC PROBLEM 2 CARD D

Smith, Brown, Jones, and Robinson are told that they have each won one of four academic prizes, but none of them knows which, so they are speculating.

Smith thinks that Robinson has won the logic prize.

LOGIC PROBLEM 3 CARD A

The four academic prizes are for outstanding work in mathematics, English, French, and logic.

Jones feels confident that Smith has not won the mathematics prize.

Who won each prize?

LOGIC PROBLEM 3 CARD B

Robinson is of the opinion that Brown has won the French prize.

The winners of the mathematics and logic prizes were correct in their speculations.

LOGIC PROBLEM 3 CARD C

The English and French prize winners were not correct in their predictions of the prize winners.

Brown thinks that Jones has won the English prize.

The math prize was awarded to Robinson.

LOGIC PROBLEM 3 CARD D

Copyright © 2001 by The McGraw-Hill Companies, Inc. All Rights Reserved.

Logic Problem Clue Cards

PROBLEMS 4 AND 5 (Activity Set 2.3)

MATERIAL CARD 20

On a plane, Smith, Jones, and Robinson are pilot, copilot and engineer, but not necessarily in that order.

The copilot has the same name as the singer living in the same town.

LOGIC PROBLEM 4 CARD A

On the plane there is a singing group, The 3-Ms, whose names are M. Smith, M. Robinson, and M. Jones.

The copilot lives in the city between that of M. Jones and that of the engineer.

LOGIC PROBLEM 4 CARD B

M. Robinson lives in New York City.

Jones consistently beats the engineer in racquetball.

LOGIC PROBLEM 4 CARD C

The singers each reside in a different city, New York, San Francisco and Chicago, as do the members of the plane crew.

What is the pilot's name.

LOGIC PROBLEM 4 CARD D

The Chevrolet owner lives in the house next door to the house where the horses are kept.

The person in the red house is English.

The Japanese drives a Ford.

Each house is occupied by a person of different nationality.

The person in the ivory house does not drink milk.

LOGIC PROBLEM 5 CARD A

The Plymouth driver lives next door to the person with the fox.

The Honda owner drinks orange juice.

The Norwegian lives next to the blue house.

Coffee is drunk in the green house.

Who owns the skunk?

LOGIC PROBLEM 5 CARD B

The Chevrolet owner lives in the yellow House.

Milk is drunk in the middle house.

The Norwegian lives in the first house.

The Spaniard owns a dog.

If each person has one home, one pet, one car, a different nationality and a different drink, who drinks the water?

LOGIC PROBLEM 5 CARD C

The Ukrainian drinks tea.

The green house is immediately to the right of the ivory house.

The Mercedes driver owns snails.

The fox lives with the owner of the house farthest from the green house.

The five houses are in a row.

LOGIC PROBLEM 5 CARD D

Copyright © 2001 by The McGraw-Hill Companies, Inc. All Rights Reserved.

Object Pieces for Logic Problem 5 (Activity Set 2.3)

MATERIAL CARD 21

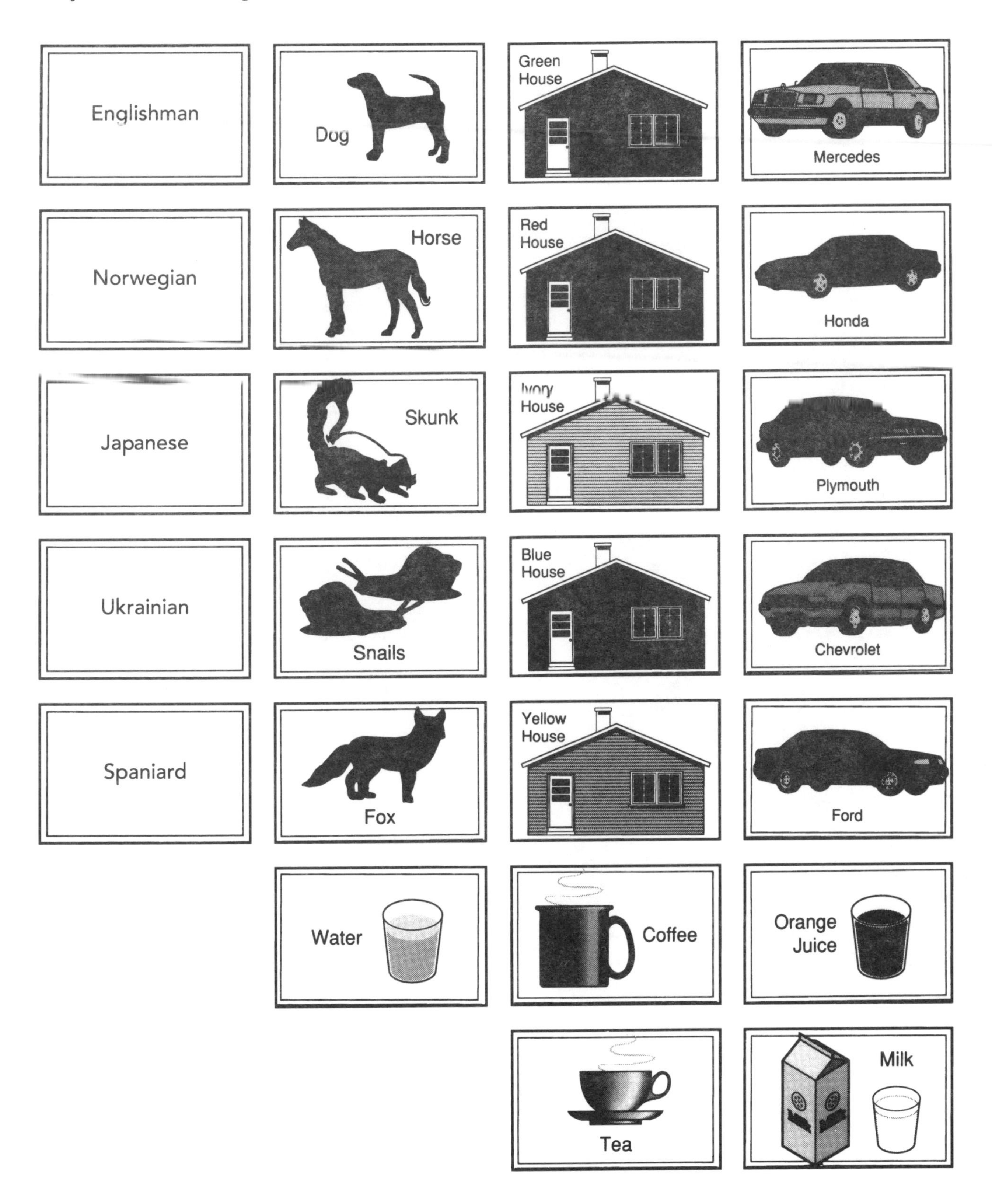

Copyright © 2001 by The McGraw-Hill Companies, Inc. All Rights Reserved.

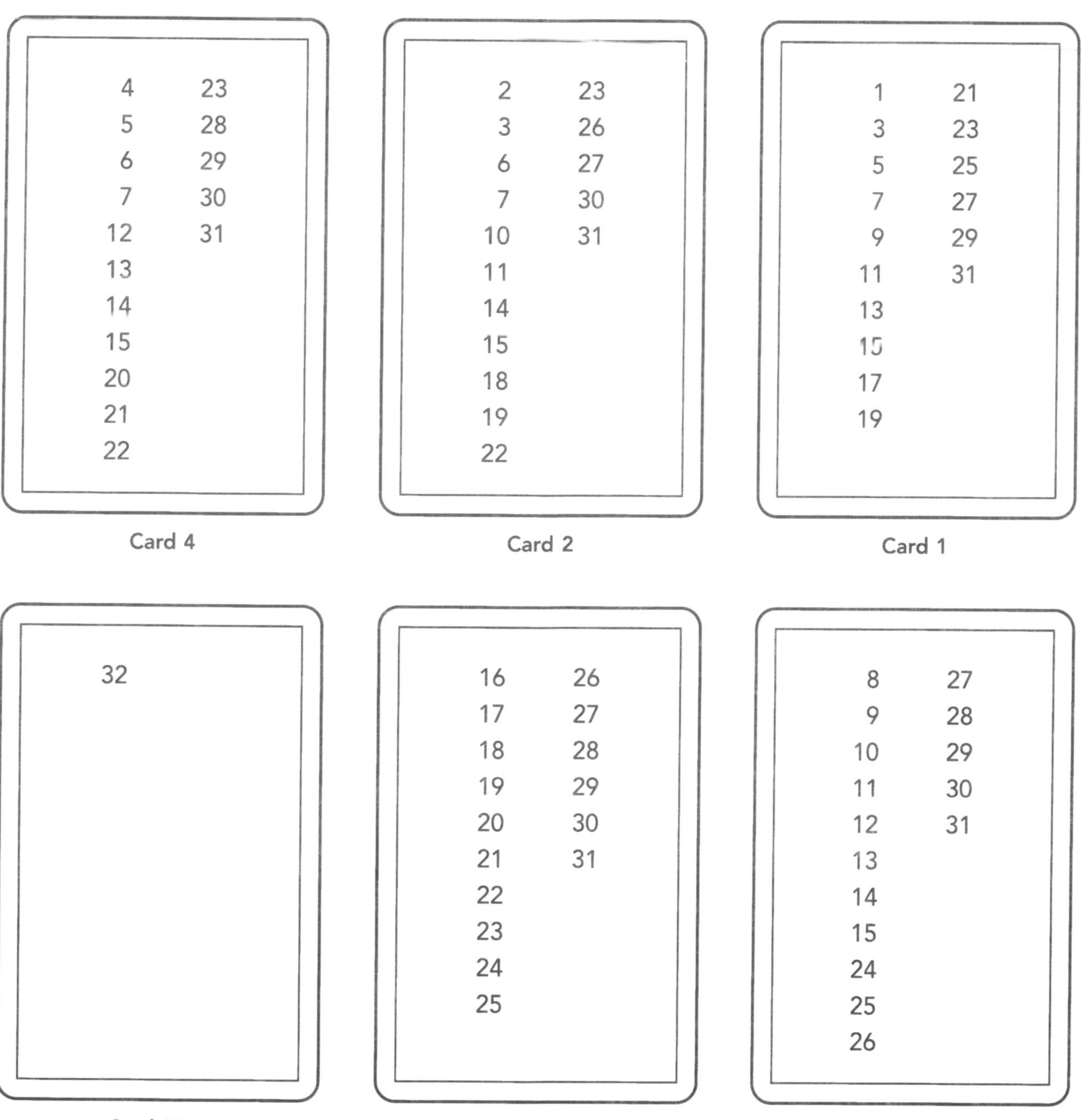

Card 32

Copyright © 2001 by The McGraw-Hill Companies, Inc. All Rights Reserved.

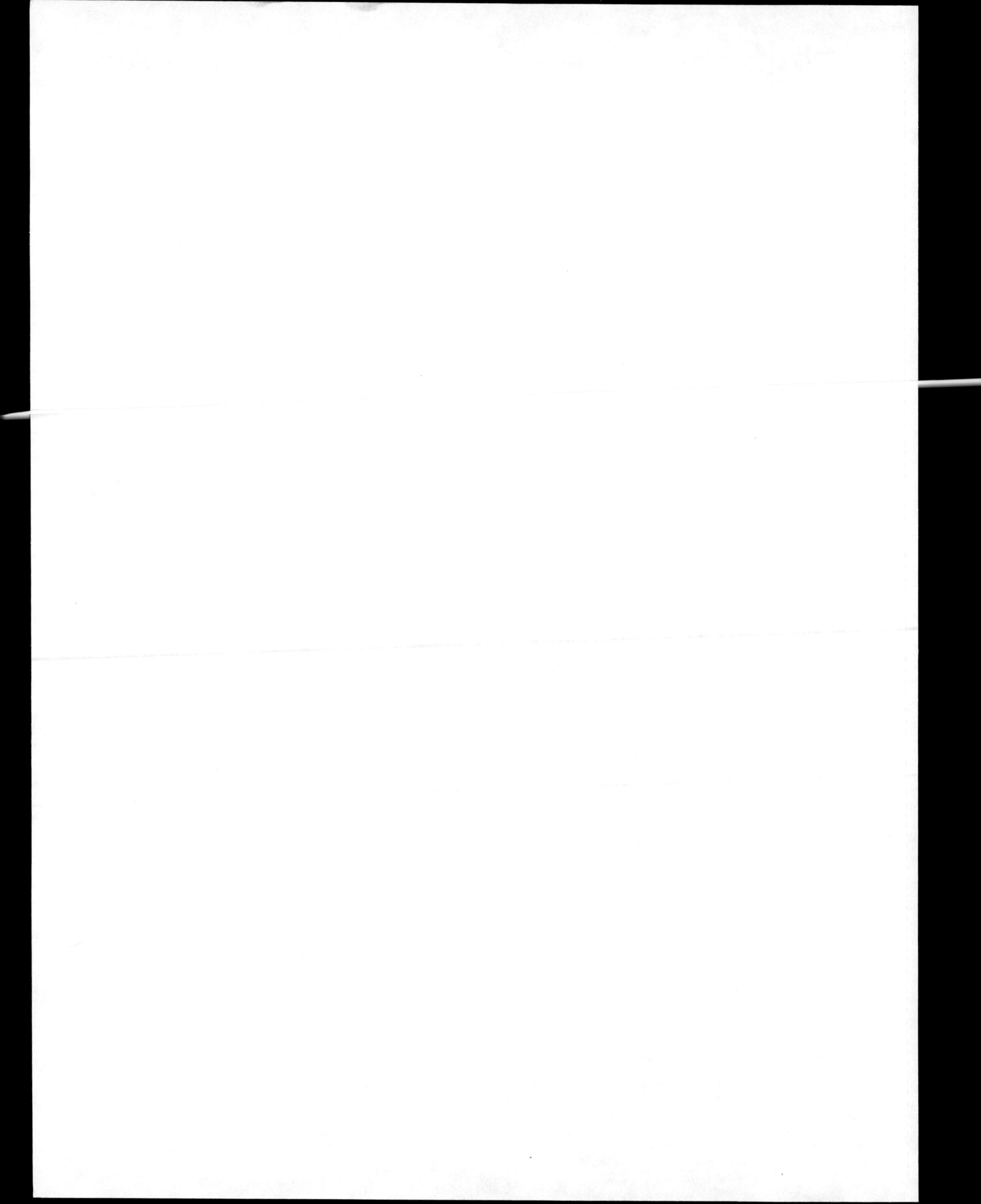

Decimal Squares* (Activity Sets 6.1 and 6.2)

MATERIAL CARD 23

*Decimal Squares is a registered trademark of Scott Resources.

Copyright © 2001 by The McGraw-Hill Companies, Inc. All Rights Reserved.

*Decimal Squares is a registered trademark of Scott Resources.

Copyright © 2001 by The McGraw-Hill Companies, Inc. All Rights Reserved.

MATERIAL CARD 25

*Decimal Squares is a registered trademark of Scott Resources.

Copyright © 2001 by The McGraw-Hill Companies, Inc. All Rights Reserved.

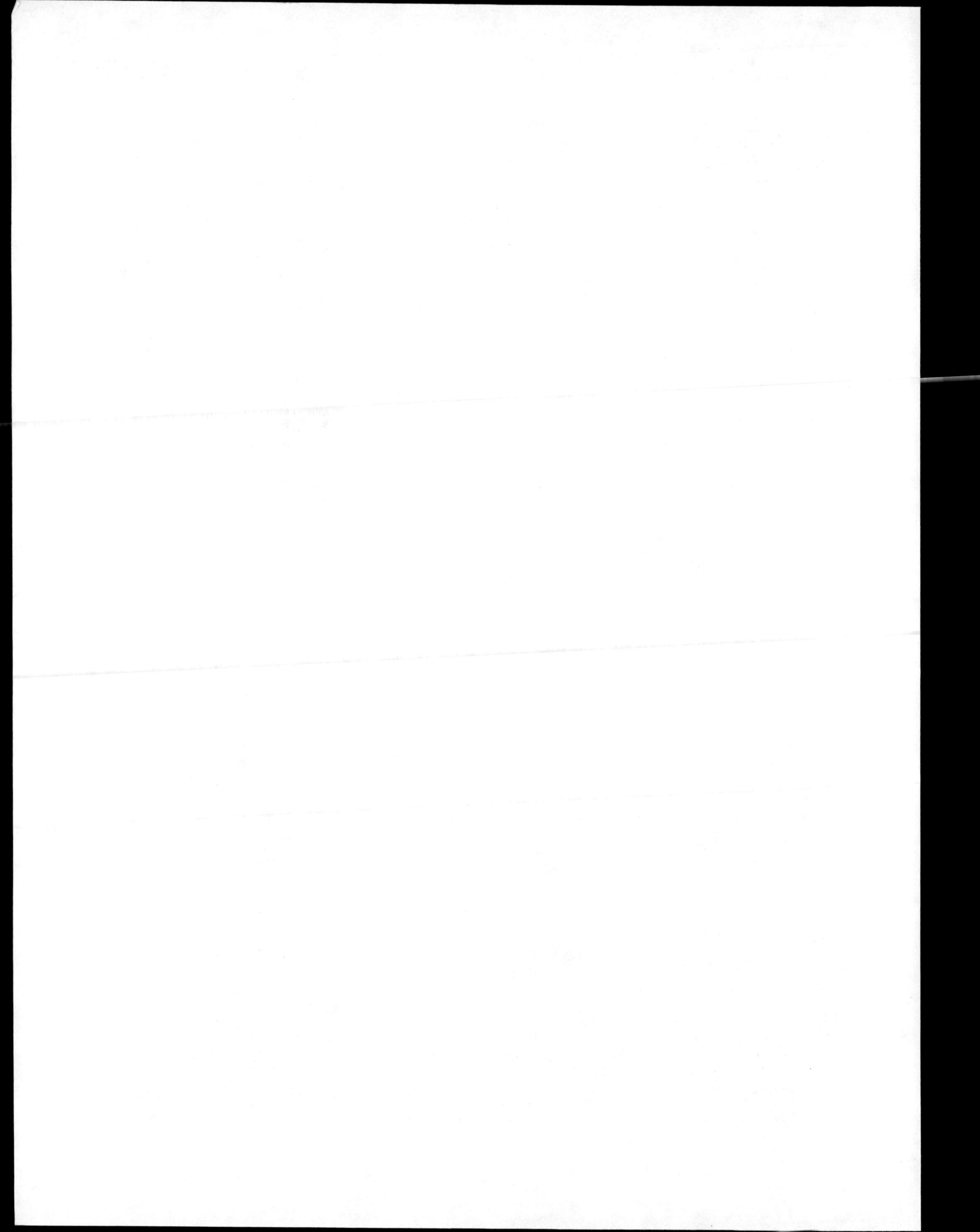

Decimal Squares* (Activity Sets 6.1 and 6.2)

MATERIAL CARD 26

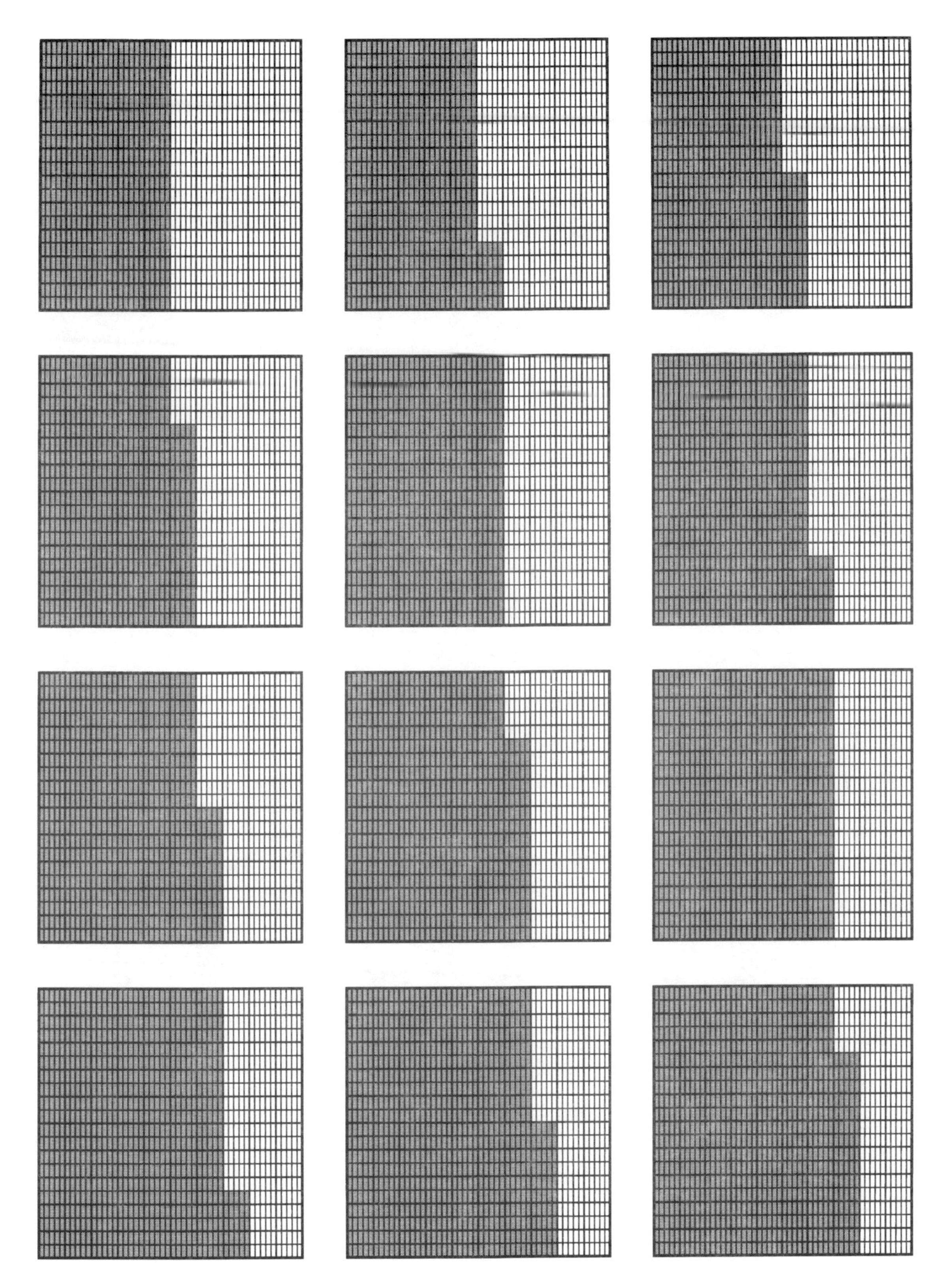

*Decimal Squares is a registered trademark of Scott Resources.

Copyright © 2001 by The McGraw-Hill Companies, Inc. All Rights Reserved.

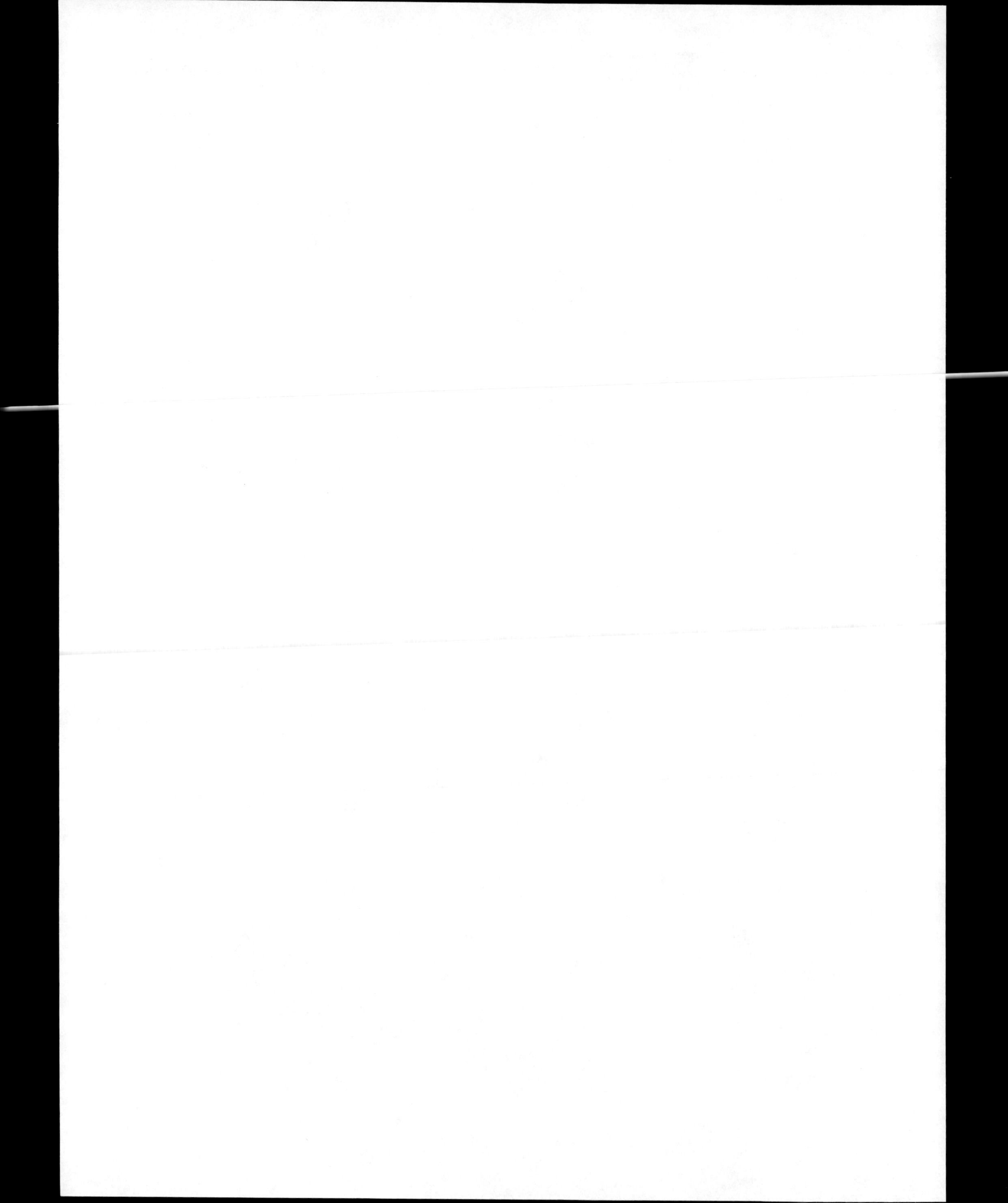

Rectangular Geoboard Template

(Activity Sets 6.4, 9.1, and 10.2)

MATERIAL CARD 27

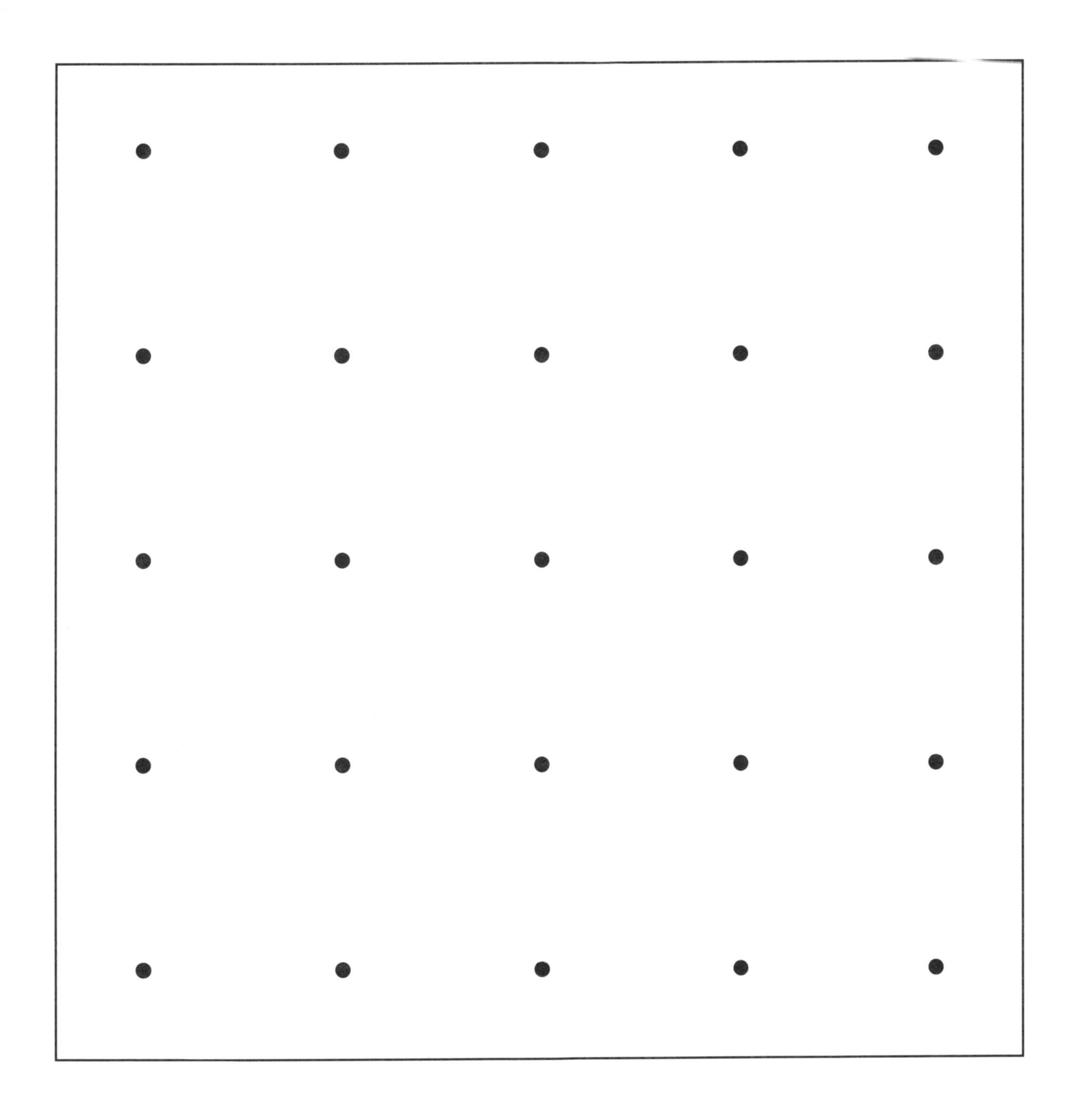

Copyright © 2001 by The McGraw-Hill Companies, Inc. All Rights Reserved.

MATERIAL CARD 28

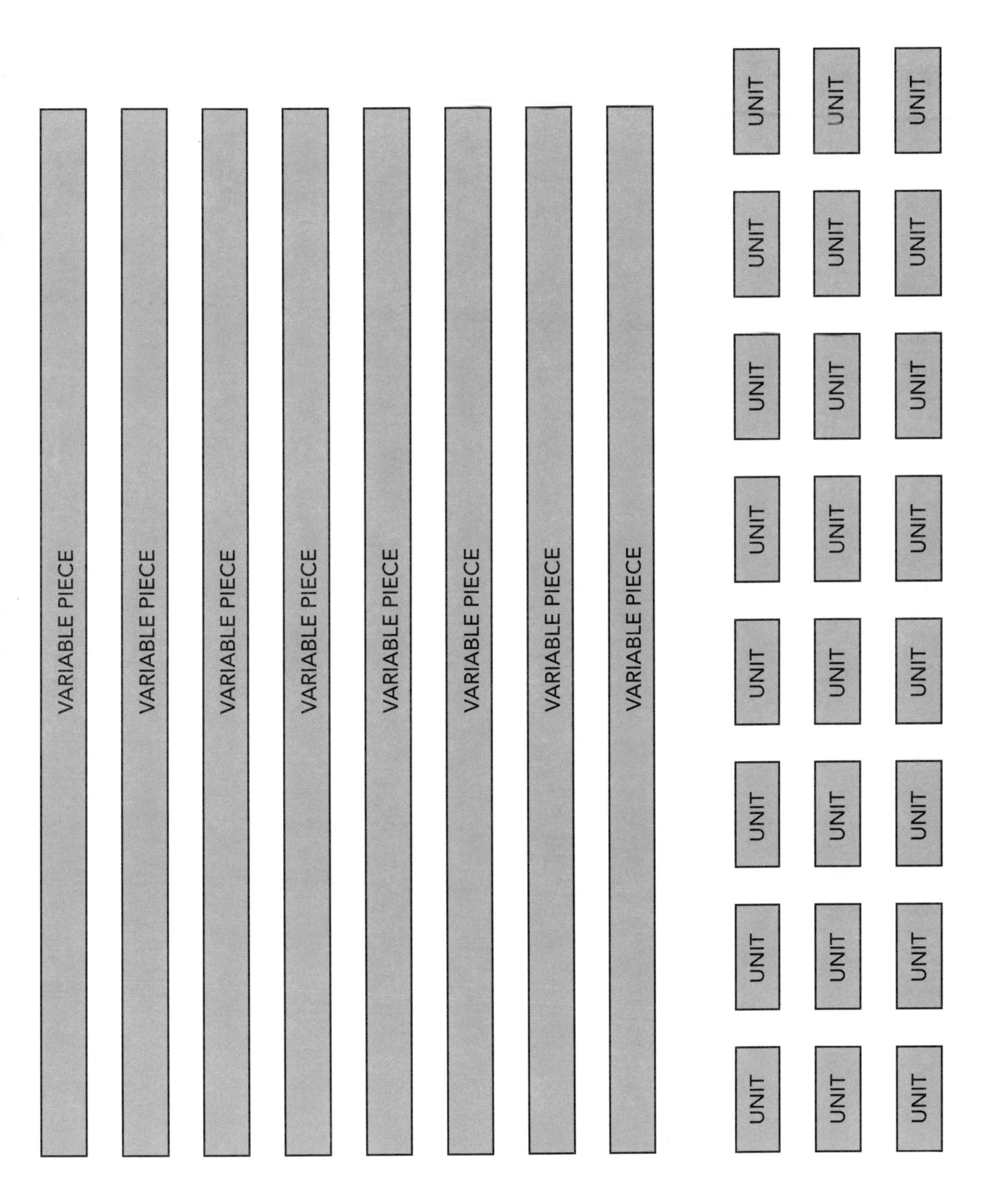

Copyright © 2001 by The McGraw-Hill Companies, Inc. All Rights Reserved.

Algebraic Expression Cards (Activity Set 1.3 JUST FOR FUN)

MATERIAL CARD 29

Team A		Team B	
I have $n+1$ **Who has** two less than a number	**I have** $4n+5$ **Who has** two more than the square root of a number	**I have** $n+2$ **Who has** five less than a number	**I have** $5n+3$ **Who has** one more than three times the square of a number
I have $n-2$ **Who has** one more than two times a number	**I have** $\sqrt{n}+2$ **Who has** five less than eight times a number	**I have** $n-5$ **Who has** two more than three times a number	**I have** $3n^2+1$ **Who has** two more than one-third of a number
I have $2n+1$ **Who has** five less than seven times a number	**I have** $8n-5$ **Who has** one more than four times the square of a number	**I have** $3n+2$ **Who has** five less than six times a number	**I have** $\frac{n}{3}+2$ **Who has** six less than the cube of a number
I have $7n-5$ **Who has** seven more than the square of a number	**I have** $4n^2+1$ **Who has** four more than one-third of a number	**I have** $6n-5$ **Who has** eight more than the square of a number	**I have** n^3-6 **Who has** nine more than four times a number
I have n^2+7 **Who has** three more than half of a number	**I have** $\frac{n}{3}+4$ **Who has** seven less than the cube of a number	**I have** n^2+8 **Who has** five less than half of a number	**I have** $4n+9$ **Who has** seven less than half of a number
I have $\frac{n}{2}+3$ **Who has** five more than four times a number	**I have** n^3-7 **Who has** one more than a number	**I have** $\frac{n}{2}-5$ **Who has** three more than five times a number	**I have** $\frac{n}{2}-7$ **Who has** two more than a number

Copyright © 2001 by The McGraw-Hill Companies, Inc. All Rights Reserved.

Simulation Spinners (Activity Set 8.2)

MATERIAL CARD 30

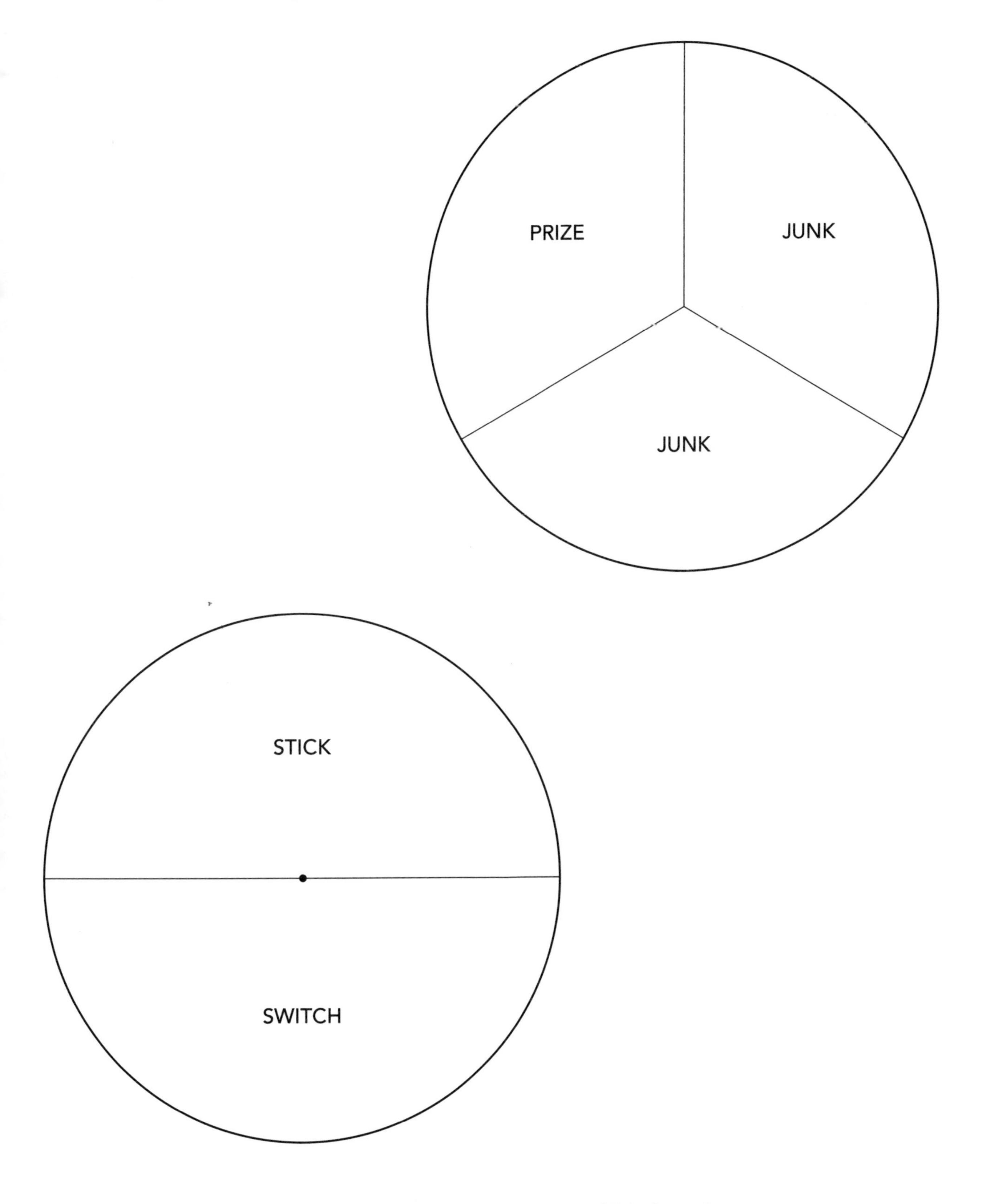

Copyright © 2001 by The McGraw-Hill Companies, Inc. All Rights Reserved.

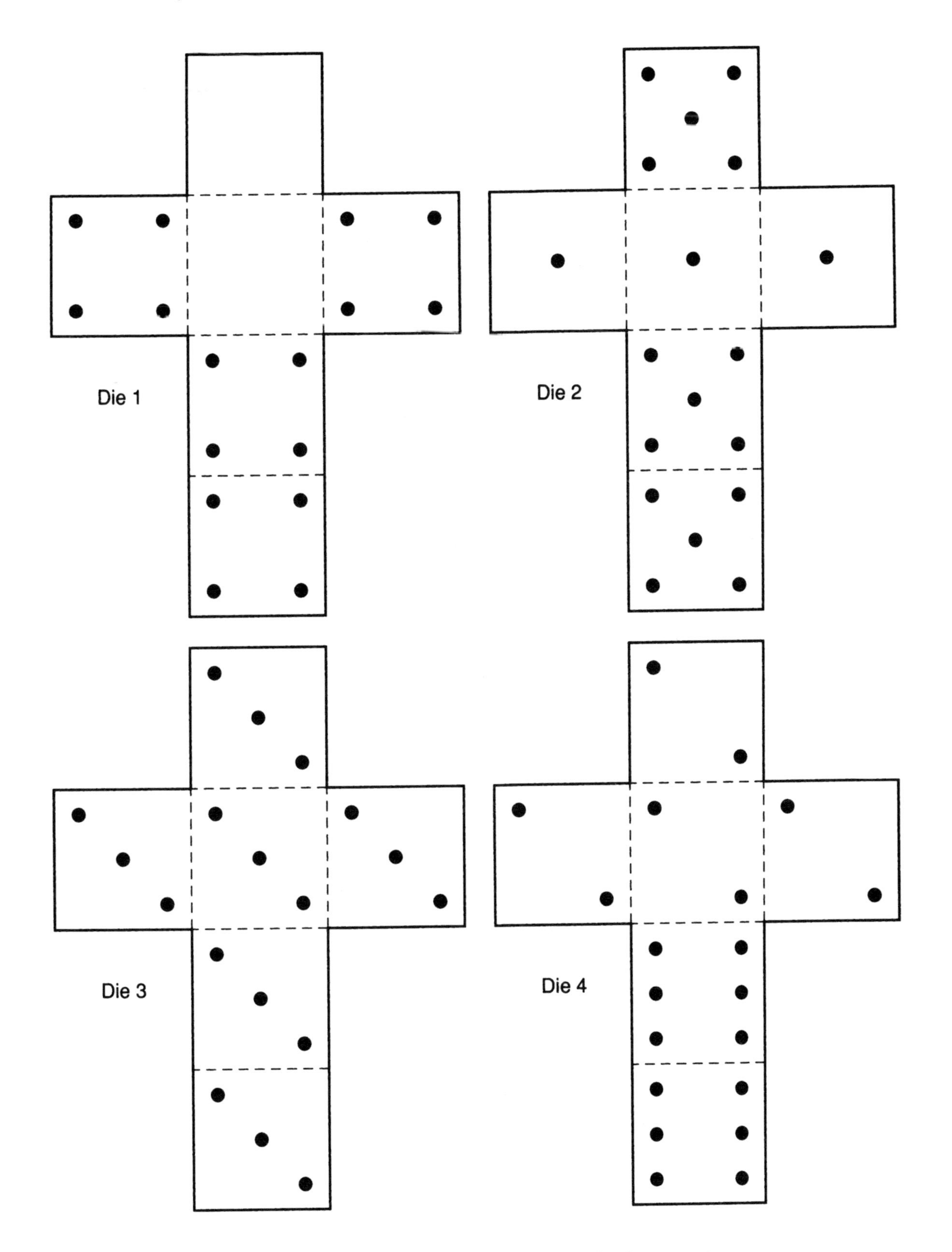

Copyright © 2001 by The McGraw-Hill Companies, Inc. All Rights Reserved.

Metric Ruler, Protractor, and Compass

MATERIAL CARD 32

(Activity Sets 9.1, 10.1, 10.3, 11.1, 11.2, and 11.3)

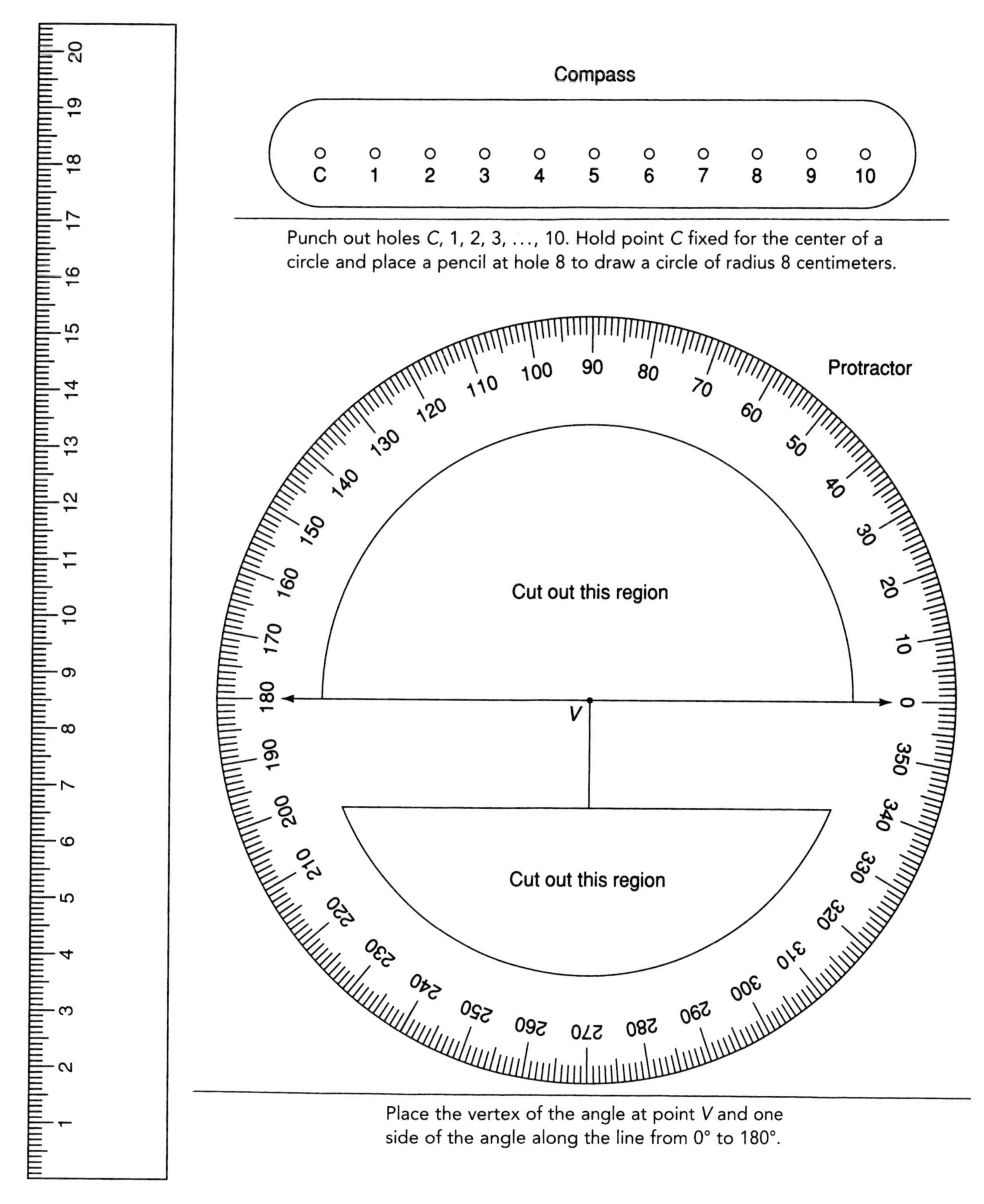

Compass
C 1 2 3 4 5 6 7 8 9 10
Punch out holes *C*, 1, 2, 3, . . ., 10. Hold point *C* fixed for the center of a circle and place a pencil at hole 8 to draw a circle of radius 8 centimeters.
Protractor
Cut out this region
Cut out this region
V

Place the vertex of the angle at point *V* and one side of the angle along the line from 0° to 180°.

Copyright © 2001 by The McGraw-Hill Companies, Inc. All Rights Reserved.

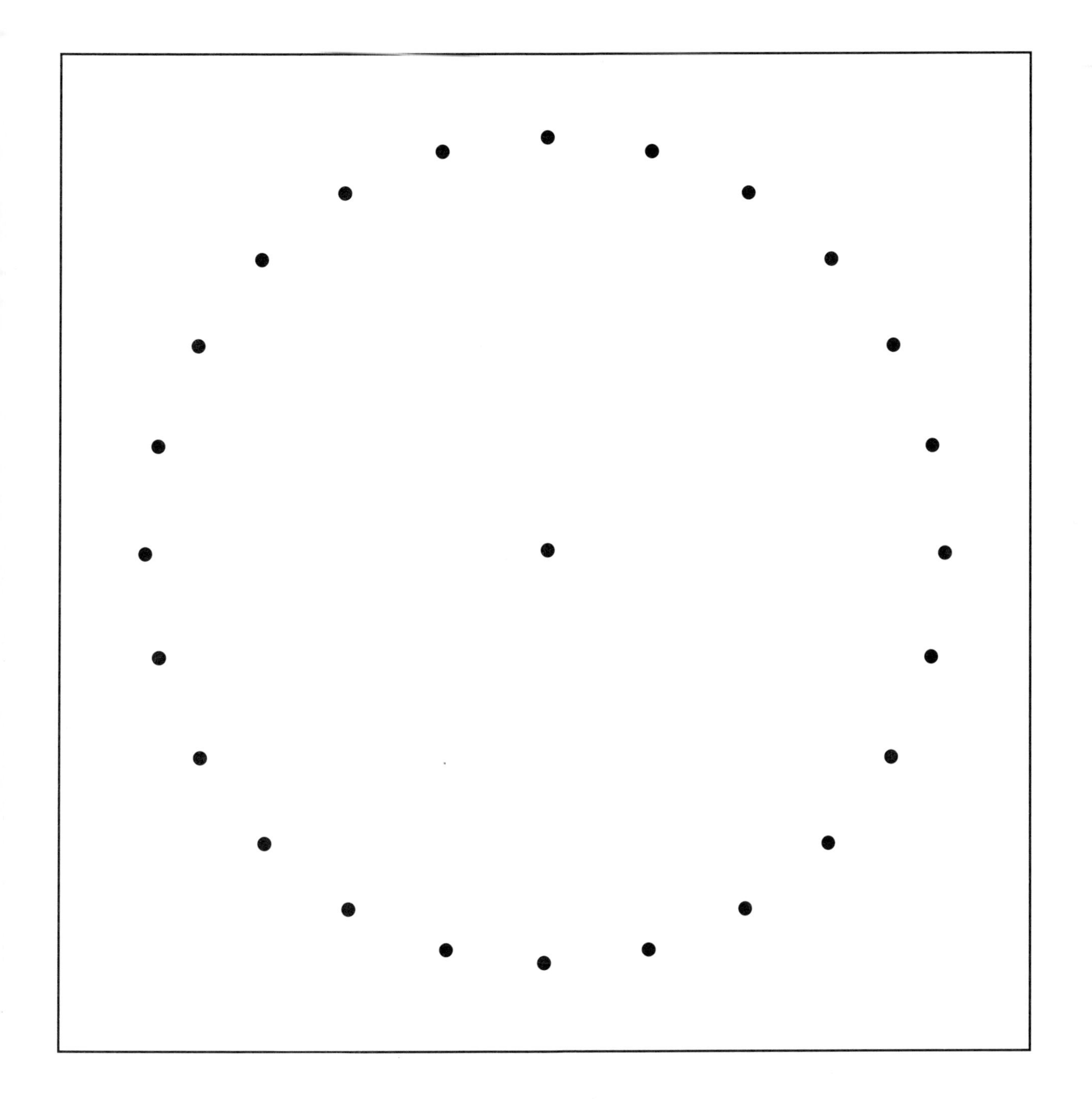

Copyright © 2001 by The McGraw-Hill Companies, Inc. All Rights Reserved.

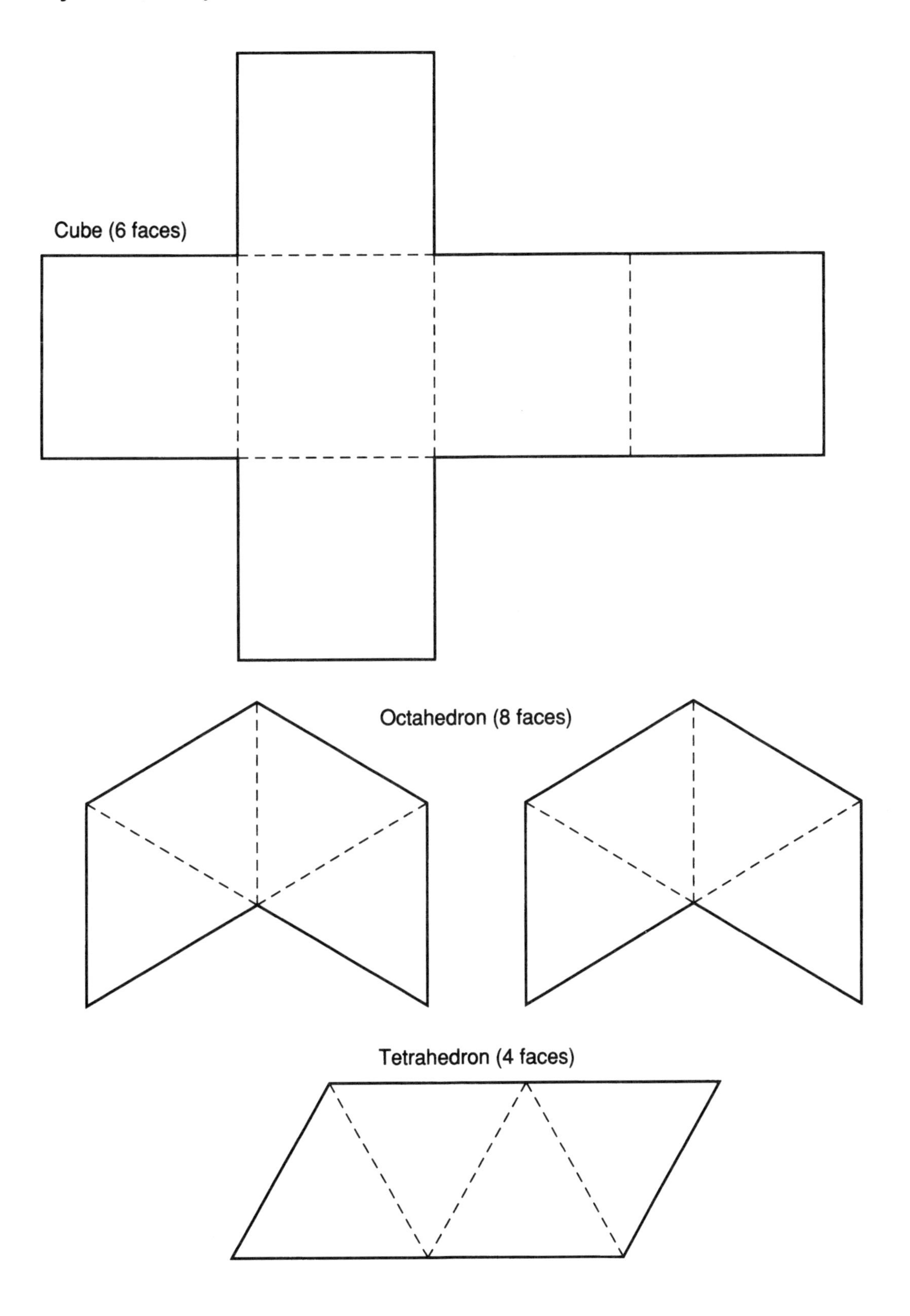

Copyright © 2001 by The McGraw-Hill Companies, Inc. All Rights Reserved.

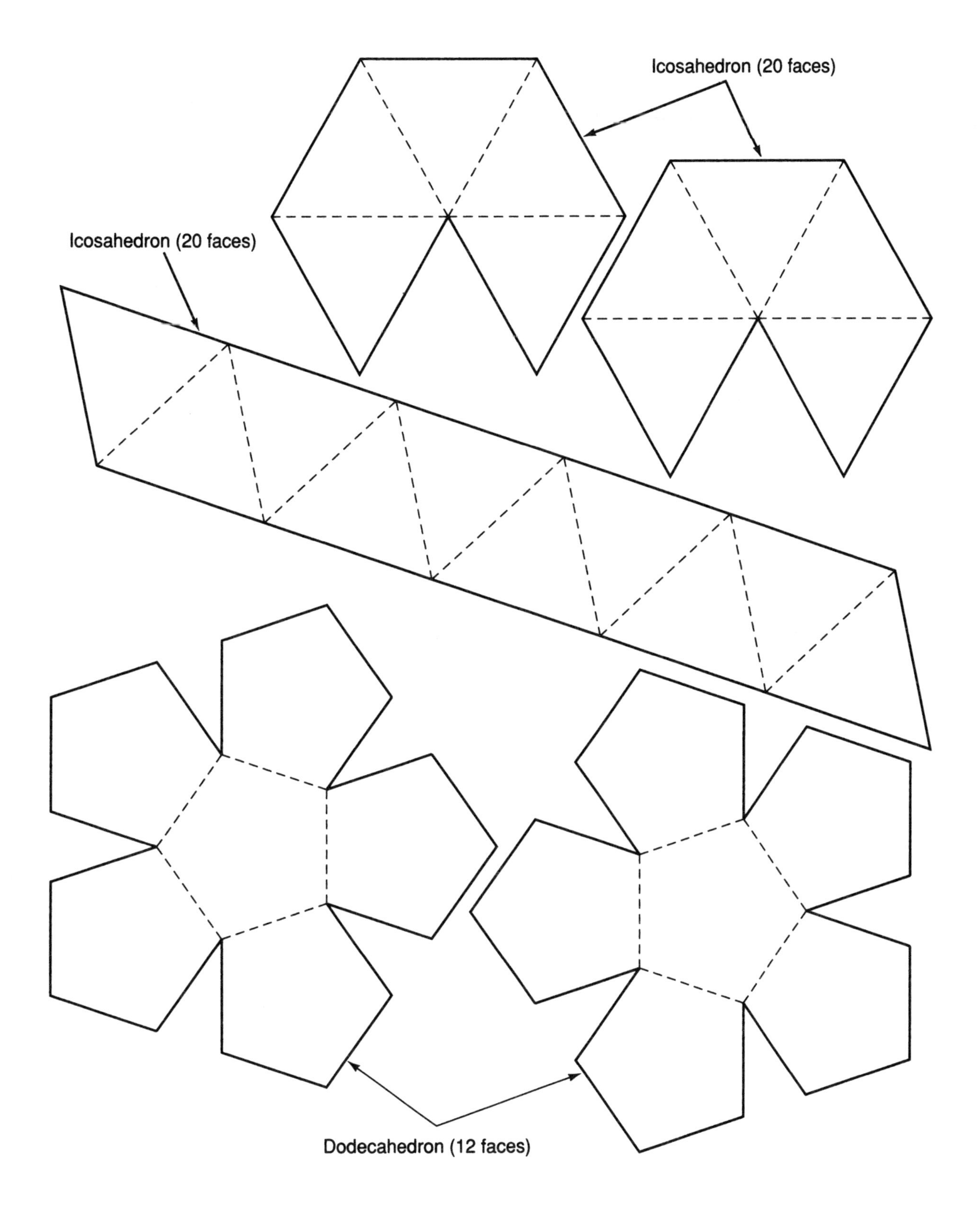

Copyright © 2001 by The McGraw-Hill Companies, Inc. All Rights Reserved.

Cube Patterns for Instant Insanity

(Activity Set 9.3 JUST FOR FUN)

MATERIAL CARD 36

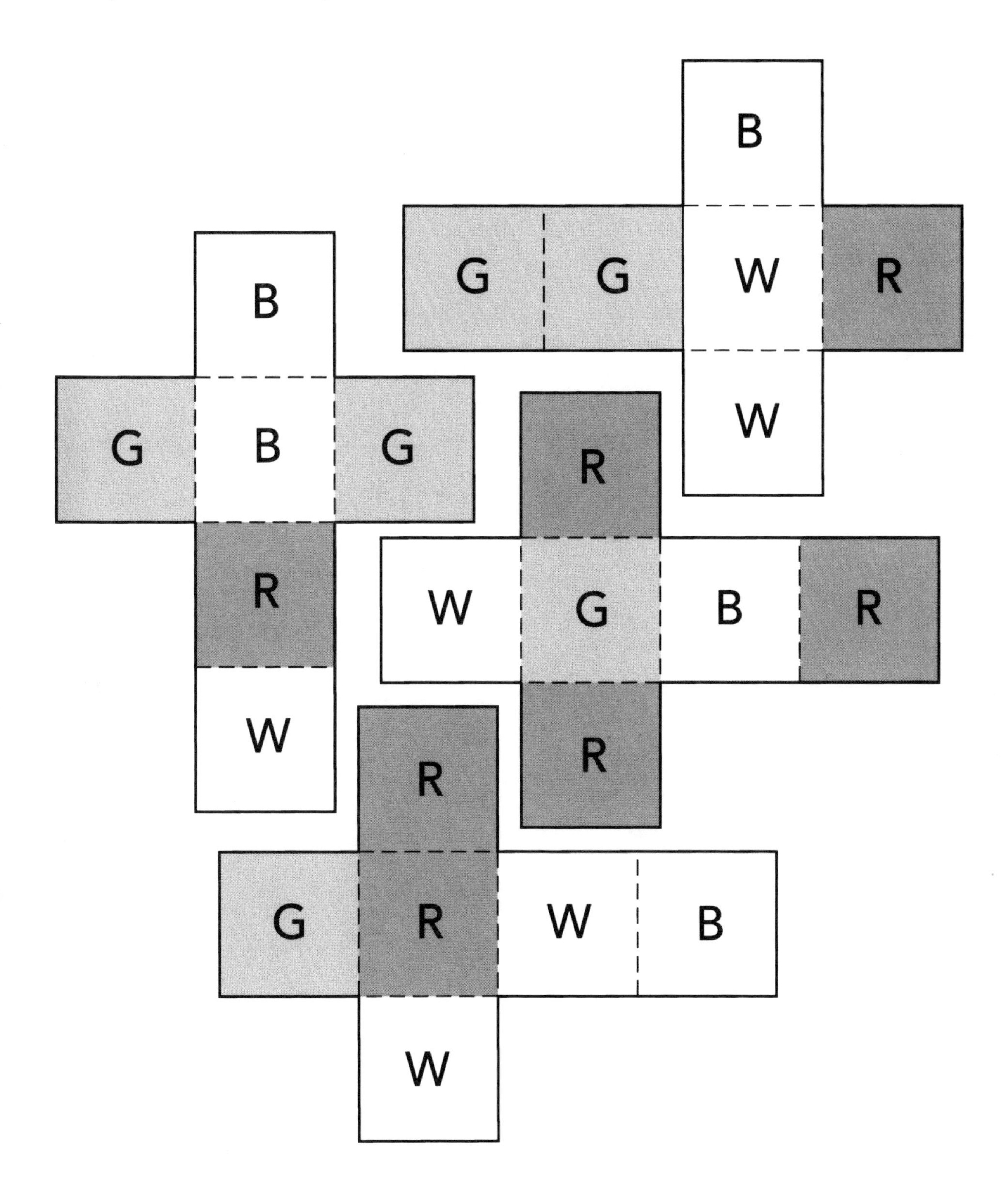

Copyright © 2001 by The McGraw-Hill Companies, Inc. All Rights Reserved.

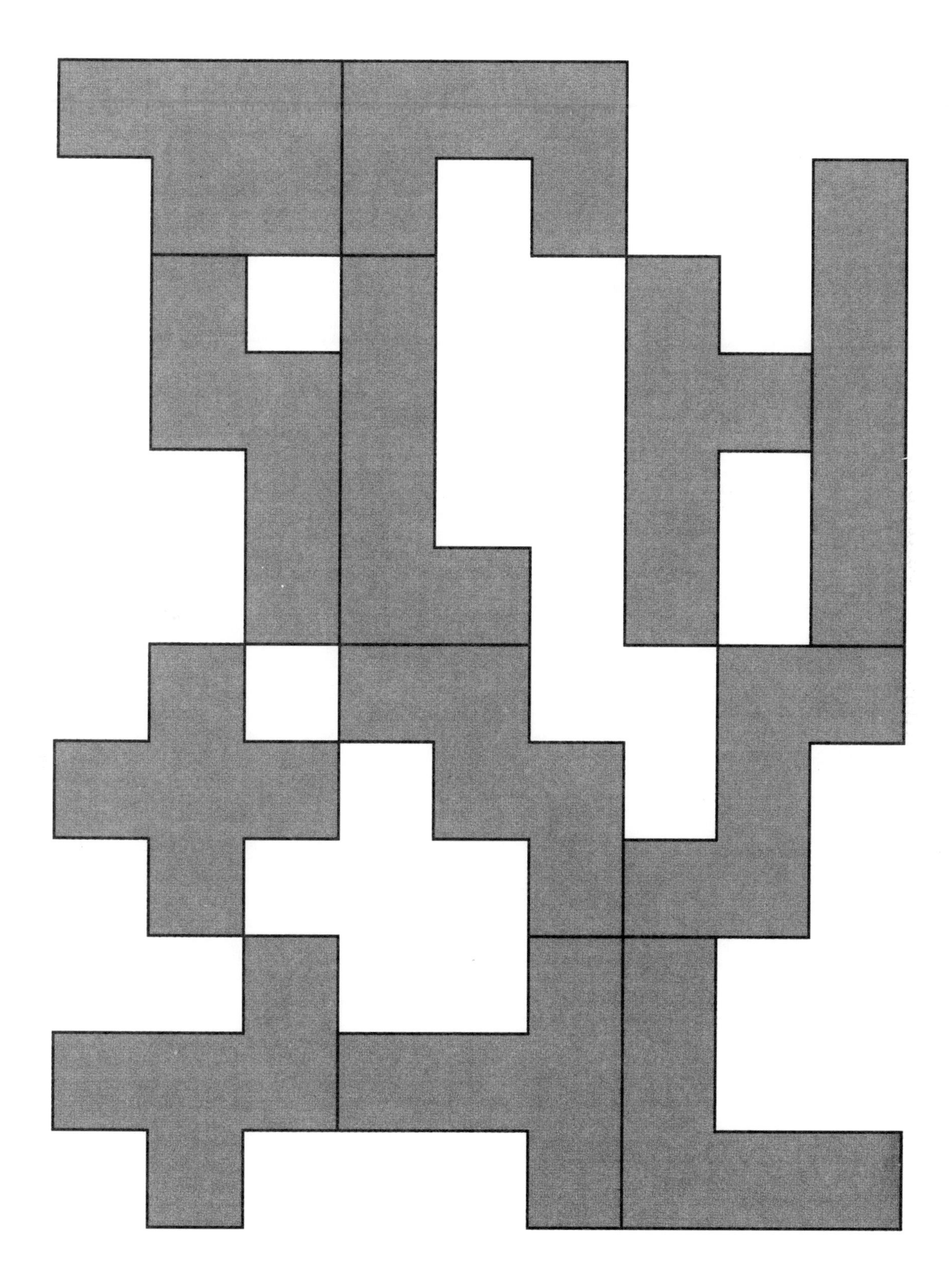

Copyright © 2001 by The McGraw-Hill Companies, Inc. All Rights Reserved.

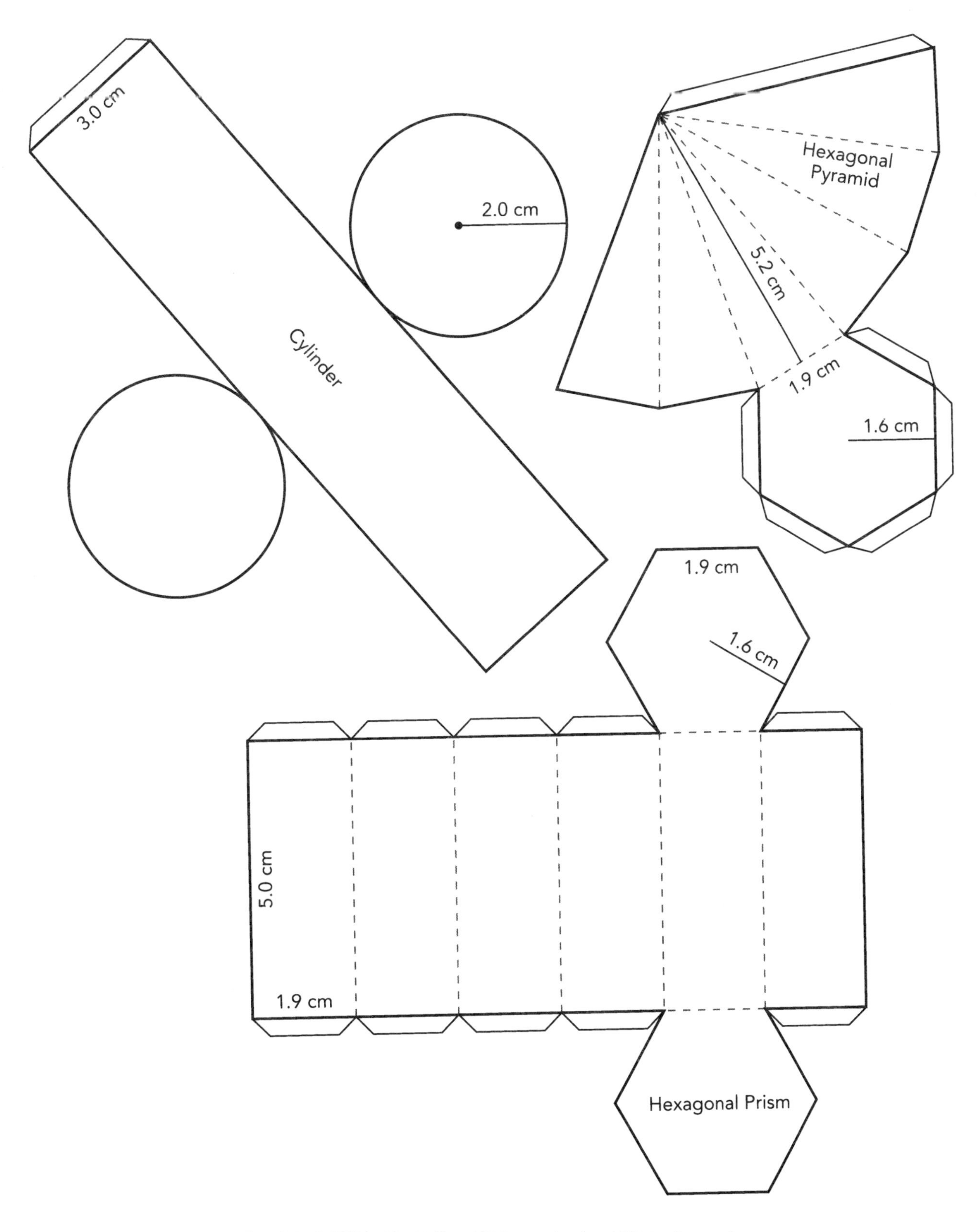

Copyright © 2001 by The McGraw-Hill Companies, Inc. All Rights Reserved.

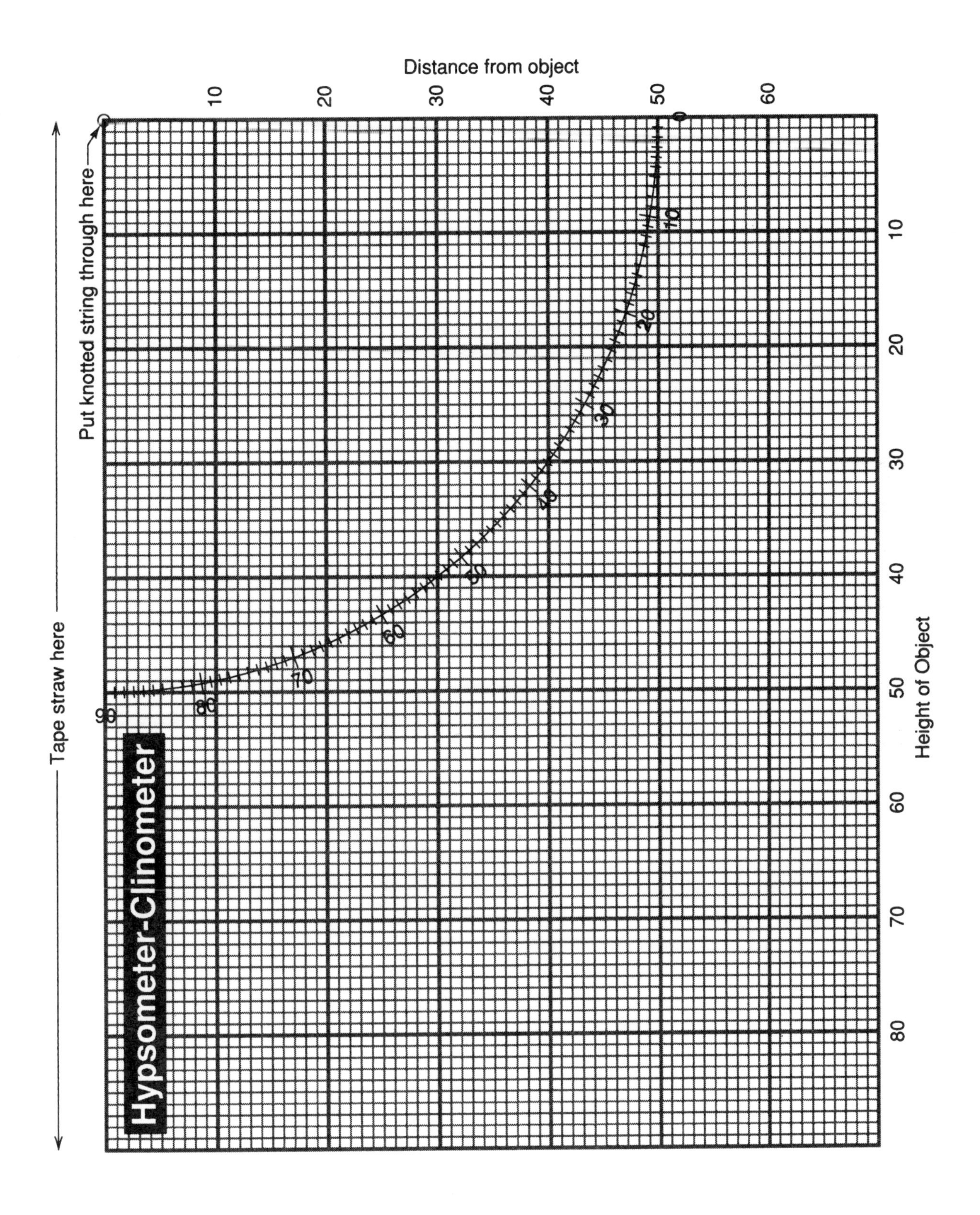

Copyright © 2001 by The McGraw-Hill Companies, Inc. All Rights Reserved.

INTEREST GAMEBOARD

$50 $100 $200 $300 $400 $500 $600

.5	1	1.5	2	2.5	3	3.5	4	4.5	5
1	2	3	4	5	6	7	8	9	10
2	4	6	8	10	12	14	16	18	20
3	6	9	12	15	18	21	24	27	30
4	8	12	16	20	24	28	32	36	40
5	10	15	20	25	30	35	40	45	50
6	12	18	24	30	36	42	48	54	60

Copyright © 2001 by The McGraw-Hill Companies, Inc. All Rights Reserved.